ANNUAL REVIEW OF NEUROSCIENCE

EDITORIAL COMMITTEE (1992)

ANNUAL REVIEW OF NEUROSCIENCE

VOLUME 15, 1992

W. MAXWELL COWAN, *Editor*
Howard Hughes Medical Institute

ERIC M. SHOOTER, *Associate Editor*
Stanford University School of Medicine

CHARLES F. STEVENS, *Associate Editor*
Salk Institute for Biological Studies

RICHARD F. THOMPSON, *Associate Editor*
University of Southern California

ANNUAL REVIEWS INC   4139 EL CAMINO WAY   P.O. BOX 10139   PALO ALTO, CALIFORNIA 94303-0897

ANNUAL REVIEWS INC.
Palo Alto, California, USA

International Standard Serial Number: 0147-006X
International Standard Book Number: 0-8243-2415-3

Annual Review and publication titles are registered trademarks of Annual Reviews Inc.

∞ The paper used in this publication meets the minimum requirements of American National Standard for Information Sciences—Permanence of Paper for Printed Library Materials, ANSI Z39.48-1984.

Annual Reviews Inc. and the Editors of its publications assume no responsibility for the statements expressed by the contributors to this *Review*.

TYPESET BY BPCC-AUP GLASGOW LTD., SCOTLAND
PRINTED AND BOUND IN THE UNITED STATES OF AMERICA

Annual Review of Neuroscience
Volume 15, 1992

CONTENTS

SOME RELATED ARTICLES IN OTHER *ANNUAL REVIEWS*

From the *Annual Review of Cell Biology*, Volume 7 (1991):

Cell and Substrate Adhesion Molecules in Drosophila, Michael Hortsch and Corey S. Goodman

Mechanisms and Functions of Cell Death, Ronald E. Ellis, Junying Yuan, and H. Robert Horvitz

The Notch *Locus and the Cell Biology of Neuroblast Segregation*, Spyros Artavanis-Tsakonas, Christos Delidakis, and Richard G. Fehon

Laminin Receptors, Robert P. Mecham

Signal Transduction in the Visual System of Drosophila, Dean P. Smith, Mark A. Stamnes, and Charles S. Zuker

From the *Annual Review of Immunology*, Volume 9 (1991):

The SCID-HU Mouse: A Small Animal Model for HIV Infection and Pathogenesis, Joseph McCune, Hideto Kaneshima, John Krowka, Reiko Namikawa, Henry Outzen, Bruno Peault, Linda Rabin, Chu-Chih Shih, Edwin Yee, M. Lieberman, I. Weissman, and L. Shultz

MHC Class-II Molecules and Autoimmunity, Gerald T. Nepom and Henry Erlich

From the *Annual Review of Pharmacology & Toxicology*, Volume 32 (1992):

Mutagenesis of the Human Beta-2 Adrenergic Receptor: How Structure Elucidates Function, Jacek Ostrowski, Michael A. Kjelsberg, Marc G. Caron, and Robert J. Lefkowitz

From the *Annual Review of Physiology*, Volume 54 (1992):

Ionic Current Mechanisms Generating Vertebrate Primary Cardiac Pacemaker Activity at the Single Cell Level: An Integrative View, Donald L. Campbell, Randall L. Rasmusson, and Harold C. Strauss

Charge Movement and the Nature of Signal Transduction in Skeletal Muscle Excitation-Contraction Coupling, Eduardo Riós, Gonzalo Pizarro, and Enrico Stefani

Mechanotransduction, Andrew S. French

Role of cGMP and Ca^{2+} in Vertebrate Photoreceptor Excitation and Adaption, U. B. Kaupp and K.-W. Koch

From the *Annual Review of Psychology*, Volume 43 (1992):

ANNUAL REVIEWS INC. is a nonprofit scientific publisher established to promote the advancement of the sciences. Beginning in 1932 with the *Annual Review of Biochemistry,* the Company has pursued as its principal function the publication of high quality, reasonably priced *Annual Review* volumes. The volumes are organized by Editors and Editorial Committees who invite qualified authors to contribute critical articles reviewing significant developments within each major discipline. The Editor-in-Chief invites those interested in serving as future Editorial Committee members to communicate directly with him. Annual Reviews Inc. is administered by a Board of Directors, whose members serve without compensation.

Annu. Rev. Neurosci. 1992. 15:1–29

THE EVOLUTION OF EYES

Michael F. Land

Neuroscience Interdisciplinary Research Centre, School of Biological Sciences, University of Sussex, Brighton BN1 9QG, United Kingdom

Russell D. Fernald

Programs of Human Biology and Neuroscience and Department of Psychology, Stanford University, Stanford, California 94305

KEY WORDS: vision, optics, retina

INTRODUCTION: EVOLUTION AT DIFFERENT LEVELS

Since the earth formed more than 5 billion years ago, sunlight has been the most potent selective force to control the evolution of living organisms. Consequences of this solar selection are most evident in eyes, the premier sensory outposts of the brain. Because organisms use light to see, eyes have evolved into many shapes, sizes, and designs; within these structures, highly conserved protein molecules for catching photons and bending light rays have also evolved. Although eyes themselves demonstrate many different solutions to the problem of obtaining an image—solutions reached relatively late in evolution—some of the molecules important for sight are, in fact, the same as in the earliest times. This suggests that once suitable biochemical solutions are found, they are retained, even though their "packaging" varies greatly. In this review, we concentrate on the diversity of eye types and their optical evolution, but first we consider briefly evolution at the more fundamental levels of molecules and cells.

Molecular Evolution

The opsins, the protein components of the visual pigments responsible for catching photons, have a history that extends well beyond the appearance of anything we would recognize as an eye. Goldsmith (1990) recently

1

0147–006X/92/0301–0001$02.00

compared the opsins' evolutionary lineages in detail. These molecules consist of seven transmembrane helices with short loops on both sides of the membrane. Covalently attached to the molecules is a highly conjugated molecule, the chromophore, which is one of a family of only four close relatives of vitamin A. The chromophore accepts the photon of light; as a result, the molecule flips from the 11-*cis* to the all-*trans* form, which in turn triggers a biochemical cascade that leads to excitation of the receptor cell. These features are common to all metazoan opsins. Based on the degree of similarity in their DNA, they must share a common ancestry. Two regions of the molecule, particularly the cytoplasmic loop between helices 1 and 2 and the site of attachment of the chromophore molecule in helix 7, show very close similarity in opsins from vertebrates, insects, and *Octopus*, whose ancestries diverged in the Cambrian. Within the vertebrates, the primary structure of the rod and cone opsins clearly has a phylogeny that maps onto the phylogeny of the parent species. With the recent work by Nathans and coworkers (e.g. Nathans 1987) we have begun to understand the molecular and phylogenetic basis of color vision at a molecular level.

Although metazoan opsins have apparently evolved along several separate lines from a common ancient ancestor, what happened earlier is not so clear. Interestingly, "bacteriorhodopsin" from *Halobacterium* does not show significant amino acid similarity with cattle rhodopsin, and it is double-bond 13 of the chromophore, rather than 11, that is altered by light. Nevertheless, like metazoan opsins, bacteriorhodopsin seems to belong to a larger superfamily of proteins, all of which have seven transmembrane helices and operate by activating second-messenger cascades (e.g. Hall 1987). These proteins include the β-adrenergic receptor and the muscarinic acetylcholine receptor. Whether these similarities indicate a very ancient common ancestry or a more recent appropriation of one protein to another function, is not yet clear. For further discussion and detailed references, consult Goldsmith's excellent review (1990).

Opsins are not the only visual proteins with an interesting history. Vertebrate lenses are formed from modified epithelial cells, which contain high concentrations of soluble proteins known as "crystallins" (e.g. Bloemendal 1981) because of their highly organized packing. The distribution of these proteins is responsible for the remarkable refractive index gradients in these lenses, which underlie their optical properties (see below). Of the ten crystallins now known, α-A is the most ubiquitous, as it is only missing in bony fishes. Once thought to be "dull proteins" (de Jong et al 1988), the production and arrangement of lens proteins instead offers a tantalizing glimpse at the versatile bag of tricks used during the evolution of optical structures.

Phylogenetic trees based on DNA sequences of α-A crystallin reveal strongly directional selection in vertebrates that require increasingly flexible lenses. However, how the observed amino acid substitutions contribute to flexibility is unknown (de Jong et al 1988). The consequences of relaxed selection can be seen in the crystallins of the subterranean rodent *Spalax ehrenbergi*, which is completely blind and has only rudimentary eyes. The gene for α-A crystallin changes four times as fast as in sighted rodent relatives, but at only one fifth the rate for neutral evolution found in pseudogenes (Hendriks et al 1987). Because *Spalax* still responds to photoperiods for thermoregulation (Haim et al 1983), perhaps some feature of α-A is required for functionality, or, as Hendriks et al (1987) suggest, α-A may play a role in the development of the eye. In any event, the severity of the constraints on lens construction are mirrored in the extent of evolutionary selection for α-A lens protein. For other lens proteins, there are other themes and variations, but these give no clear idea of how selection is acting.

Until recently, all crystallins were thought to be unique to lens tissue and to have evolved for this specialist function. However, this is apparently only partly true, as crystallins fall into two distinct groups. One group, the Alpha and Beta-gamma crystallins, are indeed specialized lens proteins. Each is the product of gene duplication and divergent evolution from distinct ancestors with a role in stress responses. Alpha crystallins are related to ubiquitous and ancient heat-shock proteins and to a schistosome egg antigen; Beta-gamma crystallins are relatives of a bacterial spore coat calcium-binding protein (Wistow & Piatigorski 1988). The other group of lens proteins are enzymes or their close relatives, which are often used as enzymes elsewhere in the animal (e.g. Wistow et al 1988a). Moreover, some taxon-specific lens proteins are actually products of the same genes as the enzymes (Hendriks et al 1988; Wistow et al 1988b). One gene coding for a protein with two entirely different functions has been called "gene sharing" (Wistow et al 1990) and is considered an evolutionary strategy that preceeded gene duplication and specialization (Piatigorski & Wistow 1989). Such shared genes are subjected to two or more different selective constraints, which makes evolutionary change complicated at best.

Why have enzymes been recruited as vertebrate lens tissue, which comprises up to 40% of the lens? There is considerable speculation on this point (e.g. Wistow & Kim 1991), but no compelling insight. Whatever the reason, this molecular opportunism in vertebrates is apparently such a good idea that molluscs independently evolved the same strategy (Doolittle 1988). Squid eyes, which are interesting for their convergence with fish eyes at the organ level (see below), have lenses whose protein content is almost 100% the enzyme glutathione S-transferase (Tomarev & Zinovieva

1988). This convergence of molecular strategy hints that "enzyme as lens" might have a deeper structural basis. Or, as Piatigorski et al (1988) suggest, gene regulation of enzymes conveniently lends itself to exploitation for tissue-specific expression. Thus, enzymes may be used to make lenses because either that type of protein makes good lenses, or it may just be easy to get lens cells to make a lot of enzyme, or there may be less obvious reasons for this strategy.

Edwards & Meyer (1990) reported that conservation of a crystalline cone constituent exists throughout the insects and crustacea. This suggests that just as α-A crystallins in vertebrates show a monophyletic line, so do at least some lens proteins in arthropods, even though at the organ level arthropods show every known type of optical structure. We end this section with a remark of Goldsmith (1990), which summarizes the multilevel nature of eye evolution: "The eyes of cephalopods, arthropods, and vertebrates are not homologous, yet at the molecular level some of their constituents are."

Cellular Evolution

After molecules, receptor cells are at the next level of organization. At the receptor cell level, there are interesting indications of ancient common ancestries. As Eakin (1963) first pointed out, the protostomes (the annelid-mollusk-arthropod line) tend to have receptors in which the expanded photopigment-bearing membrane is composed of microvilli; in the deuterostomes (the echinoderm-chordate line), however, the receptors have membranes derived from cilia, which are expanded into plate- or disklike structures. Eakin's analysis, which suggests a simple dual phylogeny for photoreceptor cells, ran into difficulties because of many anatomic exceptions: Many of the microvillous (rhabdomeric) structures contained at least part of a cilium, for example (Eakin 1972). In terms of physiology, however, our knowledge tends to support Eakin's original scheme. The microvillous receptors of mollusks and arthropods depolarize to light, and the eventual result of photon absorptions is the opening of Na^+ channels in the receptor membrane (e.g. Fuortes & O'Bryan 1972). The ciliary receptors of vertebrates, on the other hand, hyperpolarize to light by the closing of Na^+ channels (e.g. Yau & Baylor 1989). An exception to this scheme seems to be the receptors in the primitive chordate Salpa, which are microvillous, but hyperpolarize (Gorman et al 1971).

One other physiologic class of receptor, which is found infrequently in the mollusks and annelids, hyperpolarizes to light like vertebrate receptors (Leutscher-Hazelhoff 1984). Here, however, the mechanism is different, as it involves an increase in the conductance of K^+ ions (Gorman & McReynolds 1978). Anatomically, these receptors are characterized by the presence of many expanded cilia. Functionally, they all seem to be concerned with simple defensive "off" responses to shadow or motion (Land 1968; Salvini-Plawen & Mayr 1977). Thus, a case can be made on both anatomic and physiologic grounds that there may be a rather small number of anatomic and physiologic types of photoreceptor cell, possibly with independent evolutionary origins. Alternatively, they may have repeatedly arisen because they represent the only ways to utilize the available membrane machinery in the service of photoreception. The answer may lie in the relationship of the physiologic response to the cells' specific microanatomy, which at present remains unexplored and enigmatic.

The rest of this review addresses evolution at the level of the whole organ, the eye itself. Here, we are largely concerned with the ways that the preexisting molecular and cellular building blocks have been assembled to provide various solutions to the problem of obtaining and transmitting an image.

THE EVOLUTION OF OPTICAL MECHANIMS

Parallel Evolution in Optical Design

When we trace the evolution of different kinds of eye, the greatest problem lies in deciding whether similarity in structure is due to evolutionary convergence or to common descent. There is a relatively small number of ways to produce an eye that gives a usable image, and most have been "discovered" more than once, thus giving rise to similar structures in unrelated animals. Citing the most notorious example, the phylogenetically unrelated eyes of squid and fish are similar in a great many details, presumably because the logic of the production of large, camera-type eyes necessitates a spherical lens, iris, eye-muscles, etc. (Packard 1972). By contrast, human and fish eyes are related by common descent, although optically they are rather different from each other. A superficial study of the eyes does not always allow such a distinction to be made, and lineages in eyes must be traced by either knowing the phylogeny of the animals in advance, or looking at other characters that are related less to optical "design principles." In the case of fish and cephalopods, the inverted and multilayered structure of the fish retina, compared with the simpler, noninverted retina of octopus and squid, demonstrates most clearly the unrelatedness of the eyes themselves.

Eye evolution has proceeded in two stages. In almost all the major animal groups, one finds simple eye-spots that consist of a small number of receptors in an open cup of screening pigment cells (Figure 1a). In an impressive analysis of the detailed structure, anatomic origins, and phylogenetic affinities of these eye-spots, Salvini-Plawen & Mayr (1977)

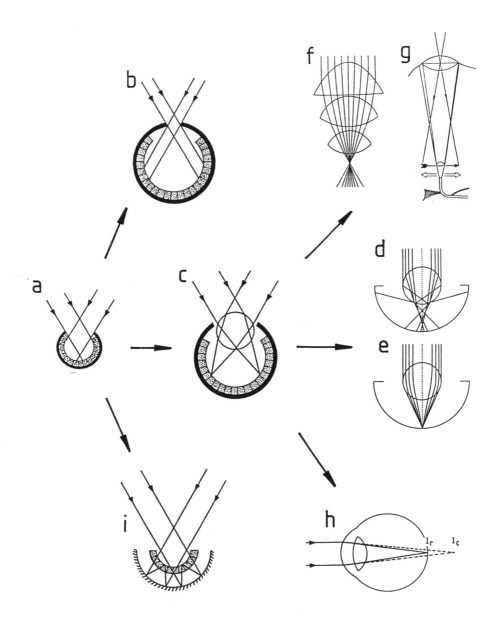

concluded that such structures had evolved independently at least 40 times, and probably as many as 65 times. These eye-spots are useful in selecting a congenial environment, as they can tell an animal a certain amount about the distribution of light and dark in the surroundings. However, with only shadowing from the pigment cup to restrict the acceptance angle of individual receptors, the resolution is much too poor for the eye to detect predators or prey, or to be involved in pattern recognition or the control of locomotion. All these tasks require the eye to have an optical system that can restrict receptor acceptance angles to a few degrees or better. This second stage in eye evolution, the provision of a competent optical system, has occurred much less frequently than the first, in only six of the 33 metazoan phyla listed by Barnes (1987): the Cnidaria, Mollusca, Annelida, Onychophora, Arthropoda, and Chordata. These are, however, the most successful phyla, as they contribute about 96% of known species. Perhaps the attainment of optical "lift-off" has contributed to this success.

The most exciting feature of the later stages of optical evolution has been the diversity of mechanisms that have been tried out in various parts of the animal kingdom. At last count, there were ten optically distinct ways of producing images (Figures 1 and 2). These include nearly all those known from optical technology (the Fresnel lens and the zoom lens are two of the few exceptions that come to mind), plus several solutions involving array optics that have not really been invented. Some of these solutions, such as the spherical graded-index lens (Figure 1e), have evolved many times; others, such as the reflecting superposition eyes of shrimps and lobsters (Figure 2j), have probably only evolved once. Four or five of these mechanisms have only been discovered in the last 25 years, which is remarkable given that excellent anatomic descriptions of most of these eyes have been available since the 1900s or earlier.

As most textbooks continue to refer to "the lens eye" or "the compound eye," as though these represented the totality of optical types, it seems appropriate to provide a brief review of all known mechanisms of image formation in eyes. We concentrate here on the new mechanisms, but do

---

Figure 1 The evolution of single-chambered eyes. The arrows indicate developments, rather than specific evolutionary pathways. See text for details and references. Compiled from many sources.

(a) Pit eye, common throughout the lower phyla. (b) Pinhole eye of *Haliotis* or *Nautilus*. (c) Eye with lens. (d) Homogeneous lens. (e) Inhomogeneous "Matthiessen" lens. (f) Multiple lens eye of male *Pontella*. (g) Two-lens eye of *Copilia*. Solid arrow shows image position; open arrow, the movement of the second lens. (h) Terrestrial eye of man with cornea and lens; *Ic*, image formed by cornea alone; *Ir*, final image on retina. (i) Mirror eye of the scallop *Pecten*.

not omit those that have been understood for much longer. Thus, the remainder of this chapter is mostly devoted to the mechanisms, capabilities, evolutionary origins, and affinities of the many kinds of "advanced" image-forming eye. The conventional division of eyes into "simple," i.e. single-chambered or camera-like, and "compound" is retained, because the mechanisms involved really are very different and represent topologically "con-cave" and "convex" solutions to the problem of image formation (Gold-smith 1990). Useful, supplementary accounts of optical mechanisms in the invertebrates are given by Land (1981a), and in several chapters in Ali (1984). Nilsson (1989) gives an excellent account of compound eye optics and evolution. Walls (1942) still provides the best comparative account of vertebrate eye optics, but Hughes (1977, 1986) and Sivak (1988) offer important new perspectives.

Simple Eyes

PIT EYES   These eye-spots are of interest here only because they must have provided the ancestors for optically more advanced eyes (Figure 1a). These eyes are typically less than 100 μm in diameter and contain from only 1 to about 100 receptors. They are found in all but five of the 33 metazoan phyla. They may be derived from ciliated ectodermal cells or, less commonly, from nonciliated ganglionic cells. The eyes may be "everse," i.e. the receptors are directed towards the light, and the nerve fibers pass through the back of the eye-cup. Or, they may be "inverse," i.e. the nerve fibers emerge from the front of the cup (Salvini-Plawen & Mayr 1977). Burr (1984) has reviewed behaviors that these eyes can mediate. There are three ways to improve the performance of an eye-spot. An enlarged cup and reduced aperture produces a pinhole eye (Figure 1b). The incor-poration of a refractile structure into the eye sharpens the retinal image and thus improves directionality (Figure 1c). And, the provision of a reflecting layer behind the receptors has two effects: First, it increases the amount of light available to the receptors; second, if the receptors move forward in the eye, it throws an image on them (Figure 1f). One can discern the beginnings of all these processes in the eye-spots of different invertebrate groups (reviews in Ali 1984).

PINHOLES   The only one good example of a pinhole eye is found in the ancient cephalopod mollusk *Nautilus* (Figure 1b). A few other mollusks have what one might describe as "improved pits." In the abalone *Haliotis*, the eye-cup is 1 mm long with a 0.2 mm pupil, and perhaps 15,000 receptors (Tonosaki 1967; Messenger 1981). The *Nautilus* eye, however, is quite different. Except for the absence of a lens, it is an advanced eye in all respects. It is large, almost 1 cm in diameter; it has an aperture that can be expanded from 0.4 to 2.8 mm (Hurley et al 1978); and it has extraocular muscles that mediate a response to gravity, thus stabilizing the eye against the rocking motion of the swimming animal (Hartline et al 1979).

Optically, however, this is a poor eye. The point-spread function (blur circle) on the retina cannot be smaller than the pupil, which limits resolution to several degrees at best. Muntz & Raj (1984) used the animal's optomotor response to test resolution and found that the minimum effective grating period was 11–22.5°, which is worse than expected. The real problem with this eye is that a reduction of the pupil diameter to improve resolution means a serious loss of retinal illuminance, and vice versa. Even at full aperture, the image is six times dimmer than in the eye of an octopus or fish, and the resolution is awful. The real mystery is that the pinhole has been retained. Almost any lens-like structure, however crude, placed in the aperture would improve resolution, sensitivity, or both. Thus, it must remain an evolutionary conundrum that this simple modification has not occurred here, when it has so often elsewhere.

SPHERICAL LENSES   In aquatic animals, the most common optical system in single-chambered eyes is based on a spherical lens (Figures 1c–e). Initially, such a lens would have arisen by an increase in the refractive index of the material within the eye-cup, brought about by the addition of protein or carbohydrate. Eyes with such undifferentiated (or "Fullmasse") lenses can still be found in some gastropod mollusks and annelids (see Land 1981a for earlier references). However, such a lens can only reduce the diameter of the blur circle on the retina, not form a sharp image, because the focal length cannot be shortened enough to fit the eye. In more advanced lens eyes, the required reduction in focal length is achieved because the lens has a special inhomogeneous construction, with dense, high refractive index material in the center, and a gradient of decreasing density and refractive index toward the periphery. In 1877, Matthiessen discovered this gradient in fish lenses (see Pumphrey 1961; Hughes 1986; Axelrod et al 1988). He was struck by the short focal length (about 2.5 radii, known as "Matthiessen's ratio"); if the lens were homogeneous, the refractive index would be 1.66, an unattainable value. In fact, the central refractive index is about 1.52, which falls to less than 1.4 at the periphery. The effect of the gradient is twofold. First, the focal length is reduced (and, concomitantly, the relative aperture increased) because light is continuously bent within the lens, not just at its surfaces. Second, with the correct gradient the lens can be made aplanatic, i.e. free from the spherical aberration, which makes homogeneous spherical lens virtually unusable (Pumphrey 1961) (Figure 1d and e). The exact form of the gradient that permits this condition was not achieved theoretically until quite recently

(Luneberg 1944; Fletcher et al 1954), although Matthiessen had proposed a parabolic gradient that was very similar. In spite of a recent suggestion by Fernald & Wright (1983) that fish lenses might have a substantial homogeneous core, it now seems that a continuous gradient, like that of Matthiessen or Luneberg, must be present to account for the observed ray paths (Axelrod et al 1988).

By measuring the focal length, it is easy to tell whether a particular group of animals has "discovered" how to make this kind of lens. If the focal length is around 2.5 radii, then the lens must have a gradient construction. A homogeneous lens with the same central refractive index would have a focal length of 4 radii. By this criterion, "Matthiessen" lenses have evolved at least eight times: in the fish, in the cephalopod mollusks (excluding *Nautilus*), at least four times in the gastropod mollusks (littorinids, strombids, heteropods, and some pulmonates), in the annelids (alciopid polychaetes), and once in the copepod crustaceans (*Labidocera*). Details are given in Land (1981a, 1984a). The remarkable lens eyes of cubomedusan jellyfish (Piatigorsky et al 1989) are not included here, as their optical properties have not been examined. Interestingly, the above list does include all aquatic lens eyes of any size; none have homogeneous lenses. One can conclude that there is one right way of producing such lenses, and that natural selection always finds it. Matthiessen lenses are indeed of excellent optical quality, as they offer high resolution with high light-gathering power. Their only residual defect is chromatic aberration (Fernald 1990), which need not have a serious effect.

Lens construction accounts for one aspect of the remarkable convergence between fish and cephalopod eyes. The identity of Matthiessen's ratio in the two groups, itself a result of the refractive index of the dry material of the lens center, and the inevitable spherical symmetry of the image effectively dictate the eyes' shape and proportions. The presence of eye muscles can be explained from the need to stabilize the image. This need grows with image quality, if that quality is not to be compromised by blur. Similarly, the need for an accommodation mechanism is determined by eye size, in the same way that focusing becomes more critical for camera lenses as the focal length increases. Thus, many of the convergent features that seem so remarkable (Packard 1972) are inevitable, given a particular type and size of eye.

MULTIPLE LENSES   Among aquatic eyes (Figure 1*f* and *g*), there are alternatives to the single spherical lens, but they are certainly not common. Two of the most interesting are found in copepod crustaceans, in which they are derived from parts of the single median eye. In *Pontella*, the lens is a triplet (Figure 1*f*); two elements are actually outside the eye in the animal's rostrum, and a third element is close to the retina of only six (!) receptors (Land 1984a). The eyes are sexually dimorphic—the females only have a doublet—and the animals themselves are conspicuously marked in blue and silver, which suggests a role for the eyes in the recognition of species and potential mates. Optically, the intriguing feature of the eyes is the first surface, which is parabolic. Ray tracing shows that this configuration can correct the spherical aberration of the other five interfaces in the optical system to provide a point image. This seems to be an interesting alternative solution, as an aspheric surface achieves the same result as the inhomogeneous optics of the Matthiessen lens.

Another copepod, *Copilia*, has fascinated biologists for more than a century. Its eyes are constructed strangely (Figures 1g and 3b), and they move to and fro in the longitudinal plane, thus scanning the water in front of the animal (Exner 1891). Each eye has two lenses that are arranged like a telescope: A large, long focal length "objective" lens forms an image on or close to a second, short focal length "eyepiece" lens immediately in front of the cluster of five to seven photoreceptors (Wolken & Florida 1969; Downing 1972). The second lens and receptors move together as a unit during scanning. The function of this astonishing system is still not well understood, but we discuss the possible role of scanning later in this review.

CORNEAL REFRACTION   In our own eyes, two thirds of the optical power lies in the cornea (Figure 1h). The lens, which is entirely responsible for image formation in our aquatic ancestors, is now mainly concerned with adjustments of focus. The use of a curved air/tissue interface for image formation is limited to terrestrial animals, and is actually a rather uncommon optical mechanism. Apart from the land vertebrates, the only other large group to use corneal refraction are the spiders (Land 1985a), whose eyesight can be remarkably acute. Williams & McIntyre (1980) estimate that the interreceptor angle in the jumping spider *Portia* is only 2.5 arc min. Considering the size of the animal (1 cm), this compares quite favorably with 0.5 arc min in the human fovea. The larvae of some insects also have simple eyes that form an image by using the cornea; the most impressive are the eyes of tiger beetle larvae (*Cicindela*), in which the interreceptor angle is about 1.8°. This is quite comparable in performance to the compound eyes of the adults that supplant them (Friederichs 1931; Land 1985b). The dorsal ocelli of adult insects are of the same general design, but are profoundly out of focus. They are concerned with stabilizing flight relative to the sky, and not with imaging.

For an eye of the corneal type to realize its maximum possible (diffraction limited) acuity, it must be corrected for spherical aberration. There are two ways this might be done: The cornea itself might be aspherical, as the surface that directs all parallel rays to a single point is not spherical, but elliptical; alternatively, an inhomogeneous lens might be used to produce the correction. According to Millodot & Sivak (1979), the cornea of the human eye is aspheric and thus corrected; the lens corrects itself by being inhomogeneous. The penalty of an aspheric correction is that the eye loses its radial symmetry, and thus has one "good" axis and reduced resolution elsewhere. Where all-round vision is needed, it may be better to go for the other solution. In the rat eye, which has a nearly spherical cornea, the lens is in fact overcorrected for spherical aberration, thus compensating for the cornea (Chaudhuri et al 1983). One further trick that seems to obtain a little more resolution from the eye is the inclusion of a negative lens, which is formed from the retinal surface, into the fovea, immediately in front of the receptors. This produces a system with telephoto properties and a locally enlarged image. Snyder & Miller (1978) first described this arrangement in an eagle, in which the eye's focal length effectively increased by 50%; a similar mechanism also occurs in some jumping spider eyes (Williams & McIntyre 1980).

The transition from lens-based to cornea-based optics, which accompanies the evolution of terrestrial life, must have involved a weakening of the power of the lens as the cornea became effective, much as happens today during metamorphosis in anurans (Sivak & Warburg 1983). Greater problems arise when an animal needs to operate effectively in both media at the same time. There seem to be two solutions. One solution is to retain all the power in the lens and have a flat cornea without power in either medium. This is approximately the situation in penguins and seals (Sivak 1988). An interesting variant of this occurs in porpoises (*Phocoena*), in which the cornea retains some power by having different inner and outer radii and an internal refractive index gradient. When the porpoise focuses in air, the cornea is flattened further (Kroeger 1989). The alternative to a nearly flat cornea is to provide the lens with huge powers of accommodation. This occurs in some diving birds, in which the powerful ciliary muscle squeezes the lens into, and partly through, the rigid iris, thus deforming the front surface into a locally very high curvature. In diving mergansers, this mechanism can produce as much as 80 diopters of accommodation, compared with 3–6 diopters in nondiving ducks (Sivak et al 1985).

CONCAVE REFLECTORS   Small eye-spots, in which the pigment cup is overlaid by a multilayer mirror, are found in some rotifers, platyhelminthes, and copepod crustaceans (see Ali 1984). However, none of these eyes are large enough to form usable images. In scallops (*Pecten*) and their relatives, the situation is different (Figure 1*i*). They have up to 100 respectable-sized

(1 mm) eyes around the edge of the mantle, each of which contains a "lens," a two-layered retina, and a reflecting tapetum. If one looks into the eye through the pupil, a bright inverted image is visible. Its location indicates that it could only have been formed by the concave reflector, not by the weak, low refractive index lens (Land 1965). The image visible to an observer is indeed the same one the animal sees. It falls onto the distal layer of the retina, where there are receptors that give "off"-responses. Thus, the animal sees moving objects—and shuts—as the image crosses successive receptors. These eyes represent an evolutionary line that is apparently quite unrelated to other molluskan eyes (Salvini-Plawen & Mayr 1977, Figure 8). The only other large eye that uses a mirror as an imaging device—rather than just a light-path doubler, as in the tapetum of a cat's eye—is in the deep-water ostracod crustacean *Gigantocypris*. These large (1 cm) animals have a pair of parabolic reflectors that focus light onto blob-like retinae at their foci. The resolution is probably very poor, but the light-gathering power is enormous, with a calculated F-number of 0.25 (Land 1984a).

Compound Eyes

In the last 25 years, we have seen a great revival of interest in compound eyes, with the discovery of three new optical types (reflecting and parabolic superposition and afocal apposition), the reinstatement of a fourth (refracting superposition), and rediscovery and naming of a fifth (neural superposition). In fact, the only type of compound eye to have avoided recent reappraisal is the classical apposition eye of diurnal insects and crustaceans, in which the erect image in the eye as a whole is built up from the elementary contributions of all the separate ommatidia (Figure 2b). Even that mechanism, proposed in the 1826 "mosaic theory" of Johannes Müller, came close to eclipse in the mid-nineteenth century and had to be revived by Sigmund Exner in his great monograph on compound eye optics (Exner 1891). Exner, the undisputed father of the subject, made two major discoveries, which we discuss below: the lens cylinder and the principle of superposition imagery. As seems to be the fate of ideas about compound eye function, they also came close to abandonment in the 1960s (see Land 1981a; Nilsson 1989), but survived the challenge undamaged. For readers interested in the history of the subject, Hardie's new (1989) translation of Exner's monograph, which includes a modern appendix, is a feast.

APPOSITION EYES   These are the best-known and most common compound eyes, and their relative simplicity strongly suggests that they are the ancestral type in each lineage. Each unit, or ommatidium, consists of a lens that forms an image onto the tip of the rhabdom, a light-guiding structure of photopigment-containing membrane formed from the contributions of a small number of receptor cells. The presence of the small, inverted image behind each facet caused confusion in the nineteenth century, but its role here is only to delineate each rhabdom's field of view and increase its brightness; the image is not resolved within the rhabdom. The animal itself sees the overall erect image across the eye, which is formed by the apposed "pixels" contributed by the individual ommatidia.

Apposition eyes are found in all three arthropod subphyla; the Chelicerata, Crustacea, and Uniramia (myriapods and insects). There is, however, no universal agreement regarding the number of times they evolved. Manton & Anderson (1979) favor separate origins of the three main groups, whereas Paulus (1979) believes that the groups are monophyletic, and that the original arthropods possessed faceted eyes. Among chelicerates, the horseshoe crabs (*Limulus*) have apposition eyes, and the prevailing view is that the simple eyes of scorpions and spiders are derived from these by reorganization under single lenses. The opposite appears to have occurred in the centipede *Scutigera*, in which a compound eye has apparently reevolved from scattered single elements. In insects and crustaceans, the compound eyes take many forms. However, there are sufficient detailed similarities in the way that individual ommatidia are constructed for a common ancestry to be a distinct possibility (Paulus 1979), although Nilsson (1989) takes the opposite view. Outside the arthropods, there are two remarkable examples of independently evolved apposition eyes, one in the annelids (on the tentacles of sabellid tube worms) and one in the mantle eyes of bivalve mollusks of the family *Arcacae* (see Salvini-Plawen & Mayr 1977; Land 1981a). In both cases, the eyes' function is to detect the movements of predators. In some of the tube worms, the eyes are little more than collections of pigmented tubes with receptors at the bottom.

Figure 2 The evolution of compound eyes. Arrows indicate developments, rather than specific evolutionary pathways, which are more complex. For further details and references see text. Compiled from many sources.

(a) Hypothetical ancestor with receptors in pigmented tubes. (b) Apposition eye. (c) Focal apposition ommatidium with image at rhabdom tip. (d) Multiinterface lens (*Notonecta*). (e) Lens-cylinder (*Limulus*); numbers in d and e are refractive indices. (f) Neural superposition in a dipteran fly; the numbers indicate the receptors and laminar structures that view the same directions in space. (g) Afocal apposition optics with intermediate image and collimated exit beam. (h) Superposition eye with deep-lying image. (i) Refracting superposition; inset shows axial and oblique ray paths. (j) Reflecting superposition; inset shows two views of ray paths through mirror box. (k) Parabolic superposition (*Macropipus*); inset shows focused beam recollimated by parabolic mirror.

This was probably how compound eyes originated in the mainstream of the Arthropoda (Figure 2a).

The image in each ommatidium may be produced in three different ways. In terrestrial insects, the curved cornea nearly always forms the image (Figure 2c). This mechanism is not available underwater; the

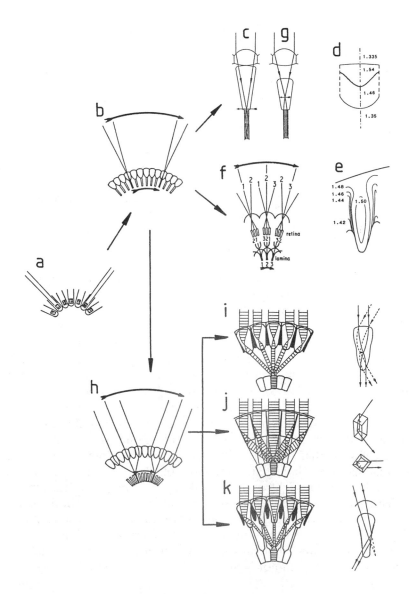

alternatives are the use of other lens surfaces (*Notonecta*, Schwind 1980), or a lens with a variable refractive index (*Limulus*, Exner 1891) (Figure 2*d* and *e*). Exner discovered the latter mechanism, which has affinities with the Matthiessen lens. He described it as a lens cylinder and showed that such a cylinder would form an image if the gradient of refractive index fell in an approximately parabolic fashion from the axis to the circumference. Eighty years later, interference microscopy made it possible to confirm Exner's farsighted conjecture in *Limulus* (Land 1979) and in the superposition eyes of many species (Figure 2*i*, see below).

The most serious limitation to the resolving power of apposition eyes, and of compound eyes in general, is diffraction (Mallock 1894; Barlow 1952; Snyder 1979). Image quality depends on lens diameter; the smaller the lens, the more blurred the image. The half-width of the diffraction image of a point source is given by λ/D radians. Thus, for green light ($\lambda = 0.5$ μm) and a lens diameter D of 25 μm, the diffraction image is 1.1° wide. The minimum angle that separates ommatidial axes cannot usefully be much smaller than this, which severely limits the quality of compound eye vision. By comparison, humans resolve 100 times better, as they have a single lens and a daylight pupil 2.5 mm in diameter. An improvement in the resolution of a compound eye requires an increase in both the sizes and number of the facets, which quickly results in structures of absurd dimensions. This is beautifully illustrated in Kirschfeld (1976).

NEURAL SUPERPOSITION EYES   In the dipteran flies, there is a variant of the apposition eye in which the elements (rhabdomeres) that comprise the rhabdom are not fused, but separated from each other (Figure 2*f*). In these insects, each inverted image is really resolved by the seven receptive elements in the focal plane, which raises all the problems of how the many inverted images are put together to form the overall erect image. The solution to this was first proposed by Vigier (1908, translation in Braitenberg & Strausfeld 1973), and rediscovered and proved by Kirschfeld (1967). The angle between the visual directions of the rhabdomeres within an ommatidium is the same as that between the ommatidial axes themselves, so that the six eccentric rhabdomeres in one ommatidium all have fields of view that coincide with the central rhabdomeres in adjacent ommatidia. Beneath the retina, the axons of all the retinula cells that view the same direction (eight, as the central rhabdomere is double) from seven adjacent ommatidia, collect up into the same "cartridge" in the lamina, after an impressively complicated piece of neural rewiring. Therefore, there is no difference between these eyes and ordinary apposition eyes at the level of the lamina. Dipterans thereby gain a sevenfold increase in the effective size of the photon signal and do not have to sacrifice resolution by increasing rhabdom size and, hence, acceptance angle. Thus, the flies have about an extra 15 minutes of useful vision at dawn and sunset.

AFOCAL APPOSITION   Butterflies have apposition eyes, but with an unusual construction (Figure 2g). The cornea forms an image, just as in the eye of a bee or grasshopper. Unlike those insects, however, the image is not at the rhabdom tip, but at the front focus of a second lens contained (as a lens cylinder) in the crystalline cone. This lens, which is of very short focal length, then recollimates the light, so that it emerges into the rhabdom as a parallel beam, not a focused spot (Nilsson et al 1984, 1988). This construction is basically the same as in a two-lens astronomical telescope, with an angular magnification of about ×6. It is considered "afocal" because there is no external focus, in contrast to the "focal" arrangement of an ordinary apposition eye (Figure 2c). As far as the resolution of the eye is concerned, the acceptance angle of each ommatidium is not determined by the angular subtense of the rhabdom tip, but by the critical angle for total internal reflection, which is set by the refractive index of the rhabdom itself. In practice, the situation is a little more complicated because the narrowness of the rhabdoms (ca. 2 μm) means that waveguide effects are important (Nilsson et al 1988; van Hateren 1989). Overall, the performance of the afocal apposition eye is marginally better than its focal equivalent (van Hateren & Nilsson 1987).

The afocal apposition eye is an important link between the apposition and superposition types, which we discuss next. It can be derived from an ordinary apposition eye by assuming that the second lens arises as a waveguide "funnel," which improves the transfer of light into the rhabdom. According to van Hateren & Nilsson (1987), such a structure can evolve into a lens without impediment. Once the second lens is present, and the system is afocal, it can further evolve into a superposition eye by an increase in the focal length of the second lens and a sinking of the retina to a more proximal position (Figure 2h). This type of transformation has apparently occurred several times in both the Lepidoptera and Coleoptera.

REFRACTING SUPERPOSITION EYES   In the eyes of many nocturnal insects and crustaceans, the rhabdom tips are not immediately behind the facet lenses, as they are in apposition eyes, but lie much deeper, with a zone of clear material that separates them from the optics (Figure 2i). Exner (1891) demonstrated that in the eye of the firefly Lampyris, a real, erect image is formed at the level of the retina. This image is produced by the super-position of rays from many elements across the eye surface. Exner also showed that such imagery is not possible if the optical elements behave as simple lenses. However, a single image will be produced by the array if each element behaves as a two-lens telescope that inverts the light path, but (unlike afocal apposition) has little actual magnification. A problem with this mechanism seemed to be that in *Lampyris*, and in other eyes of this type, the optical elements do not have sufficient optical power in their curved surfaces to function as telescopes. Exner's solution was again to postulate the presence of lens cylinder optics (see Apposition Eyes, above). These lens cylinders differ from those of *Limulus*, however, as they are twice the length, with a focus in the middle, not at the tip. Each half of the structure then behaves like one lens of a telescope, and overall the system becomes an afocal inverter, with a parallel output beam. Disbelief in both lens cylinders and superposition optics arose during the 1960s, and the modern reinstatement of Exner's ideas followed accurate refractive index measurements by Seitz (1969) and Hausen (1973) (see also Kunze 1979).

The feature crucial to the optical performance of all types of super-position eye is the accuracy with which the beams from each telescopic element coincide at the deep focus. In spite of an historic belief that there cannot be perfect coincidence, we now know that the superposition is so good in some diurnal moths that the eye operates at the diffraction limit for a single facet, which means that optically these eyes are as acute as equivalent apposition eyes (Land 1984b). By having a large effective pupil and large receptors, superposition eyes gain a 100-fold, or even a 1000-fold, increase in sensitivity; hence, their popularity in dim-light situations. McIntyre and colleagues have published a particularly fine series of studies, which explores all these issues, on the design of scarab beetle eyes (McIntyre & Caveney 1985; Warrant & McIntyre 1990a,b).

REFLECTING SUPERPOSITION EYES   Exner (1891) was actually wrong once. He thought that the eyes of long-bodied decapod crustaceans (shrimps, crayfish, lobsters) had superposition eyes of the refracting kind discussed above. However, attempts in the 1950s and 1960s to demonstrate lenses or lens cylinders in these eyes failed. Instead, these studies, which found square, homogeneous, low refractive-index, box-like structures, caused considerable confusion because no optical function could be ascribed to such elements. Thus, shrimps were blind, for about 20 years. This serious problem was resolved by Vogt (1975), who studied crayfish, and Land (1976), who studied shrimp. They discovered that the ray-bending was not done by lenses, but by mirrors in the walls of each "box." A comparison of Figures 2*i* and 2*j* shows that both telescopes and mirrors have the ability to invert the direction of a beam of light, so both can give rise to a superposition image. In many ways, the mirror solution seems more straightforward than the complicated telescope arrangement. This, however, is only true for the rather idealized case of Figure 2*j*, which illustrates rays in a section along a perfect row of mirrors in the center of the eye. Most rays away from the eye's plane of symmetry do not encounter a single mirror, but are reflected from two sides of the mirror "box" that makes up each optical element. There are, then, two important questions: What is the fate of these doubly reflected rays? Do all initially parallel rays reach a common focus? Here, the square arrangement of the facet array—almost unique to the decapod crustaceans—turns out to be crucial. Image formation is only possible if most rays encounter a "corner-reflector."

Consider first a simpler arrangement for producing a point image by reflection. This consists of a series of concentric "saucer rims," each angled to direct rays to a common focus; Figure 2*j* would then be any radial section through this array. The problem here is that such a stack has a single axis, and only rays nearly parallel to that axis form an image; other rays are reflected chaotically around the stack. The alternative is to replace the single reflecting strips with an array of mirror-pairs set at right angles. This substitution is possible because rays reflected from a corner go through two right angles and leave in a plane parallel to the incident rays (Figure 2*j*, inset). In other words, the rays behave almost as though they had encountered a single mirror at normal incidence, as in the saucer rim array. The beauty of the corner-reflector arrangement is that the orientation of each mirror pair is no longer important, unlike the situation in the single mirror array. Thus, the structure as a whole no longer has a single axis and can be used to make a wide-angle eye (Vogt 1977; Land 1981a). Clearly, this mirror-box design only works with right-angle corners and not hexagons, which accounts for the square facets. Various other features of these eyes are important for their function. The mirror boxes must be the right depth, about twice the width, so that most rays are reflected from two faces, but not more. Rays that pass straight through are intercepted by the unsilvered "tail" of the mirror boxes, whose refractive index decreases proximally to provide the appropriate critical angle for reflexion (Vogt 1980). Finally, there is a weak lens in the cornea of the crayfish. This lens "pre-focuses" the light that enters the mirror box, thus giving a narrower beam at the retina (Bryceson 1981). All these features provide an image comparable in quality to that produced by refracting superposition optics (Bryceson & McIntyre 1983; Nilsson 1989).

Reflecting superposition eyes, which are only found in the decapod crustaceans, presumably evolved within that group back in the Cambrian. The nearest relatives of the decapods, the euphausiids (krill), have refracting superposition eyes. The larval stages of decapod shrimps have apposition eyes with hexagonal facets, which change at metamorphosis into superposition eyes with square facets (Land 1981b; Fincham 1984; Nilsson 1983, 1989). Presumably, this transformation would have been no more difficult in evolution than in ontogeny. Interestingly, most of the true crabs (Brachyura), normally regarded as "advanced" decapods, have retained the apposition eyes into adult life. Undoubtedly, this reflects the crabs' littoral or semiterrestrial environment, in which light levels are high compared with the benthic or pelagic environment of shrimps and lobsters.

PARABOLIC SUPERPOSITION EYES   This final type of eye is the most recently discovered (Nilsson 1988) and the most difficult to understand. From an evolutionary viewpoint, it is also the most interesting because it has some characteristics of apposition eyes, as well as both other types of superposition eye (Figure 2k). It was first discovered in a swimming crab (*Macropipus* = *Portunus*). Each optical element consists of a corneal lens, which on its own focuses light close to the proximal tip of the crystalline cone, as in an apposition eye. Rays parallel to the axis of the cone enter a light-guiding structure that links the cone to the deep-lying rhabdom. Oblique rays, however, encounter the side of the cone, which has a reflecting coating and a parabolic profile. The effect of this mirror surface is to recollimate the partially focused rays, so that they emerge as a parallel beam that crosses the eye's clear-zone, as in other superposition eyes. This relatively straightforward mechanism is complicated because rays in the orthogonal plane (perpendicular to the page) encounter rather different optics. For these rays, the cone behaves as a cylindrical lens, thus creating a focus on the surface of the parabolic mirror. It then recollimates the rays on their reverse passage through the cone. This mechanism has more in common with refracting superposition. Thus, this eye uses lenses and mirrors in both apposition and superposition configurations and it would be the ideal ancestor of most kinds of compound eye. Sadly, the evidence is against this, as all the eyes of this kind discovered to date are from the brachyuran crabs or the anomuran hermit crabs, neither of which is an ancestral group to other crustaceans (Nilsson 1989). However, this eye does demonstrate the possibility of mixing mirrors and lenses, thus providing a viable link between the refracting and reflecting superposition types. This is important because such transitions do appear to have occurred. The shrimp *Gennadas*, for example, has a perfectly good refracting superposition eye, whereas its ancestors presumably had reflecting optics as in other shrimps (Nilsson 1990).

Zero-, One-, and Two-Dimensional Eyes

A novel classification of eye types that cuts across the scheme just presented is worth a brief comment. Most eyes, of whatever type, resolve surrounding space onto a two-dimensional retinal sheet; the third dimension is added by the brain on the basis of further clues, such as binocular disparity, and motion parallax. However, there are a few eyes in which the retina is essentially one-dimensional: a line of receptors, rather than a sheet. For a century, we have known that the eyes of heteropod sea-snails have linear retinae, three to six receptors wide and several hundred receptors long. A study of one of these, *Oxygyrus*, demonstrated that the retina works by scanning, and the eye tilts through 90° in about a second (Land 1982). In this way, the receptor row samples the surrounding water and presumably detects food particles (Figure 3*a*). A rather similar scanning system is found in the copepod *Labidocera*. Here, however, there are only ten receptors in the line, and only the males have the specialized eyes, which implies a role in mate detection (Land 1988). The principal eyes of jumping spiders (Salticidae) provide a third example. These have complex layered retinae, five to seven receptors wide by about 50 long (Land 1972; Blest 1985). They move the retinae from side to side (at right angles to the long dimension) and rotate them while examining novel stimuli (Land 1969). Interestingly, this scanning system operates in parallel with the fixed two-dimensional retinae of the antero-lateral eyes, which detect movement and act as "viewfinders" for the principal eyes. The last example is found in

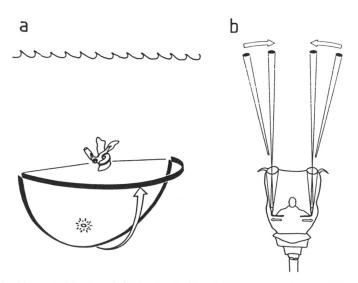

Figure 3   Alternatives to the two-dimensional retina. (*a*) The heteropod sea-snail *Oxygyrus* has a linear retina that scans the upwelling light in the ocean. (*b*) The copepod *Copilia* has retinae that only subtend 3° and scan through a total of 14°.

the apposition compound eyes of mantis shrimps (Stomatopoda). These have conventional, two-dimensional retinae, except that each eye has a strip of six rows of enlarged ommatidia that passes through the center. The eye keeps its remarkable octochromatic color-vision system in this strip, whose field of view is less than 5° by 180° (Cronin & Marshall 1989). This arrangement means that for the animal to determine the color of the stimulus, the eye must center it on the band and scan across it, which is indeed what happens. The eyes show both targeting movements and slow, small scanning movements (Land et al 1990). Whether the rest of the eye can function during scanning is not clear; the mantis shrimps do not have the extra stationary pair of eyes available to the jumping spiders.

The only convincing example of a zero-dimensional eye is the copepod *Copilia*, whose optical system was discussed earlier (Figure 1*g*). Each eye has a field of view only about 3° wide, and within this the image is probably not further resolved. Although the eyes move back and forth through about 14° (Downing 1972), they still only scan a minute fraction of the space around the animal (Figure 3*b*), which raises the obvious question, What can *Copilia* find to look at? Moray (1972) points out that although the scanning movements are horizontal, the predominant movements of the plankton on which *Copilia* feeds are vertical. Thus, the prey itself may provide a second dimension to the scanning. Without such additional motion, it is hard to imagine that the eyes are of much value.

Evolutionary Fine Tuning: Sampling the Environment

Thus far, we have only considered the broad types of eye and their macro-evolution, but not the way that each eye is adjusted to particular conditions. In general, one finds only a weak relationship between eye-type and the ecological niche of the animal; no one type is obviously more useful than the others in a given set of conditions. This is because all eye types can be built with higher or lower resolution, or absolute sensitivity, by varying the sizes of the eye itself, the optical components, the receptors, or the ganglion cell pools (see Land 1981a). The only real limitations on the uses of the different types are that compound eyes are diffraction limited to a resolution of about one degree and that superposition compound eyes are intrinsically more sensitive, size for size, than apposition ones, and so are more common in nocturnal or deep water arthropods.

One can often find a detailed correspondence between eye structure and niche in the way the image is sampled by the retina. The best-known example of such a relationship is in the distribution of ganglion cells in mammals. Animals that inhabit flat, open environments usually have "visual streaks" of high ganglion cell density that correspond approximately to the horizon (cheetah, plains kangaroo, rabbit); however, arboreal species, and those whose lateral view is generally obscured by vegetation (human, tree kangaroo, rat), typically have a radially symmetric pattern (Hughes 1977). Similar conclusions apply in birds (Meyer 1977; Martin 1985; Hayes & Brooke 1990). Many sea-birds, such as the manx shearwater and fulmar petrel, whose important visual world is the few degrees around and below the horizon, tend to have a horizontal, ribbon-like area of high ganglion cell density. Woodland birds, however, have varying arrangements of nonlinear areas that contain one or often two foveas. In reef fish, the same patterns occur. Collin & Pettigrew (1988a,b) found that fish that swam over sandy bottoms or below the open surface tend to have elongated horizontal high density areas, but those that live in holes and cracks in the reef itself had more circular areas (Figure 4a and b). One particularly interesting case is the surface-feeding fish *Aplocheilus*, which has two horizontal streaks separated by about 40°. One views the surface from below; the other, which views the surface from above, looks out of the water just above the edge of Snell's window (Munk 1970).

Interestingly, the visual system's priorities and interests in vertebrates show up at the level of the ganglion cells—the third-order neurons at the point where information must be compressed before transmission to the brain. There are differences in the densities of the receptors, the ultimate sampling stations, but they are much less striking than those seen in the ganglion cell layer. By contrast, in arthropods with compound eyes, variations in sampling related to the structure of the environment are found in the layout of the most peripheral structures, the ommatidia (Land 1989). Presumably, this difference arises because the information bottleneck in compound eyes is at the periphery, as the relatively low optical resolution is provided by the diffraction limited lenses. In any event, the equivalents of visual streaks and acute zones of various kinds appear when one maps the directions of the ommatidial axes onto a sphere around the animal (Figure 4c and d). Some common patterns include a region of enhanced vertical resolution around the horizon in many flying insects and in crabs that inhabit flat sandy or muddy shores (Zeil et al 1986); forward-pointing acute zones related to the velocity flow field in forward flight (Land & Eckert 1985; Land 1989); and special regions of high acuity, which usually point forwards and upwards, that are concerned either with sexual pursuit, in which case they are confined to males (Collett & Land 1975; van Praagh et al 1980), or with the capture of other insect prey on the wing, as in the eyes of dragonflies (Sherk 1978) (Figure 4d). There is even a remarkable analogue of the two horizontal streaks of *Aplocheilus*,

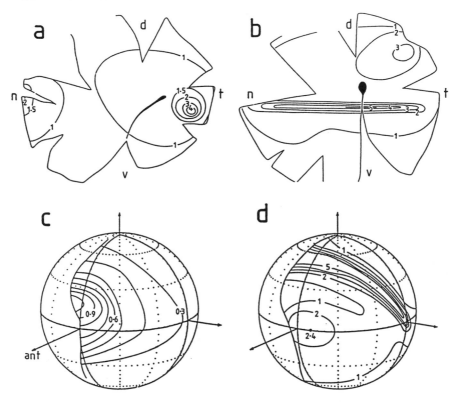

Figure 4   Adjustments of retina to habitat in fish (*a* and *b*) and insects (*c* and *d*). (d = dorsal, v = ventral, n = nasal, t = temporal.)

(*a*) Ganglion cell density map of the left retina of the reef rish *Cephalopholis miniatus*, which lives in cavities in the coral and has a forward-pointing acute zone. (*b*) A similar map for *Lethrinus chrysostomas*, which lives in open water and has a "visual streak" that images the underwater horizon. Numbers are thousands of ganglion cells per mm². Modified from Collin & Pettigrew (1988a,b). (*c*) Map of visual field of the left eye of a male blow-fly *Calliphora erythrocephala*, which shows an acute zone pointing forward and slightly dorsal. (*d*) Similar map for the dragonfly *Anax junius*, which has a streak of elevated resolution crossing the dorsal quadrant. Figures show the numbers of ommatidial axes per square degree of field. ant; anterior direction. (From Land 1989.)

in the eye of the surface-feeding backswimmer, *Notonecta* (Schwind 1980). In many instances, the provision of an acute zone in a compound eye necessitates an improvement in the diffraction limit, which can only be achieved by locally increasing the sizes of the facet lenses. In dipteran flies and dragonflies, this leads to the paradoxical situation in which the region of highest acuity appears superficially to have the coarsest mosaic.

CONCLUSIONS

1. At the molecular level, vision has an ancestry that predates the appearance of recognizable eyes. Thus, all metazoan opsins are sufficiently similar at the amino acid level to be regarded as having a single ancient origin. They may have originated from an earlier family of transmembrane proteins. In contrast, the many proteins responsible for refraction in vertebrate lenses are neither related to one another, nor have a single origin. Two, which appear in all vertebrates, are relatives of different stress response proteins of ancient origin, whereas most others double as enzymes. Various amounts of these different proteins are used to make lenses following rules that have thus far eluded discovery.

2. The number of anatomically and physiologically distinct receptor types seems to be small and may represent ancient lineages. As a general rule, the microvillous receptors of Protostomes depolarize by an increase in Na^+ conductance, whereas vertebrate ciliary receptors hyperpolarize to light by a Na^+ conductance decrease. A further class of multiciliate receptors also hyperpolarizes, but via an increase in K^+ conductance.

3. Eyes with well-developed optical systems evolved many times at the end of the Cambrian period. There are now about ten optically distinct mechanisms. These include pinholes, lenses of both multi-element and inhomogeneous construction, aspheric surfaces, concave mirrors, apposition compound eyes that employ a variety of lens types, and three kinds of superposition eye that utilize lenses, mirrors, or both.

4. Because the number of physical solutions to the problem of forming an image is finite, convergent evolution has been very common. The best example is the inhomogeneous Matthiessen lens, which has evolved independently in the vertebrates, several times in the mollusks and annelids, and once in the crustaceans. Similar cases of convergence can be found among compound eyes.

5. Not all eyes employ two-dimensional retinae to receive the image. In one crustacean (*Copilia*), the surroundings are scanned with a pair of receptors that have point-like fields. A number of mollusks and crustaceans have one-dimensional strip-like retinae, which scan at right angles to their long dimension.

6. Both simple and compound eyes may show very specific adaptations to environment and way of life. In vertebrates, these are usually manifest as variations in ganglion cell density across the retina. However, in the compound eyes of insects, in which the optics are limiting, these variations are seen in the size and disposition of the ommatidia.

Literature Cited

Ali, M. A., ed. 1984. *Photoreception and Vision in Invertebrates.* New York: Plenum. 858 pp.

Axelrod, D., Lerner, D., Sands, P. J. 1988. Refractive index within the lens of a goldfish determined from the paths of thin laser beams. *Vision Res.* 28: 57–65

Barlow, H. B. 1952. The size of ommatidia in apposition eyes. *J. Exp. Biol.* 29: 667–74

Barnes, R. D. 1987. *Invertebrate Zoology.* Philadelphia: Saunders. 893 pp.

Blest, A. D. 1985. The fine structure of spider photoreceptors in relation to function. In *Neurobiology of Arachnids,* ed. F. G. Barth, pp. 79–102. Berlin: Springer. 385 pp.

Bloemendal, H., ed. 1981. *Molecular and Cellular Biology of the Eye Lens.* New York: Wiley

Braitenberg, V., Strausfeld, N. J. 1973. Principles of the mosaic organization in the visual system's neurophil of *Musca domestica* L. In *Handbook of Sensory Physiology,* ed. H. Autrum, VII/3: 631–60. Berlin: Springer. 755 pp.

Bryceson, K. 1981. Focusing of light by corneal lenses in a reflecting superposition eye. *J. Exp. Biol.* 90: 347–50

Bryceson, K., McIntyre, P. 1983. Image quality and acceptance angle in a reflecting superposition eye. *J. Comp. Physiol. A* 151: 367–80

Burr, A. H. 1984. Photomovement behavior in simple invertebrates. See Ali 1984, pp. 179–215

Chaudhuri, A., Hallett, P. E., Parker, J. A. 1983. Aspheric curvatures, refractive indices and chromatic aberration for the rat eye. *Vision Res.* 23: 1351–63

Collett, T. S., Land, M. F. 1975. Visual control of flight behaviour in the hoverfly, *Syritta pipiens* L. *J. Comp. Physiol.* 99: 1–66

Collin, S. P., Pettigrew, J. D. 1988a. Retinal topography in reef fishes. I. Some species with well developed areae but poorly developed streaks. *Brain Behav. Evol.* 31: 269–82

Collin, S. P., Pettigrew, J. D. 1988b. Retinal topography in reef fishes. II. Some species with prominent horizontal streaks and high-density areae. *Brain Behav. Evol.* 31: 283–95

Cronin, T. W., Marshall, N. J. 1989. Multiple spectral classes of photoreceptors in the retinas of gonodactyloid stomatopod crustaceans. *J. Comp. Physiol. A* 166: 261–75

De Jong, W. W., Leunissen, J. A. M., Hendriks, W., Bloemenda. H. 1988. Evolution of alpha-crystallin: In quest of a function. In *Molecular Biology of the Eye: Genes, Vision, and Ocular Diseases,* ed. J. Piatigorski, T. Shinohara, P. Zelenka, pp. 149–58. New York: Liss

Doolittle, R. F. 1988. More molecular opportunism. *Nature* 336: 18

Downing, A. C. 1972. Optical scanning in the lateral eyes of the copepod *Copilia. Perception* 1: 193–207

Eakin, R. M. 1963. Lines of evolution in photoreceptors. In *General Physiology of Cell Specialization,* ed. D. Mazia, A. Tyler, pp. 393–425. New York: McGraw-Hill

Eakin, R. M. 1972. Structure of invertebrate photoreceptors. In *Handbook of Sensory Physiology,* ed. J. A. Dartnall, VII/1: 625–84. Berlin: Springer. 810 pp.

Edwards, J. S., Meyer, M. R. 1990. Conservation of antigen 3G6: a crystalline cone constituent in the compound eye of arthropods. *J. Neurobiol.* 21: 441–52

Exner, S. 1891. *The Physiology of the Compound Eyes of Insects and Crustaceans.* Transl. R. C. Hardie, 1989. Berlin: Springer. 177 pp. (From German)

Fernald, R. D. 1990. The optical system of fishes. In *The Visual System of Fish,* ed. R. Douglas, M. Djamgoz, pp. 45–61. London: Chapman & Hall. 526 pp.

Fernald, R. D., Wright, S. 1983. Maintenance of optical quality during crystalline lens growth. *Nature* 301: 618–20

Fincham, A. A. 1984. Ontogeny and optics of the common prawn *Palaemon (Palaemon) serratus* (Pennant, 1777). *Zool. J. Linn. Soc.* 81: 89–113

Fletcher, A., Murphy, R., Young, A. 1954. Solutions of two optical problems. *Proc. R. Soc. London Ser. A* 223: 216–25

Friederichs, H. F. 1931. Beiträge zur Morphologie und Physiologie der Sehorgane der Cicindeliden (Col.). *Z. Morphol. Okol. Tiere* 21: 1–172

Fuortes, M. G. F., O'Bryan, P. M. 1972. Generator potentials in invertebrate photoreceptors. In *Handbook of Sensory Physiology,* ed. M. G. F. Fuortes, VII/2: 321–38. Berlin: Springer

Goldsmith, T. H. 1990. Optimization, constraint, and history in the evolution of eyes. *Q. Rev. Biol.* 65: 281–322

Gorman, A. L. F., McReynolds, J. S. 1978. Ionic effects on the membrane potential of hyperpolarizing receptors in the scallop retina. *J. Physiol. (London)* 275: 345–55

Gorman, A. L. F., McReynolds, J. S., Barnes, S. N. 1971. Photoreceptors in primitive chordates: fine structure, hyperpolarizing receptor potentials, and evolution. *Science* 172: 1052–54

Haim, A., Heth, G., Pratt, H., Nevo, E. 1983. Photoperiodic effects on thermoregulation in a "blind" subterranean mammal. *J. Exp. Biol.* 107: 59–64

Hall, Z. W. 1987. Three of a kind: the Beta-adrenergic receptor, the muscarinic acetylcholine receptor, and rhodopsin. *Trends Neurosci.* 10: 99–101

Hartline, P. H., Hurley, A. C., Lange, G. D. 1979. Eye stabilization by statocyst mediated oculomotor reflex in *Nautilus*. *J. Comp. Physiol.* 132: 117–28

Hausen, K. 1973. Die Brechungsindices im Kristallkegel der Mehlmotte *Ephestia kuhniella*. *J. Comp. Physiol.* 82: 365–78

Hayes, B. P., Brooke, M. de L. 1990. Retinal ganglion cell distribution and behaviour in procellariiform seabirds. *Vision Res.* 30: 1277–90

Hendriks, W., Leunissen, J., Nevo, E., Bloemendal, H., de Jong, W. W. 1987. The lens protein alpha-A-crystallin of the blind mole rat, *Spalax ehrenbergi*: Evolutionary change and functional constraints. *Proc. Natl. Acad. Sci. USA* 84: 5320–24

Hughes, A. 1977. The topography of vision in mammals. In *Handbook of Sensory Physiology*, ed. F. Crescitelli, VII/5: 613–756. Berlin: Springer. 813 pp.

Hughes, A. 1986. The schematic eye comes of age. In *Visual Neuroscience*, ed. J. D. Pettigrew, K. J. Sanderson, W. R. Levick, pp. 60–89. Cambridge: Cambridge Univ. Press. 448 pp.

Hurley, A. C., Lange, G. D., Hartline, P. H. 1978. The adjustable "pin-hole camera" eye of *Nautilus*. *J. Exp. Zool.* 205: 37–44

Kirschfeld, K. 1967. Die Projektion der optischen Umwelt auf der Raster der Rhabdomere im Komplexauge von *Musca*. *Exp. Brain Res.* 3: 248–70

Kirschfeld, K. 1976. The resolution of lens and compound eyes. In *Neural Principles in Vision*, ed. F. Zettler, R. Weiler, pp. 354–70. Berlin: Springer. 430 pp.

Kroeger, R. H. 1989. *Dioptrik, Funktion der Pupille, und Akkommodation bei Zahnwalen*. Dissertation. University of Tübingen

Kunze, P. 1979. Apposition and superposition eyes. In *Handbook of Sensory Physiology*, ed. H.-J. Autrum, VII/6A: 441–502. Berlin: Springer. 729 pp.

Land, M. F. 1965. Image formation by a concave reflector in the eye of the scallop, *Pecten maximus*. *J. Physiol. (London)* 179: 138–53

Land, M. F. 1968. Functional aspects of the optical and retinal organization of the mollusc eye. *Symp. Zool. Soc. London* 23: 75–96

Land, M. F. 1969. Movements of the retinae of jumping spiders (Salticidae: Dendry-

phantinae) in response to visual stimuli. *J. Exp. Biol.* 51: 471–93

Land, M. F. 1972. Mechanisms of orientation and pattern recognition by jumping spiders (Salticidae). In *Information Processing in the Visual Systems of Arthropods*, ed. R. Wehner, pp. 231–47. Berlin: Springer. 334 pp.

Land, M. F. 1976. Superposition images are formed by reflection in the eyes of some oceanic decapod crustacea. *Nature* 263: 764–65

Land, M. F. 1979. The optical mechanism of the eye of *Limulus*. *Nature* 280: 396–97

Land, M. F. 1981a. Optics and vision in invertebrates. In *Handbook of Sensory Physiology*, ed. H.-J. Autrum, VII/6B: 471–592. Berlin: Springer. 629 pp.

Land, M. F. 1981b. Optical mechanisms in the higher crustacea with a comment on their evolutionary origins. In *Sense Organs*, ed. M. S. Laverack, D. J. Cosens, pp. 31–48. Glasgow: Blackie. 394 pp.

Land, M. F. 1982. Scanning eye movements in a heteropod mollusc. *J. Exp. Biol.* 96: 427–30

Land, M. F. 1984a. Crustacea. See Ali 1984, pp. 401–38

Land, M. F. 1984b. The resolving power of diurnal superposition eyes measured with an ophthalmoscope. *J. Comp. Physiol. A* 154: 515–33

Land, M. F. 1985a. The morphology and optics of spider eyes. In *Neurobiology of Arachnids*, ed. F. G. Barth, pp. 53–78. Berlin: Springer. 385 pp.

Land, M. F. 1985b. Optics of insect eyes. In *Comprehensive Insect Physiology, Biochemistry and Pharmacology*, ed. G. A. Kerkut, L. I. Gilbert, Vol. 6, pp. 225–75. Oxford: Pergamon. 448 pp.

Land, M. F. 1988. The functions of eye and body movements in *Labidocera* and other copepods. *J. Exp. Biol.* 140: 381–91

Land, M. F. 1989. Variations in the structure and design of compound eyes. In *Facets of Vision*, ed. D. G. Stavenga, R. C. Hardie, pp. 90–111. Berlin: Springer. 454 pp.

Land, M. F., Eckert, H. 1985. Maps of the acute zones of fly eyes. *J. Comp. Physiol. A* 158: 525–38

Land, M. F., Marshall, J. N., Brownless, D., Cronin, T. W. 1990. The eye-movements of the mantis shrimp *Odontodactylus scyllarus* (Crustacea: Stomatopoda). *J. Comp. Physiol. A* 167: 155–56

Leutscher-Hazelhoff, J. T. 1984. Ciliary cells evolved for vision hyperpolarize—why? The *Branchiomma* viewpoint. *Naturwissenschaften* 71: 213

Luneberg, R. K. 1944. *The Mathematical Theory of Optics*. PhD thesis. Brown

Univ., Providence, RI. Republished 1964, Berkeley: Univ. Calif. Press. 448 pp.

Mallock, A. 1894. Insect sight and the defining power of composite eyes. *Proc. R. Soc. London Ser. B* 55: 85–90

Manton, S. M., Anderson, D. T. 1979. Polyphyly and the evolution of arthropods. In *The Origin of Major Invertebrate Groups*, ed. M. R. House, pp. 269–321. London: Academic. 515 pp.

Martin, G. R. 1985. Eye. In *Form and Function in Birds*, ed. A. S. King, J. McLelland, Vol. 3, pp. 311–73. London: Academic

McIntyre, P., Caveney, S. 1985. Graded-index optics are matched to optical geometry in the superposition eyes of scarab beetles. *Philos. Trans. R. Soc. London Ser. B* 311: 237–69

Messenger, J. B. 1981. Comparative physiology of vision in molluscs. In *Handbook of Sensory Physiology*, ed. H.-J. Autrum, VII/6C: 93–200. Berlin: Springer. 663 pp.

Meyer, D. B. 1977. The avian eye and its adaptations. In *Handbook of Sensory Physiology*, ed. F. Crescitelli, VII/5: 549–611. Berlin: Springer. 813 pp.

Millodot, M., Sivak, J. 1979. Contribution of the cornea and lens to the spherical aberration of the eye. *Vision Res.* 19: 685–87

Moray, N. 1972. Visual mechanism in the copepod *Copilia*. *Perception* 1: 193–207

Munk, O. 1970. On the occurrence and significance of horizontal band-shaped retinal areae in teleosts. *Vidensk. Medd. Dan. Naturhist. Foren.* 133: 85–120

Muntz, W. R. A., Raj, U. 1984. On the visual system of *Nautilus pompilius*. *J. Exp. Biol.* 109: 253–63

Nathans, J. 1987. Molecular biology of visual pigments. *Annu. Rev. Neurosci.* 10: 163–94

Nilsson, D.-E. 1983. Evolutionary links between apposition and superposition optics in crustacean eyes. *Nature* 302: 818–21

Nilsson, D.-E. 1988. A new type of imaging optics in compound eyes. *Nature* 332: 76–78

Nilsson, D.-E. 1989. Optics and evolution of the compound eye. In *Facets of Vision*, ed. D. G. Stavenga, R. C. Hardie, pp. 30–73. Berlin: Springer. 454 pp.

Nilsson, D.-E. 1990. Three unexpected cases of refracting superposition eyes in crustaceans. *J. Comp. Physiol. A* 167: 71–78

Nilsson, D.-E., Land, M. F., Howard, J. 1984. Afocal apposition optics in butterfly eyes. *Nature* 312: 561–63

Nilsson, D.-E., Land, M. F., Howard, J. 1988. Optics of the butterfly eye. *J. Comp. Physiol. A* 162: 341–66

Packard, A. 1972. Cephalopods and fish: the limits of convergence. *Biol. Rev.* 47: 241–307

Paulus, H. F. 1979. Eye structure and the monophyly of the arthropoda. In *Arthropod Phylogeny*, ed. A. P. Gupta, pp. 299–383. New York: Van Nostrand Reinhold

Piatigorski, J., O'Brien, W. E., Norman, B. L., Kalumuk, K., Wistow, G. J., et al. 1988. Gene sharing by δ-crystallin and argininosuccinate lyase. *Proc. Natl. Acad. Sci. USA* 85: 3479–83

Piatigorsky, J., Horwitz, J., Kuwabara, T., Cutress, C. E. 1989. The cellular eye lens and crystallins of cubomedusan jellyfish. *J. Comp. Physiol. A* 164: 577–87

Piatigorski, J., Wistow, G. J. 1989. Enzyme/crystallins: gene sharing as an evolutionary strategy. *Cell* 57: 197–99

Pumphrey, R. J. 1961. Concerning vision. In *The Cell and the Organism*, ed. J. A. Ramsay, V. B. Wigglesworth, pp. 193–208. Cambridge: Cambridge Univ. Press. 350 pp.

Salvini-Plawen, L. V., Mayr, E. 1977. On the evolution of photoreceptors and eyes. *Evol. Biol.* 10: 207–63

Schwind, R. 1980. Geometrical optics of the *Notonecta* eye: adaptations to optical environment and way of life. *J. Comp. Physiol. A* 140: 59–68

Seitz, G. 1969. Untersuchungen am dioptrischen Apparat des Leuchtkäferauges. *Z. Vergl. Physiol.* 62: 61–74

Sherk, T. E. 1978. Development of the compound eyes of dragonflies (Odonata). III. Adult compound eyes. *J. Exp. Zool.* 203: 61–80

Sivak, J. G. 1988. Optics of amphibious eyes in vertebrates. In *Sensory Biology of Aquatic Animals*, ed. J. Atema, R. R. Fay, A. N. Popper, W. N. Tavolga, pp. 466–85. New York: Springer. 936 pp.

Sivak, J. G., Hildbrand, T., Lebert, C. 1985. Magnitude and rate of accommodation in diving and nondiving birds. *Vision Res.* 25: 925–33

Sivak, J. G., Warburg, M. 1983. Changes in optical properties of the eye during metamorphosis of an anuran, *Pelobates syriacus*. *J. Comp. Physiol. A* 150: 329–32

Snyder, A. W. 1979. Physics of vision in compound eyes. In *Handbook of Sensory Physiology*, ed. H.-J. Autrum, VII/6A: 225–313. Berlin: Springer. 729 pp.

Snyder, A. W., Miller, W. H. 1978. Telephoto lens system of falconiform eyes. *Nature* 275: 127–29

Tomarev, S. I., Zinovieva, R. D. 1988. Squid major lens polypeptides are homologous to glutathione S-transferase subunits. *Nature* 336: 86–88

Tonosaki, A. 1967. Fine structure of the

retina in *Haliotis discus*. *Z. Zellforsch*. 79: 469–80

Van Hateren, J. H. 1989. Photoreceptor optics, theory and practice. In *Facets of Vision*, ed. D. G. Stavenga, R. C. Hardie, pp. 74–89. Berlin: Springer. 454 pp.

Van Hateren, J. H., Nilsson, D.-E. 1987. Butterfly optics exceed the theoretical limits of conventional apposition eyes. *Biol. Cybern*. 57: 159–68

Van Praagh, J. P., Ribi, W., Wehrhahn, C., Wittmann, D. 1980. Drone bees fixate the queen with the dorsal frontal part of their compound eyes. *J. Comp. Physiol. A* 136: 263–66

Vigier, P. 1908. Sur l'existence réelle et le rôle des neurones. La neurone perioptique des Diptères. *C R Soc. Biol. (Paris)* 64: 959–61

Vogt, K. 1975. Zur Optik des Flusskrebsauges. *Z. Naturforsch*. 30c: 691

Vogt, K. 1977. Ray path and reflection mechanisms in crayfish eyes. *Z. Naturforsch*. 32c: 466–68

Vogt, K. 1980. Die Spiegeloptik des Flusskrebsauges. The optical system of the crayfish eye. *J. Comp. Physiol*. 135: 1–19

Walls, G. L. 1942. *The Vertebrate Eye and Its Adaptive Radiation*. Bloomington Hills: Cranbrook Inst. Reprinted 1963, New York: Hafner. 785 pp.

Warrant, E. J., McIntyre, P. 1990a. Limitations to resolution in superposition eyes. *J. Comp. Physiol. A* 167: 785–803

Warrant, E. J., McIntyre, P. 1990b. Screening pigment, aperture and sensitivity in the dung beetle superposition eye. *J. Comp. Physiol. A* 167: 805–15

Williams, D. S., McIntyre, P. 1980. The

principal eyes of a jumping spider have a telephoto component. *Nature* 288: 578–80

Wistow, G., Anderson, A., Piatigorski, J. 1990. Evidence for neutral and selective processes in the recruitment of enzyme-crystallins in avian lenses. *Proc. Natl. Acad. Sci. USA* 87: 6277–80

Wistow, G., Kim, H. 1991. Lens protein expression in mammals: taxon-specificity and the recruitment of crystallins. *J. Mol. Evol*. 32: 262–69

Wistow, G. J., Lietman, T., Piatigorski, J. 1988a. The origins of crystallins. In *Molecular Biology of the Eye: Genes, Vision, and Ocular Diseases*, ed. J. Piatigorski, T. Shinohara, P. Zelenka, pp. 139–47. New York: Liss

Wistow, G. J., Lietman, T., Williams, L. A., Stapel, S. O., de Jong, W. W., et al. 1988b. Tau-crystallin/alpha-enolase: one gene encodes both an enzyme and a lens structural protein. *J. Cell Biol*. 107: 2729–36

Wistow, G. J., Piatigorski, J. 1988. Lens crystallins: The evolution and expression of proteins for a highly specialized tissue. *Annu. Rev. Biochem*. 57: 479–504

Wolken, J. J., Florida, R. G. 1969. The eye structure and optical system of the crustacean copepod, *Copilia*. *J. Cell Biol*. 40: 279–85

Yau, K.-W., Baylor, D. A. 1989. Cyclic GMP-activated conductance of retinal photoreceptor cells. *Annu. Rev. Neurosci*. 12: 289–327

Zeil, J., Nalbach, G., Nalbach, H.-O. 1986. Eyes, eyestalks and the visual world of semi-terrestrial crabs. *J. Comp. Physiol. A* 159: 801–11

Annu. Rev. Neurosci. 1992. 15:31–56

DEVELOPMENT OF LOCAL CIRCUITS IN MAMMALIAN VISUAL CORTEX

Lawrence C. Katz and Edward M. Callaway

Department of Neurobiology, Duke University Medical Center, Durham, North Carolina 27710

KEY WORDS: plasticity, intrinsic circuits, pathfinding, horizontal connections, activity

INTRODUCTION

Most of the synapses in the mammalian cerebral cortex are components of local circuits. Not only do these local, or intrinsic, connections dominate in numerical terms, but such synapses are directly responsible for generating neuronal codes that convey sensory data and elicit behavior. In the mammalian primary visual cortex, the precise arrangement of local circuits in both vertical and horizontal dimensions is crucial for such fundamental properties of cortical neurons as detecting the orientation and direction of movement of visual stimuli. The elaboration of local connections, therefore, is probably a key factor in the emergence of computationally competent neural circuits.

Despite the central role of local circuits in neuronal processing, until recently several factors conspired to limit insight into how precise local excitatory and inhibitory interconnections emerge during development. First, the patterns of intrinsic circuitry in many brain areas were either unknown or known only in crude outline, which greatly hindered interpretation of developmental studies. Second, techniques used to elucidate local circuits were limited, as they relied almost exclusively on Golgi staining. In recent years, advances in techniques have improved our understanding of adult local circuits and have provided powerful tools for detailed investigations of local circuit development. These studies have been particularly

31

fruitful in the mammalian primary visual cortex, in which the arrangements of local circuits in the adult, and the response properties generated by these circuits, have been examined in considerable detail. Furthermore, the availability of at least some information concerning the emergence of specific response properties and behaviors during postnatal development offers an opportunity to correlate the elaboration of local connections to the emergence of distinct physiological properties.

In this review, we concentrate on the visual cortex to examine the emergence of intrinsic processing machinery in the mammalian neocortex. We define the general mechanisms and constraints on the elaboration of local axonal connections and focus primarily on the differentiation of excitatory links within and between the cortical layers. These excitatory synapses arise principally from spine-bearing neurons: the pyramidal cells outside of layer 4 and the spiny stellate cells within layer 4. The developmental history of subplate and layer I neurons has been reviewed by others (Marin-Padilla 1988; Shatz et al 1988). Although we emphasize development of the visual cortex, we also present additional information garnered from other brain regions that illuminate some of the basic mechanisms.

TECHNICAL ADVANCES IN THE STUDY OF LOCAL CIRCUITS

Most information about adult and developing cortical circuitry has been inferred from Golgi studies of immature animals. Although adult myelinated axons usually stain poorly, the unmyelinated processes of young animals stain well. Even in young animals, however, the Golgi technique suffers several well-known limitations. It stains capriciously, which makes it difficult to study a restricted neuronal population in isolation. Also, the completeness of axonal staining is always in question because of the pronounced age-dependence of impregnation. This is particularly troublesome in developmental studies, in which the growth, elaboration, and retraction of processes are critical events.

Intracellular Staining in Brain Slices

Intracellular staining, with either horseradish peroxidase (HRP), biocytin, or fluorescent dyes, overcomes many limitations of the Golgi method, thus allowing detailed visualization of the complete axonal arbors of selected subsets of cells in visual cortex (e.g. Gilbert & Wiesel 1979, 1983; Martin & Whitteridge 1984). Coupling intracellular staining with in vitro brain slice preparations has overcome many limitations of both Golgi staining and intracellular staining in vivo (Katz 1987). Specific neurons in specific locales (such as a cortical layer) can be targeted for injection, and the yield of filled cells is dramatically higher than with in vivo staining. Intracellular staining in brain slices has been successfully employed to visualize developing neurons in a variety of systems, including the visual cortex (Callaway & Katz 1990; Friauf et al 1990; Katz 1991), the optic tectum (Katz & Constantine-Paton 1988), the retina (Dann et al 1988; Ramoa et al 1987), and the hippocampus (Rihn & Claiborne, 1990). The high yield of filled neurons allows systematic assessment of the morphological consequences of specific manipulations to a developing system; by using directed intracellular injections, one can visualize dozens of similar neurons in each animal. For example, Katz et al (1989) examined the influence of ocular dominance column boundaries on cortical stellate cells in primate visual cortex; they required only a few animals to obtain enough cells to complete the study. The capricious nature of Golgi staining, or the restricted yield of intracellular staining in vivo, would make such investigations otherwise impossible.

Intracellular staining in slices has its own limitations, however, including the severing of long collaterals and difficulties in staining long processes in their entirety. Therefore, complementary anterograde and retrograde tracing techniques for visualizing local patterns of connections are frequently required.

New Fluorescent Tracers for Visualizing Local Circuits

The structure of the developing brain limits the usefulness of conventional tracing reagents; they usually diffuse widely in the large extracellular space in young animals and produce injection sites much larger than the entire extent of a developing local projection. Several new fluorescent tracers that overcome these problems have proven useful in the analysis of local cortical circuits. Retrogradely transported fluorescent latex microspheres, or "beads" (Katz et al 1984; Katz & Iarovici 1990) are composed of relatively large, hydrophobic particles that diffuse very little after injection into very young brains, thus providing the resolution necessary for developmental studies. Retrogradely transported beads are also retained in living cells for many months following uptake, thus permitting visualization of the eventual fate of neurons connected to particular locales in the developing animal. However, this approach does not reveal the detailed axonal or dendritic arborization of labeled neurons.

The fluorescent carbocyanine dye, 1,1'-dioctadecyl-3,3,3',3'-tetramethylindocarbocyanine perchlorate, popularly known as DiI, and DiA and DiO, which are of the same family, are also very powerful tools for analyzing developing local circuits. These dyes travel both anterogradely and retrogradely and they fill axons, including growth cones and fine branches, exceptionally well. In addition, these dyes diffuse along fibers even in fixed tissue (Godement et al 1987), which overcomes many of the problems of making tracer injections in prenatal animals or in difficult to access targets. Finally, these dyes allow detailed analysis of circuits in the human brain (Burkhalter & Bernardo 1989), which previously relied almost exclusively on the Golgi technique.

LINKING LAYERS: THE DEVELOPMENT OF VERTICAL CONNECTIONS

One of the earliest noted features of adult cortical organization was the strong tendency for connections to run vertically between cortical layers (Cajal 1911; Lorente de No 1944). For many years, vertical connections were considered the predominant, if not exclusive, dimension of connectivity in the cortex (Lorente de No 1933). The discovery of orientation columns, in which radially aligned groups of neurons shared preference for the orientation of a visual stimulus, was also attributed to the strong vertical links between cortical layers. Hubel & Wiesel (1962) observed that cells with increasingly complex response properties were located in different layers and hypothesized that interlaminar connections were used to construct more elaborate physiological properties from simpler ones. For example, they proposed that several "simple" cells in layer 4 converged on overlying cells in layer 3 to generate neurons with "complex" receptive fields. In this hierarchical scheme of cortical processing, specific intrinsic vertical connections between excitatory neurons (spiny stellate and pyramidal cells) in different layers form the neural circuitry required to analyze information from a local portion of the visual field. The use of Golgi, extracellular, and intracellular staining techniques provided an anatomic framework for the hierarchical theory and revealed the basic patterns of excitatory interlaminar connections in the adult visual cortex of cats and primates (reviewed in Gilbert 1983).

Except for the recent work of Bolz & Gilbert (1986, 1989), who tested the physiological roles of vertical connections by inhibiting the activity of one layer while recording responses in another, the exact contribution of interlaminar connections to receptive field properties remains largely unknown. In this light, developmental studies may provide some insight into the role of interlaminar connections: If the time course of the emergence of a specific vertical connection is known, the differentiation of the circuit can be correlated with the emergence of physiological response properties.

Development of Vertical Connections:
Specificity vs. Exuberance

When a developing pyramidal cell completes its migration, it has only a single process, its efferent axon, which either has reached or is en route to distant targets (Cajal 1911; Miller 1988). The transition from one simple process to the exquisite order and remarkable laminar specificity of adult vertical excitatory connections could involve several distinct cellular mechanisms. For example, consider the pattern of vertical connections of pyramidal cells in layer 2/3. In adult cats and primates, the main efferent axons of these cells descend through all the cortical layers (except layer 1) on their way to making long-distance connections to extrastriate cortical areas. Collateral branches are primarily present in layers 2/3 and 5; layer 4 contains few, if any, branches (Gilbert & Kelly 1975; Gilbert & Wiesel 1979, 1983; Lund & Boothe 1975; Tigges & Tigges 1982). The specificity of this connection, like that of any vertical connection in cortex, could be achieved in at least three ways: selective "pruning" of initially elaborate and poorly specified collaterals present in all layers; sprouting of collaterals in all layers, followed by differential growth of collaterals situated in appropriate layers; or initially specific outgrowth of collaterals exclusively in the appropriate layers. Ample precedents exist for the role of all three mechanisms in the development of the nervous system in general and the visual system in particular. Extensive pruning of geniculocortical afferents has been implicated as the mechanism that underlies the segregation of inputs to form the ocular dominance stripes in striate cortex (LeVay & Stryker 1979; LeVay et al 1980). Elimination of immature sprouts may be involved in the refinement of geniculocortical topography in hamster visual cortex (Naegele et al 1988), and differential outgrowth of appropriately situated terminal arbors (coupled with elimination of inappropriate sprouts) appears to be one mechanism by which lamina-specific arborizations are formed by retinal ganglion cell axons in the lateral geniculate nucleus (Sretevan & Shatz 1984, 1986). The formation of the specific pattern of intrinsic vertical connections in cerebral cortex, however, does not seem to involve formation of sprouts or collaterals in inappropriate layers. Instead, studies of cats, monkeys, and humans indicate that interlaminar connections are generally highly specific from the initial elaboration of local collaterals.

Lund et al (1977) first noted that specific patterns of laminar connections could be achieved by specific outgrowth, rather than by regressive events. They studied the emergence of local circuits in macaque cortex by using Golgi staining. In layer 6 of late postnatal and young adult brains, at least four cell types can be differentiated, based on highly specific and stereotyped patterns of laminar arborizations of their dendrites and axons. These patterns may be related to the patterns of inputs from magnocellular and parvocellular layers of the lateral geniculate nucleus, which subserve very different functions in visual processing. Lund et al examined these cells at embryonic day 127 (E127) and noted that the specific dendritic and axonal arborizations were already present by that age. In many cases the sprouts in the "appropriate" layers were < 50 μm long, but no sprouts or growth cones were observed in incorrect layers. Although the emergence of the laminar patterns was not followed from the very earliest stages (i.e. before the emergence of any collateral arbors), Lund et al concluded that "the specific pattern is attained in the initial growth of the neuron, not apparently by later loss of processes or 'pruning' of a more random initial growth." Thus, at E127, well before birth at E165 and certainly before the onset of any visual experience, the specificity ·of this interlaminar circuit has already been achieved.

One could argue that a period of exuberance for these intrinsic connections in the monkey occurs before E127, or that the Golgi technique might fail to impregnate very immature, fine sprouts in the inappropriate layers. These possibilities seem less likely in view of observations of layer 2/3 neurons in cat and human visual cortex, which started at younger developmental stages and used different staining techniques. In the adult cortex, layer 2/3 pyramidal cells form extensive collateral arbors within both layer 2/3 and layer 5, but they almost completely avoid the intervening layer 4. Intracellular staining with Lucifer yellow in brain slices prepared from neonatal cats demonstrated that this specific pattern of collaterals was present from the very earliest times that collateral sprouts could be detected (Katz 1991). Even extremely immature cells, in which only growth cones emerged from the main efferent axon, had sprouts exclusively in the appropriate layers. Furthermore, the number of primary collaterals remained constant throughout postnatal development, which suggests that little, if any, elimination of primary collaterals occurred. By using extracellular staining in fixed fetal human brains, Burkhalter et al (1990) also concluded that the formation of interlaminar connections of layer 2/3 cells involved specific outgrowth in the appropriate layers. At 24–26 weeks' gestation, crystals of DiI, which were placed in the upper layers, labeled only a radial band of fibers, with no evidence of collateral spouts in layers 2/3 or 5. Only at 29 weeks were collaterals first visible in layer 5; no sprouts were evident in layer 4.

In addition to the extrinsically projecting pyramidal cells, other cells have demonstrated early specificity. In the adult monkey, for example, spiny stellate cells in layer 4 (the geniculate afferent recipient layer) have stereotyped projections to layers 3 and 5. Lund et al (1977) found that the specificity of these characteristic patterns of connections developed early; by E127, the rudiments of the adult connections were already in place. No cells with widespread or unusual collateral systems were present. Investigations of the development of vertical connections by layer 4 spiny stellate cells in the cat, which used either Golgi (Meyer & Ferres-Torres 1984) or intracellular staining (Katz & Callaway 1990), similarly found that the laminar specificity characteristic of these cells was present from the earliest times that collaterals began to form.

The vertical connections formed by other nonprojecting neurons also seem to develop highly specifically. In the adult cortex, about 20% of neurons are GABA-containing inhibitory interneurons. These cells fall into an array of cell types, which are principally distinguishable by their local axonal connections (reviewed in Fairen et al 1984; Lund 1988). The laminar organization of these inhibitory connections is as precise as that of excitatory neurons. Golgi studies strongly suggest that the laminar specificity of inhibitory connections, like that of excitatory vertical connections, also arises via specific outgrowth, rather than through a process of elimination. When the development of a particular class of inhibitory neurons in monkey layer 4 was analyzed, the characteristic laminar distribution of collaterals was present by E127 (Lund et al 1977). A similar pattern of development was observed in the cat (Meyer & Ferres-Torres 1984).

Timing of Specific Collateral Outgrowth

There are large differences in the age of neurons in different cortical layers, and one might expect that the emergence of intrinsic vertical circuits would reflect these maturational differences. The cell layers in the mammalian cortex (except for layer 1 and the subplate zone) are formed in an "inside-out" pattern, in which later-generated cells migrate past postmigratory cells to occupy positions near the top of the cortical plate (Angevine & Sidman 1961). In animals with increasingly long gestation times, these differences can be dramatic. In the cat, more than one month elapses between the time that layer 6 cells reach their position and the layer 2 cells reach theirs (Luskin & Shatz 1985); in monkeys, at least 50 days separate the arrival times of layer 6 and layer 2 cells (Rakic 1974). Recent observations indicate that when timing differences do exist, arborizations sometimes form first in the more mature layers. Very young layer 4 spiny stellate cells form arbors in "older" layers 4 and 5 before extending collaterals into the "younger" layer 2/3 (Katz & Callaway 1990; Lund et al 1977; Meyer & Ferres-Torres 1984). In humans, layer 2/3 cells may form collaterals in the older layer 5, several months before forming collaterals within layer 2/3 (Burkhalter et al 1990). Some investigators have reported that inhibitory interneurons in deeper layers appear to differentiate before those located more superficially (Miller 1986). The patterns of connections between cortical areas can also reflect a similar age-dependence, as older cells form arbors in older layers first (Coogan & Burkhalter 1988).

Alternatively, local circuits in the entire visual cortex might develop more or less in concert, perhaps initiated by some cortex-wide cue (Lund et al 1977; Parnavelas et al 1978). Developing layer 6 neurons form equally extensive arbors in layer 4 and layer 3, with no obvious delay in either layer (Lund et al 1977). Similarly, layer 2/3 cells in cat form sprouts and collaterals simultaneously in layers 2/3 and 5 (Katz 1991). Rakic et al (1986) have observed that cortical synaptogenesis, which should at least partly reflect the formation of local connections, proceeds simultaneously in all cortical layers, and not in an inside-out fashion.

Although the "gradient rule" and the "simultaneity rule" are each consistent with the development of some vertical circuits, clearly no single rule applies to all. On a sufficiently coarse time scale, circuits may develop in near synchrony. However, as the time course of circuit differentiation of specific cell types is examined in more detail, discontinuities in the apparent simultaneity are revealed. When the differentiation of some nine different types of spine-free cells in the cat was examined at closely spaced intervals, different types of these inhibitory interneurons had apparently matured at dramatically different rates, even within the same cortical layer (Meyer & Ferres-Torres 1984). Distinctions in the timing of local circuit formation may be more readily apparent in animals with protracted cortical development, like cats and humans. In the next section, we consider some of the possible cues that could produce layer-specific branching patterns.

Cues for Generating Vertical Specificity

To relate the mechanisms involved in local circuit formation to the formation of other connections in the brain more closely, it is worthwhile to distinguish between activity-independent and activity-dependent events during the development of neuronal connections. In general, the pathfinding events and cues that are involved in the navigation of axons towards their correct targets elsewhere in the brain involve a complex interplay between mechanical constraints, diffusible growth factors, and specific molecules on neuronal and nonneuronal cell surfaces. Neuronal activity during pathfinding does not appear to play a significant role. In the vertebrate visual system, for example, axons of retinal ganglion cells make appropriate pathway choices at the optic chiasm, and locate their appropriate targets, even in the presence of TTX, which blocks neuronal activity (reviewed in Udin & Fawcett 1988).

In contrast, the mechanisms that determine how developing axons form specific patterns of terminal arborizations within target areas generally require either evoked or spontaneous neuronal activity (for recent reviews, see Harris & Holt 1990; Shatz 1990). Well-known examples of synaptic rearrangements within the visual system include the selection of appropriate layers in the lateral geniculate nucleus by ingrowing retinal ganglion cell axons and the formation of the segregated pattern of ocular dominance columns by geniculocortical axons in the visual cortex. In contrast to pathway finding cues, blocking neuronal activity by TTX application prevents the normal segregation of retinal ganglion cell terminals into appropriate geniculate layers (Shatz & Stryker 1988; Sretevan et al 1988), prevents the formation of ocular dominance columns in cortex (Stryker & Harris 1986), and reduces the specificity of retinotectal topography (Meyer 1983; Schmidt & Edwards 1983).

The limited evidence currently available suggests that the mechanisms for the formation of specific vertical connections have more in common with the pathway choices observed in other parts of the brain, than with activity-dependent reorganization of synapses. Unlike the formation of cortical ocular dominance columns, vertical connections develop specifically from the outset of axonal differentiation. Furthermore, the development of this specificity may not require either patterned visual experience or spontaneous retinal activity. Continuous binocular deprivation for the first few postnatal months had no discernable effect on either the pattern of interlaminar connections of layer 4 stellate cells or the specificity of the vertical connections of layer 2/3 pyramidal cells; cells still formed normal numbers of branches in the appropriate layers and avoided forming sprouts in inappropriate layers (Katz & Callaway 1990 and unpublished observations). Physiological experiments also suggest that patterned visual activity is not required for the development of specific vertical connections. End-stop inhibition, which relies on vertical connections from layer 6 to layer 4, is present in visually inexperienced animals (Braastad & Heggelund 1985). At least until the end of the first postnatal month, neither binocular deprivation nor dark rearing disrupts the appearance of oriented cells throughout the cortical layers. And, these manipulations do not degrade the columnar organization of orientation selective cells (Sherk & Stryker 1976; Braastad & Heggelund 1985).

Numerous cell surface molecules, which either stimulate or inhibit neurite outgrowth, have been described throughout the nervous system. The absence of sprouts of layer 2/3 pyramidal cells in layer 4 could result from the presence of inhibitory factors on either neuronal or nonneuronal cells, as described in optic tectum (Walter et al 1987a,b). However, even while layer 2/3 cells fail to sprout in layer 4, collaterals from both layer 4 and layer 6 cells are growing well within the layer. Therefore, if inhibitory cues are involved, they must be specific for one group of cells.

Not surprisingly, some aspects of the development of vertical circuits within the cortex display a striking similarity to the selection of specific targets by other cortical cells. Pyramidal cells in layer 5 project to the brainstem, where they form specific connections to the basal pontine nuclei. These connections are formed by "interstitial sprouting": The axon first grows past the target nucleus; at a later point in development, sprouts emerge and grow specifically into the appropriate subregion of the target. The specificity of the terminal arborization, therefore, emerges via specific outgrowth from a main axon, and not via selection of appropriate collaterals (O'Leary & Terashima 1988). The descriptions of interstitial sprouting in the corticopontine system are reminiscent of interstitial sprouting in the formation of interlaminar connections. Recent in vitro evidence suggests that this interstitial sprouting is induced by a diffusible factor secreted by the pons (Heffner et al 1990). Although the sprouting factors that might be responsible for specific interlaminar connections in cortex have not yet been identified, such factors may be responsible for targeting geniculocortical afferents to layer 4 (Bolz et al 1990).

Differentiation of Vertical Connections and the Emergence of Response Properties

Because the inputs from individual neurons in the lateral geniculate nucleus carry unoriented, monocular information, almost all of the critical cortical response properties must be synthesized within the cortex itself. From a developmental standpoint, physiological studies can answer two important questions: Are response properties innately specified, or are they modified by the animal's visual experience? What is the relationship between the emergence of specific connections and the appearance of distinct response properties?

There is now widespread agreement that the basic response property of orientation selectivity, and the arrangement of cells with similar orientations into columns, is present very early in postnatal development. Many subsequent studies in the cat have confirmed the initial observations of Hubel & Wiesel (1963), who found that orientation selective cells can be detected as soon as the eyes are open, about postnatal day 8, and that these cells are organized into columns (Albus & Wolf 1984; Blakemore & Van Sluyters 1975; Braastad & Heggelund 1985; Frégnac & Imbert 1978; Wiesel & Huble 1974). These studies suggest that at least the rudiments of the local circuitry necessary for generating orientation selectivity are present at eye opening.

Close parallels apparently exist between the emergence of specific response properties and the differentiation of vertical connections. A detailed laminar analysis of the emergence of response properties in the cat revealed that during the first two postnatal weeks, only cells that resemble simple cells were visually responsive; these were only found in layers 4 and 6, which receive direct geniculate input (Albus & Wolf 1984). Cells with complex receptive fields were only detected in layers 2/3 and 5 after the beginning of the third postnatal week (>P14). The emergence of complex cells may be related to the formation of connections from layer 4. Until about P12, the proportion of layer 4 cells that form vertical connections to the overlying layer 2/3 is very small. The number of cells that form such connections only increases to near adult values toward the beginning of the third week, even though the strength of the 4–2/3 connection is still much less than in the adult (Katz & Callaway 1990). Because complex receptive field properties emerge at approximately the same time that layer 4 cells grow into layer 2/3, this connection may be vital for conveying the information necessary to generate complex receptive fields. Furthermore, the temporal difference in the emergence of simple and complex receptive fields, coupled with the anatomic results, supports the idea that complex receptive fields may be generated, at least initially, by inputs from simple cells in layer 4 (for an alternative hypothesis for generating complex receptive fields in the adult, see Malpeli 1983). The simultaneous emergence of complex cells in layers 2/3 and 5 also seems to echo the simultaneity in the development of connections from layer 2/3 to 5, which supports the idea that the receptive fields of layer 5 cells may be at least partially generated from inputs from layer 2/3.

CONNECTING COLUMNS: DEVELOPMENT AND MODIFICATION OF HORIZONTAL CONNECTIONS

Although Golgi techniques strongly emphasized vertical connections that link different layers, newer tracing techniques have revealed extensive horizontal connections over distances of several millimeters within individual cortical layers. These connections, variously termed tangential, intralaminar, or horizontal, are especially prominent in layers 2/3 and 5 of the visual cortex in primate and nonprimate species. Individual pyramidal neurons injected with HRP form numerous periodic aggregates of synaptic terminals spaced at about 1 mm intervals in the tangential plane, over a distance of up to 6 mm from end to end (3 mm from the cell body) (Gilbert & Wiesel 1979, 1983; Martin & Whitteridge 1984). These periodic connections are reciprocal: Small cortical injections of mixed retrograde and anterograde tracers label distinct clusters of neurons and synaptic terminals, also spaced at about 1 mm intervals over similar distances, in cats, tree shrews, and primates (Burkhalter & Bernardo 1989; Gilbert & Wiesel 1989; Livingstone & Hubel 1984; Rocklund 1985; Rocklund & Lund 1982, 1983).

Several roles have been suggested for these connections (Mitchison & Crick 1982), although their precise functions are unknown. Long horizontal connections may generate the elongated receptive fields encountered in layer 6 (Bolz & Gilbert 1989). Their relationship to functional cortical organization suggests that long horizontal connections are involved in integrating visual information across columnar boundaries to form a cohesive map of the visual field. In the cat, and probably in tree shrews, horizontal connections in layer 2/3 specifically interconnect columns that share the same preferred orientation (Gilbert & Wiesel 1989). In primates, horizontal connections specifically link the cytochrome oxidase rich "blobs" to one another, and "interblob" regions to each other (Burkhalter & Bernardo 1989; Livingstone & Hubel 1984). Gray & Singer (1989) have reported that synchronized oscillatory responses to same-orientation stimuli extend over many millimeters. They also propose that the synchrony of the responses is mediated by tangential connections. Because the horizontal connections span a cortical area greater than that corresponding to the classical receptive field of an individual neuron, such connections may be important in modifying a neuron's responsivity according to the context of surrounding stimuli (e.g. Allman et al 1985; Gilbert & Wiesel 1990; Gray et al 1989; Gulyas et al 1987; von der Malsburg & Schneider 1986). Thus, the emergence of perceptual capabilities that require integration of information from distant points in the visual field may also require the development of appropriate horizontal links within a cortical layer.

Normal Development of Horizontal Connections in Visual Cortex

In very young cats and humans, clustered horizontal connections are absent (area 17: Burkhalter et al 1990; Callaway & Katz 1990; Luhmann et al 1990a; area 18: Price 1986). Until the end of the first postnatal week in cats, retrograde tracers injected into the superficial layers of cortex result in a continuous pattern of labeling over a limited tangential domain; similarly, DiI injections in neonatal human brains revealed no clustering, even at four months postnatal. By the end of the first postnatal week in the cat, tracer injections reveal crude clusters of retrogradely labeled cells over a somewhat greater tangential extent than seen only a few days earlier. These crude clusters emerge simultaneously in layers 2/3 and 5. There is some controversy over the subsequent process of cluster development.

Luhmann et al (1986, 1990a) claim that the number, spacing, and tangential extent of clusters, which result from a single retrograde tracer injection, all increase steadily during the first postnatal month, until the most distant clusters are up to 10 mm from an injection site. After the fourth postnatal week, both the number and tangential extent of labeled clusters decline to adult levels. Thus, according to these investigators, the development of early postnatal clusters involves reductions in both the length of tangential fibers and the number of clusters linked to a single site. In contrast, Callaway & Katz (1990), who used sequential tracer injections at precisely the same cortical locus, observed that the number, tangential extent, and location of clusters labeled from a single site remained unchanged between P15 and P30, at which time labeling was indistinguishable from that in adult animals (Figure 2). The tangential extent of clusters reached its adult values (3–4.5 mm from the injection site) by eight days postnatal, and there was no subseqent increase or reduction. However, Callaway & Katz found that the early clusters were "crude;" as many retrogradely labeled cells were present in the spaces between clusters. As animals matured, the proportion of labeled cells between clusters gradually diminished, until the adult organization—distinct clusters of labeled cells with few cells between clusters—appeared by the end of the first postnatal month (see Figure 1).

Some of the differences between the observations of the two groups are probably methodological. The failure to differentiate between crude and refined clusters may be attributable to the use of a neuronal tracer (WGA-HRP) that produces injection sites in excess of 1 mm in diameter. These injections are far larger than the distance between adjacent clusters; therefore, labeling between clusters would be expected regardless of the precision of connections between them. Luhmann et al (1986, 1990a) also analyzed their data with an image processor that employed a band-pass filter designed to remove contributions from labeling between clusters.

Differences related to the tangential extent of retrograde label are more difficult to reconcile. In more than 20 experiments involving microsphere injections in animals aged P12–P21, Katz & Callaway never observed intrinsic label extending more than 4.5 mm from an injection site. Luhmann et al (1990a) reported label extending more than 10 mm from a microsphere injection in a few cases, but considerably less in others. Using current source density analysis, these investigators report that they saw no direct electrophysiological evidence for such extensive horizontal collaterals (Luhmann et al 1990b). Indirect evidence for changes in the length of tangential arbors was obtained by using extracellular single-unit recordings, which occasionally revealed cells with ectopic receptive fields. The distance of such fields from the receptive field center was comparable to

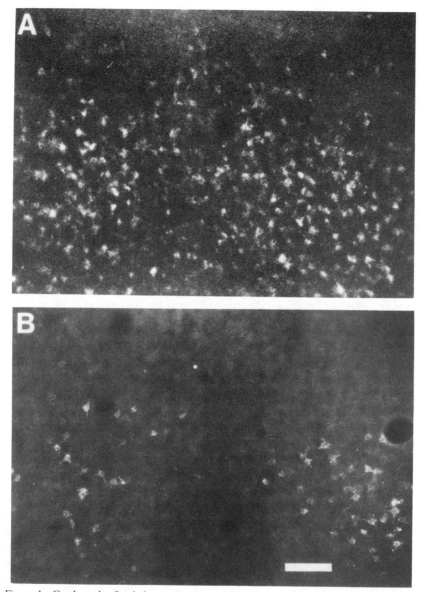

Figure 1 Crude and refined clusters in tangential sections of developing cat visual cortex. (*A*) Part of the pattern of retrograde labeling following a single microsphere injection in a postnatal day 12 cat. Two crude clusters are visible as zones of dense, bright retrograde labeling; between these zones are numerous labeled cells that have made inappropriate connections to the injection site. (*B*) At postnatal day 38, a similar injection results in highly refined clusters of labeled cells; clusters are smaller and few cells between clusters are labeled. Scale bar: 100 μm. (From Callaway & Katz 1990, with permission.)

the tangential spread of collaterals. Ectopic fields were extremely rare in the adult (Luhmann et al 1990c). The loss of ectopic receptive fields could represent loss of tangential arbors, but could also result from other maturational changes in striate cortex during the first months of life.

Based on this work, it is not clear whether a postnatal reduction in the length of horizontal connections actually occurs. This issue was directly addressed by making closely spaced or overlapping injections of red microspheres at an early age (P14–15), followed by green microspheres at a later age (P29 or P38). The retrograde labeling that resulted from the 2 injections was invariably coextensive, and always less than 4.5 mm from the injection sites. When the injection sites were superimposed directly, the refined clusters caused by the later injection were located in the middle of crude clusters from the earlier injection (Callaway & Katz 1990) (Figure 2). These data argue against a postnatal reduction in horizontal axon length, or change in the number or position of clusters between P14 and P38. Single time point experiments indicate that the adult extent and pattern of clusters are attained even before P38, which suggests that there is no overall change in these parameters in the developmental history of clusters.

Mechanisms of Cluster Refinement

Over the past decade it has become clear that regressive phenomena, such as cell death, process elimination, and synapse elimination, play a major role in shaping the patterns of connectivity in the developing nervous system (for review, see Purves & Lichtman 1985). In the development of specificity in several corticocortical projections, both selective cell death and process elimination have been implicated (Innocenti & Caminiti 1980; O'Leary et al 1981; Price & Blakemore 1985a,b). In the normal development of clustered intrinsic horizontal connections, at least two transitions potentially mediated by regressive phenomena have been identified: the change from an unclustered distribution of retrograde labeling to a crudely clustered distribution and the change from crude to refined clusters. The use of persistent cellular labels to follow developing horizontal connections indicates that neither transition involves the selective death of incorrectly situated cells (Callaway & Katz 1990).

The use of the above-mentioned double-injection paradigm, in which different tracers were injected at different times to label crude and refined clusters, demonstrated that selective process elimination is responsible for cluster refinement. Neurons located between clusters, which are labeled by early injections, are still clearly visible after clusters have refined, but no longer make connections to the original injection site. This indicates that

P15,29,31

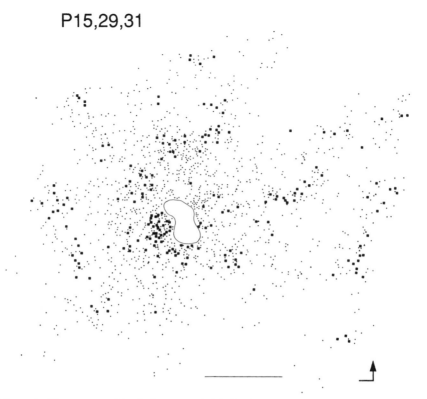

Figure 2 The patterns of intrinsic retrograde labeling in a tangential section of layer 2/3 of cat striate cortex, following injections of fluorescent microspheres at a single site at two different times. The injection site is within the central blank area. Red microspheres were injected on postnatal day 15, followed by green microspheres on day 29. The animal was perfused on day 31. Dots mark cells labeled only by the P15 injection, and squares indicate cells labeled by both the P15 injection and the P29 injection. Triangles indicate cells labeled only at P29. Cells labeled by the P15 injection were arranged in the crude clusters typical of this age, whereas the double labeled cells—those that made an appropriate connection at P15 and maintained it until P29—are arranged in refined clusters. This indicates that the refinement of clusters results from elimination of collaterals, and not cell death. The double-injection also reveals that the number, position, and tangential extent of clusters does not change between P15 and P29. Scale bar: 1 mm. (From Callaway & Katz 1990, with permission.)

those cells that made "inappropriate" connections have selectively lost certain horizontal collaterals (Figure 2).

Retrograde labeling experiments can indicate the presence or absence of some sort of connection between two points, but leave a great deal unanswered about the types of modifications that horizontal axonal arbors undergo during the emergence and refinement of clusters. Reconstructions of individual layer 2/3 pyramidal cells from cat striate cortex show that early in development, cells extend long, relatively unbranched horizontal axon collaterals several millimeters from the cell body (Figure 3, *top*). These branches lack the clusters of fine distal branches characteristic of adult cells (Callaway & Katz 1990, 1991). During the time at which retrograde tracing experiments show no clusters, the directions of collateral growth appear random. Even three weeks postnatally, when crude clusters are clearly present, cells still lack obviously clustered horizontal collaterals. This apparent contradiction between the appearance of single cells and the results of retrograde tracing studies suggests that the differentiation of crude clusters may result from a rearrangement of synapses along axon collaterals without any major redistribution in the positions of the collaterals themselves. Reorganization of axon collaterals is observed between three and six weeks postnatally, as the pattern of retrograde labeling refines: The long horizontal collaterals of a single cell become grouped, and clusters of distal branches are elaborated at the 1 mm intervals typical of adult cells (Figure 3).

Role of Neuronal Activity in Cluster Emergence and Refinement

The initial emergence of crude clusters does not require patterned visual experience. Neither binocular deprivation, nor dark rearing, nor binocular intraocular injections of TTX prevent crude clusters from forming (Luhmann et al 1986, 1990a; Callaway & Katz 1991 and unpublished observations). Apparently, patterned visual activity does not carry the critical cues for establishing crude clusters. This result does not imply that neuronal activity is not involved in the emergence of clusters, as it has become increasingly apparent that even intrinsic activity can provide the cues necessary for the refinement of connections in the visual and other systems. Spontaneous activity of retinal ganglion cells during prenatal life appears to be sufficient and necessary to allow retinal afferents to segregate into eye-specific layers in the lateral geniculate nucleus (Shatz & Stryker 1988; Sretavan & Shatz 1984, 1986; Sretavan et al 1988). Geniculate afferents in the developing monkey segregate into ocular dominance patches before birth and before the onset of patterned visual input (Rakic 1976). It is certainly possible that spontaneous activity in either the geniculate itself, or perhaps within the cortex, is sufficient to organize the pattern of crude clusters. At this point, however, it is equally possible that activity-independent cues in the cortex provide the basis for the organization of a crude system of orientation columns and clusters.

The refinement of clusters, on the other hand, shows a clear dependence

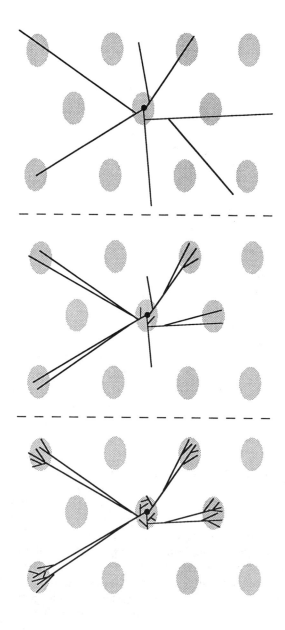

on patterned visual activity. Prolonged binocular deprivation or dark rearing severely degrades or eliminates the clustered organization of horizontal connections. Deprivation during the first two postnatal months does not cause degeneration, but may alter the number and extent of clusters (Luhmann et al 1990a). Deprivation also prevents the normal progression from "crude" to "refined" clusters (Callaway & Katz 1991). Thus, the pattern of clusters that results from a small tracer injection in a P38 binocularly deprived cat resembles, at least superficially, a similar injection in a normal P14 animal. Nevertheless, binocular deprivation does not simply arrest the development of these local connections at a less mature state. Intracellular due injections reveal that individual cells reorganize their axonal arborizations, much like normal cells: The long horizontal collaterals group together, and aggregates of distal branches are elaborated. However, the precision with which these changes occur is much reduced from normal. Thus, each cell makes connections over a region that represents a larger than normal range of orientations.

As with the establishment of ocular dominance columns, there may be a "critical period" in early postnatal life during which the precision of horizontal connections is modifiable by visual experience (Hubel & Wiesel 1970). In the cat, the critical period for response to monocular lid suture in layer 4 extends from about three weeks to three months postnatally. Within layer 2/3, which is rich in long distance horizontal connections, the critical period extends considerably longer (Daw & Fox 1991). This may reflect continued plasticity of intrinsic circuits in these layers for some time after plastic changes in thalamic afferents are no longer possible.

Although a critical period for alterations in horizontal connections apparently exists, the boundaries of this period are still poorly defined. If regular vision is restored to a binocularly deprived animal with crude clusters at six weeks of age, the clusters will refine to a normal adult state by three months of age (E. M. Callaway and L. C. Katz, unpublished observations). Because prolonged binocular deprivation causes degener-

---

Figure 3 Schematic diagram depicting the development of the horizontal axonal arbor of a layer 2/3 pyramidal neuron, as seen in the tangential plane at three postnatal ages. Stippled areas represent iso-orientation columns, lines represent axon collaterals, and the dot in the center stippled area represents the neuron's cell body. (*Top*) 0–2 weeks postnatal. The initial outgrowth of horizontal axon collaterals is random, and these collaterals project for several millimeters to both correct and incorrect orientation columns. At this time, axonal arbors lack distal collateral branches. (*Middle*) 2–4 weeks postnatal. Axon collaterals projecting to incorrect orientation columns have been selectively eliminated. For at least some neurons, long collaterals projecting to correct columns are added. (*Bottom*) > 5 weeks postnatal. Distal collateral branches have been added selectively within correct orientation columns. At 6 weeks, axonal arbors are indistinguishable from adult arbors.

ative changes in horizontal connections, manipulations that change the pattern of visual input but not the overall level, such as induced strabismus, (Hubel & Wiesel 1965), may be more appropriate for determining the period during which horizontal collaterals can be modified.

Cluster Refinement and the Development of Orientation Columns

Because of the direct relationship between orientation columns and clustered connections (Gilbert & Wiesel 1989), developmental studies of cluster development have provided several insights into orientation column development. Although there is abundant electrophysiological evidence that oriented cells are arranged in columns even early in development, it has proven difficult to visualize the overall organization of columns in developing animals. Studies with 2-deoxyglucose (Thompson et al 1983) revealed patches of labeling in layer 4 in 21-day-old cats, but the adult columnar organization was not apparent until about 35 days of age. The sequential retrograde tracer experiments described above have important implications for the development of the overall pattern of orientation columns. Because the positions of clusters reflect the positions of iso-orientation domains, the fact that the positions of clusters do not seem to change during postnatal development (Figure 2) implies that at least a crude overall map of orientation columns emerges by P8.

The postnatal refinement of clusters may be related to the increased precision of orientation tuning. If, as outlined above, we accept the premise that adult clusters in the cat label iso-orientation columns, then the cells located between the developing clusters represent inputs from other orientation domains. Thus, during the time that crude clusters are present, cells in layer 2/3 are much more likely to receive input from "incorrect" orientations than cells in the adult cortex. This could account for the consistent observations that the orientation tuning in young cats is considerably broader than in older animals. Adult values for tuning emerge at about one month postnatal (Albus & Wolf 1984; Braastad & Heggelund 1985; Imbert & Buisseret 1975), the time at which clusters achieve their adult level of refinement. Alternatively, cluster refinement could result from increased precision of orientation tuning following changes in other circuitry.

For many years after the discovery of plasticity in the ocular dominance column system, a sometimes acrimonious debate raged as to whether orientation specificity was innately determined or, like ocular dominance specificity, subject to modification by visual experience (reviewed in Frégnac & Imbert 1984). There is now general agreement that the basic outlines of orientation processing—including orientation specific cells, the precision of orientation tuning, and the organization of orientation columns—do not depend on patterned visual experience (Sherk & Stryker 1976; Stryker et al 1978). However, results showing that binocular deprivation reduces the specificity of horizontal connections indicate that some aspects of orientation-dependent visual processing may depend on patterned visual experience for their development. In this context, it would be of considerable interest to test deprived animals either psychophysically or with more complex visual stimuli to determine whether their ability to process more global stimuli has been affected.

Cross Columnar Correlations and the Development of Horizontal Connections

Mechanisms based on correlations between pre- and postsynaptic activity (Hebb 1949) have been strongly implicated in the development of ocular dominance columns in visual cortex (Hubel & Wiesel 1965; Stryker & Strickland 1984) and in the development of retinotectal circuitry in frogs and goldfish (reviewed in Constantine-Paton et al 1990). Hebb-like mechanisms have also been proposed to guide the development of clustered intrinsic connections (Callaway & Katz 1990; Luhmann et al 1990c). Some evidence suggests that such a mechanism may in fact be employed, particularly during the refinement of clustered horizontal connections (Callaway & Katz 1991).

The functional relationships between source and target populations suggest that correlated activity could regulate the development of horizontal connections. Because horizontal connections link columns with similar stimulus preference—iso-orientation columns in cats; colour-specific blobs versus interblobs in primates—the shared response properties could allow for the requisite correlation of activity between axons and their targets. Such correlations have been demonstrated in adult animals for both same-orientation columns (Gray et al 1989; Gray & Singer 1989; T'so et al 1986) and in the blob system in primates (T'so & Gilbert 1988). Furthermore, the correlations in primates appear to be not only for blob versus interblob regions, but also for color opponentcy and ocular dominance within the blob system, and again for orientation preference and ocular dominance in nonblob regions. If these correlations were present between the crude columns in young animals, they could mediate the subsequent refinement of clusters. This hypothesis is consistent with the available evidence; the absence of cluster refinement in lid-sutured cats could be explained by the fact that these animals do not experience oriented visual stimuli that would be necessary to drive the appropriately correlated responses.

CONCLUDING REMARKS AND
FUTURE DIRECTIONS

We have attempted to define some of the basic developmental rules that guide the formation of the exquisitely precise, three-dimensional lattice of the mature cortex. Patterns of vertical connections, which give rise to the basic processing capabilities of cortex, appear to be highly specified, possibly by molecular cues within the cortex. On the other hand, the links that unite cortical modules in the tangential plane are only crudely specified and subject to activity-dependent modifications. Despite recent progress, we are still woefully ignorant of the cues or mechanisms used to establish or modify local cortical circuits. We suspect that in vitro approaches—especially long-term cultures of developing cortical slices—will provide the control and accessibility over molecular cues and activity to enable a detailed understanding of the emergence and refinement of circuits. In addition to in vitro analyses, far more insight into the workings of developing cortical machinery is required. Conventional, microelectrode-based extracellular recording techniques are limited in developing systems, in which the primary mode of synaptic communication may be via subthreshold electrical events. New recording techniques, such as optical recording with voltage- or ion-sensitive dyes, may provide a new perspective on the emergence of cortical processing machinery in the intact animal.

Considering the cortical lattice from an evolutionary perspective, vertical connections may represent the primordial axis of cortical intrinsic circuits, whereas horizontal connections may represent relatively recent additions necessitated by increasingly complex forms of processing. Although stereotyped vertical connections are observed in all vertebrate cortices, highly specific horizontal connections are especially prominent in the primate cortex. The crude specification and subsequent activity-dependent modification of horizontal connections may offer opportunities for forging novel links between processing modules, thus providing an avenue for generating new perceptual or cognitive capabilities.

ACKNOWLEDGMENTS

We wish to thank Drs. A. Burkhalter, D. Fitzpatrick, D. Iarovici, and D. Purves for their helpful comments on this manuscript. We are also grateful to Dr. T. Wiesel for his insight and encouragement of the work in the authors' laboratory. We have been supported by the National Institutes of Health grants EY07960 (Katz) and EY06128 (Callaway). Lawrence Katz is a Lucille P. Markey Scholar, and this work was supported in part by a grant from the L. P. Markey Charitable Trust.

Literature Cited

Albus, K., Wolf, W. 1984. Early post-natal development of neuronal function in the kitten's striate cortex: a laminar analysis. *J. Physiol.* 348: 153–85

Allman, J., Meizen, F., McGuinness, E. L. 1985. Stimulus specific responses from beyond the classical receptive field: neurophysiological mechanisms for local-global comparisons in visual neurons. *Annu. Rev. Neurosci.* 8: 407–30

Angevine, J. B., Sidman, R. L. 1961. Autoradiographic study of cell migration during histogenesis of cerebral cortex of the mouse. *Nature* 192: 766–68

Blakemore, C., Van Sluyters, R. C. 1975. Innate and environmental factors in the development of the kitten's visual cortex. *J. Physiol.* 248: 663–716

Bolz, J., Gilbert, C. D. 1986. Generation of end-inhibition in the visual cortex via interlaminar connections. *Nature* 320: 362–65

Bolz, J., Gilbert, C. D. 1989. The role of horizontal connections in generating long receptive fields in the cat visual cortex. *Eur. J. Neurosci.* 1: 263–68

Bolz, J., Novak, N., Gotz, M., Bonhoeffer, T. 1990. Formation of target-specific neuronal projections in organotypic slice cultures from rat visual cortex. *Nature* 346: 359–62

Braastad, B. O., Heggelund, P. 1985. Development of spatial receptive-field organization and orientation selectivity in kitten striate cortex. *J. Neurophysiol.* 53: 1158–78

Burkhalter, A., Bernardo, K. L. 1989. Organization of corticocortical connections in human visual cortex. *Proc. Natl. Acad. Sci. USA* 86: 1071–75

Burkhalter, A., Bernardo, K. L., Charles, V. 1990. Postnatal development of intracortical connections in human visual cortex. *Soc. Neurosci. Abstr.* 16: 1129

Cajal, S. R. 1911 (1972). *Histologie du Système Nerveux de l'Homme et des Vertèbres.* Madrid: CSIC

Callaway, E. M., Katz, L. C. 1990. Emergence and refinement of clustered horizontal connections in cat striate cortex. *J. Neurosci.* 10: 1134–53

Callaway, E. M., Katz, L. C. 1991. Effects of binocular deprivation on the development of clustered horizontal connections in cat striate cortex. *Proc. Natl. Acad. Sci. USA* 88: 745–49

Constantine-Paton, M., Cline, H. T., Debski, D. 1990. Patterned activity, synaptic convergence, and the NMDA receptor in developing visual pathways. *Annu. Rev. Neurosci.* 13: 129–54

Coogan, T. A., Burkhalter, A. 1988. Sequential development of connections between striate and extrastriate visual cortical areas in the rat. *J. Comp. Neurol.* 278: 242–52

Dann, J. F., Buhl, E. H., Peichle, L. 1988. Postnatal dendritic maturation of alpha and beta ganglion cells in cat retina. *J. Neurosci.* 8: 1485–99

Daw, N. W., Fox, K. 1991. Function of NMDA receptors in the developing visual cortex. In *Development of the Visual System*, ed. C. J. Shatz, D. M.-K. Lam, pp. 243–52. Cambridge: MIT Press

Fairen, A., DeFelipe, J., Regidor, J. 1984. Nonpyramidal neurons: general account. In *Cerebral Cortex*, ed. A. Peters, E. G. Jones, 1: 201–54. New York: Plenum

Frégnac, Y., Imbert, M. 1978. Early development of visual cortical cells in normal and dark-reared kittens: relationship between orientation selectivity and ocular dominance. *J. Physiol.* 278: 27–44

Frégnac, Y., Imbert, M. 1984. Development of neuronal selectivity in primary visual cortex of cat. *Physiol. Rev.* 64: 325–434

Friauf, E., McConnell, S. K., Shatz, C. J. 1990. Functional synaptic circuits in the subplate during fetal and early postnatal development of cat visual cortex. *J. Neurosci.* 10: 2601–13

Gilbert, C. D. 1983. Microcircuitry of the visual cortex. *Annu. Rev. Neurosci.* 6: 217–47

Gilbert, C. D., Kelly, J. P. 1975. The projection of cells in different layers of the cat's visual cortex. *J. Comp. Neurol.* 163: 81–106

Gilbert, C. D., Wiesel, T. N. 1979. Morphology and intracortical projections of functionally characterized neurones in the cat visual cortex. *Nature* 280: 120–25

Gilbert, C. D., Wiesel, T. N. 1983. Clustered intrinsic connections in cat visual cortex. *J. Neurosci.* 3: 1116–33

Gilbert, C. D., Wiesel, T. N. 1989. Columnar specificity of intrinsic horizontal and cortico-cortical connections in cat visual cortex. *J. Neurosci.* 9: 2432–42

Gilbert, C. D., Wiesel, T. N. 1990. The influence of contextual stimuli on the orientation selectivity of cells in primary visual cortex of the cat. *Vision Res.* 11: 1689–1701

Godement, P., Vanselow, J., Thanos, S., Bonhoeffer, F. 1987. A study in developing visual systems with a new method of staining neurones and their processes in fixed tissue. *Development* 101: 697–713

Gray, C. M., Konig, P., Engel, A. K., Singer, W. 1989. Oscillatory responses in cat visual cortex exhibit inter-columnar synchronization which reflects global stimulus properties. *Nature* 338: 334–37

Gray, C. M., Singer, W. 1989. Stimulus-specific neuronal oscillations in orientation columns of cat visual cortex. *Proc. Natl. Acad. Sci. USA* 86: 1698–1702

Gulyas, B., Orban, G. A., Duysens, J., Maes, H. 1987. The suppressive influence of moving textured backgrounds on responses of cat striate neurons to moving bars. *J. Neurophysiol.* 57: 1767–91

Harris, W. A., Holt, C. E. 1990. Early events in the embryogenesis of the vertebrate nervous system: Cellular determination and pathfinding. *Annu. Rev. Neurosci.* 13: 155–69

Hebb, D. O. 1949. *The Organization of Behavior. A Neuropsychological Theory*, p. 62. London: Chapman & Hall. 5th ed.

Heffner, C. D., Lumsden, A. G. S., O'Leary, D. D. 1990. Target control of collateral extension and directional axon growth in the mammalian brain. *Science* 247: 217–20

Hubel, D. H., Wiesel, T. N. 1962. Receptive fields, binocular interaction and functional architecture in the cat's visual cortex. *J. Physiol.* 160: 106–54

Hubel, D. H., Wiesel, T. N. 1963. Receptive fields of cells in striate cortex of very young, visually inexperienced kittens. *J. Neurophysiol.* 26: 994–1002

Hubel, D. H., Wiesel, T. N. 1965. Binocular interaction in striate cortex of kittens reared with artificial squint. *J. Neurophysiol.* 28: 1041–59

Hubel, D. H., Wiesel, T. N. 1970. The period of susceptibility to the physiological effects of unilateral eye closure in kittens. *J. Physiol.* 206: 419–36

Imbert, M., Buisseret, P. 1975. Receptive field characteristics and plastic properties of visual cortical cells in kittens reared with or without visual experience. *Exp. Brain Res.* 22: 25–36

Innocenti, G. M., Caminiti, R. 1980. Postnatal shaping of callosal connections from sensory areas. *Exp. Brain Res.* 38: 381–94

Katz, L. C. 1987. Local circuitry of identified projection neurons in cat visual cortex brain slices. *J. Neurosci.* 7: 1223–49

Katz, L. C. 1991. Specificity in the development of vertical connections in cat striate cortex. *Eur. J. Neurosci.* 3: 1–9

Katz, L. C., Burkhalter, A., Dreyer, W. J. 1984. Fluorescent latex microspheres as a retrograde neuronal marker for in vivo and in vitro studies of visual cortex. *Nature* 310: 498–500

Katz, L. C., Callaway, E. M. 1990. Development of interlaminar connections of layer 4 neurons in cat striate cortex. *Soc.*

Neurosci. Abstr. 16: 1129

Katz, L. C., Constantine-Paton, M. 1988. Relationships between segregated afferents and postsynaptic neurons in the optic tectum of three-eyed frogs. *J. Neurosci.* 8: 3160–80

Katz, L. C., Gilbert, C. D., Wiesel, T. N. 1989. Local circuits and ocular dominance columns in monkey striate cortex. *J. Neurosci.* 9: 1389–99

Katz, L. C., Iarovici, D. M. 1990. Green fluorescent latex microspheres: a new retrograde tracer. *Neuroscience* 34: 511–20

LeVay, S., Stryker, M. P. 1979. The development of ocular dominance columns in the cat. *Soc. Neurosci. Symp.* 4: 83–98

LeVay, S., Wiesel, T. N., Hubel, D. H. 1980. The development of ocular dominance columns in normal and visually deprived monkeys. *J. Comp. Neurol.* 191: 1–51

Livingstone, M. S., Hubel, D. H. 1984. Specificity of intrinsic connections in primate primary visual cortex. *J. Neurosci.* 4: 2830–35

Lorente de No, R. 1933. Studies on the structure of the cerebral cortex. *J. Psychol. Neurol.* 45: 382–438

Lorente de No, R. 1944. Cerebral cortex: Architecture, intracortical connections, motor projections. In *Physiology of the Nervous System*, ed. J. F. Fulton, pp. 291–325. London: Oxford Univ. Press

Luhmann, H. J., Martinez-Millan, L., Singer, W. 1986. Development of horizontal intrinsic connections in cat striate cortex. *Exp. Brain Res.* 63: 443–48

Luhmann, H. J., Singer, W., Martinez-Millan, L. 1990a. Horizontal interactions in cat striate cortex: I. Anatomical substrate and postnatal development. *Eur. J. Neurosci.* 2: 344–57

Luhmann, H. J., Gruel, J. M., Singer, W. 1990b. Horizontal interactions in cat striate cortex: II. A current source-density analysis. *Eur. J. Neurosci.* 2: 358–68

Luhmann, H. J., Gruel, J. M., Singer, W. 1990c. Horizontal interactions in cat striate cortex: III. Ectopic receptive fields and transient exuberance of tangential interactions *Eur. J. Neurosci.* 2: 369–77

Lund, J. S. 1988. Anatomical organization of macaque monkey striate visual cortex. *Annu. Rev. Neurosci.* 11: 253–88

Lund, J. S., Boothe, R. 1975. Interlaminar connections and pyramidal neuron organization in the visual cortex, area 17, of the macaque monkey. *J. Comp. Neurol.* 159: 305–34

Lund, J. S., Boothe, R. G., Lund, R. D. 1977. Development of neurons in the visual cortex of the monkey (*Macaca nemestrina*): a Golgi study from fetal day 127 to postnatal maturity. *J. Comp. Neurol.* 176: 149–88

Luskin, M. B., Shatz, C. J. 1985. Neurogenesis of the cat's primary visual cortex. *J. Comp. Neurol.* 242: 611–31

Malpeli, J. G. 1983. Activity of cells in area 17 of the cat in absence of input from layer A of the lateral geniculate nucleus. *J. Neurophysiol.* 49: 595–610

Marin-Padilla, M. 1988. Early ontogenesis of the human cerebral cortex. See Peters & Jones 1988, pp. 1–34

Martin, K. A. C., Whitteridge, D. 1984 Form, function and intracortical projections of spiny neurons in the striate cortex of the cat. *J. Physiol.* 353: 463–504

Meyer, R. L. 1983. Tetrodotoxin inhibits the formation of refined retinotopography in goldfish. *Dev. Brain Res.* 6: 293–98

Meyer, G., Ferres-Torres, R. 1984. Postnatal maturation of neurons in the visual cortex of the cat. *J. Comp. Neurol.* 228: 226–44

Miller, M. W. 1986. Maturation of rat visual cortex. III. Postnatal morphogenesis and synaptogenesis of local circuit neurons. *Dev. Brain. Res.* 25: 271–85

Miller, M. W. 1988. Development of projection and local circuit neurons in neocortex. See Peters & Jones 1988, pp. 133–75

Mitchison, G., Crick, F. 1982. Long axons within the striate cortex: their distribution, orientation, and patterns of connection. *Proc. Natl. Acad. Sci. USA* 79: 3661–65

Naegele, J. R., Jhaveri, S., Schneider, G. E. 1988. Sharpening of topographic projections and maturation of geniculocortical axon arbors in the hamster. *J. Comp. Neurol.* 277: 593–607

O'Leary, D. D. M., Stanfield, B. B., Cowan, W. M. 1981. Evidence that the early postnatal restriction of the cells of origin of the callosal projection is due to the elimination of axonal collaterals rather than to the death of neurons. *Dev. Brain Res.* 1: 607–17

O'Leary, D. D. M., Terashima, T. 1988. Cortical axons branch to multiple subcortical targets by interstitial axon budding: implications for target recognition and "waiting periods." *Neuron* 4: 901–10

Parnavelas, J. G., Bradford, R., Mounty, E. J., Lieberman, A. R. 1978. The development of non-pyramidal neurons in the visual cortex of the rat. *Anat. Embryol.* 155: 1–14

Peters, A., Jones, E. G., eds. 1988. *Cerebral Cortex: Development and Maturation of Cerebral Cortex.* New York: Plenum

Price, D. J. 1986. The postnatal development of clustered intrinsic connections in area 18 of the visual cortex in kittens. *Dev. Brain Res.* 24: 31–38

Price, D. J., Blakemore, C. 1985a. Regressive events in the postnatal development of association projections in the visual cortex. *Nature* 316: 721–24

Price, D. J., Blakemore, C. 1985b. The postnatal development of the association projection from visual cortical area 17 to area 18 in the cat. *J. Neurosci.* 5: 2443–52

Purves, D., Lichtman, J. W. 1985. *Principles of Neural Development,* Chap. 6,12. Sunderland, Mass: Sinauer

Rakic, P. 1974. Neurons in rhesus monkey visual cortex: systematic relation between time of origin and eventual disposition. *Science* 183: 425–27

Rakic, P. 1976. Prenatal genesis of connections subserving ocular dominance in the rhesus monkey. *Nature* 261: 589–91

Rakic, P., Bourgeois, J.-P., Eckenhoff, E. F., Zecevic, N., Goldman-Rakic, P. S. 1986. Concurrent overproduction of synapses in diverse regions of the primate cerebral cortex. *Science* 232: 232–35

Ramoa, A. S., Campbell, G., Shatz, C. J. 1987. Transient morphological features of identified ganglion cells in living fetal and neonatal retina. *Science* 237: 522–25

Rihn, L. L., Claiborne, B. J. 1990. Dendritic growth and regression in rat dentate granule cells during late postnatal development. *Dev. Brain Res.* 54: 115–24

Rockland, K. S. 1985. Anatomical organization of primary visual cortex (area 17) in the ferret. *J. Comp. Neurol.* 241: 225–36

Rockland, K. S., Lund, J. S. 1982. Widespread periodic intrinsic connections in the tree shrew visual cortex. *Science* 215: 1532–34

Rockland, K. S., Lund, J. S. 1983. Intrinsic laminar lattice connections in primate visual cortex. *J. Comp. Neurol.* 206: 303–18

Schmidt, J. T., Edwards, D. L. 1983. Activity sharpens the map during the regeneration of the retinotectal projection in goldfish. *Brain Res.* 209: 29–39

Shatz, C. J. 1990. Impulse activity and the patterning of connections during CNS development. *Neuron* 5: 745–56

Shatz, C. J., Stryker, M. P. 1988. Prenatal tetrodotoxin infusion blocks segregation of retinogeniculate afferents. *Science* 242: 87–89

Shatz, C. J., Chun, J. J. M., Luskin, M. B. 1988. The role of the subplate in the development of the mammalian telencephalon. See Peters & Jones 1988, pp. 35–58

Sherk, H., Stryker, M. P. 1976. Quantitative study of cortical orientation selectivity in visually inexperienced kitten. *J. Neurophysiol.* 39: 63–70

Sretevan, D. W., Shatz, C. J. 1984. Prenatal

development of individual retinogeniculate axons during the period of segregation. *Nature* 308: 845–48

Sretevan, D. W., Shatz, C. J. 1986. Prenatal development of retinal ganglion cell axons: segregation into eye-specific layers within the cat's lateral geniculate nucleus. *J. Neurosci.* 6: 234–51

Sretevan, D. W., Shatz, C. J., Stryker, M. P. 1988. Modification of retinal ganglion cell axon morphology by prenatal infusion of tetrodotoxin. *Nature* 336: 468–71

Stryker, M. P., Harris, W. A. 1986. Binocular impulse blockade prevents formation of ocular dominance columns in the cat's visual cortex. *J. Neurosci.* 6: 2117–33

Stryker, M. P., Sherk, H., Leventhal, A. G., Hirsch, H. V. B. 1978. Physiological consequences for the cat's visual cortex of effectively restricting early visual experience with oriented contours. *J. Neurophysiol.* 41: 896–909

Stryker, M. P., Strickland, S. L. 1984. Physiological segregation of ocular dominance columns depends on the pattern of afferent electrical activity. *Invest. Opthalmol. Visual Sci.* 25 (Suppl); 278

Thompson, I. D., Kossut, M., Blakemore, C. 1983. Development of orientation columns in cat striate cortex revealed by 2-deoxyglucose autoradiography. *Nature* 301: 712–15

Tigges, J., Tigges, M. 1982. Principles of axonal collateralization of laminae II-III pyramids in area 17 of squirrel monkey: a quantitative Golgi study. *Neurosci. Lett.* 29: 99–104

T'so, D. Y., Gilbert, C. D. 1988. The organization of chromatic and spatial interactions in the primate striate cortex. *J. Neurosci.* 8: 1712–27

T'so, D. Y., Gilbert, C. D., Wiesel, T. N. 1986. Relationships between horizontal interactions and functional architecture in cat striate cortex as revealed by cross-correlation analysis. *J. Neurosci.* 6: 1160–70

Udin, S. B., Fawcett, J. W. 1988. Formation of topographic maps. *Annu. Rev. Neurosci.* 11: 289–327

von der Malsburg, C., Schneider, W. 1986. A neural cocktail-party processor. *Biol. Cybern.* 54: 29–40

Walter, J., Kern-Veits, B., Huf, J., Stolze, B., Bonhoeffer, F. 1987a. Recognition of position-specific properties of tectal cell membranes by retinal axons in vitro. *Development* 101:685–96

Walter, J., Henke-Fahle, S., Bonhoeffer, F. 1987b. Avoidance of posterior tectal membranes by temporal retinal axons. *Development* 101: 909–13

Wiesel, T. N., Hubel, D. H. 1974. Ordered arrangement of orientation columns in monkeys lacking visual experience. *J. Comp. Neurol.* 158: 307–18

Annu. Rev. Neurosci. 1992. 15:57–85

THE BIOSYNTHESIS OF NEUROPEPTIDES: PEPTIDE α-AMIDATION

Betty A. Eipper, Doris A. Stoffers, and Richard E. Mains

Department of Neuroscience, The Johns Hopkins University School of Medicine, Baltimore, Maryland 21205

KEY WORDS: monooxogenase, alternate splicing, ascorbate, copper, bifunctional enzyme

INTRODUCTION

The prevalence, diversity, and multiplicity of functions of peptides in the nervous system are well known. However, we are only now approaching the point at which our knowledge about the biosynthesis of neuropeptides can provide us with insights into nervous system function, just as our knowledge of the enzymes involved in the biosynthesis of conventional neurotransmitters has so effectively done. To this end, we focus our review on one of the steps in peptide biosynthesis, peptide α-amidation. We summarize information obtained primarily over the past few years, because several previous reviews comprehensively cover earlier work on peptide α-amidation (Bertelsen et al 1990; Bradbury & Smyth 1987a,b; Eipper & Mains 1988; Kreil 1985; Mains et al 1990). Several recent reviews have dealt more broadly with peptide processing (Chretien et al 1989; Harris 1989; Mains et al 1990; Mains & Eipper 1990; Schwartz 1990; Sossin et al 1989; Steiner et al 1989). Recent rapid progress on serine proteases identified by homology to the yeast endoprotease *Kex2*, which is known to play a role in α-mating factor biosynthesis, is also central to our developing understanding of neuropeptide biosynthesis (Bresnahan et al 1990; Hatsuzawa et al 1990; Seidah et al 1990, 1991; Smeekens & Steiner 1990).

57

0147–006X/92/0301–0057$02.00

α-AMIDATION IS A CRITICAL DETERMINANT OF BIOLOGIC ACTIVITY FOR MANY NEUROPEPTIDES

Examination of any list of neuropeptides reveals that at least half have at their COOH-terminus an α-amide moiety. Instead of terminating with an ionizable, free carboxylic acid (-COOH ↔ -COO⁻), these peptides terminate with a nonionizable α-amide (-CONH₂). Some of the α-amidated

Table 1 Some α-amidated peptides found in the nervous and endocrine systems

α-amidated residue	
A alanine	b,o CRH; p Galanin; μ-Conotoxin
C cysteine	crustacean cardioactive peptide; conotoxins G1, M1, S1
D aspartic	deltorphin[a]
E glutamic	joining peptide
F phenylalanine	FMRF-NH₂; gastrin; cholecystokinin; CGRP; γ₁MSH
G glycine	oxytocin; vasopressin; GnRH; pancreastatin; leucokinin I, II; *Manduca* adipokinetic hormone; leucokinin I, II
H histidine	Apamin; scorpion toxin II
I isoleucine	h,r CRH; PHI; *Manduca* diuretic hormone; rat neuropeptide EI (melanin concentrating hormone)
K lysine	ELH; cecropin A; PACAP38[b]; conotoxin G1A
L leucine	b,h GHRH; b-amidorphin; mastoparan; cecropin B; buccalin; myomodulin; PACAP27; proglucagon(111-123)
M methionine	Substance P; Substance K; PHM; gastrin releasing peptide; neurokinin A, B; neuromedin B, C
N asparagine	VIP (mammalian); neuromedin U; corazonin; mast cell degranulating peptide
P proline	calcitonin; TRH
Q glutamine	melittin; levitide
R arginine	preproglucagon(89-118)[c]
S serine	frog granuliberin-R
T threonine	rat galanin; avian VIP; locust adipokinetic hormone
V valine	αMSH; r,p,h secretin; metorphamide/adrenorphin
W tryptophan	cockroach myoactive peptide; sea anemone peptide; crustacean erythrophore concentrating peptide
Y tyrosine	NPY; PYY; PP; ω-conotoxin; amylin

References: Most of the peptides listed can be found in many catalogs of synthetic peptides. To keep the reference list reasonable, no reference is given for commonly available peptides.

[a] Richter et al (1987) predict the Asp-NH₂ from the cDNA sequence, but such a peptide has not actually been found yet.

[b] PACAP, pituitary adenylate cyclase activating peptide.

[c] Andrews et al 1986.

peptides found in the nervous and endocrine systems are listed in Table 1, along with selected bioactive peptides from other sources that have interesting actions in the nervous system. Although peptides terminating with α-amidated neutral amino acids have been found more frequently, neuropeptides that contain the α-amide of every amino acid have been found. Tamburini et al (1990) observed a range of kinetic parameters when they evaluated the ability of a purified α-amidating enzyme to catalyze the α-amidation of a set of model peptide substrates of the form N-dansyl-$(Gly)_4$-X-Gly, where X could be any amino acid; the peptide with Asp at position X was the only peptide for which no activity was detectable. Antisera to peptide amides often show great specificity for this modification; in some cases, cross-reactivity with other neuropeptides that terminate with the same α-amidated amino acid has been observed (Minth et al 1989; Nahon et al 1989).

For many of these peptides, the presence of the α-amide moiety is essential for biologic activity. For example, the α-amide in CRH and TRH is required for high affinity binding of the peptides to receptors, for biologic activity on the pituitary, and for behavioral actions in the whole animal (DeSouza & Kuhar 1986; Harvey 1990; Tazi et al 1987; Vale et al 1981). Neither the naturally occurring products, CRH-Gly and TRH-Gly, nor the free acid forms, CRH-OH and TRH-OH (which are not normally found in tissues), bind with high affinity to the relevant receptors. Interestingly, even peptides designed to act as antagonists to α-melanotropin, neuropeptide Y, oxytocin, and vasopressin must have the α-amide moiety (Eberle 1981; Fuhlendorff et al 1990; Sawyer & Manning 1989). However, in secretin (Gafvelin et al 1985) and growth hormone releasing hormone (Aitman et al 1989; Wehrenberg & Ling 1983), for example, presence of the α-amide moiety is unimportant for biologic activity.

Tatemoto et al (1986) successfully identified several novel bioactive peptides by identifying peptides with α-amides. Speculations regarding the utility of this modification have included a role in protecting peptides from enzymatic degradation (half-life) and increasing binding affinity (Fuhlendorff et al 1990; Kreil 1985).

As indicated in Table 1, peptides with neutral, acidic, and basic amino acid α-amides have all been identified. A similar enzymatic process is used to create every α-amidated amino acid. As we discuss in detail below, in mammals, a single gene undergoes tissue specific and developmentally regulated alternative splicing to generate multiple forms of amidating enzyme. The precursors to α-amidated peptides always contain a Gly residue to the COOH-terminal side of the residue to be α-amidated. Unless the peptide is situated at the extreme COOH-terminus of the preprohormone, as with prepromelittin (Kreil 1985), proteolysis at an endo-proteolytic cleavage site is required to reveal the α-amidation site. Thus, the signal for α-amidation is the sequence -X-Gly-Basic-Basic- or occasionally -X-Gly-Basic (as for CRH; Kreil 1985). The presence of such a sequence in a preprohormone structure elucidated by analysis of its cDNA justifies a search for the putative product peptide terminating with -X-NH_2.

EARLY CHARACTERIZATION OF THE ENZYMES INVOLVED IN PEPTIDE α-AMIDATION

The amidation assay developed by Bradbury et al (1982), which utilized $[^{125}I]$-D-Tyr-Val-Gly as the peptide substrate, permitted the first identification of an enzyme activity capable of catalyzing the α-amidation of peptidyl-Gly substrates. Isotopic labeling experiments demonstrated that the nitrogen of the α-amide was derived from the α-NH_2-group of Gly, with concomitant production of glyoxylate (Bradbury et al 1982). Soluble amidating enzymes have since been purified to homogeneity from bovine neurointermediate pituitary (Murthy et al 1986), frog skin (Mizuno et al 1986), rat medullary thyroid carcinoma (Gilligan et al 1989; Mehta et al 1988), and rat brain (Noguchi et al 1989; Takahashi et al 1990). The amidating enzymes purified from the bovine neurointermediate pituitary, frog skin, and rat brain all have apparent molecular weights of about 40 kDa. The soluble amidating enzyme purified from rat medullary thyroid carcinoma has a significantly higher apparent molecular weight, 75 kDa (Mehta et al 1988). A 92 kDa amidating enzyme was purified from porcine atrium particulate fractions extracted with 0.1% Lubrol (Kojima et al 1989). Variable extents of glycosylation can contribute to heterogeneity in molecular weight without greatly altering kinetic parameters (Bradbury & Smyth 1988a).

The activity of each of the purified amidating enzymes is dependent on copper, molecular oxygen, and a reducing cofactor. When enzyme acitivity is inhibited by addition of divalent metal ion chelators, copper is generally the only ion that effectively restores enzymatic activity (Bradbury & Smyth 1985; Murthy et al 1986). Ascorbate consumption has been measured for the reaction catalyzed by the bovine enzyme and is approximately equimolar to the amount of α-amidated product produced (Murthy et al 1987). Of particular interest for neuronal tissue is that catecholamines, such as dopamine and norepinephrine, can replace ascorbic acid and provide reducing equivalents for the reaction (Bradbury & Smyth 1985;

Murthy et al 1987). Amidating enzymes assayed in human cerebrospinal fluid and tumor tissues (Glauder et al 1990; Scott et al 1990; Wand et al 1985a,b), hypothalamus from several species (Emeson 1984; Faivre-Bauman et al 1988; Gale et al 1988), serum from several species (Eipper et al 1985; Tajima et al 1990; Wand et al 1985a), human semen (Pekary et al 1990), porcine stomach (Dickinson & Yamada 1991), and anglerfish islets (Mackin et al 1987) exhibit similar properties.

These properties of the purified enzyme all indicated that peptide α-amidation was catalyzed by a monooxygenase and led to categorization of the amidating enzyme as a peptidylglycine α-amidating monooxygenase (PAM) (EC 1.14.17.3) (Murthy et al 1986). Bradbury & Smyth (1987b) demonstrated conversion of glyoxylic acid phenylhydrazone to oxalic acid phenylhydrazide by a porcine pituitary amidating enzyme. Zabriskie et al (1991) observed incorporation of ^{18}O from $^{18}O_2$, thus demonstrating the occurrence of hydroxylation directly. As noted when the dependence of peptide α-amidation on copper and ascorbate was first reported, peptide amidation shares many similarities with the conversion of dopamine to norepinephrine by dopamine β-monooxygenase (EC 1.14.17.1) (Eipper & Mains 1988; Stewart & Klinman 1988).

Studies with purified amidating enzymes have consistently demonstrated the ability of each enzyme preparation to catalyze the α-amidation of a variety of peptidyl-Gly substrates. Structural elements important for the recognition of peptidylglycine substrates by amidating enzyme are located at the extreme COOH-terminus of the peptide, and neutral, hydrophobic amino acids are preferred in the penultimate position (Bradbury & Smyth 1988b; Tamburini et al 1988).

One of the few discrepancies of the amidating enzymes studied in various tissues has been the wide range of pH optima reported. When assayed with D-Tyr-Val-Gly as the peptide substrate, the pH optimum of crude or purified neurointermediate pituitary and brain PAM has always been alkaline (Emeson 1984; Murthy et al 1986; Takahashi et al 1990); in contrast, more acidic pH optima were observed with α-N-acetyl-Tyr-Val-Gly (Perkins et al 1990b,c) and a series of uncharged, nonpeptide substrates (Katapodis & May 1990). As Mehta et al (1988) and Tamburini et al (1990) have suggested, the presence of an ionizable, α-NH_2-group not far from the COOH-terminus of the peptide substrate is partially responsible for the pH profiles observed. The amidating enzyme purified from rat medullary thyroid carcinoma has a more acidic pH optimum (pH 5.5) (Mehta et al 1988). As discussed below, both the substrate selected for detecting activity and the source of enzyme are important determinants of the pH optimum.

cDNAs ENCODING PEPTIDYLGLYCINE α-AMIDATING MONOOXYGENASES

By purifying the bovine neurointermediate pituitary and frog skin monooxygenases, cDNAs encoding both enzymes were isolated (Eipper et al 1987; Mizuno et al 1987). The cDNA encoding bovine PAM was identified by screening an expression library prepared with RNA from the bovine intermediate pituitary with antisera to purified bovine PAM-A and -B (Eipper et al 1987) (Figure 1). The cDNA characterized encoded a 972 amino acid protein with an NH_2-terminal signal peptide and a putative transmembrane domain near its COOH-terminus. Amino acid sequence data for purified bovine PAM-B and for several cyanogen bromide peptides derived from PAM-A and -B were all located within the NH_2-terminal third of this larger protein (Figure 1). Paired basic amino acids were located at positions compatible with their use in the generation of soluble PAM-A and PAM-B from a membrane associated PAM precursor protein. A second cDNA, which differed from the first by the deletion of a 54 bp segment encoding an 18 amino acid peptide within the putative cytoplasmic domain of the PAM precursor, was also described (see below).

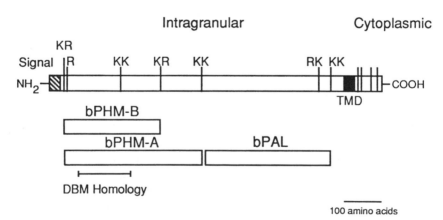

Figure 1 The bifunctional bovine PAM precursor. The protein encoded by bPAM-1 cDNA is shown (Eipper et al 1987). The position of the signal peptide and putative transmembrane domain (TMD) are indicated. Paired basic potential endoproteolytic cleavage sites are indicated by vertical lines. The NH_2-termini identified by amino acid sequence analysis of purified bPAM-A/B and bPAL (see below) are shown (Eipper et al 1987, 1991). The region exhibiting homology to bovine dopamine-β-monooxygenase (DBM) is indicated (Southan & Kruse 1989). As described below, the abbreviation PHM replaces PAM when describing the indicated region of the PAM protein.

The first amidating enzyme cDNA characterized from frog skin, AE-I, encoded a soluble 42 kDa protein highly homologous (66% identical) to the NH_2-terminal third of the bovine PAM precursor (Figure 2) (Mizuno et al 1987). A second cDNA isolated from frog skin encoded a larger, transmembrane protein highly homologous to the entire protein encoded by the bovine PAM cDNA (Figure 2) (Ohsuye et al 1988). Because of the many scattered differences in sequence between frog AE-I and AE-II, Ohsuye et al (1988) concluded that the two frog cDNAs must be derived from separate genes. *Xenopus laevis* are tetraploid and express two genes for many proteins encoded by a single gene in other species (Fritz et al 1989; Mohun et al 1988).

Verification that these cDNAs encoded peptide α-amidating enzymes came from production of enzyme activity upon expression of the cDNAs in mammalian (Mains et al 1988; Perkins et al 1990a) and bacterial (Mizuno et al 1987) systems. Extracts of AtT-20 corticotrope tumor cells transfected with a cDNA encoding bPAM-1 exhibited a ten-fold increase

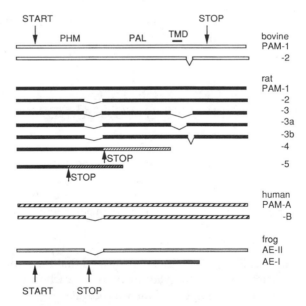

Figure 2 Forms of PAM cDNAs identified in various species. The relationship between PAM cDNAs from bovine neurointermediate pituitary (Eipper et al 1987), rat atrium, and anterior pituitary (Stoffers et al 1989; Stoffers et al 1991, unpublished), human thyroid carcinoma (Glauder et al 1990), and frog skin (Mizuno et al 1987; Ohsuye et al 1988) are shown; start and stop sites for protein translation are indicated. The various crosshatching patterns indicate identity; lines indicate sites of alternative splicing. PHM, PAL, and TMD as above.

in amidation activity when assayed with a series of substrates of the D-Tyr-X-Gly type (Perkins et al 1990a); this observation supported the idea that a single amidating enzyme is involved in production of the wide variety of α-amidated peptides observed (Table 1).

Elucidation of the structure of cDNAs encoding bovine and human dopamine β-monooxygenase permitted comparison of the monooxygenase domain of PAM to dopamine-β-monooxygenase (Figure 1) (Southan & Kruse 1989; Wang et al 1990). As indicated in Figure 1, the region from amino acids 61 to 317 of bovine PAM shows 31% identity to amino acids 202 to 481 of bovine dopamine β-monooxygenase. This region of the bovine PAM precursor contains seven Cys residues; five are in similar or identical position in bovine dopamine β-monooxygenase. Histidine clusters considered important in the binding of copper to dopamine β-monooxygenase are conserved in PAM. Mechanistic studies by Katapodis & May (1988) demonstrated that bovine neurointermediate pituitary PAM, like dopamine β-monooxygenase, readily catalyzes several alternate monooxygenase reactions, including sulfoxidation, amine N-dealkylation, and O-dealkylation. Thus, PAM and dopamine β-monooxygenase, which are important determinants of catecholaminergic and peptidergic function, appear to be members of a family of type II copper (non blue, EPR-detectable) monooxygenases (Southan & Kruse 1989).

THE TWO-STEP CONVERSION OF PEPTIDYL-GLYCINE SUBSTRATES INTO α-AMIDATED PRODUCTS

While examining the reaction mechanism and the properties of both purified and recombinant enzymes, it became clear that the conversion of peptidyl-X-Gly substrates into peptidyl-X-NH_2 products involves a relatively stable reaction intermediate and two enzyme activities.

The Stable Reaction Intermediate

When Bradbury & Smyth (1987b) demonstrated that the initial step in the amidation reaction was hydroxylation of the α-carbon of the COOH-terminal Gly, the resulting intermediate was postulated to exist transiently, thus rapidly undergoing dismutation to the α-amidated product and gly-oxylate. Subsequent studies have demonstrated that peptidyl-α-hydroxy-glycine is quite stable under the conditions that might be encountered in the secretory pathway. When Young & Tamburini (1989) demonstrated that the 75 kDa amidating enzyme purified from medullary thyroid carcinoma catalyzed both the copper- and ascorbate-dependent stereospecific production of peptidyl-α-hydroxyglycine and its subsequent conversion to an α-amidated peptide, they noted the relative stability of their model peptidyl-α-hydroxylglycine compounds.

Tajima et al (1990) used a preparation of amidating enzyme partially purified from horse serum to identify peptidyl-α-hydroxyglycine as a stable reaction product. The nonenzymatic conversion of the peptidyl-α-hydroxyglycine into α-amidated product plus glyoxylate was accelerated by alkaline pH (pH > 7.5), high temperature (T > 37°), and high concentrations of copper ([Cu^{2+}] > 0.2 mM). Similarly, Takahashi et al (1990) used a purified, soluble 36 kDa amidating enzyme prepared from rat brain to demonstrate production of a stable reaction intermediate. Katapodis & May (1990) used PAM purified from bovine neurointermediate pituitary to form an intermediate species that could be converted into α-amidated product by a second protein factor. Expression of the cDNA encoding frog AE-I yielded an enzyme activity that formed a stable intermediate identified as peptidyl-α-hydroxylglycine; base catalyzed conversion of the peptidyl-α-hydroxyglycine compound into α-amidated product occurred rapidly (Suzuki et al 1990).

The Second Protein Factor

The stability of the peptidyl-α-hydroxyglycine intermediate at pH values found in the secretory pathway necessitates participation of a second enzyme activity in peptide α-amidation. Such an activity was found by several experimental approaches. Noguchi et al (1989) purified a soluble α-amidating enzyme from rat brain and noted that the pH dependence of the purified enzyme differed from that exhibited by crude extracts; pursuing this observation, they purified a 41 kDa factor responsible for restoring peptide α-amidation at low pH (Takahashi et al 1990). By using a non-peptide substrate (benzoyl-α-hydroxyglycine), Katapodis et al (1990) demonstrated that bovine neurointermediate pituitary contained a soluble 45 kDa protein (α-hydroxyglycine amidating dealkylase) capable of converting α-hydroxybenzoylglycine to benzamide; when added to purified bovine neurointermediate pituitary PAM or the stable intermediate formed by this enzyme, this preparation increased the rate of formation of α-amidated product tenfold. Interestingly, the insect cells used for expression of frog AE-1 contained an enzymatic activity capable of converting the peptidyl-α-hydroxyglycine intermediate into α-amidated product (Suzuki et al 1990).

While examining stable AtT-20 cell lines that express cDNAs encoding bPAM, we began to utilize the assay system developed by Mizuno et al (1986) (Perkins et al 1990a,b,c). In optimizing assay conditions that utilize α-N-acetyl-Tyr-Val-Gly instead of D-Tyr-Val-Gly, it became clear that secretory granules contained a protein factor capable of stimulating the α-amidation of α-N-acetyl-Tyr-Val-Gly by purified bovine PAM-A and PAM-B at low pH; this protein factor was referred to as a stimulator of PAM, or SPAM (Perkins et al 1990c). We noted that cells and tissues containing high levels of PAM activity also contained high levels of SPAM activity. In comparing AtT-20 lines that express cDNAs encoding the full-length 108 kDa bovine PAM precursor, a protein that corresponds to the soluble 38 kDa PAM-B protein and antisense RNA to reduce levels of PAM activity, we found that expression of SPAM activity was clearly linked to expression of the full-length bovine PAM precursor.

We purified this stimulatory activity to homogeneity from bovine neuro-intermediate pituitary extracts. By using rabbit polyclonal antisera to bPAM (561–579), we could demonstrate that the stimulatory activity contained this antigenic determinant and was derived from the PAM precursor (Eipper et al 1991; Perkins et al 1990b). Sequence analysis revealed that the first 12 amino acids of this 50 kDa protein were identical with residues 434 to 445 of bovine PAM (Figure 1). Thus, endoproteolytic cleavage at Lys^{432}–Lys^{433} generates PAM-A and this stimulatory factor. Expression in hEK293 cells of a cDNA encoding the signal peptide, PAL (see below), transmembrane, and cytoplasmic domains of rat PAM leads to production of membrane-associated PAL activity (Eipper et al 1991). Expression of truncated cDNAs encoding only the soluble PHM or PAL domain led to secretion of the corresponding activity (Eipper et al 1991; Kato et al 1990a). The protein factors responsible for stimulation of peptide α-amidation at low pH are enzymes that catalyze the conversion of the stable peptidyl-α-hydroxyglycine intermediate into product α-amidated peptide.

The Two-Step Amidation Reaction

At physiologic pH, peptide α-amidation proceeds as follows:

$$\textit{Step I}: \quad \text{Peptidyl-CH(R)}-\overset{\displaystyle O}{\overset{\|}{C}}-\text{NH}-\text{CH}_2-\text{COO}^- + \text{ascorbate} + \text{O}_2 \xrightarrow{\text{(PHM)}}$$

$$\text{Peptidyl-CH(R)}-\overset{\displaystyle O}{\overset{\|}{C}}-\text{NH}-\text{CH(OH)COO}^- + \text{dehydroascorbate} + \text{H}_2\text{O}$$

$$\textit{Step II}: \quad \text{Peptidyl-CH(R)}-\overset{\displaystyle O}{\overset{\|}{C}}-\text{NH}-\text{CH(OH)COO}^- \xrightarrow{\text{(PAL)}}$$

$$\underset{\text{α-amidated peptide}}{\text{Peptidyl-CH(R)}-\overset{\displaystyle O}{\overset{\|}{C}}-\text{NH}_2} + \underset{\text{glyoxylate}}{\text{H}\overset{\displaystyle O}{\overset{\|}{C}}\text{COO}^-}$$

The enzyme catalyzing the first step in the process of peptide amidation is thus a peptidylglycine α-hydroxylating monooxygenase (PHM) (EC 1.4.17.3) (Eipper et al 1991; Perkins et al 1990b; Suzuki et al 1990). The enzyme catalyzing the second step in peptide amidation is a peptidyl-α-hydroxyglycine α-amidating lyase (PAL) (EC 4.3.2). The soluble amidating enzymes purified from bovine neurointermediate pituitary, PAM-A and PAM-B, contain only the monooxygenase activity and would more properly be referred to as PHM-A and PHM-B. Although the PHM domain resembles dopamine β-monooxygenase in structure, striking similarities between the PAL domain and other enzymes of the same class (human arginosuccinate lyase and B. subtilis adenylosuccinate lyase) were not apparent. The reaction catalyzed by PAL is formally similar to the reaction catalyzed by ureidoglycollate lyase (EC 4.3.2.3; Takada & Noguchi 1986).

PAM: A Bifunctional Enzyme and Enzyme Precursor

When intact, the PAM precursor (Figure 1) is thus a bifunctional enzyme that catalyzes sequential steps in a biosynthetic pathway. The existence of an enzymatic function for the entire intragranular domain of the PAM precursor is consistent with its high degree of conservation across species. Multifunctional enzymes of this type are commonly found in amino acid and nucleotide biosynthesis (Bazan et al 1989; Coggins & Hardie 1986; DiPatti et al 1990). However, PAM is the first multifunctional enzyme shown to participate in peptide biosynthesis. The functional independence of the two domains is demonstrated by expression of the individual catalytic activities by using expression vectors that contain truncated cDNAs (Eipper et al 1991; Kato et al 1990a; Perkins et al 1990a,c). The bifunctional PAM protein is subject to tissue specific endoproteolysis that generates independent, soluble forms of PHM and PAL, as well as soluble bifunctional enzymes, such as the 75 kDa medullary thyroid carcinoma amidating enzyme (Mehta et al 1988). Functional differences that distinguish the intact PAM protein and the soluble enzymes produced from it have not yet been investigated.

Assay for Peptide α-Amidation

With the realization that peptide amidation is a two-step process, modified assay systems that reflect both catalytic steps were needed. Because crude extracts generally contain both enzymatic activities, as either separate entities or part of a bifunctional protein, assays specific for each enzyme and assays reflecting overall conversion of peptidylglycine substrate to α-amidated product were needed. Suzuki et al (1990) documented a simple way to convert peptidyl-α-hydroxyglycine into α-amidated product with base, thus allowing the assay of PHM activity in the absence of PAL. Coupled with simple phase separation systems, like that devised by Mizuno et al (1986) for α-N-acetyl-Tyr-Val-Gly and α-N-acetyl-Tyr-Val-NH$_2$ or α-amide specific immunoassays (Dickinson & Yamada 1991; Hilsted & Hansen 1988; Scott et al 1990), this approach makes it feasible to assay PHM activity accurately. Nonpeptide substrates developed by Bradbury & Smyth (1987b) and Katapodis et al (1990) also make possible the independent measurement of PHM and PAL, respectively. In addition, these nonpeptide substrates have opened the way to develop specific inhibitors of both enzymes (Bradbury et al 1990; Katapodis & May 1990). Purified PHM has also been used to produce enough of the stable radiolabeled and unlabeled peptidyl-α-hydroxyglycine peptide intermediate to use as the substrate to assay PAL (Eipper et al 1991).

ALTERNATIVE SPLICING GENERATES MULTIPLE FORMS OF PAM

Major Forms of PAM mRNA in Several Species

With the recognition that atrium contains very high levels of PAM activity and mRNA, two types of cDNA encoding rat PAM were isolated from an atrial cDNA library (Figure 2) (Stoffers et al 1989). rPAM-1 encodes a 976 amino acid putative transmembrane protein that contains both PHM and PAL catalytic domains. When a 315 bp segment (optional exon A) is deleted from rPAM-2, an 871 amino acid putative transmembrane protein is encoded. The 105 amino acid domain present in rPAM-1 separates the PHM and PAL catalytic domains and contains one potential paired basic endoproteolytic cleavage site and a potential site for N-glycosylation. As shown in Figure 2, deletion of optional exon A makes rPAM-2 homologous to frog AE-II. cDNAs encoding two forms of human PAM were isolated from a medullary thyroid carcinoma cDNA library and characterized (Glauder et al 1990). The two types of human PAM cDNA are very similar to those characterized in rat atrium; they differ by the presence (hPAM-A) or absence (hPAM-B) of a 321 bp segment between the PHM and PAL domains (Figure 2).

The proteins encoded by the longest forms of rat, bovine, human, and frog PAM cDNA are compared in Figure 3. The amino acid sequence predicted for rat PAM-1 is shown in its entirety. No further data are shown if the amino acid present in all four species is identical. Of the 976 positions in rPAM-1, 635 (65%) are identical in frog, beef, and man. A period is shown if the other species have a highly conservative substitution; inclusion of highly conservative substitutions raises the homology to 76%, and well over 90% if only the mammalian molecules are considered. The actual amino acids present are in bold for positions in which the substitutions are clearly not conservative.

Although the precise sequence is poorly conserved in the signal peptide, the properties considered essential to recognition by the signal peptidase are conserved (von Heijne 1987). The site of signal peptide cleavage has been verified for recombinant rat (Beaudry et al 1990) and frog PAM (Suzuki et al 1990). The sequence of the ten amino acid putative propeptide is identical in all four species. Within the PHM domain [rPAM(42–382)], 255 of the 341 positions (75%) are identical or highly homologous. Within the PAL domain [rPAM(498–820)], 247 of the 323 positions (76%) exhibit identity or high homology. Within optional exon A, 92 of the 105 positions are conserved, thus giving the same degree of identity as that found among the mammalian PHM and PAL domains (a homologous frog sequence has not been identified yet). Large regions of the cytoplasmic domain show marked conservation across species, which points to an important functional role for this domain, perhaps in interaction with clathrin through the mediation of adaptor proteins (Payne & Schekman 1989). Nonconservative substitutions are clustered in a few regions of the PAM precursor. In particular, the region that follows the monooxygenase domain and precedes optional exon A, the region that follows the COOH-terminus of the PAL domain and precedes the transmembrane domain, and the 18 amino acid region that distinguishes bPAM-1 and -2 (and rPAM-2 and 3b) are especially poorly conserved.

Additional Forms of PAM mRNA

By rescreening the rat atrial cDNA library, additional alternatively spliced forms of PAM mRNA were identified (Figure 2). In rPAM-3, removal of the 258 bp optional exon B leads to loss of the transmembrane domain (Stoffers & Eipper 1989; Stoffers et al 1991). Other forms of PAM mRNA were found in pituitary cDNA libraries; these cDNAs encode proteins in which subregions of optional exon B are deleted: rPAM-3a and 3b. The region deleted in rPAM-3a includes the transmembrane domain, whereas the region deleted in rPAM-3b corresponds to the region deleted in bPAM-2 (Figure 2). Similar data were obtained by Kato et al (1990b). All of these forms of PAM mRNA encode both the PHM and PAL domains, although they differ in predicted subcellular localization and endoproteolytic cleavage pattern.

Rescreening the atrial cDNA library also yielded a cDNA of the rPAM-4 type and permitted identification of rPAM-4 RNA in atrial samples (Stoffers et al 1991). This type of PAM cDNA is identical to rPAM-1 through optional exon A and diverges from the sequence of rPAM-1 at

```
                    SIGNAL              PRO ●
           MAGRARSGLLLLLGLLALQSSCLAFRSPLSVFKRFKETTRSFSNECLGTIGPVTPLDASDFALDIRMPGVTPKESDTYF    80
RAT
FROG       MD  .-LI .- .  .-F IF N YC      .E.        L .   TR MSPG.    .T      T  L
BOVINE     --  -F  -- .   .-.-FP                              R  I .          Q
HUMAN      --  VP  -  .   .-F---P                P          TR  V I .         Q

           CMSMRLPVDEEAFVIDFKPRASMDTVHHMLLFGCNMPSSTGSYWFCDEGTCTDKANILYAWARNAPPTRLPKGVGFQVGGET   162
RAT
FROG       KY  . . . . . N A          V     DD  D SA  N  .S M      .    E   R  K.
BOVINE                       N                    .    N                     R
HUMAN      .                      G.

           GSKYFVLQVRYGDISAFRDNHKDCSGVSVHLTRVPQPLIAGMYLMMSVDTVIPPGEKVVNADISCQYKMYPMHVFAYRVHTH   244
RAT                                                                          ●
FROG       .   .K Q K . .  . PEK    I S  .N .   QE  . L NRPTI P
BOVINE          .          N  . . G.                 G  . . H K
HUMAN                                                  A    . . H N

           HLGKVVSGYRVRNGQWTLIGRQNPQLPQAFYPVEHPVDVTFGDILAARCVFTGEGRTEATHIGGTSSDEMCNLYIMYYMEAK   326
RAT                                                                              . A
FROG       Q Q  . H K.  S        ....P  . T .  K MS .    .K
BOVINE                 S                .  . .         V
HUMAN                                    .    .

           YALSFMTCTKNVAPDMFRTIPAEANIPIVKPDMVMM----HGHHKEAENKEKSALMQQPKQGEEVLEQGDFYSLLSKLLG    404
RAT                                                                     ●
FROG       .T  . VQTGN KL EN EI   S  M  MMMG    ..EAEA.TNTAL   .        RE .
BOVINE     .  .        Q  I  P     S                ----  T  .  L        RE  G
HUMAN      .  .        Q     P     S            ---- E  TY . IP L        RE  D

           ERED-VHVHKYNPTEKTESGSDLVAEIANVQKKDLGRSDAREGAEHEEWGNAILVRDRIHRFHQLESTLRPAESRAFSFQQ    485
RAT        V                           T  .  -QR    .     R  V   VL L
FROG                                                                     ---------------
BOVINE         A   E                      .     R     .   R     P  V L
HUMAN          A   E                                                     ---------------
```

```
                                                                                              565
RAT      --PGEGPWEPEPSGDFHVEEELDWPRVYLLPGQVSGVALDSKNNLVIFHRGDHVWDGNSFDSKFVVQQRGLGPIEEDTILVI
FROG     -----------     .LDT  N KV   .  P                       E  RN   .        Q S   .
BOVINE   PL  T  H.  A                 PQ                                                   I
HUMAN    PP  T  H.  M A               P

                                                                                              647
RAT      DPNNAEILQSSGKNLFYLPHGLSIDTDGNYWVTDVALHQVFKLDPHSKEGPLLILGRSMQPGSDQNHFCQPTDVAVEPSTGA
FROG     T.K. K. Q  .      R           -VGAE T  .F  RK                              I  N
BOVINE   A.            K               K    T                                      . D  T
HUMAN    A.     E      K              NN    .                                      .  .  .

                                                                                              729
RAT      VFVSDGYCNSRIVQFSPSGKFVTQWGEESSGSSPRPGQFSVPHSLALVPHLDQLCVADRENGRIQCFKTDTKEFVREIKHAS
FROG     F .       M   N M .M    A..NL   R.   TM.SDQG                       .AK G   .Q  QE
BOVINE   : .                     A LE  . L    R          P G                            .   P
HUMAN    : :                                            L G

                                                                                              810
RAT      FGRNVFAISYIP-GFLFQLNGKPYFGDQEPVQGFVMNFSSGEIIDVFKPVRKHFDMPHDIVASEDGTVYIGDAHTNTVWKFT
FROG     E .  AGV.A.          ST      E            ML  N..TIA N.       A..   .      A A     -
BOVINE   - L  A.                                        N            A
HUMAN    - L  A.           .

                                                                                              890
                                                                     TMD
RAT      LTEKMEHRSVKKAGIEVQEIKEAEAVVEPKV-ENKPTSSELQKMQEKQKLSTEPGSGVSVVLITTLLVIPVLVLLAIVMFI
FROG     SPS A        E TT-.F T..RSRP TNE VG QT   PS VQ S..   I   .   .     AI
BOVINE   S            E TT   S   T M--   A I          VK    PA        .   .     AL
HUMAN    L               T M--                       IK    P                     AI

                                                                                              969
RAT      RWKKSR-AFG-DHDRKLESSSGRVLGFRGKSGGGLNLGNFFASRKGYSRKGFDRVSTEGSDQEK-DEDDGTESEEEYSAPL
FROG     . V MYG   IG. S   G. .L          T    .              T          .       D  .  .      P
BOVINE   -         SF: A    L.                                             -  -  -            P
```

RAT / FROG / BOVINE / HUMAN (repeated for each block)

Figure 3 Comparison of proteins encoded by PAM precursors. The amino acid sequence of rPAM-1 is given in its entirety in the single letter code (defined in Table 1). The sequences of the longest proteins encoded by PAM cDNAs from bovine, human, and frog are compared. When complete identity is observed, no further information is given. A single period is shown when the amino acids present at that position are highly conserved: aliphatic, aromatic, positively charged, negatively charged, or very small side chain (Taylor 1987). When the amino acid similarities only meet less stringent criteria (hydrophobicity, polarity, size), the amino acid replacements are shown in regular type. When there was no homology (Taylor 1987), the residue present is given in bold type. Major structural features are indicated: the signal peptide, pro-region, exon A (first set of brackets), exon B (second set of brackets), transmembrane domain (TMD; crosshatched), pairs of basic amino acids (●).

this site; the novel 1.2 kb sequence at the 3′ end of the rPAM-4 mRNA encodes a hydrophilic, basic 20 amino acid segment before reaching an in-frame stop codon. cDNAs of the rPAM-4 type encode only the PHM domain and thus resemble frog AE-I; nevertheless, the 3′-region of rPAM-4 was not homologous to the 3′-region of frog AE-I, and frog AE-I does not contain a region homologous to optional exon A.

By screening rat anterior and neurointermediate pituitary cDNA libraries, still another type of PAM mRNA was identified. mRNAs of the rPAM-5 type are identical to rPAM-1 until position 1216 (amino acid 307 of rPAM-1); a novel 1.2 kb 3′-region encodes only five amino acids before reaching a stop codon. Based upon homology to dopamine β-mono-oxygenase and the smallest forms of PHM, rPAM-5 is not expected to encode an active protein.

Two forms of PAM cDNA have been characterized from rat medullary thyroid carcinoma (Bertelsen et al 1990). Type B mRNA is essentially identical to rPAM-3b, whereas Type A mRNA is identical to rPAM-1 through most of optional exon B, at which point it diverges. Three of the five forms of PAM cDNA described in cDNA libraries prepared from Wistar rats (Kato et al 1990b) correspond exactly to the forms of PAM cDNA diagramed in Figure 2: clone 201 corresponds to rPAM-1, clone 202 to rPAM-2, and clone 205 to rPAM-3b. Although clones 203 and 204 correspond to rPAM-3a and -3b in the region of optional exon B, they contain optional exon A.

The PAM Gene

Analysis of Southern blots has consistently indicated that mammals con-tain a single gene encoding PAM (Stoffers et al 1991; Kato et al 1990b). Genomic clones encoding rat PAM were isolated from a λ_{FIX} genomic library. Exon-containing fragments were subcloned and sequenced. The gene encoding rat PAM encompasses at least 150 kb of genomic DNA and is subdivided into more than 26 exons. The PHM and PAL domains are each encoded by multiple exons, whereas the untranslated regions are contained within single exons. Figure 4 illustrates the structure of the rat PAM gene in the region around optional exon A. Generation of rPAM-4 involves the use by the cell of an alternative poly(A) addition signal in the 1.2 kb region of the gene that follows optional exon A. In contrast to the suggestion by Bertelsen et al (1990), the 315 nts of optional exon A are not part of a retained intron.

Tissue Specific Processing of the Bifunctional PAM Protein

The proteins encoded by the alternatively spliced PAM mRNAs, which we have identified, are illustrated in Figure 5 and vary in size from 108

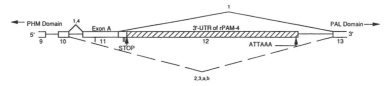

Figure 4 The exon A and rPAM-4 specific region of the rat PAM gene. The PHM domain ends with a 72 nt cassette exon. The 315 nt of exon A are contiguous with the 1207 nt specific to the 3'-end of rPAM-4. Paired basic amino acid potential endoproteolytic cleavage sites are indicated by tick marks below the open boxes. The 3'-untranslated region (3'-UTR) of rPAM-4 is indicated by diagonal hatching. Splicing choices made predominantly in brain and heart are indicated by solid lines; choices more prevalent in pituitary and submaxillary gland are marked by dashed lines.

kDa to 35 kDa. These proteins would be expected to have very different cellular dispositions. Proteins encoded by mRNAs of the rPAM-1, -2 or 3b type would be integral membrane proteins, whereas proteins encoded by rPAM-3, -3a, -4, and -5 would be soluble intragranular proteins.

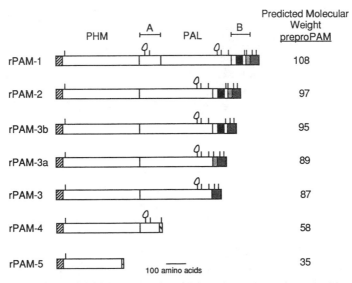

Figure 5 Proteins produced by alternative splicing of rat PAM RNA. The forms of rat PAM protein encoded by the cDNAs shown in Figure 2 are compared (Stoffers et al 1989, 1991; Stoffers & Eipper 1989, unpublished). Signal peptide, crosshatching; transmembrane domain, solid; 18 amino acid domain absent from rPAM-3b, light stippling; remainder of rPAM-1 cytoplasmic domain, darker stippling; pairs of basic amino acids, vertical line; potential N-linked glycosylation site, irregular closed shape. The PHM and PAL domains and the locations of exons A and B are shown.

Although the soluble proteins would be expected to undergo secretion upon exocytosis, intact rPAM-1, -2 and -3b would remain membrane associated; endoproteolytic processing would be required to generate secreted enzyme activity.

Tissue-specific, post-translational processing of the PAM proteins encoded by these cDNAs generates further diversity. In amidating enzymes prepared from heart atrium and rat medullary thyroid carcinoma, the PHM and PAL domains remain associated (Eipper et al 1988; Husten & Eipper 1991; Kojima et al 1989; Mehta et al 1988). In contrast, endoproteolytic cleavage separates the majority of the PHM and PAL into separate catalytic units in bovine neurointermediate pituitary and frog skin (Eipper et al 1991; Katapodis et al 1990; Mizuno et al 1986; Murthy et al 1986), whereas in anterior pituitary, both membrane and soluble forms occur in substantial quantity (May et al 1988). The PAM precursor, like a typical prohormone, has a set of paired basic amino acid sequences that are used in a tissue-specific manner to generate smaller, soluble forms of enzyme. Because optional exon A contains the only paired basic sequence that separates the PHM and PAL domains of rat (but not bovine or human) PAM, the presence or absence of this domain should greatly affect processing.

For both bPHM-B (Eipper et al 1987) and the 75 kDa rat CA-77 or MTC amidating enzyme (Beaudry et al 1990), comparison of the experimentally determined NH_2-terminal sequence and the sequence of the protein encoded by the corresponding cDNA indicates that endoproteolytic cleavage occurs following Lys^{29}-Arg^{30}, and to a greater extent following Arg^{36}. The several forms of amidating enzyme purified from frog skin are all derived by endoproteolytic cleavage following the corresponding single Arg residue (Mizuno et al 1986). Expression in C127 cells of a cDNA equivalent to rPAM-2, but truncated to delete the transmembrane and cytoplasmic domains, leads to secretion of enzyme from which only the signal peptide was lacking (Beaudry et al 1990). Similarly, analysis of the NH_2-terminus of frog AE-I expressed in insect cells indicates removal of only the signal peptide (Suzuki et al 1990). Because these expressed enzymes are active, presence of the propeptide does not completely inhibit activity, and the role of this fully conserved region is unclear.

Consistent with an important role for cleavage at paired basic amino acid sites, the NH_2-terminus of bovine neurointermediate pituitary PAL arises through endoproteolytic cleavage after a Lys-Lys sequence (Eipper et al 1991), and the COOH-terminus of the 75 kDa MTC amidating enzyme is thought to arise by cleavage after a Lys-Lys sequence (Bertelsen et al 1990). CA-77 cells express forms of PAM mRNA equivalent to rPAM-3b (Figure 5) and rPAM-1 (except for an altered COOH-terminus) (Bertelsen et al 1990). Pulse-chase studies with these cells demonstrate conversion of newly synthesized 94 kDa Type A (rPAM-3b) protein into a secreted 75 kDa protein and conversion of newly synthesized 105 kDa Type B protein into secreted 41 and 43 kDa proteins that differ at their NH_2-termini (Beaudry & Bertelsen 1989). Expression of bPAM-1 cDNA in AtT-20 cells leads to increased production of soluble PHM and PAL, which demonstrates that the bPAM precursor undergoes endoproteolytic processing into smaller, soluble enzymes (Mains et al 1991; Perkins et al 1990a). The ability of cells to process PAM precursors is cell-type specific; hEK293 cells transfected with a cDNA encoding rPAM-1 from which the PHM domain had been deleted do not carry out extensive endoproteolytic processing, and most of the PAL activity remains membrane associated (Eipper et al 1991). Expression of a cDNA equivalent to rPAM-3a, but with optional exon A in COS-7 cells, resulted in secretion of a 37 kDa PHM protein and a 53 kDa PAL protein, which demonstrates the occurrence of endoproteolytic cleavage by this fibroblast line (Kato et al 1990a).

PAM IN THE NERVOUS SYSTEM

Amidation Activity in the Nervous System

Levels of amidation activity in the central nervous system are highest in the hypothalamus (Braas et al 1989; Gale et al 1988; Meng & Tsou 1988). Approximately 50% of the amidating activity is membrane associated (Braas et al 1989). Although soluble forms of PHM (36 kDa) and PAL (41 kDa) have been purified from a crude granule fraction prepared from male Wistar rat brain (Takahashi et al 1990), membrane associated forms remain to be characterized. Subcellular fractionation studies have consistently revealed an association of amidation activity with crude synaptosomal and neurosecretory granule fractions (Emeson 1984; Gale et al 1988).

Forms of PAM mRNA in the Nervous System

Based on Northern blot analysis, the hypothalamus also contains the highest levels of PAM mRNA in the central nervous system (Braas et al 1989); levels in the anterior pituitary are threefold higher, and levels in the atrium approximately 15-fold higher. Other central nervous system regions (olfactory bulb, cerebral cortex, hippocampus, striatum, thalamus, brainstem) contain levels of PAM mRNA approximately threefold below those in hypothalamus, whereas PAM mRNA is barely detectable in the cerebellum. As in the rat atrium, rat brain contains 4.2 and 3.8 kb forms of PAM mRNA; the relative amounts vary in different brain regions. When visualized with a cDNA probe specific for optional exon A, the 4.2 kb

PAM mRNA in the central nervous system appeared to correspond to rPAM-1 in the heart.

Although cDNA clones encoding PAM have not yet been isolated from neural tissue, reverse transcription, coupled with the polymerase chain reaction, confirms the presence of similar forms of PAM mRNA in the heart and the central nervous system (Stoffers et al 1991). Oligonucleotide primers spanning the PHM domain revealed a single form of mono-oxygenase RNA in all tissues. Primers spanning optional exon A only revealed significant amounts of the longest form, rPAM-1, in the heart, hypothalamus, and cerebral cortex. Similarly, oligonucleotides spanning optional exon B revealed that the most prevalent PAM mRNAs in heart and brain contain optional exon B; in pituitary and other tissues, PAM mRNAs lacking exon B or a subregion were more prevalent. Thus, the PAM gene product is subjected to a similar set of RNA splicing events in brain and heart (depicted by solid lines in Figure 4). This situation is similar to the calcitonin/calcitonin gene-related peptide gene, which is also subjected to one splicing pattern in brain and heart and a different pattern in other tissues (Amara et al 1982; Crenshaw et al 1987). The nucleotide sequence upstream of the calcitonin-specific splice acceptor inhibits the production of calcitonin transcripts in calcitonin gene-related peptide-producing cells (Emeson et al 1989). It is intriguing to speculate that the same tissue-specific splicing factors may operate on both the calcitonin and PAM genes.

Localization of PAM in the Nervous System

Immunocytochemical studies by Rhodes et al (1990) demonstrated a wide-spread distribution of PAM protein in the rat brain; the highest levels were in the periventricular and supraoptic nuclei, neocortex, and sensory ganglia. PAM immunoreactivity was also observed in neurons not known to produce α-amidated peptides, as well as in ependyma, choroid plexus, oligodendroglia, and Schwann cells. Because levels of PAM activity did not drop dramatically in the distal segment of the sciatic nerve after sectioning, Schwann cells were thought to produce the enzyme (Rhodes et al 1990).

Riboprobes prepared from rat PAM-1 cDNA were used to localize PAM mRNA in the adult male rat brain (Figure 6) (Schafer et al 1991). Consistent with a role for PAM in the α-amidation of all neuropeptides, PAM transcripts were identified in virtually all major brain areas except the cerebellum. Levels of PAM mRNA were highest in the hippocampus, supraoptic, and paraventricular nuclei and piriform cortex. As observed by immunocytochemistry (Rhodes et al 1990), PAM transcripts were also identified in nonneuronal cells, including some ependymal cells, and

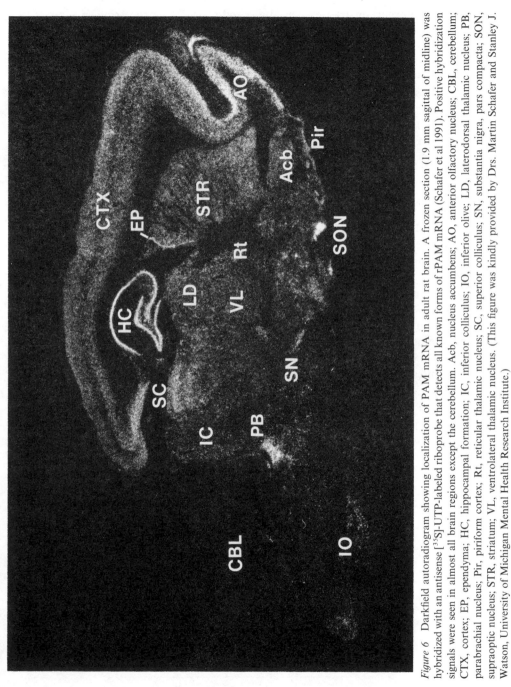

Figure 6 Darkfield autoradiogram showing localization of PAM mRNA in adult rat brain. A frozen section (1.9 mm sagittal of midline) was hybridized with an antisense [^{35}S]-UTP-labeled riboprobe that detects all known forms of rPAM mRNA (Schafer et al 1991). Positive hybridization signals were seen in almost all brain regions except the cerebellum. Acb, nucleus accumbens; AO, anterior olfactory nucleus; CBL, cerebellum; CTX, cortex; EP, ependyma; HC, hippocampal formation; IC, inferior colliculus; IO, inferior olive; LD, laterodorsal thalamic nucleus; PB, parabrachial nucleus; Pir, piriform cortex; Rt, reticular thalamic nucleus; SC, superior colliculus; SN, substantia nigra, pars compacta; SON, supraoptic nucleus; STR, striatum; VL, ventrolateral thalamic nucleus. (This figure was kindly provided by Drs. Martin Schafer and Stanley J. Watson, University of Michigan Mental Health Research Institute.)

surrounding certain arteries. The ability to localize this processing enzyme to particular cells types may serve as a morphologic assay for the occurrence of α-amidation and thus aid in the identification of novel α-amidated peptides. However, the existence of peptide precursors and other peptide processing enzymes, such as carboxypeptidase E in glial cells, may call for an expansion of the cell types usually considered as sources of bioactive peptides (Shinoda et al 1989; Vilijn et al 1989).

PHYSIOLOGIC STUDIES OF PEPTIDE α-AMIDATION IN THE NERVOUS AND ENDOCRINE SYSTEMS

Amidation as a Rate-Limiting Step

One of the difficulties in understanding peptidergic neurons has been the lack of information on the rate limiting steps in peptide processing. In the biosynthetic pathway that leads from tyrosine to dopamine, norepinephrine, and epinephrine, levels of tyrosine, tyrosine hydroxylase, and tetahydrobiopterin can be rate limiting (Fillenz 1990). In peptidergic neurons, levels of PAM can clearly be rate limiting under physiologic conditions. The observation that Gly-extended forms of many peptides are found in tissue extracts first indicated that PAM might play a rate limiting role (Table 2). For example, the rat ventral prostate contains concentrations of TRH-Gly that exceed those of TRH by a factor of 100 (Pekary et al 1989). Although levels of SP-Gly in the central nervous system are generally only 2–3% of the levels of SP, levels of SP-Gly-Lys are 14-fold below the level of SP-Gly (Marchand et al 1990). Similarly, significant levels of gastrin-Gly, compared with gastrin, are found in the antrum of the stomach, whereas gastrin-Gly-Arg(-Arg) could not be detected (Dickinson et al 1990; Hilsted et al 1988).

Table 2 Peptides that occur in significant amounts in the Gly-extended form under physiologic conditions

α-melanotropin-Gly (Fender 1988)
gastrin-Gly (Hilsted & Hansen 1988; Varro et al 1990)
joining peptide-Gly (Fender 1988)
NPY-Gly (Andrews et al 1985)
oxytocin-Gly (Amico et al 1986)
PHI-Gly (Cauvin et al 1989)
SP-Gly (Kream et al 1985)
TRH-Gly (Pekary et al 1989)
vasotocin-Gly (Chauvet et al 1990)

We utilized AtT-20 corticotrope tumor cells as a model system to investigate rate limiting steps in peptide production (Mains et al 1991; Perkins et al 1990a). Stable cell lines expressing increased levels of the full-length bPAM precursor or the PHM domain alone were generated. Stable cell lines with reduced levels of PAM activity were generated by transfection with vector that expressed antisense PAM RNA; in these cells, levels of PAM mRNA, PHM activity, PAL activity, and PAM protein were reduced. AtT-20 cells produce α-amidated-joining peptide (JP-NH$_2$) from their endogenous pro-ACTH/endorphin precursor. The ability of these cell lines to produce JP-NH$_2$ was assessed by biosynthetic labeling and separation of newly synthesized JP-Gly and JP-NH$_2$. Under conditions in which wild type AtT-20 cells α-amidate approximately half of the JP produced, cell lines expressing increased levels of PHM alone or PHM and PAL produced almost entirely JP-NH$_2$. Cell lines exhibiting a three fold reduction in PHM and PAL activity produced approximately half as much JP-NH$_2$ as wild type cells. Thus, the monooxygenase step appears to be rate limiting in peptide amidation.

Regulation of PAM Expression

As might be expected of a rate limiting enzyme in a biosynthetic pathway, levels of PAM are subject to regulation. AtT-20 cells exhibit increased levels of PAM mRNA when stimulated with cAMP or CRH and decreased levels of PAM mRNA when inhibited with dexamethasone (Eipper et al 1987; Thiele et al 1989). These effects appear slowly and are dependent upon protein synthesis. The effects are also tissue specific; in cultures of rat atrial myocytes, dexamethasone increases PAM expression (Thiele et al 1989). In CA-77 rat medullary thyroid carcinoma cells (Birnbaum et al 1989a), as in AtT-20 corticotropes, dexamethasone treatment decreases PAM secretion. By using in situ hybridization histochemistry, Grino et al (1990) demonstrated an increase in levels of PAM mRNA in the magnocellular neurons of the paraventricular nucleus following adrenalectomy.

PAM expression is also responsive to thyroid hormone. Levels of PAM mRNA in the anterior pituitary of the adult male rat rise several-fold following chemical or surgical thyroidectomy (Ouafik et al 1990). Levels of several α-amidated peptides (NPY, SP, VIP) also rise dramatically in hypothyroidism (references in Ouafik et al 1990). In the prostate gland, the ratio of TRH-Gly to TRH increases with age or following castration (Pekary et al 1989), which suggests that some positive signal is required to maintain full PAM activity. Likewise, the ratio of gastrin-Gly to gastrin-NH$_2$ in the antrum is highest in the unstimulated stomach (Varro et al 1990). There may also be some control over whether active (α-amidated) or inactive (Gly-extended) peptides are released in vivo. Both α-amidated and Gly-extended gastrin occur in the blood during fasting and both are secreted by antral tissue in response to a meal (Hilsted & Hansen 1988).

Quantitation of gastrin and gastrin-Gly during development in neonatal rats indicates that the naturally occurring surge in corticosteroids associated with weaning increases the extent of gastrin α-amidation, whereas dietary changes stimulate gastrin synthesis (Marino et al 1988). Several studies have documented changes in PAM mRNA, PAM activity, and the levels of α-amidated peptides during the development of the hypothalamus, cerebral cortex, pancreas, stomach, heart, and bone (Birnbaum et al 1989b; Faivre-Bauman et al 1988; Hilsted et al 1988; Maltese et al 1989; Marino et al 1988; Mogensen et al 1990; Ouafik et al 1989; Scharfmann et al 1988). How developmental cues regulate these changes in PAM expression, and which developmental processes depend on the timely appearance of α-amidated peptides are fascinating issues to be addressed in the future.

The fact that PHM and PAL are secreted along with their α-amidated peptide products raises the question of whether amidation occurs extracellularly. Certainly, potential substrates, such as gastrin-Gly, are co-localized with gastrin in G-cells and cosecreted in response to stimulation (Dickinson et al 1990; Hilsted & Hansen 1988). Similarly, integral membrane protein forms of PAM could reach the cell surface and function while attached to the cell surface.

Role of Copper

The importance of copper availability for peptide α-amidation in vivo and in cell culture has been demonstrated by using disulfiram (Antabuse™), a copper chelator for the treatment of chronic alcoholism. Disulfiram treatment of rats for a few days at doses as low as the human therapeutic amounts obliterated the ability of the intermediate pituitary to α-amidate newly synthesized α-melanotropin and joining peptide (Mains et al 1986). Marchand et al (1990) treated rats with disulfiram and observed a tenfold increase in the amount of Substance P-Gly compared with Substance P-NH$_2$. Functional studies in these same animals indicated that the disulfiram treatment induced a significant increase in pain threshold, consistent with a role for Substance P in pain sensation. Similar treatments of rats lowered brain and pituitary levels of α-amidated peptides without affecting levels of other peptides; a compensatory increase in PHM activity was observed in pituitary extracts assayed in a test tube under optimal conditions (Mueller et al 1991). Antral extracts from animals treated with diethyldithiocarbamate had levels of gastrin-Gly that exceeded the levels of gastrin-NH$_2$. The extracts exhibited a major increase in amidation activity when assayed under optimal conditions, which again suggests the occurrence of a compensatory increase in enzyme production (Dickinson et al 1990; Hilsted 1990). When Bradbury et al (1990) treated CA-77 cells with inhibitors of PHM, they observed a transient decrease in activity followed by a compensatory increase. Treatment of neonatal mice with copper-chelating agents, such as D-penicillamine, also causes serious neurologic abnormalities (Yamamoto et al 1990). Possible contributions by disulfiram-induced inhibition of PAM to the peripheral neuropathy and other undesirable side effects associated with long-term treatment with disulfiram have not yet been investigated.

The average daily intake of copper in humans is 1 mg, which is below the estimated daily requirement of 1.5 to 3 mg. This low intake of copper is thought to affect the brain, heart, pancreas, pituitary, adrenal, thyroid, and gonad (Bhathena & Recant 1990; Lukaski et al 1988). In Menkes' disease, which is an X-linked recessive trait that exhibits severe neurologic-complications, serum copper levels are low due to impaired gut absorption of copper (Waldrop & Ettinger 1990). Cells from individuals with Menkes' disease have increased intracellular metallothionein levels (Waldrop & Ettinger 1990) and, thus, may have decreased availability of Cu^{2+} for PAM. Wilson's disease, a disorder in intrahepatic copper metabolism that is inherited as an autosomal recessive trait and leads to elevated tissue levels of copper, often presents with neurologic complications that appear subsequent to hepatic injury (Woods & Colon 1989). A role for altered peptide amidation in the psychiatric disorders, tremor, increased salivation, speech abnormalities, and cardiac involvement sometimes associated with Wilson's disease remains to be investigated (Kuan 1987). A possible role for altered peptide α-amidation in the pathology of these diseases needs to be investigated, as well.

Role of Ascorbate

The contribution of ascorbic acid (vitamin C) to peptide α-amidation has also been investigated in vivo and in cell culture, thus capitalizing on the fact that ascorbate is made in the liver and delivered to the pituitary and nervous tissue by the blood in species such as rodents and that guinea pigs (like humans) are dependent on dietary sources for their total supplies of ascorbate. The ability of mouse pituitary corticotrope tumor cells to synthesize α-amidated joining peptide is a reflection of the availability of ascorbic acid in the culture medium (Mains et al 1991; May et al 1989). However, some amidation occurs even in the total absence of ascorbate, just as other ascorbate dependent enzymes, such as prolyl hydroxylase, still show residual function with other sources of reducing equivalents (Chauhan et al 1985). Although anterior pituitary corticotropes, inter-mediate pituitary melanotropes, and AtT-20 corticotrope tumor cells all perform α-amidation well in the presence of normal supplies of ascorbate, the ability of the three different cell types to utilize alternate cofactors differs. Although melanotropes exhibit a stringent requirement for ascorbate, AtT-20 cells can utilize catecholamines to support peptide amidation (May et al 1989). The central nervous system is very well protected against ascorbate deficiency and maintains its ascorbate supplies in the face of nearly fatal ascorbate deprivation (Eipper & Mains 1988; Mains et al 1990). However, in severely scorbutic guinea pigs, there is a significant elevation in the ratio of Glycine-extended to α-amidated gastrin (Fenger & Hilsted 1988; Hilsted et al 1986). Peptide α-amidation in many peripheral tissues may be much more sensitive to decreased ascorbate supplies than peptide α-amidation in the central nervous system and pituitary.

CONCLUSIONS AND PERSPECTIVES

Peptide α-amidation is clearly important for the functioning of many biologic communication networks. Although recent studies have delineated the structure of the proteins that mediate α-amidation, the cofactors required, and some details of the reaction mechanisms, much remains to be learned in these areas. The gene encoding PAM is complex, which gives rise to a family of bifunctional proteins that mediate α-amidation and raises questions about factors that regulate tissue-specific mRNA and protein processing and the developmental and hormonal control of expression of this gene. A better understanding of this processing enzyme will provide insight into the functioning of peptidergic neurons and better tools for physiologic studies of these neurons.

ACKNOWLEDGMENTS

This review was supported by grants from the Public Health Service (DK-32949, DK-32948, DA-00266, DA-00098, DA-00097) and from the American Heart Association. We thank Drs. Martin Schafer and Stanley J. Watson for kindly providing the photograph in Fig. 6.

Literature Cited

Aitman, T. J., Rafferty, B., Coy, D., Lynch, S. S., Clayton, R. N. 1989. *Peptides* 10: 1–4

Amara, S. G., Jonas, V., Rosenfeld, M. G., Ong, E. S., Evans, R. M. 1982. *Nature* 298: 240–44

Amico, J. A., Ervin, M. G., Finn, F. M., Leake, R. D., Fisher, D. A., Robinson, A. G. 1986. *Metabolism* 35: 596–601

Andrews, P. C., Hawke, D. H., Lee, T. D., Legesse, K., Noe, B. D., Shively, J. E. 1986. *J. Biol. Chem.* 261: 8128–33

Andrews, P. C., Hawke, D., Shively, J. E., Dixon, J. E. 1985. *Endocrinology* 116: 2677–81

Bazan, J. F., Fletterick, R. J., Pilkis, S. J. 1989. *Proc. Natl. Acad. Sci. USA* 86: 9642–46

Beaudry, G. A., Bertelsen, A. H. 1989. *Biochem. Biophys. Res. Commun.* 163: 959–66

Beaudry, G. A., Mehta, N. M., Ray, M. L., Bertelsen, A. H. 1990. *J. Biol. Chem.* 265: 17694–99

Bertelsen, A. H., Beaudry, G. A., Galella, E. A., Jones, B. N., Ray, M. L., Mehta, N. 1990. *Arch. Biochem. Biophys.* 279: 87–96

Bhathena, S. J., Recant, L. 1990. In *Metal Ions in Biology and Medicine*, ed. P. Collery, L. A., Poirier, H. Manfait, J. C. Etienne, pp. 84–88. Paris: Libbey Eurotext

Birnbaum, R. S., Bertelsen, A. H., Roos, B. A. 1989a. *Mol. Cell. Endocrinol.* 61: 109–16

Birnbaum, R. S., Howard, G. A., Roos, B. A. 1989b. *Endocrinology* 124: 3134–36

Braas, K. M., Stoffers, D. A., Eipper, B. A., May, V. 1989. *Mol. Endocrinol.* 3: 1387–98

Bradbury, A. F., Smyth, D. G. 1985. In *Biogenetics of Neurohormonal Peptides*, ed. R. Hakanson, J. Thorell, pp. 171–86. New York: Academic

Bradbury, A. F., Finnie, M. D. A., Smyth, D. G. 1982. *Nature* 298: 686–88

Bradbury, A. F., Mistry, J., Roos, B. A., Smyth, D. G. 1990. *Eur. J. Biochem.* 189: 363–68

Bradbury, A. F., Smyth, D. G. 1987a. *Biosci. Rep.* 7: 907–16

Bradbury, A. F., Smyth, D. G. 1987b. *Eur. J. Biochem.* 169: 579–84

Bradbury, A. F., Smyth, D. G. 1988a. *Biochem. Biophys. Res. Commun.* 154: 1293–1300.

Bradbury, A. F., Smyth, D. G. 1988b. *Physiol. Bohemoslov.* 37: 267–74

Bresnahan, P. A., Leduc, R., Thomas, L., Thorner, J., Gibson, H. L., et al. 1990. *J. Cell Biol.* 111: 2851–59

Cauvin, A., Vandermeers, A., Vandermeers-Piret, M. C., Robberecht, P., Christophe, J. 1989. *Endocrinology* 125: 1296–1302

Chauhan, U., Assad, R., Peterkofsky, B. 1985. *Biochem. Biophys. Res. Commun.* 131: 277–83

Chauvet, J., Rouille, Y., Chauvet, M. T., Acher, R. 1990. *Neuroendocrinology* 51: 233–36

Chretien, M., Sikstrom, R. A., Lazure, C., Mbikay, M., Seidah, N. G. 1989. *Biosci. Rep.* 9: 693–700

Coggins, J. R., Hardie, D. G. 1986. *Multidomain Proteins—Structure and Evolution.* New York: Elsevier. 344 pp.

Crenshaw, E. B., Jonas, V., Swanson, L. W., Rosenfeld, M. G. 1987. *Cell* 49: 389–98

DeSouza, E. B., Kuhar, M. J. 1986. *Methods Enzymol.* 124: 560–90

Dickinson, C. J., Marino, L., Yamada, T. 1990. *Am. J. Physiol.* 258: G810–14

Dickinson, C. J., Yamada, T. 1991. *J. Biol. Chem.* 266: 334–38

Di Patti, M. C. B., Musci, G., Giartosio, A., Dalessio, S., Calabrese, L. 1990. *J. Biol. Chem.* 265: 21016–22

Eberle, A. N. 1981. *Ciba Found. Symp.* 81: 13–31

Eipper, B. A., Mains, R. E. 1988. *Annu. Rev. Physiol.* 50: 333–44

Eipper, B. A., May, V., Braas, K. M. 1988. *J. Biol. Chem.* 263: 8371–79

Eipper, B. A., Myers, A. C., Mains, R. E. 1985. *Endocrinology* 116: 2497–2504

Eipper, B. A., Park, L. P., Dickerson, I. M., Keutmann, H. T., Thiele, E. A., et al. 1987. *Mol. Endocrinol.* 1: 777–90

Eipper, B. A., Perkins, S. N., Husten, E. J., Johnson, R. C., Keutmann, H. T., Mains, R. E. 1991. *J. Biol. Chem.* 266: 7827–33

Emeson, R. B. 1984. *J. Neurosci.* 4: 2604–13

Emeson, R. B., Hedjran, F., Yeakley, J. M., Guise, J. W., Rosenfeld, M. G. 1989. *Nature* 341: 76–80

Faivre-Bauman, A., Loudes, C., Barret, A., Patte, C., Tixier-Vidal, A. 1988. *Dev. Brain Res.* 40: 261–67

Fenger, M. 1988. *Regul. Pept.* 20: 345–57

Fenger, M., Hilsted, L. 1988. *Acta Endocrinol.* 118: 119–24

Fillenz, M. 1990. *Noradrenergic Neurons.* Cambridge: Cambridge Univ. Press. 238 pp.

Fritz, A. F., Cho, K. W., Wright, C. V., Jegalian, B. G., DeRobertis, E. M. 1989. *Dev. Biol.* 131: 584–88

Fuhlendorff, J., Gether, U., Aakerlund, L., Langeland-Johansen, N., Thogersen, H., et al. 1990. *Proc. Natl. Acad. Sci. USA* 87: 182–86

Gafvelin, G., Carlquist, M., Mutt, V. 1985. *FEBS Lett.* 184: 347–50

Gale, J. S., McIntosh, J. E. A., McIntosh, R. P. 1988. *Biochem. J.* 251: 251–59

Gilligan, J. P., Lovato, S. J., Mehta, N. M., Bertelsen, A. H., Jeng, A. Y., Tamburini, P. P. 1989. *Endocrinology* 124: 2729–36

Glauder, J., Ragg, H., Rauch, J., Engels, J. W. 1990. *Biochem. Biophys. Res. Commun.* 169: 551–58

Grino, M., Guillaume, V., Boudouresque, F., Conte-Devoix, B., Maltese, J. Y., Oliver, C. 1990. *Mol. Endocrinol.* 4: 1613–19

Harris, R. B. 1989. *Arch. Biochem. Biophys.* 275: 315–33

Harvey, S. 1990. *J. Endocrinol.* 125: 345–58

Hatsuzawa, K., Hosaka, M., Nakagawa, T., Nagase, M., Shoda, A., et al. 1990. *J. Biol. Chem.* 265: 22075–78

Hilsted, L. 1990. *Regul. Pept.* 29: 179–87

Hilsted, L., Bardram, L., Rehfeld, J. F. 1988. *Biochem. J.* 255: 397–402

Hilsted, L., Hansen, C. P. 1988. *Am. J. Physiol.* 255: G665–69
Hilsted, L., Rehfeld, J. F., Schwartz, T. W. 1986. *FEBS Lett.* 196: 151–54
Husten, E. J., Eipper, B. A. 1991. *J. Biol. Chem.* 266: In press
Katapodis, A. G., May, S. W. 1988. *Biochem. Biophys. Res. Commun.* 151: 499–505
Katapodis, A. G., May, S. W. 1990. *Biochemistry* 29: 4541–48
Katapodis, A. G., Ping, D., May, S. W. 1990. *Biochemistry* 26: 6115–20
Kato, I., Yonekura, H., Tajima, M., Yanagi, M., Yamamoto, H., Okamoto, H. 1990a. *Biochem. Biophys. Res. Commun.* 172: 197–203
Kato, I., Yonekura, H., Yamamoto, H., Okamoto, H. 1990b. *FEBS Lett.* 269: 319–23
Kojima, M., Mizuno, K., Kangawa, K. Matsuo, H. 1989. *J. Biochem.* 105: 440–43
Kream, R. M., Schoenfeld, T. A., Mancuso, R., Clancy, A. N., El-Bermani, W., Macrides, F. 1985. *Proc. Natl. Acad. Sci. USA* 82: 4832–36
Kreil, G. 1985. *Enzymol. Post-Trans. Modif. Proteins* 2: 41–51
Kuan, P. 1987. *Chest* 91: 579–83
Lukaski, H. C., Klevay, L. M., Milne, D. B. 1988. *Eur. J. Appl. Physiol.* 58: 74–80
Mackin, R. B., Flacker, J. M., Mackin, J. A., Noe, B. D. 1987. *Gen. Comp. Endocrinol.* 67: 263–69
Mains, R. E., Bloomquist, B. T., Eipper, B. A. 1991. *Mol. Endocrinol.* 5: 187–93
Mains, R. E., Dickerson, I. M., May, V., Stoffers, D. A., Perkins, S. N., et al. 1990. *Front. Neuroendocrinol.* 11: 52–89
Mains, R. E., Eipper, B. A. 1990. *Trends Endocrinol. Metab.* 1: 388–94
Mains, R. E., May, V., Cullen, E. I., Eipper, B. A. 1988. In *Molecular Biology of Brain and Endocrine Peptidergic Systems*, ed. M. Chretien, K. W., McKerns, pp. 201–13. New York: Plenum
Mains, R. E., Park, L. P., Eipper, B. A. 1986. *J. Biol. Chem.* 261: 11938–41
Maltese, J. Y., Giraud, P., Kowalski, C., Ouafik, L. H., Salers, P., et al. 1989. *Biochem. Biophys. Res. Commun.* 158: 244–50
Marchand, J. E., Hershman, K., Kumar, M. S. A., Thompson, M. L., Kream, R. M. 1990. *J. Biol. Chem.* 265: 264–73
Marino, L. R., Sugano, K., Yamada, T. 1988. *Am. J. Physiol.* 254: G87–92
May, V., Cullen, E. I., Braas, K. M., Eipper, B. A. 1988. *J. Biol. Chem.* 263: 7550–54
May, V., Mains, R. E., Eipper, B. A. 1989. *Horm. Res.* 32: 18–21
Mehta, N. M., Gilligan, J. P., Jones, B. N.,

Bertelsen, A. H., Roos, B. A., Birnbaum, R. S. 1988. *Arch. Biochem. Biophys.* 261: 44–54
Meng, F., Tsou, K. 1988. *J. Neurochem.* 50: 1352–55
Minth, C. D., Qiu, H., Akil, H., Watson, S. J., Dixon, J. E. 1989. *Proc. Natl. Acad. Sci. USA* 86: 4292–96
Mizuno, K., Ohsuye, K., Wada, Y., Fuchimura, K., Tanaka, S., Matsuo, H. 1987. *Biochem. Biophys. Res. Commun.* 148: 546–52
Mizuno, K., Sakata, J., Kojima, M., Kangawa, K., Matsuo, H. 1986. *Biochem. Biophys. Res. Commun.* 137: 984–91
Mogensen, N. W., Hilsted, L., Bardram, L., Rehfeld, J. F. 1990. *Devl. Brain Res.* 54: 81–86
Mohun, T., Garrett, N., Stutz, F., Spohr, G. 1988. *J. Mol. Biol.* 202: 67–76
Mueller, G.. P., Husten, E. J., Eipper, B. A. 1991. *Endocrinology* 128: Abstr. 1021
Murthy, A. S. N., Keutmann, H. T., Eipper, B. A. 1987. *Mol. Endocrinol.* 1: 290–99
Murthy, A. S. N., Mains, R. E., Eipper, B. A. 1986. *J. Biol. Chem.* 261: 1815–22
Nahon, J. L., Presse, F., Bittencourt, J. C., Sawchenko, P. E., Vale, W. 1989. *Endocrinology* 125: 2056–65
Noguchi, M., Takahashi, K., Okamoto, H. 1989. *Arch. Biochem. Biophys.* 275: 505–13
Ohsuye, K., Kitano, K., Wada, Y., Fuchimura, K., Tanaka, S., et al. 1988. *Biochem. Biophys. Res. Commun.* 150: 1275–81
Ouafik, L. H., May, V., Keutmann, H. T., Eipper, B. A. 1989. *J. Biol. Chem.* 264: 5839–45
Ouafik, L. H., May, V., Saffen, D. W., Eipper, B. A. 1990. *Mol. Endocrinol.* 4: 1497–1505
Payne, G. S., Schekman, R. 1989. *Science* 245: 1358–65
Pekary, A. E., Knoble, M., Garcia, N. 1989. *Endocrinology* 125: 679–85
Pekary, A. E., Reeve, J. R., Smith, V. A., Friedman, S. 1990. *Int. J. Androl.* 13: 169–79
Perkins, S. N., Eipper, B. A., Mains, R. E. 1990a. *Mol. Endocrinol.* 4: 132–39
Perkins, S. N., Husten, E. J., Eipper, B. A. 1990b. *Biochem. Biophys. Res. Commun.* 171: 926–32
Perkins, S. N., Husten, E. J., Mains, R. E., Eipper, B. A. 1990c. *Endocrinology* 127: 2771–78
Rhodes, C. H., Xu, R. Y., Angeletti, R. H. 1990. *J. Histochem. Cytochem.* 38: 1301–11
Richter, K., Egger, R., Kreil, G. 1987. *Science* 238: 200–2
Sawyer, W. H., Manning, M. 1989. *Trends Endocrinol. Metab.* 1: 48–50

Schafer, M. K. H., Stoffers, D. A., Eipper, B. A., Watson, S. J. 1991. *J. Neurosci.* In press

Scharfmann, R., Leduque, P., Aratan-Spiref, S., Dubois, P., Basmaciogullari, A., Czernichow, P. 1988. *Endocrinology* 123: 1329–34

Schwartz, T. W. 1990. In *Molecular Biology of the Islets of Langerhans*, ed. H. Okamoto, pp. 153–205. Cambridge: Cambridge Univ. Press

Scott, F. M., Treston, A., Avis, I., Kasprzyk, P., Eipper, B. A., et al. 1990. *Clin. Res.* 38: 248A

Seidah, N. G., Gaspar, L., Mion, P., Marcinkiewicz, M., Mbikay, M., Chretien, M. 1990. *DNA Cell Biol.* 9: 415–24

Seidah, N. G., Marcinkiewicz, M., Benjannet, S., Gaspar, L., Beaubien, G., et al. 1991. *Mol. Endocrinol.* 5: 111–22

Shinoda, H., Marini, A. M., Cosi, C., Schwartz, J. P. 1989. *Science* 245: 415–17

Smeekens, S. P., Steiner, D. F. 1990. *J. Biol. Chem.* 265: 2997–3000

Sossin, W. S., Fisher, J. M., Scheller, R. H. 1989. *Neuron* 2: 1407–17

Southan, C., Kruse, L. I. 1989. *FEBS Lett.* 255: 116–20

Steiner, D. F., Chan, S. J., Smeekens, S. P., Bell, G. I., Emdin, S., Falkmer, S. 1989. *Biol. Bull.* 177: 172–75

Stewart, L. C., Klinman, J. P. 1988. *Annu. Rev. Biochem.* 57: 551–92

Stoffers, D. E., Eipper, B. A. 1989. In *ICSU Short Reports, Advances in Gene Technology: Molecular Neurobiology and Neuropharmacology*, ed. R. L. Rotundo, F. Ahman et al., 9: 120. Oxford: IRL

Stoffers, D. A., Green, C. B. R., Eipper, B. A. 1989. *Proc. Natl. Acad. Sci. USA* 86: 735–39

Stoffers, D. A., Ouafik, L. H., Eipper, B. A. 1991. *J. Biol. Chem.* 266: 1701–7

Suzuki, K., Shimoi, H., Iwasaki, Y., Kawahara, T., Matsuura, Y., Nishikawa, Y. 1990. *EMBO J.* 9: 4259–65

Tajima, M., Iida, T., Yoshida, S., Komatsu, K., Namba, R., et al. 1990. *J. Biol. Chem.* 265: 9602–5

Takada, Y., Noguchi, T. 1986. *Biochem. J.* 235: 391–97

Takahashi, K., Okamoto, H., Seino, H.,

Noguchi, M. 1990. *Biochem. Biophys. Res. Commun.* 169: 524–30

Tamburini, P. P., Jones, B. N., Consalvo, A. P., Young, S. D., Lovata, S. J., et al. 1988. *Arch. Biochem. Biophys.* 267: 623–31

Tamburini, P. P., Young, S. D., Jones, B. N., Palmesino, R. A., Consalvo, A. P. 1990. *Int. J. Pept. Protein Res.* 35: 153–56

Tatemoto, K., Efendic, S., Mutt, V., Makk, G., Feistner, G. J., Barchas, J. D. 1986. *Nature* 234: 476–78

Taylor, W. R. 1987. In *Nucleic Acid and Protein Sequence Analysis*, ed. M. J. Bishop, C. J. Rawlings, pp. 285–322. Oxford: IRL

Tazi, A., Dantzer, R., LeMoal, M., Rivier, J., Vale, W., Koob, G. F. 1987. *Regul. Pept.* 18: 37–42

Thiele, E. A., Marek, K. L., Eipper, B. A. 1989. *Endocrinology* 125: 2279–88

Vale, W., Spiess, J., Rivier, J. 1981. *Science* 213: 1394–97

Varro, A., Nemeth, J., Bridson, J., Lee, C., Moore, S., Dockray, G. J. 1990. *J. Biol. Chem.* 265: 21476–81

Vilijn, M. H., Das, B., Kessler, J. A., Fricker, L. D. 1989. *J. Neurochem.* 53: 1487–93

von Heijne, G. 1987. *Sequence Analysis in Molecular Biology*. San Diego: Academic

Waldrop, G. L., Ettinger, M. J. 1990. *Biochem. J.* 267: 417–12

Wand, G. S., Ney, R. L., Baylin, S., Eipper, B. A., Mains, R. E. 1985a. *Metabolism* 34: 1044–52

Wand, G. S., Ney, R. L., Mains, R. E., Eipper, B. A. 1985b. *Neuroendocrinology* 41: 482–89

Wang, N., Southan, C., DeWolf, W. E. Jr., Wells, T. N. C., Kruse, L. I., Leatherbarrow, R. J. 1990. *Biochemistry* 29: 6466–74

Wehrenberg, W. B., Ling, N. 1983. *Biochem. Biophys. Res. Commun.* 115: 525–30

Wood, S. E., Colon, V. F. 1989. *Am. Fam. Physician* 40: 171–78

Yamamoto, M., Akiyama, C., Aikawa, H. 1990. *Dev. Brain Res.* 55: 51–55

Young, S. D., Tamburini, P. P. 1989. *J. Am. Chem. Soc.* 111: 1933–34

Zabriskie, T. M., Cheng, H., Vederas, J. C. 1991. *J. Chem. Soc. Chem. Commun.*, pp. 571–72

Annu. Rev. Neurosci. 1992. 15:87–114

ADRENERGIC RECEPTORS AS MODELS FOR G PROTEIN-COUPLED RECEPTORS

Brian Kobilka

Howard Hughes Medical Institute, Departments of Cardiology and Molecular and Cellular Physiology, Stanford University Medical Center, Stanford, California 94305

KEY WORDS: signal transduction, desensitization

INTRODUCTION

Approximately 80% of known hormones and neurotransmitters activate cellular signal transduction mechanisms by activating G protein-coupled receptors (Birnbaumer et al 1990). Studying the structure and function of these receptors has been challenging because they are not naturally abundant proteins and they require a lipid environment to be fully active. The cDNA or genomic clones for many members of this class of receptors have recently been obtained. Applying the techniques of molecular biology to the study of G protein-coupled receptors has helped circumvent many of the technical difficulties that had complicated previous efforts to characterize these proteins biochemically.

The adrenergic receptors have been one of the most extensively studied classes of G protein-coupled receptors. Genomic and/or cDNA clones for nine types of adrenergic receptors have been obtained: two types of α_1 receptors (Cotecchia et al 1988; Schwinn et al 1990); three types of α_2 receptors (Kobilka et al 1987; Lomasney et al 1990; Regan et al 1988; Weinshank et al 1990); the β_1 receptor (Frielle et al 1987); the β_2 receptor (Dixon et al 1986); the β_3 receptor (Emorine et al 1989); and the avian β receptor (Yarden et al 1986). Significant progress has been made in elucidating the structural domains involved in ligand binding, G protein activation, and desensitization, and much of what is learned about the mechanism of action of adrenergic receptors will likely apply to the other

87

G protein-coupled receptors. This review focuses on recent advances in understanding the structure and function of adrenergic receptors and attempts to integrate developments in the study of other G protein-coupled receptors.

RECEPTOR-G PROTEIN INTERACTIONS

Adenylyl cyclase is activated by G protein-coupled receptors such as the β_2 receptor. Figure 1 illustrates some common features of the interactions

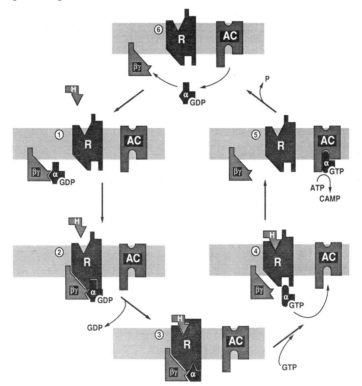

Figure 1 Outline of a cycle of signal transduction for the β_2 adrenergic receptor. (1) The components include the hormone (H) epinephrine, the β_2 receptor (R), the subunits of G_s (α, β, and γ), and adenylyl cyclase (AC). (2) Receptors bound to G_s have a higher affinity for the hormone. (3) The binding of the hormone to the β_2 receptor leads to a structural change in the receptor, which facilitates the release of GDP and the binding of GTP to the α subunit of G_s. (4) The binding of GTP to the α subunit results in a release of the α subunit from both the receptor and the β-γ subunit. (5) The GTP-$G_{s\alpha}$ subunit binds to and activates adenylyl cyclase, which catalyzes the conversion of ATP to cAMP. cAMP activates PKA, which phosphorylates a variety of cytosolic and nuclear proteins (not shown). (6) The activation of adenylyl cyclase continues until GTP is hydrolyzed to GDP and the $G_{s\alpha}$ subunit is released from the enzyme and complexes with the β-γ subunit.

between receptors and G proteins (based on reviews by Birnbaumer et al 1990; Bourne et al 1991; Freismuth et al 1988; Ross 1989). Receptors form complexes with heterotrimeric GTP binding proteins that consist of α, β, and γ subunits. Formation of this complex alters the properties of both the receptor and the G protein. When complexed with a G protein, the receptor often has a higher affinity for its agonist. When the G protein forms a complex with an agonist occupied receptor, the G protein α subunit exhibits a more rapid rate for the dissociation of GDP and the binding of GTP. The $G_{s\alpha}$-GTP complex dissociates from the receptor and from the β and γ subunits. The free $G_{s\alpha}$-GTP subunit can then continue to activate adenylyl cyclase until the bound GTP is hydrolysed to GDP. The $G_{s\alpha}$-GDP reassociates with the β-γ subunits, and the cycle can begin again.

Six classes of G protein α subunits have been characterized either biochemically or by cloning: G_s, the G_i proteins, G_o, G_{olf}, G_t, and $G_{Z/X}$. The effector systems activated by these proteins have been reviewed elsewhere (Birnbaumer et al 1990), but are briefly summarized here. G_s activates adenylyl cyclase and Ca^{2+} channels. Members of the G_i class of G proteins (G_{i1}, G_{i2}, and G_{i3}) inhibit the activity of adenylyl cyclase and activate phospholipase C, phospholipase A2, and K^+ channels. G_o proteins may be involved in modulating the activity of voltage sensitive Ca^{2+} channels and phospholipase C. G_{olf} is found in nasal epithelium and may mediate signal transduction by olfactory receptors. G_t proteins (transducins) mediate signal transduction by the opsins. The function of $G_{Z/X}$ is not known.

The β and γ subunits are tightly, but noncovalently, bound to each other. The β-γ dimer is essential for effective activation of the α subunit by a receptor. Little is known, however, about the physical interaction between the receptor and the β-γ dimer. Evidence suggests that mammalian β-γ subunits play a direct role in activating phospholipase A2 (Kim et al 1989).

Expression of the cloned adrenergic receptors in cultured cells facilitates the study of receptor-G protein interactions. β adrenergic receptors (β_1, β_2, and β_3) all activate G_s. Stimulation of the α_1 adrenergic receptor leads to the activation of phospholipase C (Cotecchia et al 1990b). Stimulation of α_2 receptors expressed in cultured cells leads to the inhibition of adenylyl cyclase and the activation of phospholipase C (Cotecchia et al 1990b). This may be due to either the ability of a receptor to activate more than one type of G protein, or the ability of the G_α subunit to activate more than one effector system. The cellular response to stimulation of a specific G protein-coupled receptor may also differ for different cells. Agonist stimulation of dopaminergic D_2 receptors expressed in mouse fibroblasts (Ltk-cells) leads to inhibition of adenylyl cyclase and activation of phospholipase C (Vallar et al 1990). Although activation of the same receptor expressed in GH_4C_1 pituitary cells produces inhibition of adenylyl cyclase, it fails to activate phospholipase C; instead, a decrease of intracellular calcium is observed. Thus, receptor-effector coupling in highly differentiated cells in vivo might differ from what is observed when these receptors are expressed in tissue culture cells.

COMMON STRUCTURAL FEATURES OF G PROTEIN-COUPLED RECEPTORS

The human β_2 receptor can be used to illustrate the common structural features shared by members of the G protein-coupled receptor family (Figure 2). These receptors have seven hydrophobic domains that are thought to span the membrane. The receptor's amino terminus is extracellular, and the carboxyl terminus is intracellular. Usually one or more

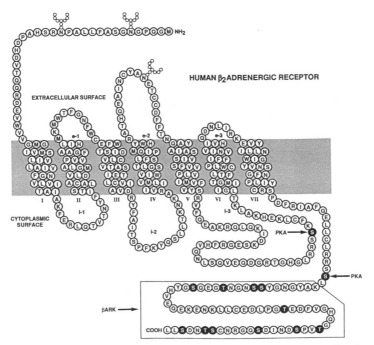

Figure 2 Diagram of the proposed membrane topology for the human β_2 adrenergic receptor. Circles with letters represent amino acids identified by the single letter amino acid code. Proposed membrane spanning domains are numbered with roman numerals. Extracellular domains are designated by a prefix e-. Intracellular domains are designated by a prefix i-. Black circles with white letters indicate potential sites for phosphorylation by PKA or βARK.

sites for asparagine-linked glycosylation are found in the amino terminal domain. The cytoplasmic domains often contain potential sites for phosphorylation by one or more cytosolic kinases. High-resolution structural analysis is not yet available for any of the G protein-coupled receptors. Although bacteriorhodopsin is not itself a G protein-coupled receptor, analysis of its primary amino acid sequence indicates it shares a similar structure with members of this receptor family. The structure of bacteriorhodopsin, recently determined by high-resolution electron cryomicroscopy (Henderson et al 1990), then provides our best current picture of the G protein-coupled receptors. The seven hydrophobic regions of bacteriorhodopsin form membrane spanning domains, which all contribute to the retinal binding pocket; five of these are involved in the formation of the proton pump.

Biochemical (Dohlman et al 1987) and immunologic techniques have been used to study the topology of the β_2 adrenergic receptor (Wang et al 1989b) and evidence from these studies supports the existence of seven membrane spanning domains. When the sequences of the various cloned receptors are compared, the greatest degree of homology is found in these hydrophobic transmembrane domains. Functionally similar receptors, such as the subtypes of the α_2 adrenergic receptor, have 75% identical amino acid composition in the hydrophobic domains. The greatest degree of diversity in amino acid composition and size is found in the amino terminus, the third intracellular domain, and the carboxyl terminus. The amino terminus ranges from 12 amino acids for the α_2 receptor (human chromosome 2), which lacks asparagine-linked sugars (Lomasney et al 1990), to 394 amino acids for the TSH receptor (Parmentier et al 1989). The third intracellular loop ranges in size from 29 amino acids for the substance P receptor (Yokota et al 1989) to 242 amino acids for the human M3 cholinergic receptor (Peralta et al 1987). The carboxyl terminus ranges from 21 amino acids for the α_2 receptor (human chromosome 2) to 164 amino acids for the hamster α_1 receptor (Cotecchia et al 1988). Except for the extremely long amino terminal domain in all of the glycoprotein hormone receptors (TSH, FH, LH), no clear correlation is apparent between the size of these hydrophilic domains and the functional properties of ligand binding and G protein activation. As discussed below, the amino terminal domain of the glycoprotein receptors is involved in ligand binding.

Not all receptors that activate G proteins are members of the seven membrane spanning family. For example, the receptor for insulin-like growth factor II, has a single membrane spanning domain yet it directly activates G_{i-2}, which in turn activates a calcium-permeable channel in a mouse fibroblast cell line (Okamoto et al 1990). G protein activation by this receptor is discussed further below.

THE LIGAND BINDING DOMAIN

In contrast to the overall structural similarities seen among the G protein-coupled receptors, the natural agonists for these receptors are structurally diverse. These agonists range in size from a single photon of light to 28 kd glycoprotein hormones. Within this spectrum are a diverse set of odorants for olfaction and small hormone and neurotransmitter molecules, such as catecholamines, acetylcholine, serotonin, histamine, and tachykinins. Detailed analysis of the physical interaction between receptors and their cognate ligands has been limited to a few of the G protein-coupled receptors. Most of the evidence to date strongly supports the hypothesis that the ligand binding site for adrenergic and muscarinic receptors is formed by the membrane spanning domains. Removal of most of the amino terminus by proteolysis (Rubenstein et al 1987; Wong et al 1988) or by deletion mutations (Dixon et al 1987) has little effect on the ability of β-adrenergic receptors to bind ligands. Biophysical analysis of the ligand binding site for the fluorescent β receptor antagonist carazolol indicates that this compound is buried at least 10.9 Å into the hydrophobic core of the receptor (Tota & Strader 1990).

Photoaffinity labeling studies have been used to identify receptor domains that are in close proximity to the ligand binding site in the turkey β receptor (Wong et al 1988), the hamster β_2 adrenergic receptor (Dohlman et al 1988), and the human platelet α_2 adrenergic receptor (Matsui et al 1989). In two cases, identification of the labeled peptide by obtaining amino acid sequence was possible (Dohlman et al 1988; Wong et al 1988). In the turkey β receptor, label was associated with the seventh hydrophobic domain. In the hamster β_2 receptor, the antagonist was bound to a peptide in the second hydrophobic domain. In the human α_2 receptor, identification of the labeled peptide by sequence analysis was not possible. However, the amino acid sequence of this receptor was used to predict proteolytic cleavage sites and identify the labeled peptide by its physical characteristics (Matusi et al 1989). The studies found that the labeled α_2 receptor peptide is from the third hydrophobic domain. Each of these studies observed a different site for labeling of the receptor with a photoactivatable ligand, which may be caused by differences in the binding site for each receptor. More likely, this suggests that several hydrophobic domains participate in the formation of the ligand binding pocket. The latter interpretation is consistent with data from mutagenesis studies, which are discussed below.

The Catecholamine Binding Domain

Mutagenesis studies have provided extremely valuable information regarding the role of specific amino acids in adrenergic and muscarinic receptors in ligand binding. Figure 3 illustrates key interactions between the β_2 adrenergic receptor and the β receptor agonist isoproterenol. The importance of these residues has been demonstrated in a series of experiments that combine site directed mutagenesis with pharmacophore analysis (Strader et al 1989a,b, 1991). The carboxylate group on the aspartate residue in the third hydrophobic domain is believed to act as a counterion for the catecholamine nitrogen. When this aspartate residue is changed to a serine, catecholamines are no longer effective agonists (Strader et al 1991). Catecholesters that may form a hydrogen bond to the serine hydroxyl group are capable, however, of acting as full agonists for this mutant receptor (Strader et al 1991). These studies support the hypothesis that the chemical nature of the aspartate residue in the third membrane spanning domain is important in direct interactions with the amine of catechol agonists. An aspartate is also present in the same position of the third membrane spanning domain of the muscarinic receptor. Mutation of this aspartate also interferes with agonist binding, which suggests that this residue may interact with the amine nitrogen of acetylcholine (Fraser et al 1989).

A similar approach was used to demonstrate the role of serine 204 and 207 in the fifth hydrophobic domain in interacting with the meta- and para-hydroxyls on the catechol ring (Strader et al 1989b). β_2 adrenergic receptor mutants were made in which either serine 204 or serine 207 were replaced by alanine. These mutant receptors were expressed in cells and analyzed for their ability to activate adenylyl cyclase following stimulation

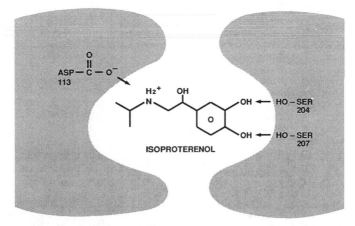

Figure 3 Sites of interaction between the β_2-adrenergic receptor and the agonist isoproterenol. (Based on work described in Strader et al 1989b.)

by isoproterenol or derivatives of isoproterenol, which lacked either the para- or the meta-hydroxyl on the catechol ring. Isoproterenol was able to activate both mutant receptors, and the isoproterenol derivatives were both effective in activating the wild type β_2 receptor. The β_2 receptor mutant lacking serine 204 could only be activated by isoproterenol and the isoproterenol derivative lacking the para-hydroxyl, but not by the isoproterenol derivative lacking the meta-hydroxyl. Conversely, the β_2 receptor mutant lacking serine 207 could only be activated by isoproterenol and the isoproterenol derivative lacking the meta-hydroxyl, but not by the derivative lacking the para-hydroxyl. These studies support the hypothesis that serine 204 forms a hydrogen bond with the meta-hydroxyl of the catechol ring and that serine 207 forms a hydrogen bond with the para-hydroxyl. Mutation of several other amino acids in different membrane spanning domains also affects ligand binding (Strader et al 1989c); however, it is not possible to determine if these mutations affect the binding domain or the overall structure of the protein.

Determinants of Subtype Specific Ligand Binding

The aspartate and serine residues indicated in Figure 3 are found in all of the adrenergic receptors. These residues are, therefore, likely to be important components of the catechol binding pocket of all of the adrenergic receptors, but not likely to be important for determining differences in binding to subtype specific ligands. Evidence from chimeric receptor studies have suggested that determinants of subtype specific ligand binding are found on several of the membrane spanning domains. The seventh hydrophobic domain is important in determining differences in antagonist binding specificity between the β_2 and α_2 adrenergic receptors (Kobilka et al 1988). More recently, a substitution of the phenylalanine at position 412 in the human platelet α_2 adrenergic receptor for asparagine (the amino acid found in the homologous region of the β_2 receptor) has been observed to result in loss of binding to the α_2 receptor antagonist yohimbine and acquisition of high affinity for the β receptor antagonists alprenolol, propanolol, and pindolol (S. Suryanaryana and B. Kobilka 1991, in press). These studies suggest a direct interaction between subtype specific ligands and specific amino acids in the seventh membrane spanning domain of the α_2 and β_2 adrenergic receptor. Nevertheless, recent biochemical studies have shown that a proteolytic product of the porcine α_2 adrenergic receptor, which contains only the first five membrane spanning domains, is capable of binding antagonists (Wilson et al 1990). The authors suggest that the seventh membrane spanning domain may be essential in directing the proper folding of the binding pocket. Once the pocket is formed, the seventh membrane spanning domain can be proteolyticically removed without affecting ligand binding.

Studies of chimeric receptors formed from pharmacologically closely related β_1 and β_2 receptors suggest that most of the membrane spanning domains contribute to determining ligand binding specificity (Dixon et al 1989; Frielle et al 1988). This may occur by contributing either through allosteric effects or directly to the binding pocket. Figure 4 shows antagonist binding properties for a series of β_1-β_2 receptor chimers. The β_1 specific antagonist betaxolol and the β_2 specific antagonist ICI118551 were used to compete for binding sites with the nonspecific β antagonist [^{125}I]Cyanopindolol. There is a progressive change found in the relative potency of these two antagonists as β_1 receptor sequence is replaced by β_2 receptor sequence.

Role of Extracellular Domains in Ligand Binding

There is no evidence for the direct involvement of extracellular domains in ligand binding in adrenergic receptors. Mutation of cysteine residues in the hydrophilic domains e-1 and e-2 (Figure 2) adversely affects ligand binding (Strader et al 1989c), but these effects are probably caused by changes in the structure of the receptor.

As indicated above, the G protein-coupled receptors for the glycoprotein hormones TSH, FSH, and LH have a distinctive structure with the amino terminal domain constituting more than 50% of the protein. Studies of chimeric receptors constructed from the TSH receptor and the receptor for LH and hCG indicate that this domain confers specificity for binding (Nagayama et al 1991). A chimeric receptor that has the amino terminal domain from the LH/hCG receptor and the remainder of the receptor derived from the TSH receptor could bind hCG with high affinity. Furthermore, hCG stimulated adenylyl cyclase activity was observed in cells that express this chimeric receptor. These studies confirm the importance of this large hydrophilic domain in hormone binding, especially in dictating binding specificity. The possibility exists, however, that the membrane spanning domains may participate in ligand binding, but are not important for conferring binding specificity. The glycoprotein hormones share a common β chain and such common structural domains on the glycoprotein hormones may be involved in interacting with the membrane spanning domains.

G PROTEIN COUPLING

G Protein Coupling Domains

Mutagenesis studies on adrenergic and muscarinic receptors have revealed that several cytosolic domains are involved in the functional coupling of

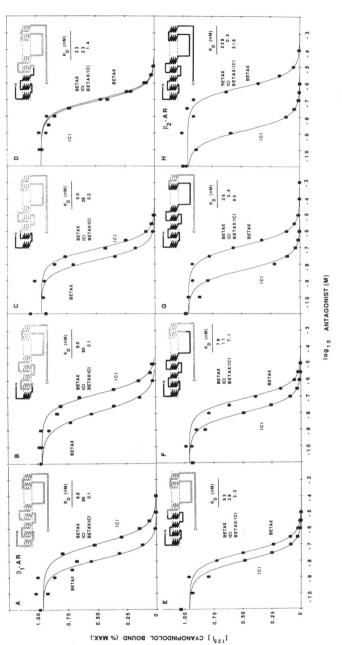

Figure 4 Competition binding studies on β_1, β_2, and chimeric β_1/β_2 adrenergic receptors. The nonselective antagonist $[^{125}I]$iodocyanopindolol competes for binding sites with the β_1 selective antagonist betaxolol (BETAX, ●) or the β_2 selective antagonist ICI118551 (ICI, ■). The composition of the chimeric receptors are indicated in the inset. β_1 sequence is indicated by an open line, and β_2 sequence is indicated by a solid line. (From Frielle et al 1988.)

receptors and G proteins. The study of chimeric receptors constructed from receptors with different G protein-coupling specificities has been valuable in discovering that the third cytoplasmic domain is responsible for determining the specificity of receptor-G protein interactions (Cotecchia et al 1990; Kobilka et al 1988; Wong et al 1990). The exchange of a 12 amino acid sequence in the amino terminal segment of the third cytoplasmic domain of the muscarinic (M1) receptor with the sequence from the homologous domain of the turkey β receptor altered G protein coupling specificity (Wong et al 1990). This chimeric receptor could activate phospholipase C (the property of the muscarinic receptor), as well as adenylyl cyclase (the property of the β receptor), when bound to the muscarinic agonist acetylcholine, but not to adrenergic receptor agonists. When the second cytoplasmic domain of this chimeric receptor is replaced with the homologous sequence from the turkey β receptor, the new chimeric receptor fully activates adenylyl cyclase when occupied by acetylcholine, but activation of phospholipase C is reduced. These data indicate that this 12 amino acid peptide of the third intracellular loop, along with regions of the second intracellular loop, is important for defining G protein-coupling specificity.

Domains within the carboxyl terminus of the third intracellular domain and the amino terminal portion of the carboxyl terminus may also participate in interactions with G proteins (Cotecchia et al 1990a; O'Dowd et al 1988; Strader et al 1989c). Although these domains apparently do not determine specificity of interactions, mutations in them modify the efficiency of coupling, as determined by the maximal stimulation of adenylyl cyclase and the concentration of agonist required to produce half-maximal stimulation.

Mutations may also produce receptors that are more efficient in G protein activation (Cotecchia et al 1990a). When the amino acid sequence REKKAA from the carboxyl end of the third intracellular loop of the α_1 receptor is replaced by KEHKAL from the β_2 receptor, the resulting chimeric receptor exhibited a 100-fold increase in affinity for norepinephrine in ligand binding assays. Furthermore, the EC 50 for activation of phospholipase C is 100-fold lower than the EC 50 for the wild type α_1 receptor; the basal level of inositol phosphate turnover in cells that express this mutant α_1 receptor is twofold higher than the basal level in cells that express the wild type α_1 receptor (Cotecchia et al 1990a). These results indicate that natural mutations in G protein-coupled receptors transform these receptors into oncogenes. Chronic stimulation of serotonin 5HT1c receptors expressed in NIH 3T3 cells leads to the generation of transformed foci in culture (Julius et al 1989). Injection of these foci into nude mice leads to the formation of tumors. The mass oncogene is an example of a seven membrane spanning domain protein that can induce tumors in nude mice (Young et al 1986).

G Protein Activation

Although specific domains involved in receptor mediated activation of G proteins have been identified by mutagenesis studies, the mechanism by which activation is accomplished has not been determined. Insight into this mechanism may be provided from studies of the receptor for insulin-like growth factor II. This receptor, which appears to activate a G_{i-2} in mouse fibroblast cells with the consequent activation of a calcium channel (Nishimoto et al 1987), is structurally very different from the adrenergic receptors; it has only a single membrane spanning domain. The extracellular amino terminus of the human IGF-II receptor is 2266 amino acids in length, and the carboxyl terminus is 163 amino acids. Although the overall structure of this receptor is markedly different from the adrenergic receptors, analysis of the cytoplasmic carboxyl terminus identified several peptide sequences that have structural similarity with domains in the adrenergic receptors, which have been implicated in G protein interactions (Okamoto et al 1990). These peptide sequences are predicted to form amphipathic α helices with one face charged, and the other hydrophobic. A synthetic peptide representing one of these domains activated purified G_{i-2} protein reconstituted into phospholipid vesicles (Okamoto et al 1990). These studies suggest that specific domains of the receptor can directly activate G proteins. In the unoccupied state, these domains may be masked from the G protein by other receptor domains. Conformational changes induced by agonist binding will probably make these domains accessible to the G protein.

G Protein Mediated Changes in Agonist Affinity

So far, we have only considered receptor activation of G proteins. As indicated above (Figure 1), receptors that are coupled to G proteins often have a higher affinity for agonists than do uncoupled receptors. This effect of G proteins on agonist affinity is observed by performing binding experiments in the absence or presence of nonhydrolyzable GTP analogues, such as GTP-γ-S. Dixon et al (1987) have observed that most mutations that diminish G protein activation also interfere with GTP mediated changes in affinity for agonists. Interestingly, in some of the β_2 receptor mutants that are unable to activate adenylyl cyclase, there is no GTP dependent change in agonist affinity; yet, these receptors exhibit an agonist affinity comparable to that of a wild type β receptor coupled to G_s (Dixon et al 1987; Strader et al 1987). However, Hausdorff et al (1990b) have reported that a mutation in the carboxyl terminal portion of the third cytoplasmic loop of the human β_2 receptor has impaired ability to activate adenylyl cyclase, but exhibits normal GTP dependent changes in agonist binding. This suggests that the mechanism by which the G protein alters the affinity of receptors for agonists may not require the G protein to be activated by the agonist bound receptor. The turkey β receptor is one of the few G protein-coupled receptors that does not normally show a significant G protein mediated change in agonist affinity. Truncation of 71 amino acids from the carboxyl terminus of this receptor results in a significant increase in GTP mediated changes in agonist affinity for the receptor (Hertel et al 1990). This mutant also appeared to be more efficient in activating adenylyl cyclase than the wild type receptor expressed in the same cell line.

DESENSITIZATION

Changes in receptor function following prolonged exposure to agonists have been observed for several G protein-coupled receptors, including the adrenergic receptors, the muscarinic receptors, and the tachykinin receptors. This process has been most thoroughly characterized for the β adrenergic receptor (reviewed by Hausdorff et al 1990a; Lefkowitz et al 1990). Desensitization of β adrenergic receptors is manifested by an increase in the concentration of agonist required to produce half maximal stimulation of adenylyl cyclase and by a decrease in the maximum amount of cAMP produced. Cellular processes that contribute to desensitization include phosphorylation of the receptor and the physical removal of the receptor from the plasma membrane.

Role of Receptor Phosphorylation

PROTEIN KINASE A   Phosphorylation of the β_2 adrenergic receptor plays an important role in desensitization. Two different kinases phosphorylate this receptor: protein kinase A (PKA) (Benovic et al 1985) and β adrenergic receptor kinase (βARK) (Benovic et al 1987b). Protein kinase A is positively regulated by cAMP and is thus activated by stimulation of either the β_2 receptor or other receptors that activate adenylyl cyclase. Phosphorylation by PKA therefore provides a mechanism by which the β_2 receptor can be regulated by other adenylyl cyclase coupled receptors. The potential sites for PKA phosphorylation are found in third intracellular loop and the proximal carboxyl terminus (see Figure 2). Removal of these sites by mutagenesis (Hausdorff et al 1989) or inhibition of PKA (Lohse et al 1990a) almost completely prevents desensitization by low (nanomolar) concentrations of agonist, such as would be found circulating in plasma. Biochemical studies with synthetic peptides indicate that the PKA consensus site within the third loop is a better substrate for PKA than is the site in the carboxyl terminus (Blake et al 1987). This is in agreement with mutagenesis studies, which show that removal of the site within the third cytoplasmic loop can prevent desensitization caused by exposure to low concentration of agonists, whereas removal of the site within the carboxyl terminus has no effect (Clark et al 1989). The PKA site within the third cytoplasmic loop is close to domains that influence coupling of the receptor to G_s. The phosphorylation of β_2 receptor by PKA appears to interfere directly with receptor mediated G_s activation. Purified β_2 receptor can be phosphorylated by purified kinase and reconstituted into phospholipid vesicles with purified G_s. In these experiments, phosphorylation by PKA impairs receptor-G protein coupling (Benovic et al 1985).

β ADRENERGIC RECEPTOR KINASE   Desensitization is also observed when cells expressing β_2 receptors that lack PKA sites are exposed for brief periods (minutes) to higher concentrations of agonist (micromolar), such as might be found within a synapse (Clark et al 1989; Hausdorff et al 1989). These receptors are phosphorylated, even though they lack sites for PKA (Hausdorff et al 1989). Furthermore, this phosphorylation occurs in the presence of inhibitors of PKA (Lohse et al 1990a) and in cells that lack functional PKA (Strasser et al 1986). The phosphorylation observed under these conditions is mediated by βARK. This kinase, which is functionally related to rhodopsin kinase, has recently been cloned (Benovic et al 1989). Sites for βARK mediated phosphorylation of the β_2 adrenergic receptor appear to be serine and threonine residues in the carboxyl terminus (Dohlman et al 1987), but the precise consensus sequence of amino acids identified by this kinase has not been determined. In contrast to PKA, βARK only phosphorylates agonist occupied receptors (Benovic et al 1987b). Therefore, phosphorylation of significant numbers of β receptors only occurs when cells are exposed to concentrations of agonist that are high enough to saturate receptor binding sites. The high concentration of catecholamines needed to saturate receptors probably occurs at a nerve synapse. Therefore, Benovic et al (1989) have proposed that desensitization mediated by βARK may be important in regulating neural transmission. The finding that mRNA for βARK is most abundant in tissues that are richly innervated by sympathetic nerves is consistent with this hypothesis (Benovic et al 1989).

Phosphorylation of β_2 adrenergic receptors by βARK does not directly interfere with the activation of G_s (Benovic et al 1987a). A recently cloned cytosolic protein, β-arrestin, interacts with βARK phosphorylated receptor and disrupts the activation of G_s by the β receptor (Lohse et al 1990b). The existence of this protein was proposed based on the similarity between βARK mediated desensitization of the β_2 receptor and rhodopsin kinase

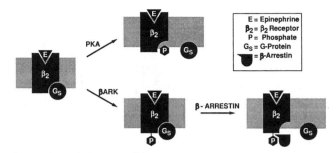

Figure 5 Comparison of desensitization mediated by PKA and BARK. Activation of the
β_2 adrenergic receptor by epinephrine (E) can result in phosphorylation of the receptor by
PKA and by βARK. Phosphorylation of the receptor by PKA interferes with the interaction
of the receptor and G_s. Phosphorylation of the receptor by βARK is not sufficient to interfere
with the activation of G_s by the receptor. β-arrestin recognizes βARK phosphorylated
receptor and blocks activation of G_s.

mediated desensitization of rhodopsin (light adaptation) (Benovic et al
1987a). Figure 5 summarizes the functional differences between PKA and
βARK mediated phosphorylation.

The effects of βARK and PKA appear to be additive in the presence of
high agonist concentrations in which both kinases appear to be active
(Figure 6) (Hausdorff et al 1989). Receptors that lack either the PKA or
the βARK sites become desensitized following a 15-minute exposure to
two micromolar isoproterenol, but to a lesser degree than does the wild
type receptor. β_2 receptors in which both PKA and βARK phosphorylation sites have been removed by mutagenesis exhibit only a slight
desensitization. None of the changes in desensitization seen in the mutant
receptors can be attributed to changes in ligand binding or G-protein
coupling, which are unaffected by the mutations. These mutagenesis studies
have been confirmed by studies that use inhibitors of both PKA and βARK
(Lohse et al 1989, 1990a). The small amount of desensitization observed
when sites for phosphorylation by both PKA and βARK are absent may
be caused by sequestration (Hausdorff et al 1989).

Redistribution of Receptors

β_2 adrenergic receptors are functionally removed from the plasma membrane following agonist exposure. This form of receptor regulation occurs
through two distinguishable processes (Figure 7): sequestration, the rapid
(within five minutes), reversible removal of functional receptors from the
plasma membrane, and down-regulation, the removal and subsequent
destruction of receptors. Down-regulation occurs more slowly than seques-

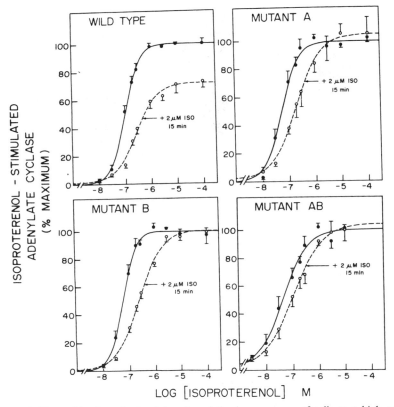

Figure 6 Desensitization of adenylyl cyclase following exposure of cells to a high concentration of agonist. Each panel is a dose/response relationship for activation of adenylyl cyclase by isoproterenol in membranes from cells that express wild type or mutant β receptors. The solid line indicates data from control cells, and the dotted line indicates data from cells that were exposed to two micromolar isoproterenol for 15 minutes. Mutant A is a β receptor that lacks sites for phosphorylation by PKA. Mutant B is a β receptor that lacks sites for phosphorylation by βARK. Mutant AB is a β receptor that lacks sites for phosphorylation by both kinases. (From Hausdorff et al 1989.)

tration; usually, several hours of agonist exposure is required to see significant loss of receptors, and the process continues for up to 24 hours.

SEQUESTRATION   The process of sequestration and its functional significance in desensitization is poorly understood. Sequestered receptors in intact cells are identified as inaccessible to hydrophilic ligands, but accessible to hydrophobic ligands. In some cases, membrane fractionation can be used to isolate the sequestered receptors in a light vesicle fraction (Waldo et al 1983). The mechanism by which receptors are internalized

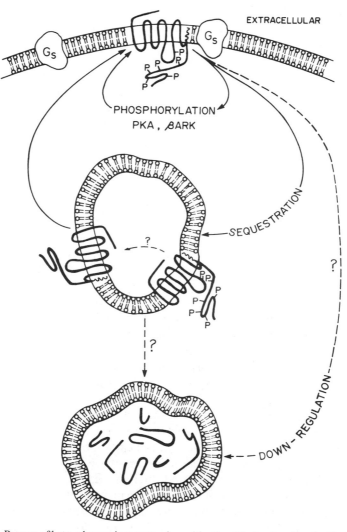

Figure 7 Process of beta-adrenergic receptor desensitization. Mechanisms involved in desensitization include phosphorylation of the receptor by PKA and βARK, and the removal of receptors from the plasma membrane by sequestration or down-regulation. Sequestration is a rapid, reversible process that may be necessary for the removal of phosphates and reactivation of receptors. Down-regulation leads to the irreversible loss of receptors by an unknown process. (From Hausdorff et al 1990a.)

after a short exposure to agonist is poorly understood. This process has been examined by immunocytochemistry in A431 cells that were exposed to isoproterenol for 30 minutes (Raposo et al 1989). A change in the cellular distribution of receptors was observed by immunofluorescence. The process was further studied by electron microscopy. A431 cells were cooled to 4^{oc} and incubated with a monoclonal antibody to an extracellular domain of the β_2 adrenergic receptor and subsequently with antimouse-IgG antibody labeled with gold. The cells were then warmed to 37^{oc} and incubated in the presence or absence of isoproterenol for 30 minutes. In cells treated with isoproterenol, 48% of the gold particles were localized in nonclathrin coated vesicles. In untreated cells, only 14% of the gold was internalized. The histochemical findings were accompanied by loss of cell surface receptors, as determined by whole cell binding with the hydrophilic β receptor antagonist [^{3}H]CGP. No attempt was made, however, to determine if the internalized receptors were capable of ligand binding by using a hydrophobic antagonist. Therefore, it is not clear if the histochemical findings represent sequestration or the process of down-regulation (discussed below). Furthermore, as pointed out by Raposo et al (1989) the binding of the gold tagged antibody complex to the receptor may have influenced receptor redistribution. For example, after 60–90 minutes, no significant difference was observed between control cells and cells treated with isoproterenol. The cell biology of the sequestration process remains fertile ground for investigation.

Phosphorylation of the β_2 adrenergic receptor is not required for sequestration as mutants lacking phosphorylation sites are sequestered normally (Hausdorff et al 1989). Activation of G_s is apparently not required for sequestration (Cheung et al 1990), although the structural domains on the β_2 adrenergic receptor that mediate sequestration are close to those that mediate G protein coupling (Cheung et al 1989). The participation of other G proteins in the process of sequestration cannot be ruled out.

Sequestered receptors are expectedly ineffective in signal transduction; they are inaccessible to agonist, and the light vesicles are deficient in G_s (Waldo et al 1983). The role of sequestration in desensitization is uncertain. Although sequestration is relatively rapid, phosphorylation precedes sequestration, and inhibitors of sequestration do not prevent desensitization (Waldo et al 1983). Hausdorff et al (1990c) have proposed that sequestration is a part of the process by which phosphates are removed from receptors and their function is restored (Figure 7). In some tissue culture cells, up to 60% of receptors may be sequestered within five minutes of agonist exposure (Cheung et al 1990). It seems unlikely that such a massive redistribution of receptors is devoid of functional significance.

DOWN-REGULATION   Down-regulation of β_2 adrenergic receptors is a complex process that occurs more slowly than sequestration and leads to the irreversible loss of receptors from the plasma membrane. Apparently, several different mechanisms contribute to the process of down-regulation. The rate of receptor loss is greatest during the first four hours of down-regulation (Bouvier et al 1989). Within 24 hours, the process usually approaches steady state. The early phase appears to require PKA mediated receptor phosphorylation. β_2 receptor mutants that lack PKA sites do not undergo this initial phase of down-regulation (Bouvier et al 1989). Valiquette et al (1990) suggested that the loss of receptors during this initial phase is mediated by clathrin coated vesicles. Mutation of two tyrosine residues on the carboxyl terminus of the β_2 adrenergic receptor led to a decrease in the early phase of down-regulation (Valiquette et al 1990). Evidence suggests that tyrosine residues play a role in the interaction of membrane proteins with clathrin (Davis et al 1986). The later phase of down-regulation, which extends from 4 to 24 hours, is apparently due to a reduction in new receptor biosynthesis. This reduction is predominantly caused by more rapid degradation of the β_2 receptor mRNA (Bouvier et al 1989). The late phase of down-regulation does not require phosphorylation of the β_2 receptor. It is not clear, however, if the gradual loss of receptor during this late phase of down-regulation is due to natural receptor turnover or to a more accelerated process that does not require receptor phosphorylation. The cellular distribution of β_2 receptors following down-regulation has been studied by two groups of investigators with different results. Wang et al (1989a) studied the process of down-regulation of β_2 receptors expressed in A431 cells. Although ligand binding studies demonstrated 80% reduction of receptors over 24 hours, no change in the amount or distribution β_2 receptor antigen could be detected. These investigators suggest that the loss of ligand binding is due to a conformational change in the receptor and not to proteolysis of receptor. Zemcik & Strader (1988) studied down-regulation in DDT-1 cells and observed a loss of antigen by immunofluorescence, which corresponded to the loss of binding activity within 30 minutes of treatment with the β_2 receptor agonist isoproterenol. These conflicting results may be due to differences in the antibodies or in the cells used in these studies. Further work is required to clarify the cell biology of down-regulation.

Desensitization of Other G Protein-Coupled Receptors

The mechanisms contributing to desensitization of the β adrenergic receptor likely apply to other G protein-coupled receptors. βARK phosphorylates the human α_2 adrenergic receptor (Benovic et al 1987c), as well as the muscarinic receptor, in an agonist dependent manner (Kwatra et al 1989). Thus, βARK or closely related kinases may be involved in agonist mediated regulation of several G protein-coupled receptors. The specificity of phosphorylation is determined by the dependence of the kinase on a receptor that is occupied by its cognate agonist.

The tachykinin receptors [substance P receptor (SPR), substance K receptor (SKR), neuromidin K receptor (NKR)] exhibit differences in the extent to which they are desensitized following agonist exposure (SPR > NKR > SKR). These differences correlate with the number of potential phosphorylation sites (serine and threonine residues) in the third intracellular loop and the carboxyl terminus (SPR − 5, NKR − 2, SKR − 1) (Shigemoto et al 1990).

Other kinases may also be involved in desensitization. Protein kinase C appears to be involved in modulating the activity of receptors that activate phospholipase C, such as the α_1 adrenergic receptor (Bouvier et al 1987) and the muscarinic receptors (Liles et al 1986). Protein kinase C may also play a role in the regulation of β_2 adrenergic receptor function by receptors that activate phospholipase C (Johnson et al 1990).

BIOSYNTHESIS AND PROCESSING OF G PROTEIN-COUPLED RECEPTORS

In addition to the general functional properties typically associated with receptors, such as ligand binding, signal transduction, and regulation, the primary amino acid sequence of receptors contain all of the essential information for the proper folding, posttranslational processing, and cellular targeting of the receptor. As integral membrane proteins, G protein-coupled receptors are likely to be cotranslationally translocated into the endoplasmic reticulum, which has been demonstrated for the β_2 adrenergic receptor expressed in a cell-free system capable of producing functional receptor protein (Kobilka 1990). In this cell-free expression system, translation and glycosylation were complete within 20 minutes; however, the receptor could not bind ligands. Additional posttranslational processing, which required approximately 30 minutes, was needed to produce functional receptor. This processing required ATP, intact microsomal membranes, and a high molecular weight cytosolic factor. Inhibitors of glycosylation and nonhydrolyzable GTP analogues did not affect posttranslational processing. Therefore, the β_2 receptors apparently acquire the ability to bind ligands within the endoplasmic reticulum.

The nature of the posttranslational modifications to the protein structure have not been determined. Covalent modification of specific amino acids may be required. Two posttranslational modifications occur in the β_2

adrenergic receptor: asparagine-linked glycosylation and palmitoylation of a cysteine residue on the carboxyl terminus (O'Dowd et al 1989). Glycosylation is not essential for processing of the protein to a functional receptor (Kobilka 1990; Rands et al 1990), but may influence the efficiency of targeting the receptors to the plasma membrane (Rands et al 1990). Mutation of the cysteine involved in palmitoylation does not affect protein processing in terms of ligand binding, but these receptors are less efficient in activating adenylyl cyclase (O'Dowd et al 1989). In the case of the β_2 receptor, disulfide bond formation is necessary for proper function (Dohlman et al 1990; Strader et al 1989c). Protein disulfide isomerase possibly plays a key role in posttranslational processing.

Posttranslational processing of the receptor may require the aid of a foldase or chaperon protein to assemble the membrane spanning domains properly. Bonifacino et al (1988) have suggested this for multisubunit membrane proteins, such as the T cell antigen receptor. The β_2 adrenergic receptor can be synthesized as a two subunit protein (Kobilka et al 1988): The first five membrane spanning domains is one subunit, and the last two membrane spanning domains is the second subunit. Expressing these domains separately does not produce a functional receptor. When expressed together, β_2 receptors that can bind ligands and activate adenylyl cyclase are expressed. Photoaffinity labeling studies show that both subunits of this "split receptor" are specifically labeled by the β receptor antagonist [^{125}I]cyanopindolol-diazerine (B. Kobilka 1988, unpublished). Thus, the hydrophilic domain that separates transmembrane domains 1–5 from 6–7 apparently does not dictate assembly or conformation. The first five hydrophobic domains may associate with domains 6 and 7 by random, fortuitous collision. More likely, this association is aided by a protein within the endoplasmic reticulum or in the cytoplasm. The few amino acid residues that are completely conserved among all G protein-coupled receptors might serve as recognition sites for the interaction between these receptors and a protein that assists in folding.

Following biosynthesis, a functional receptor is transported to a specific cellular domain, but little is known about the subcellular localization of most G protein-coupled receptors. Immunochemical localization has indicated that, in some regions of the central nervous system, β_2 receptors are concentrated at synaptic clefts (Aoki et al 1989). Even more interesting is the hypothesis that receptors are targeted to specific microdomains within the plasma membrane (Neer & Clapham 1988). Within these microdomains, a receptor may be associated with a specific G protein and components of a specific signal transduction pathway. The structural domains involved in the targeting of receptors to specific locations within cells have not been identified.

MULTIPLE ADRENERGIC RECEPTOR SUBTYPES

One of the most exciting outcomes of the application of the molecular biology techniques to the study of G protein-coupled receptors has been the confirmation, and in some cases the discovery of, multiple closely related receptor subtypes. For example, before the cloning of muscarinic receptors, two subtypes could be identified by ligand binding assays based on differences in affinity for the antagonist pirenzapine. To date, five subtypes of muscarinic receptors have been cloned (Bonner 1989). Similarly, four subtypes of adrenergic receptors (α_1, α_2, β_1, β_2) could be clearly distinguished by pharmacologic criteria in 1985; however, evidence of additional subtypes was beginning to accumulate (Bylund 1985). To date, eight subtypes of mammalian adrenergic receptors have been cloned.

An important area of future investigation is the identification of the physiologic role played by these functionally similar proteins. Consider the human α_2 adrenergic receptors. These receptors have been implicated in the control of a diverse spectrum of physiologic responses. Sympathetic tone is modulated by central nervous system α_2 adrenergic receptors; activation of these receptors by agonists leads to a reduction of sympathetic impulses to peripheral tissues. α_2 adrenergic receptors are found pre-synaptically at sympathetic nerve terminals; stimulation of these receptors leads to a reduction of neurotransmitter release. Postsynaptic α_2 receptors and extra-synaptic α_2 receptors are involved in the modulation of smooth muscle tone in arterioles, in the control of Na^+ excretion by the kidney, in platelet aggregation, and in the modulation of fat and carbohydrate metabolism. To date, three genes encoding α_2 receptors have been identified. However, we do not know which physiologic effects are mediated by each of these receptors.

The α_2 receptors are products of three genes located on human chromosome 2, 4, and 10 and are called α_2 C2, α_2 C4, and α_2 C10 for the purpose of this discussion. These receptors share approximately 75% identity within their membrane spanning domains. The binding properties of these clones expressed in Cos-7 cells are listed in Table 1 (Lomasney et al 1990). The most useful drug in discriminating among the three cloned receptors in ligand binding assays is the agonist oxymetazoline. The development of more highly subtype selective agonist and antagonist ligands is important for determining the physiologic role of these different α_2 receptor subtypes in vivo. These in vivo studies will probably be difficult to interpret, even with such compounds as oxymetazoline, however, as tissue levels of drugs are often difficult to predict because of such variables as lipid solubility of the drug, metabolism to compounds that may be

Table 1 Competition by α-adrenergic ligands for the binding of [^{3}H]yohimbine to membranes from COS-7 cells transfected with the indicated α_2-adrenergic receptors

	Ki, nM		
	α_2C2	α_2C4	α_2C10
Agonists			
(−) Epinephrine	1851	318	1671
(+) Epinephrine	8422	ND	ND
(−) Norepinephrine	1265	606	3677
Oxymetazoline	1506	125	13.2
p-Aminoclonidine	120	97	31
Antagonists			
Corynanthine	1002	182	1188
Phentolamine	9.2	14.4	6.2
Prazosin	293	67.7	2237
Rauwolscine	11	2.1	7.1
SKF 104078	105	41	97
WB 4101	132	13	47

From Lomasney et al 1990.

less selective, and renal clearance. Another confounding variable is the involvement of α_2 receptors throughout the sympathetic nervous system, which will make it difficult to distinguish central effects of drugs from peripheral effects.

Several hypotheses address a possible physiologic role for different α_2 receptor subtypes. If some of the receptors have arisen recently by gene duplication, these genes may be biologically redundant. In this case, no significant difference in tissue specific expression of the different receptors would be observed. Although the coding sequences of α_2 receptor genes may be functionally identical, the regulatory sequences may be distinct. In this case, different genes probably provide a molecular basis for expression of α_2 adrenergic receptors in specific tissues or at specific stages of development. The expression different α_2 subtypes has been studied by northern blot analysis of rat tissue RNA (Lorenz et al 1990). The results of this study indicate that expression of different subtypes is apparently tissue specific. RNA from the pituitary, aorta, lung spleen, and skeletal muscle hybridized only with the probe for the α_2 C10, whereas RNA from liver hybridized only with a probe for the α_2 C2. However, expression of RNA for the α_2 C10 and the α_2 C4 overlap significantly. RNA for these two subtypes are found throughout the brain (cortex, cerebellum, hippocampus, brainstem), whereas mRNA for α_2 C10 and C2 are found in the kidney. Also, the absence of a signal on a northern blot cannot be considered conclusive evidence that a specific subtype is not expressed in a specific tissue. Interestingly, mRNA for α_2 C4 was not detected in the kidney, even though this receptor was cloned from a cDNA library made from human kidney mRNA (Regan et al 1988). Expression of a particular subtype in an organ or tissue might be limited to a small population of cells, thus making detection of the mRNA for those receptors by northern blot analysis or the receptor protein by ligand binding assays difficult. Furthermore, different cells within a tissue might express different receptor subtypes. In situ hybridization with oligonucleotide probes, highly selective ligands, or subtype specific antibodies may provide information about the location and type of cells that express specific receptor subtypes. Such information would be useful in understanding the physiologic role of each subtype.

Another hypothesis regarding the physiologic significance of different α_2 receptor subtypes postulates that the different α_2 receptors may have distinct functions with regard to signal transduction, regulation, or cellular localization. The functional properties of α_2 C4 and α_2 C10 have been compared by expressing them in the same cell line (Cotecchia et al 1990b). The differences that were observed in inhibition of adenylyl cyclase and polyphosphatidyl inositol metabolism were small and may not have physiologic significance. These receptors may differ in other respects, such as agonist mediated regulation, or functional differences may only be observed when these receptors are expressed in highly differentiated cells in vivo, and not in tissue culture cells. Further work is needed to understand the physiologic basis for these different α_2-adrenergic receptor subtypes.

CONCLUSION

Molecular biology approaches have been used to overcome some of the technical difficulties encountered in studying such G protein-coupled receptors as the adrenergic receptors. Tremendous progress has been made over the past five years towards understanding the structure, function, and regulation of these receptors. Of particular importance has been the identification of genes that encode subtypes of α_1 and α_2 receptors. Emphasis should be placed on characterizing the cellular responses following stimulation of each receptor subtype and the role each subtype plays in normal physiologic processes. Much of what we learn from the study of adrenergic receptors will apply to other G protein-coupled receptors. Continued success towards understanding the molecular basis of adrenergic receptor function in health and disease will require creative use of techniques of molecular biology, molecular genetics, pharmacology, biochemistry, and cell biology.

Acknowledgments

The author wishes to thank Dr. Thomas Frielle for providing Figure 4 and Dr. William Hausdorff for providing Figures 6 and 7.

Literature Cited

Aoki, C., Zemcik, B. A., Strader, C. D., Pickel, V. M. 1989. Cytoplasmic loop of β-adrenergic receptors: synaptic and intracellular localization and relation to catecholaminergic neurons in the nuclei of the solitary tracts. *Brain Res.* 493: 331–47

Benovic, J. L., DeBlasi, A., Stone, W. C., Caron, M. G., Lefkowitz, R. J. 1989. β-adrenergic receptor kinase: primary structure delineates a multigene family. *Science* 246: 235–46

Benovic, J. L., Kuhn, J., Weyland, I., Codina, J., Caron, J. G., Lefkowitz, R. J. 1987a. Functional desensitization of the isolated beta-adrenergic receptor by the beta-adrenergic receptor kinase: potential role of an analog for the retinal protein arrestin (48-kDa protein). *Proc. Natl. Acad. Sci. USA* 84: 8879–82

Benovic, J. L., Mayor, F., Staniszewski, E., Lefkowitz, R. J., Caron, M. G. 1987b. Purification and characterization of beta-adrenergic receptor kinase. *J. Biol. Chem.* 262: 9026–32

Benovic, J. L., Pike, L. J., Cerione, R. A., Staniszewski, C., Yoshimasa, T., et al. 1985. Phosphorylation of the mammalian β-adrenergic receptor by cyclic AMP-dependent protein kinase. *J. Biol. Chem.* 260: 7094–7101

Benovic, J. L., Regan, J. W., Matsui, H., Mayor, F., Cotecchia, S., et al. 1987c. Agonist-dependent phosphorylation of the α₂-adrenergic receptor by the β-adrenergic receptor kinase. *J. Biol. Chem.* 262: 17251–53

Birnbaumer, L., Abramowitz, J., Brown, A. M. 1990. Receptor-effector coupling by G proteins. *Biochem. Biophys. Acta* 1031: 163–224

Blake, A. D., Mumford, R. A., Strout, H. V., Slater, E. E., Strader, C. D. 1987. Synthetic segments of the mammalian beta adrenergic receptor are preferentially recognized by cAMP-dependent protein kinase and protein kinase C. *Biochem. Biophys. Res. Commun.* 147: 168–73

Bonifacino, J. S., Lippincott-Schwartz, J., Chen, C., Antusch, D., Samelson, L. E., Klausner, R. D. 1988. Association and dissociation of the murine T cell receptor associated protein (TRAP). *J. Biol. Chem.* 263: 8965–71

Bonner, T. I. 1989. The molecular basis of muscarinic receptor diversity. *Trends Neurosci.* 12: 148–52

Bourne, H., Sanders, D. A., McCormick, F. 1991. The GTPase superfamily: conserved structure and molecular mechanism. *Nature* 349: 117–27

Bouvier, M., Collins, S., O'Dowd, B. F., Campbell, P. T., DeBlasi, A., et al. 1989. Two distinct pathways for cAMP-mediated down-regulation of the β₂-adrenergic receptor. *J. Biol. Chem.* 264: 16786–92

Bouvier, M., Leeb-Lundberg, L. M. F., Benovic, J. L., Caron, M. G., Lefkowitz, R. J. 1987. Regulation of adrenergic receptor function by phosphorylation. *J. Biol. Chem.* 262: 3106–13

Bylund, D. B. 1985. Heterogeneity of alpha-2 adrenergic receptors. *Pharmacol. Biochem. Behav.* 22: 835–43

Cheung, A. H., Dixon, R. A. F., Hill, W. S., Sigal, I. S., Strader, C. D. 1990. Separation of the structural requirements for agonist-promoted activation and sequestration of the β-adrenergic receptor. *Mol. Pharmacol.* 37: 775–79

Cheung, A. H., Sigal, I. S., Dixon, R. A. F., Strader, C. D. 1989. Agonist-promoted sequestration of the β₂-adrenergic receptor requires regions involved in functional coupling with G_s. *Mol. Pharmacol.* 35: 132–38

Clark, R. B., Friedman, J., Dixon, R. A. F., Strader, C. D. 1989. Identification of a specific site required for rapid heterologous desensitization of the β-adrenergic receptor by cAMP-dependent protein kinase. *Mol. Pharmacol.* 36: 343–48

Cotecchia, S., Exum, S., Caron, M. G., Lefkowitz, R. J. 1990a. Regions of the α₁-adrenergic receptor involved in coupling to phosphatidylinositol hydrolysis and enhanced sensitivity of biological function. *Proc. Natl. Acad. Sci. USA* 87: 2896–2900

Cotecchia, S., Kobilka, B. K., Daniel, K. W.,

Nolan, R. D., Lapetina, E. Y., et al. 1990b. Multiple second messenger pathways of α-adrenergic receptor subtypes expressed in eukaryotic cells. *J. Biol. Chem.* 265: 63–69

Cotecchia, S., Schwinn, D. A., Randall, R. R., Lefkowitz, R. J., Caron, M. G., et al. 1988. Molecular cloning and expression of the cDNA for the hamster α_1-adrenergic receptor. *Proc. Natl. Acad. Sci. USA* 85: 7159–63

Davis, C. G., Lehrman, M. A., Russell, D. W., Anderson, R. G. W., Brown, M. S., Goldstein, J. L. 1986. The J.D. mutation in familial hypercholesterolemia: amino acid substitution in cytoplasmic domain impedes internalization of LDL receptors. *Cell* 45: 15–24

Dixon, R. A. F., Hill, W. S., Candelore, M. R., Rands, E., Diehl, R., et al. 1989. Genetic analysis of the molecular basis for β-adrenergic receptor subtype specificity. *Proteins Struct. Funct. Genet.* 6: 267–74

Dixon, R. A. F., Kobilka, B. K., Strader, D. J., Benovic, J. L., Dohlman, H. G., et al. 1986. Cloning of the gene and cDNA for the mammalian beta-2-adrenergic receptor and homology with rhodopsin. *Nature* 321: 75–79

Dixon, R. A. F., Sigal, I. S., Candelore, M. R., Register, R. B., Scattergood, W., et al. 1987. Structural features required for ligand binding to the β-adrenergic receptor. *EMBO J.* 6: 3269–75

Dohlman, H. G., Bouvier, M., Benovic, J. L., Caron, M. G., Lefkowitz, R. J. 1987. The multiple membrane spanning topography of the beta 2-adrenergic receptor. *J. Biol. Chem.* 262: 14282–88

Dohlman, H. G., Caron, M. G., DeBlasi, A., Frielle, T., Lefkowitz, R. J. 1990. Role of extracellular disulfide-bonded cysteines in the ligand binding function of the β_2-adrenergic receptor. *Biochemistry* 29: 2335–42

Dohlman, H. G., Caron, M. G., Strader, C. D., Amlaiky, N., Lefkowitz, R. J. 1988. Identification and sequence of a binding site peptide of the β_2-adrenergic receptor. *Biochemistry* 27: 1813–17

Emorine, L. J., Marullo, S., Briend-Sutren, M. M., Patey, G., Tate, K., et al. 1989. Molecular characterization of the human β_3-adrenergic receptor. *Science* 245: 1118–21

Fraser, C. M., Wang, C. D., Robinson, D. A., Gocayne, J. D., Venter, C. 1989. Site-directed mutagenesis of m^1 muscarinic acetylcholine receptors: conserved aspartic acids play important roles in receptor function. *Mol. Pharmacol.* 36: 840–47

Freismuth, M., Casey, P. J., Gilman, A. G. 1988. G proteins control diverse pathways of transmembrane signaling. *FASEB J.* 3: 2125–31

Frielle, T., Collins, S., Daniel, K. W., Caron, M. G., Lefkowitz, R. J., Kobilka, B. K. 1987. Cloning of the cDNA for the human beta-one-adrenergic receptor. *Proc. Natl. Acad. Sci. USA* 84: 7920–24

Frielle, T., Daniel, K. W., Caron, M. G., Lefkowitz, R. J. 1988. Structural basis of β-adrenergic receptor subtype specificity studied with chimeric β_1/β_2-adrenergic receptors. *Proc. Natl. Acad. Sci. USA* 85: 9494–98

Hausdorff, W. P., Bouvier, M., O'Dowd, B. F., Irons, G. P., Caron, M. G., Lefkowitz, R. J. 1989. Phosphorylation sites on two domains of the β_2-adrenergic receptor are involved in distinct pathways of receptor desensitization. *J. Biol. Chem.* 264: 12657–65

Hausdorff, W. P., Caron, M. G., Lefkowitz, R. J. 1990a. Turning off the signal: desensitization of β-adrenergic receptor function. *FASEB J.* 4: 2881–89

Hausdorff, W. P., Hnatowich, M., O'Dowd, B. F., Caron, M. G., Lefkowitz, R. J. 1990b. A mutation of the β_2-adrenergic receptor impairs agonist activation of adenylyl cyclase without affecting high affinity agonist binding. *J. Biol. Chem.* 265: 1388–93

Henderson, R., Baldwin, J. M., Cesca, T. A., Zemlin, F., Beckmann, E., et al. 1990. Model for the structure of bacteriorhodopsin based on high-resolution electron cryo-microscopy. *J. Mol. Biol.* 213: 899–929

Hertel, C., Nunnally, M. H., Wong, S. K.-F., Murphy, E. A., Ross, E. M., Perkins, J. P. 1990. A truncation mutation in the avian β-adrenergic receptor causes agonist-induced internalization and GTP-sensitive agonist binding characteristic of mammalian receptors. *J. Biol. Chem.* 265: 17988–94

Johnson, J. A., Clark, R. B., Friedman, J., Dixon, R. A. F., Strader, C. D. 1990. Identification of a specific domain in the β-adrenergic receptor required for phorbol ester-induced inhibition of catecholamine-stimulated adenylyl cyclase. *Mol. Pharmacol.* 38: 289–93

Julius, D., Livelli, T. J., Jessell, T. M., Axel, R. 1989. Ectopic expression of the serotonin lc receptor and th triggering of malignant transformation. *Science* 244: 1057–62

Kim, D., Lewis, D. L., Graxialei, L., Neer, E. J., Gar-Sagi, D., Clapham, D. E. 1989. G-protein beta-gamma subunits activate the cardiac muscarinic K^+ channel via phospholipase A_2. *Nature* 337: 557–60

Kobilka, B. K. 1990. The role of cytosolic and membrane factors in processing of the human β_2 adrenergic receptor following translocation and glycosylation in a cell-free system. *J. Biol. Chem.* 265: 7610–18

Kobilka, B. K., Kobilka, T. S., Daniel, K., Regan, J. W., Caron, M. G., Lefkowitz, R. J. 1988. Chimeric α_2-, β_2-adrenergic receptors: delineation of domains involved in effector coupling and ligand binding specificity. *Science* 240: 1310–16

Kobilka, B. K., Matsui, H., Kobilka, T. S., Yang-Feng, T. L., Francke, U., et al. 1987. Cloning, sequencing, and expression of the gene coding for the human platelet α_2-adrenergic receptor. *Science* 238: 650–56

Kwatra, M. M., Benovic, J. L., Caron, M. G., Lefkowitz, R. J., Hosey, M. M. 1989. Phosphorylation of chick heart muscarinic cholinergic receptors by the beta-adrenergic receptor kinase. *Biochemistry* 28: 4543–47

Lefkowitz, R. J., Hausdorff, W. P., Caron, M. G. 1990. Role of phosphorylation in desensitization of the β-adrenoceptor. *Trends Pharmacol. Sci.* 11: 190–94

Liles, W. C., Hunter, D. D., Meier, K. E., Nathanson, N. M. 1986. Activation of protein kinase C induces rapid internalization and subsequent degradation of muscarinic acetylcholine receptors in neuroblastoma cells. *J. Biol. Chem.* 261: 5307–13

Lohse, M. J., Benovic, J. L., Caron, M. G., Lefkowitz, R. J. 1990a. Multiple pathways of rapid β_2-adrenergic receptor desensitization. *J. Biol. Chem.* 265: 3202–9

Lohse, M. J., Benovic, J. L., Codina, J., Caron, M. G., Lefkowitz, R. J. 1990b. β-arrestin: a protein that regulates β-adrenergic receptor function. *Science* 248: 1547–50

Lohse, M. J., Lefkowitz, R. J., Caron, M. G., Benovic, J. D. 1989. Inhibition of β-adrenergic receptor kinase prevents rapid homologous desensitization of β_2-adrenergic receptors. *Proc. Natl. Acad. Sci. USA* 86: 3011–15

Lomasney, J. W., Lorenz, W., Allen, L. F., King, K., Regan, J. W., et al. 1990. Expansion of the α-adrenergic receptor family: cloning and characterization of a human α_2-adrenergic receptor subtype, the gene for which is located on chromosome 2. *Proc. Natl. Acad. Sci. USA* 87: 5094–98

Lorenz, W., Lomasney, J. W., Collins, S., Regan, J. W., Caron, M. G., Lefkowitz, R. J. 1990. Expression of three α_2-adrenergic receptor subtypes in rat tissues: implications for α_2 receptor classification. *Mol. Pharmacol.* 38: 599–603

Matsui, H., Lefkowitz, R. J., Caron, M. G.,

Regan, J. W. 1989. Localization of the fourth membrane spanning domain as a ligand binding site in the human platelet α_2-adrenergic receptor. *Biochemistry* 28: 4125–30

Nagayama, Y., Wadsworth, H. L., Chazenbalk, G. D., Russo, D., Seto, P., Rapoport, B. 1991. TSH-LH/CG receptor extracellular domain chimeras as probes for TSH receptor function. *Proc. Natl. Acad. Sci. USA* 88: 902–5

Neer, E. J., Clapham, D. E. 1988. Roles of G protein subunits in transmembrane signalling. *Nature* 333: 129–34

Nishimoto, I., Hata, Y., Ogata, E., Kojima, I. 1987. Insulin-like growth factor II stimulates calcium influx in competent BALB/c 3T3 cells primed with epidermal growth factor. *J. Biol. Chem.* 25: 12120–26

O'Dowd, B. F., Hnatowich, M., Caron, M. G., Lefkowitz, R. J., Bouvier, M. 1989. Palmitoylation of the human β_2-adrenergic receptor. *J. Biol. Chem.* 264: 7564–69

O'Dowd, B. F., Hnatowich, M., Regan, J. W., Leader, W. M., Caron, M. G., Lefkowitz, R. J. 1988. Site-directed mutagenesis of the cytoplasmic domains of the human β_2-adrenergic receptor. *J. Biol. Chem.* 263: 15985–92

Okamoto, T., Katada, T., Murayama, Y., Ui, M., Ogata, E., et al. 1990. A simple structure encodes G protein-activating function of the IGF-II/mannose 6-phosphate receptor. *Cell* 62: 709–17

Parmentier, M., Libert, F., Maenhaut, C., Lefort, A., Gerard, C., et al. 1989. Molecular cloning of the thyrotropin receptor. *Science* 246: 1620–22

Peralta, E. G., Ashkenazi, A., Winslow, J. W., Smith, D. H., Ramachandran, J., Capon, D. J. 1987. Distinct primary structures, ligand-binding properties, and tissue-specific expression of four human muscarinic acetylcholine receptors. *EMBO J.* 6: 3923–29

Rands, E., Candelore, M. R., Cheung, A. H., Hill, W. S., Strader, C. D., Dixon, R. A. F. 1990. Mutational analysis of β-adrenergic receptor glycosylation. *J. Biol. Chem.* 265: 759–64

Raposo, G., Dunia, I., Delavier-Klutchko, C., Kaveri, S., Strosberg, D., et al. 1989. Internalization of β-adrenergic receptor in A431 cells involves non-coated vesicles. *Eur. J. Cell Biol.* 50: 340–52

Regan, J. W., Kobilka, T. S., Yang-Feng, T. L., Caron, M. G., Lefkowitz, R. J., Kobilka, B. K. 1988. Cloning and expression of a human kidney cDNA for an α_2-adrenergic receptor subtype. *Proc. Natl. Acad. Sci. USA* 85: 6301–5

114   KOBILKA

Ross, E. M. 1989. Signal sorting and amplification through G protein-coupled receptors. *Neuron* 3: 141–52

Rubenstein, R. C., Wong, S. K. F., Ross, E. M. 1987. The hydrophobic tryptic core of the β-adrenergic receptor retains G_s regulatory activity in response to agonists and thiols. *J. Biol. Chem.* 262: 16655–62

Schwinn, D. A., Lomasney, J. W., Lorenz, W., Szklut, P. M., Fremeau, R. T. Jr., et al. 1990. Molecular cloning and expression of the cDNA for a novel α_1-adrenergic receptor subtype. *J. Biol. Chem.* 265: 5183–89

Shigemoto, R., Yokota, H., Tsuchida, K., Nakanishi, S. 1990. Cloning and expression of a rat neuromedin K receptor cDNA. *J. Biol. Chem.* 263: 623–28

Strader, C. D., Candelore, M. R., Hill, W. S., Dixon, R. A. F., Sigal, I. S. 1989a. A single amino acid substitution in the β-adrenergic receptor promotes partial agonist activity from antagonists. *J. Biol. Chem.* 264: 16470–77

Strader, C. D., Candelore, M. R., Hill, W. S., Sigal, I. S., Dixon, R. A. F. 1989b. Identification of two serine residues involved in agonist activation of the β-adrenergic receptor. *J. Biol. Chem.* 264: 13572–78

Strader, C. D., Dixon, R. A. F., Cheung, A. H., Candelore, M. R., Blake, A. D., Sigal, I. S. 1987. Mutations that uncouple the β-adrenergic receptor from G_s and increase agonist affinity. *J. Biol. Chem.* 262: 16439–43

Strader, C. D., Gaffney, T., Sugg, E. E., Candelore, M. R., Keys, R., et al. 1991. Allele-specific activation of genetically engineered receptors. *J. Biol. Chem.* 266: 5–8

Strader, C. D., Sigal, I. S., Dixon, R. A. F. 1989c. Structural basis of β-adrenergic receptor function. *FASEB J.* 3: 1825–32

Strasser, R. H., Sibley, D. R., Lefkowitz, R. J. 1986. A novel catecholamine-activated adenosine cyclic 3′,5′-phosphate independent pathway for beta-adrenergic receptor phosphorylation in wild-type and mutant S_{49} lymphoma cells: mechanism of homologous desensitization of adenylate cyclase. *Biochemistry* 25: 1371–77

Tota, M. R., Strader, C. D. 1990. Characterization of the binding domain of the β-adrenergic receptor with the fluorescent antagonist carazolol. *J. Biol. Chem.* 265: 16891–97

Valiquette, M., Bonin, H., Hnatowich, M., Caron, M. G., Lefkowitz, R. J., et al. 1990. Involvement of tyrosine residues located in the carboxyl tail of the human β_2-adrenergic receptor in agonist-induced downregulation of the receptor. *Proc. Natl. Acad. Sci. USA* 87: 5089–93

Vallar, L., Muca, C., Magni, M., Albert, P., Bunzow, J., et al. 1990. Differential coupling of dopaminergic D_2 receptors expressed in different cell types. *J. Biol. Chem.* 265: 10320–26

Waldo, G. L., Northup, J. K., Perkins, J. P., Harden, T. K. 1983. Characterization of an altered membrane form of the beta-adrenergic receptor produced during agonist-induced desensitization. *J. Biol. Chem.* 258: 13900–8

Wang, H., Berrios, M., Malbon, C. C. 1989a. Localization of β-adrenergic receptors in A431 cells in situ. *Biochem. J.* 263: 533–38

Wang, H., Lipfert, L., Malbon, C. C., Bahouth, S. 1989b. Site-directed anti-peptide antibodies define the topography of the β-adrenergic receptor. *J. Biol. Chem.* 264: 14424–31

Weinshank, R. L., Zgombick, J. M., Macchi, M., Adham, N., Lichtblau, H., et al. 1990. Cloning expression, and pharmacological characterization of a human α_{2B}-adrenergic receptor. *Mol. Pharmacol.* 38: 681–88

Wilson, A. L., Guyer, C. A., Cragoe, E. J. Jr., Limbird, L. E. 1990. The hydrophobic tryptic core of the porcine α_2-adrenergic receptor retains allosteric modulation of binding by Na^+, H^+, and 5-amino-substituted amiloride analogs. *J. Biol. Chem.* 265: 17318–22

Wong, S. K.-F., Claughter, C., Ruoho, A. E., Ross, E. M. 1988. The catecholamine binding site of the β-adrenergic receptors formed by juxtaposed membrane-spanning domains. *J. Biol. Chem.* 263: 7925–28

Wong, S. K.-F., Parker, E. M., Ross, E. M. 1990. Chimeric muscarinic cholinergic: β-adrenergic receptors that activate Gs in response to muscarinic agonists. *J. Biol. Chem.* 265: 6219–24

Yarden, Y., Rodriguez, H., Wong, S. K.-F., Brandt, E. R., May, D. C., et al. 1986. The avian β-adrenergic receptor: primary structure and membrane topology. *Proc. Natl. Acad. Sci. USA* 83: 6795–99

Yokota, Y., Sasai, Y., Tanaka, K., Fujiwara, T., Tsuchida, K., et al. 1989. Molecular characterization of a functional cDNA for rat substance P receptor. *J. Biol. Chem.* 264: 17649–52

Young, D., Waitches, G., Birchmeier, C., Fasano, O., Wigler, M. 1986. Isolation and characterization of a new cellular oncogene encoding a protein with multiple potential transmembrane domains. *Cell* 45: 711–19

Zemcik, B., Strader, C. D. 1988. Fluorescent localization of the β-adrenergic receptor on DDT-1 cells. *Biochem. J.* 251: 333–39

Annu. Rev. Neurosci. 1992. 15:115–37

MANIPULATING THE GENOME BY HOMOLOGOUS RECOMBINATION IN EMBRYONIC STEM CELLS

Andreas Zimmer

Unit of Developmental Biology, Laboratory of Cell Biology,
National Institute of Mental Health, Bethesda, Maryland 20892

KEY WORDS: gene targeting, chimeric mice, development

INTRODUCTION

The number of cloned genes has increased exponentially in the last decade. Often, the function of those genes is unknown. And, considering the enormous efforts to map and sequence the human and mouse genome, the gap between the availablity of structural information and functional assignments is likely to increase.

The recent development of gene targeting, however, now enables us to study gene function in the living animal. Once a gene is isolated, a mutation can first be introduced into the cloned DNA and then, by homologous recombination, into the genome of mammalian cells. This procedure is generally referred to as gene targeting. If the cell in which gene targeting occurs is pluripotent [e.g. an oocyte or an embryonic stem (ES) cell], animals that bear this mutation in their germ cell lineage can be generated.

Gene targeting probably has a large variety of applications in many different fields, including agricultural production and human gene therapy. In this review, however, I only evaluate gene targeting as a means to generate mice with specific mutations. I describe some technical aspects of gene targeting and then summarize the progress that has been made.

115

THE INTRODUCTION OF MUTATIONS BY HOMOLOGOUS RECOMBINATION

When linear DNA molecules are introduced into the nuclei of mammalian cells, about 20% of these cells stably integrate the exogenous DNA into the genome (Capecchi 1980). This usually happens by joining the ends of the exogenous DNA to chromosomal sequences (Folger et al 1982). The chromosomal integration sites are apparently distributed randomly over the whole genome (Murnane et al 1989). This type of DNA integration is called random integration, or illegitimate recombination.

If the incoming DNA bears homology to endogenous sequences, it can be integrated via homologous recombination with endogenous DNA. Cells can mediate reciprocal and nonreciprocal exchange of DNA, although there is a distinct preference for the latter (Smith & Berg 1984; Thomas & Capecchi 1987). Hence, homologous recombination can result in a replacement of the endogenous sequence by the incoming DNA (gene conversion) or in the simple integration of the exogenous DNA into the homologous chromosomal site. Formally, gene conversion involves two crossover events, whereas only a single crossover is involved in the integration type recombination.

Thus, it is possible to design two different types of vectors for gene targeting as outlined in Figure 1. These constructs have been termed sequence replacement or sequence integration vectors. Because both events occur with approximately the same frequency (Thomas & Capecchi 1987), the investigator can introduce a large spectrum of different mutations, from single base pair changes to large duplications.

Homologous recombination can also occur between two extra-chromosomal DNAs or between homologous sequences that reside in the genome. These different types of recombinatorial mechanisms have already been discussed in several excellent reviews (Bollag et al 1989; Capecchi 1989b; Kucherlapati 1986).

Smithies et al (1985) provided the first example of "targeting" into an endogenous gene. They disrupted the β-globin gene with the neomycin (neo^r) and bacterial suppressor ($supF$) gene and used a "phage rescue assay" to detect and clone targeted cells. This assay was based on complementation of a bacteriophage λ mutation in genomic libraries of DNA from transfected cell pools.

Despite great efforts, our current insight into the mechanism of homologous recombination is limited, and the elucidation of the molecular basis of homologous recombination is still in its infancy. A better understanding could greatly facilitate the application of homologous recombination, because the integration of exogenous DNA by random integra-

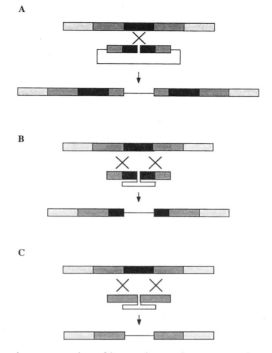

Figure 1   Schematic representation of integration- and gene conversion-type vectors. The region of homology is uninterrupted in integration-type vectors. Homologous recombination involves a single crossover event and results in duplication of the homologous region (*A*). The region of homology is interrupted by heterologous sequences in the gene conversion-type construct (*B*, *C*). Homologous recombination involves two crossovers and results in the integration of the heterologous sequence (*B*) or the replacement of target sequences by heterologous sequences (*C*).

tion is usually two or three orders of magnitude higher than homologous recombination. Hence, the low frequency is a major technical problem.

In most cases, phenotypic selection of the mutant is not possible. Thus, successful isolation of targeted cells depends on the availability of sensitive screening techniques and/or the ability to enrich for homologously recombined cell clones.

Screening for a rare genotype in a large pool of cells became a relatively easy task with the development of the "polymerase chain reaction" (PCR), a primer directed method to amplify DNA fragments in vitro (Saiki et al 1985). To detect homologous recombination events, the primers are selected so that one is complementary to heterologous sequences in the incoming DNA and the other to target endogenous DNA sequences. After homologous recombination, both primer binding sites are juxtaposed, and the amplification reaction yields a specific DNA fragment (Kim & Smithies 1988; Joyner et al 1989; Zimmer & Gruss 1989). With PCR, it is feasible to identify targeting events in as few as 1 in 10,000 cells (Kim & Smithies 1988). Furthermore, this method is sensitive enough to permit the analysis of only a portion of a single cell colony (Joyner et al 1989).

Even though PCR can detect rare events, the isolation of a specific cell clone is very difficult if the background of nontargeted cells is high. Hence, researchers have focused much attention on the development of procedures to reduce the background and enrich those cells that have undergone homologous recombination. An enrichment can be achieved by either increasing the targeting frequency or eliminating nontargeted cells.

Efforts to Increase the Targeting Frequency

The parameters that influence the frequency of homologous recombination have primarily been established by targeting natural and synthetic loci, the deletion of which allows direct selection for the mutant phenotype. Synthetic targets were created by transfecting cells with a defective neo^r (Smithies et al 1984; Song et al 1987; Thomas & Capecchi 1986; Thomas et al 1986) or herpes simplex virus thymidine kinase gene (HSV-tk) (Lin et al 1985) to obtain stable cell lines with these transgenes integrated in random chromosomal sites. These cell lines were subsequently transfected with another construct that contained a mutation at a different position. Homologous recombination between the two defective genes created one functional neo^r or HSV-tk gene, thus rendering these cells resistant to the antibiotic G418 or supporting growth in HAT medium (hypoxanthine, aminopterine, thymidine), respectively.

These experiments suggested that ratios of 10^{-2} to 10^{-5} could be expected for homologous recombination, compared with random integration. They also showed that sequence replacement and insertion events occur with approximately the same frequency (Song et al 1987). Moreover, the targeting frequency was increased with linearized templates (Thomas et al 1986). In contrast to experiments performed with yeast cells (Orr-Weaver et al 1981), however, the frequency of recombination in ES cells was not affected when nonhomologous sequences were placed at the ends of the constructs (Mansour et al 1988). This indicates that the efficiency of homologous recombination is stimulated by the linear topology of the substrate, rather than by the provision of homologous ends.

Surprisingly, neither the number of targets loci nor the number of exogenous molecules influences homologous recombination (Thomas et al 1986). Cell lines were generated with one, four, or five copies of a defective neo^r gene. After microinjection of a second defective neo^r gene, G418 resistant cell lines were obtained at the same rate in all target cell lines used. Furthermore, varying the number of microinjected molecules between 5 and 100 per cell had no effect on the homologous recombination frequency (Thomas et al 1986). Lack of dependence on the target gene copy number was also shown by Zheng & Wilson (1990). They measured the targeting frequency in normal CHO cells, which carry two copies of the dihydro-folate reductase (*DHFR*) gene, and in an amplified cell line with 800 gene copies. The frequency of homologous recombination was similar in these two cell lines. These results strongly suggest that identification of ho-mologous sequences by the cellular recombination machinery is not rate limiting. This is surprising, because the mammalian genome contains approximately 3×10^9 bases. Consequently, either a step before the align-ing of the homologous sequences or a later step must be rate limiting. To determine which step is involved, the formation of recombinational inter-mediates would have to be measured. This is not feasible at present, but circumstantial evidence suggests that intermediates may form more often than homologous recombination occurs. Thus, when Thomas & Capecchi (1986) attempted to repair an amber mutation, they observed some G418 resistant colonies in which the repair could not be accounted for by hom-ologous recombination. Instead, they found that these cell lines contained *neo*^r genes with a compensating mutation, which was apparently induced by a heteroduplex that must have formed between the chromosomal and the incoming DNA. They termed this phenomenon "heteroduplex induced mutagenesis." This phenomenon occurred with approximately the same frequency as the planned modification. Because only a small number of possibilities exist to introduce compensatory mutations, an even larger number of cells possibly formed heteroduplexes that resolved in a non-productive manner.

Most experiments that targeted synthetic genes indicate that the tar-geting frequency is independent of the integration site of the target. The one exception to this is a study by Lin et al (1985), who attempted to target a defective *tk* gene in ten different cell lines and found productive targeting events in only one.

An ideal model system to establish the condition for targeting into a natural locus is the gene encoding the enzyme hypoxanthine-guanine phosphorybosyltransferase (*hprt*). The enzyme catalyzes the conversion of guanine to guanylic acid and hypoxanthine to inosinic acid and plays an important role in the purine salvage pathway. *Hprt* is a single copy gene, which is located on the X chromosome in rodents and humans. Thus, only a single recombination event is necessary to yield a selectable phenotype in diploid male cell lines. *Hprt*⁺ cells can be selected in HAT medium, and *hprt*[−] cells in medium containing 6-thioguanine.

Capecchi and coworkers have used this model system to test the homology requirements for efficient recombination (Capecchi 1989a; Thomas & Capecchi 1987). They constructed sequence insertion and sequence replacement vectors to disrupt the *hprt* gene with a *neo*[r] cassette in exon 8. The length of homologous DNA 3′ to the *neo*[r] was constant, whereas the length 5′ to *neo*[r] varied. The constructs contained between 2.9 and 14.3 kb of *hprt* sequence. After electroporation into ES cells, they divided the cells and subjected them to different growth conditions: nonselective conditions to estimate the number of cells that survived electroporation; G418 selection to estimate the total number of stably transformed, G418 resistant cells; and G418 plus 6-thioguanine selection to detect homologously recombined cells. They found that the frequency of homologous recombination increased exponentially with increasing length of homology in both vector types. By extending the length of homologous DNA fivefold, the frequency was enhanced about 100-fold. Interestingly, they also found that the recombination frequency did not vary with the length of nonhomologous sequences (Mansour et al 1990).

Doetschman et al (1988) also mutated the *hprt* gene by inserting a *neo*[r] gene without its promotor (as I discuss later) into exon 3. In these experiments, the lengths of the homologous sequences were only 132 bp 5′ and 1.2 kb 3′ to the *neo*[r]. Surprisingly, the targeting frequency was approximately 100-fold higher than that observed by Capecchi et al, who used a much longer homologous sequence in their constructs.

Doetschman et al (1987) and Thompson et al (1989) have both reported the repair of a defective *hprt* gene in the ES cell line E14TG2a. This ES cell line carries a *hprt* with a deletion of the 5′ region, including exons 1 and 2. The constructs used to restore a functional gene contained approximately 2–4 kb of homologous sequences. In both studies, the frequency of homologous recombination was at least ten times higher than that obtained by Capecchi et al. Although it is difficult to compare the results of the above experiments directly, the discrepancy between Capecchi's and Doetschman's results might arise from a difference in the targeted genomic regions. These experiments point out that although the length of homology is important, other parameters (which are unknown) might be even more crucial for the targeting frequency.

Eliminating Nontargeted Cells

Several attempts have been made to establish conditions that would specifically eliminate nontargeted cells. One approach is based on the activation of a defective marker gene after its integration into the chromosomal target. Hence, the *neo*[r] gene, which is used to disrupt the target, lacks the *cis* regulatory sequences required for efficient expression. After homologous recombination, these *cis* elements are provided by the target gene; random integration would only rarely juxtapose the neo^r to the missing sequence. For example, efficient processing of RNA transcripts requires the presence of a DNA sequence that initiates the polyadenylation of the primary transcript. Lack of this signal in the marker gene of the targeting construct decreased the number of neo^r colonies approximately fourfold (Joyner et al 1989; Zijlstra et al 1989). In homologously recombined cells, this polyadenylation signal is provided by the target gene that guarantees appropriate processing of the primary transcripts.

In another approach, a neo^r coding region without its promoter was introduced into the target sequence, so that expression was controlled by the target gene promoter after homologous recombination (Jasin & Berg 1988; Schwartzberg et al 1990; Sedivy & Sharp 1989; Stanton et al 1990; te Riele et al 1990). Random integration infrequently resulted in activation of the neo^r, and the number of G418 resistant colonies that resulted from random integration events was reduced approximately 100-fold. Promoterless constructs can, however, only be used to target genes that are expressed in embryonic stem cells at a high enough level to activate the marker.

Another strategy for eliminating nontargeted cells is based on the fact that terminal sequences are inadvertently lost during nonreciprocal recombination (as outlined in Figure 2), whereas they are relatively unaffected by random integration. Mansour et al (1988) placed an *HSV-tk* gene at the end of a construct to target the *hprt* gene. Homologous recombination with the cellular target should result in $hprt^-neo^rHSV\text{-}tk^-$ cells, whereas random integration should yield $hprt^+neo^rHSV\text{-}tk^+$ cells. After electroporation, the cells were subjected to selection in G418 (neo^r) and gangcyclovir ($HSV\text{-}tk^-$).

This combination of positive (neo^r) and negative ($HSV\text{-}tk^-$) selections resulted in a 2000-fold enrichment for homologous recombinants over G418 selection alone. The technique, which was referred to as positive/negative selection (PNS), has the major advantage of being independent of the target gene locus. It has subsequently been used to target a large variety of genes (DeChiara et al 1990; Johnson et al 1989; McMahon & Bradley 1990; Thomas & Capecchi 1990) and is now the most widely used technique to target nonselectable genes. However, it typically gives a 10–20-fold enrichment for homologous recombinants, a figure substantially lower than that reported originally.

Gene Targeting Without Selection

All of the above-outlined gene targeting methods rely on the integration of a selectable marker gene into the target locus. Recently, we described a

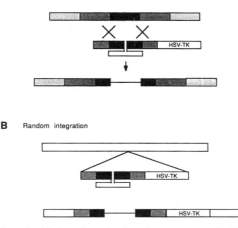

A   Homologous recombination

B   Random integration

Figure 2   The PNS method is based on the fact that sequences flanking the homologous region are lost after gene conversion (*A*), but remain intact after random integration (*B*). Thus, if a marker for negative selection (*HSV-TK*) is placed at the end of the gene targeting construct, cells that have integrated the DNA by random integration can be eliminated.

selection-independent method to target the homeobox *hox-1.1* gene (Zimmer & Gruss 1989). In the targeting construct, we inserted a 20 base pair (bp) oligonucleotide into the homeobox that, upon homologous recombination, would disrupt the reading frame; provide a binding site for the detection of targeting events by PCR; and replace one restriction enzyme site with another, thus permitting unambiguous verification of targeting events by genomic Southern blotting.

The use of microinjection to deliver the DNA was crucial to the success of our experiments. To keep the size of the starting cell pool small, we plated only a small number of cells in a single dish and attempted to inject most or all of them (Zimmer et al 1990). Microinjection is the most efficient method for transforming cells. Only 1% of the cells integrate the DNA after electroporation under optimal conditions (Mansour et al 1988), while up to 25% can be stably transformed by microinjection (Capecchi 1980; Folger et al 1982). After microinjection, the pools were expanded and analyzed by PCR. Targeted clones were finally isolated from positive pools in a multistep cloning procedure. The targeting frequency was very high: With a 1.5 kb construct, one targeting event occurred per 150 microinjected cells; the frequency was 1 per 110 for a 8.5 kb construct. If we assume that every fifth microinjected cell had integrated the exogenous DNA, the estimated ratio of homologous to illegitimate recombination was approximately 1 to 30.

Microinjection should prove useful to introduce single bp exchanges into the chromosomal target when alterations other than the desired mutation are to be avoided. With the *hox-1.1* experiment, we were concerned that the introduction of a *neo*[r] cassette with a strong promotor might affect the expression of neighboring genes in the *Hox-1* cluster, as there is circumstantial evidence that different genes in the cluster share *cis* regulatory elements. Abnormal expression of HOX genes interferes with embryonic development (Balling et al 1989; Kessel et al 1990; Wolgemuth et al 1989), and interference with other genes would make the interpretation of the effects of a *hox-1.1* deletion difficult.

EMBRYONIC STEM CELLS CAN POPULATE THE GERMLINE

Although homologous recombination has been reported after microinjection of DNA into fertilized mouse oocytes (Brinster et al 1989), this occurs at low frequency and cannot be used for most gene targeting experiments. The use of ES cells has proven to be a much better choice, as the genotype of the cells can be characterized before using them to generate animals.

Embryonic stem cells are pluripotent and are derived from pre-implantation embryos (see Figure 3). They can be propagated and manipulated in tissue culture, but require carefully adjusted culture conditions to maintain their pluripotency (Doetschman et al 1985; Evans & Kaufmann 1981; Martin 1981; Robertson 1987). The presence of the lymphocyte inhibitory factor (LIF, also termed differentiation inhibitory activity or human interleucin for DA cells), is important to prevent the differentiation of the cells. This growth factor can be added to the medium (Smith et al 1988; Williams et al 1988) or provided by embryonal fibroblast feeder cells that secrete a matrix associated form of LIF (Rathjen et al 1990).

Embryonic stem cells can be implanted into a developing pre-implantation embryo by microinjection into morulae or into the blastocoel cavity of a blastocyst (Bradley 1987). Alternatively, ES cells can be aggregated with host morulae to form chimeras. The embryos are then implanted into the uterus of a pseudopregnant foster mother and allowed to develop to term. The ES cells mix with the host cells and participate in normal embryonic development. They can colonize every tissue, including the germ cell lineage. The contribution of ES and host cells to the chimeric animals can usually be detected by the different coat colors they contribute.

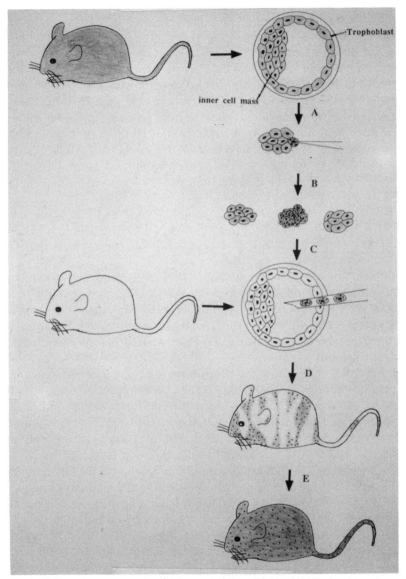

Figure 3 Schematic representation of the production of chimeric animals with embryonic stem cells. The cells are isolated from the inner cell mass of mouse blastocysts (*A*). They can be transformed and cloned by standard procedures (*B*). After implantation into host blastocysts (*C*), they contribute to all tissues in the resulting chimeras (*D*). By breeding and interbreeding, animals that are first heterozygous and eventually homozygous for the genetic alteration can be obtained (*E*).

This is very useful, because it allows one to estimate the extent with which the ES cells colonized the host embryo.

The amount of the ES cell contribution is largely dependent on the quality of the cell line and can vary significantly between individual clones. It is also dependent on the genetic background of the host embryo. Many ES cell lines colonize inbred mouse strains, such as C57 B1/6J, extensively, but poorly colonize such outbred strains as CD1. Most embryonic stem cells are derived from the mouse strain 129.

DESIGNER MICE: HOW DO THEY LOOK?

Numerous mutations have already been introduced into the mouse germ-line, a few of which have been described in the literature. By the time this manuscript is printed, the number of mutants described will probably have increased significantly. Those few mutants already described provide a preview to the surprises that may await us in the future.

Hprt-Mice: A Model for Lesh-Nyhan Syndrome?

In humans, mutations in the *hprt* gene cause the Lesh-Nyhan syndrome, a rare neurologic disorder that only affects males (for review, see Kelley & Wyngaarden 1983). This syndrome is characterized by hyperuricemia, choreoatheosis, spasticity, mental retardation, and a striking compulsive self-destructive behavior. Patients with this illness exhibit developmental retardation as early as three to four months of age. By age 2, all the neurologic manifestations of the syndrome are usually apparent. Because the mechanisms responsible for the central nervous system (CNS) disorders associated with the disease are still unknown, researchers hoped that the development of an animal model might shed some light on the etiology of the disease and contribute to the development of novel therapies. Although the *hprt* gene has been widely used as a model target for homologous recombination, the first animals with a deficient *hprt* gene were established by using a spontaneous *hprt⁻* ES cell line and a line carrying a *hprt* gene mutated by retroviral integration, respectively (Hooper et al 1987; Kuehn et al 1987). Mice hemizygous for the mutation did not exhibit any obvious behavioral abnormalities. However, close examination of the animals (Dunnett et al 1989) revealed that they had partial depletion of dopamine (DA) levels in the forebrain, which resulted from an increased DA turnover. This is intriguing, because DA depletion in the forebrain of Lesh-Nyhan patients has been correlated with many of the disease symptoms.

In rodents, the unilateral destruction of the nigrostriatal dopaminergic system with the dopamine analog 6-hydroxidopamine (6-OHDA), can lead to unilateral self-mutilation (Ungerstedt 1971). Also, extensive forebrain depletion of DA, induced by neonatal 6-OHDA injections, can result in self-mutilation and motor disorders in response to DA agonists (Breese et al 1984).

In addition to a reduction in dopamine levels, the serotonin (5-HT) concentration was significantly decreased in the basal ganglia of hemizygous animals. The 5-HT level is also altered in Lesh-Nyhan patients, but unlike the *hprt⁻* mice, the patients have an elevated level.

Altogether, these findings indicate that the *hprt* deficiency influences neurotransmitter synthesis and metabolism in mice and humans. We still do not know why the mutation selectively affects some systems and not others, and why the manifestations are so dramatically different in mice and humans.

Mutations Affecting the Development of the Cerebellum

The genetic control of the development of the mammalian CNS is poorly understood. Expression analysis of several proto-oncogenes and mammalian homologs of *Drosophila* developmental control genes suggests that they may play an important role in CNS development. In this section I discuss the results of experiments in which two genes with homology to *Drosophila* segmentation genes were disrupted.

The first example describes the targeting of the *wnt-1* (*int-1*) gene (for nomenclature, see Nusse et al 1991), which was reported independently by Thomas & Capecchi (1990) and McMahon & Bradley (1990). *Wnt-1* was first identified as a gene activated in mouse mammary tumors by proviral integrations of the mouse mammary tumor virus (MMTV) (for review, see Nusse 1988). It encodes a secreted glycoprotein of 36–42 K that is associated with the cell matrix (Bradley & Brown 1990; Papkoff & Schreyver 1990). Trangenic mouse studies showed that the expression of *wnt-1* under the control of the promotor in a MMTV long terminal repeat (LTR) can basically recapitulate the pathologic consequences of naturally occurring MMTV integrations (Tsukamoto et al 1988). Analysis of the *wnt-1* expression pattern suggested, however, that the gene has a function in spermatogenesis and neural development (Shackleford & Varmus 1987; Wilkinson et al 1987). Both groups used the PNS method to mutate the *wnt-1* gene by introducing a *neo*ʳ cassette into the second exon.

Heterozygous animals did not exhibit an obvious phenotype. Most of the homozygous animals, however, died shortly after birth. Examination of homozygous embryos at various stages showed that they failed to develop a large portion of the mesencephalon and the metencephalon; all other neuronal and nonneuronal tissues were unaffected.

This phenotype is detectable at 9.5 days post coitum (pc) (McMahon &

Bradley 1990). At this stage, the embryos are missing approximately two thirds of the mesencephalon, and by day 14.5 pc, little or no midbrain was detected. These results reflect either an additional loss of midbrain tissue between 9.5 and 14.5 days or differential growth properties of different parts of the developing mesencephalon.

Thomas & Capecchi (1990) found a malformation of the mesencephalon at day 14.5 pc and a mild midbrain hydrocephaly at day 17.5 pc. The caudal boundary of the defect was in the metencephalon. Although the dorsally derived cerebellum was missing in all examined embryos, the ventrally derived pons was unaffected.

The penetrance of the mutation seemed somewhat variable. Thomas & Capecchi examined embryos at 17.5 days pc and found that one of six homozygous embryos died in utero. On the other hand, they also obtained one homozygous animal that survived to adulthood. This mouse exhibited severe ataxia, suggestive of a cerebellar defect. A close examination of this animal's brain revealed hydrocephaly in the caudal region of the cerebral hemispheres and midbrain. Only the posterior part of the cerebellum was present. The authors suggested that the influence of *wnt-1* exhibits a rostral-caudal gradient, which begins in the mesencephalon and spreads into the metencephalon. Hence, for those individuals in whom the mutation has a lower penetrance, the anterior portion of the cerebellum would be more affected than the posterior part.

The area affected by the *wnt-1* mutation was not identical to the area of *wnt-1* expression. *Wnt-1* expression was detected at day 8.5 along the dorsal midline of the mesencephalon and rostral metencephalon and in the spinal tube. Thus, the affected area in the brain seems to extend beyond those structures that express *wnt-1*. Because the *wnt-1* product is secreted, it might be required for the normal development of a wide area. Alternatively, *wnt-1* might regulate a cascade of events, and lack of expression could interfere with the development of cells that are not directly influenced by the *wnt-1* product.

There is no correlation between *wnt-1* expression in the spinal tube and an apparent defect. *Wnt-1* belongs to a family of related genes with overlapping expression domains (McMahon & McMahon 1989; Roelink et al 1990). Hence, *wnt-1* related genes may have functionally replaced *wnt-1* in certain tissues.

The second example of targeting the homolog of a *Drosophila* segmentation gene is the mouse *en-2* gene (Joyner et al 1991). *En-2* is a homeobox-containing gene that was originally identified together with a closely related gene *en-1*, by its sequence similarity to the *Drosophila* genes engrailed (*en*) and invected (*iv*). Both mouse genes are expressed in a band of cells at the junction of the mid- and hindbrain. *En-1* is also expressed in the spinal cord, somites, and limbs. In adult animals, both *en-1* and *en-2* are expressed in some motor neurons in the pons, and some cells in the substantia nigra. *En-2* is also found in the granule cell layer of the cerebellum (Davis & Joyner 1988; Davis et al 1988).

Engrailed genes are highly conserved throughout evolution. Mutational analysis in *Drosophila* has shown that the *en* gene is required for embryonic segmentation and development of the nervous system (Lawrence & Johnston 1984; Lawrence & Struhl 1982). Based on their patterns of expression and evolutionary conservation, Davis et al (1988) postulated that the mouse engrailed genes might also be involved in the compartmentalization of the developing neural tube and the specification of differentiations of particular neuronal populations.

Joyner et al (1991) replaced 1 kb of the *en-2* gene, including 700 bp of the homeobox containing exon, by a 1.5 kb *neo^r* gene in the mouse germline. Surprisingly, mice homozygous for the *en-2* mutation survived to adulthood and did not exhibit any apparent malformation or abnormal behavior.

A close examination of the cerebellum, however, revealed subtle changes in size and folding pattern. The tuber vermis or tuber pyramis extended uninterrupted across the entire mediolateral plane of the cerebellar cortex. The folding pattern of crus I and II was altered, and these lobes were also reduced in size. Saggital sections revealed an alteration in the branching pattern of fiber tracts in addition to the externally visible structural modifications. The cellular architecture of the cerebellum was unchanged, except for a reduction in the size of the Purkinje cell layer. This observation is intriguing, because the gene is primarily expressed in the granule cell layer.

The cellular basis of foliation, that is, the characteristic folding of the cerebellar cortex, has not been established, and varies considerably among different inbred mouse strains and among individuals in a closed colony (Inouye & Oda 1980). Abnormal foliation has also been described in reeler and staggerer mutant mice (Goffinet 1984; Siedmann 1968), which exhibit a defect in Purkinje cell migration and differentiation, respectively. Neumann et al (1990) have hypothesized that a normal Purkinje cell layer is necessary for normal cerebellar foliation, but mutants like weaver or lurcher have a markedly reduced cerebellum size, abnormalities in the granule and Purkinje cell populations, and a relatively normal folial pattern (Caddy & Biscoe 1979; Herrup & Trencker 1987).

The changes in the foliation pattern are the only developmental consequences of the *en-2* mutations that have been observed. No defects were detected in the pons or the substancia nigra, where *en-2* is also expressed. Joyner et al suggest that this could be explained by a functional replacement of *en-2* by *en-1* in those cells where both genes are co-expressed. To address this question directly, it would be necessary to mutate the *en-1* gene in the mouse germline and generate mice that are homozygous for both mutations.

Disruption of the β2-Microglobulin Gene

Major histocompatibility (MHC) class I molecules consist of a highly polymorphic α-chain and a noncovalently linked β2-microglobulin (β2-m). They play an important role in the development of self-tolerance and the cytotoxic T cell response by associating with antigens and displaying them to responsive cells (Matis 1990; von Boehmer 1988; Zinkernagel & Doherty 1979). MHC class I molecules might also have functions in addition to antigen presentation, such as in the recognition of protein hormones or as olfactory cues that influence mating preference (Kittur et al 1987; Singh et al 1987; Yamazaki et al 1988). β2-m is also identical to thymotaxin, a chemotactic protein for bone marrow hematopoietic precursors (Dargemont et al 1989).

Koller & Smithies (1989) and Zijlstra et al (1989) have independently targeted the β2-m gene with a neo^r cassette inserted into exon 2. The frequency of targeting events per G418-resistant colonies was very high in both studies. Koller & Smithies (1989) found 1 per 117 with a construct containing 5 kb of homologous sequences, whereas Zijlstra et al (1989) had 1 per 25 with a construct containing 10 kb homology. The β2-m is very weakly expressed in undifferentiated ES cells. This reemphasizes that the targeting frequency is gene locus dependent, but independent of the expression of the target gene by ES cells.

Heterozygous animals did not show any phenotypic abnormalities and were bred to homozygosity (Koller et al 1990; Zijlstra et al 1990). Considering the various functions that were attributed to β2-m, Zijlstra et al (1989) thought that the mutation might interfere with embryonic development. Yet, homozygous animals survived and, at a first glimpse, appeared to be normal.

A close examination revealed that homozygous mice did not produce any β2-m protein, as expected; consequently, they did not express MHC class I molecules on their cell surfaces. Furthermore, homozygous animals had no functional Fc receptor, which indicates that β2-m is indeed essential for its expression at the cell surface. The thymus, lymph nodes, and spleen in the mutant animals were of normal size and contained a normal number of cells, which suggests that the β2-m is not the sole inducer of the migration of hematopoietic precursors from the bone marrow.

When cells from lymphoid organs were analyzed with antibodies that delineate different subsets of T cells, a 100–150-fold reduction in T-cell antigen receptor $\alpha\beta^+CD4^-8^+$ T cells became apparent. $\alpha\beta^+CD4^-8^+$ T cells develop from $\alpha\beta^{dim}CD4^+8^+$ thymocytes, and $\alpha\beta^-CD4^-8^+$ T cells are thought to represent an intermediate stage in differentiation. The distribution of $\alpha\beta^{dim}CD4^+8^+$ and $\alpha\beta^-CD4^-8^+$ T cells was however unchanged in mutant mice, which indicates that the differentiation of $\alpha\beta^-CD4^-8^+$ T cells requires the interaction with MHC class I molecules.

These mutant mice now allow us to address questions concerning the function of $\beta2$-m and class I molecules in living animals directly. They should also be valuable in analyzing the function of $CD4^-8^+$ cytolytic T cells in infections, carcinogenesis, and tissue graft rejection.

Disruption of the Insulin-Like Growth Factor II Gene

The insulin-like growth factor II (*IGF-II*) gene encodes a mitogenic polypeptide (Di Cicco-Bloom & Black 1988) with a wide expression spectrum during rodent embryogenesis (Beck et al 1987, 1988). In the adult brain, expression has been confined to the chorioid plexus and leptomeninges (Stylianopoulou et al 1988). Sara & Carlsson-Skwirut (1988) have suggested that *IGF-II* is involved in the regulation of the growth of the brain.

By using the PNS technique, this gene was disrupted by a *neo*[r] cassette in the second exon in ES cells (DeChiara et al 1990). The *IGF-II*⁻ ES cells readily colonized the germline in chimeric males to yield ES cell-derived heterozygous offspring.

Approximately 50% of the offspring had a considerably smaller body size than normal. Their body weights were about 60% of the average weight of their normal littermates. DNA analysis revealed that all mice with this growth deficiency were heterozygous for the *IGF-II* mutation. Inspection of embryos showed that the phenotype was already apparent 16 days pc. In situ hybridization showed that all components that form the chorioallantoic placenta express the *IGF-II* gene (DeChiara et al 1990); thus, the observed phenotype is conceivably an indirect consequence of an impaired placental function.

The phenotype suggested a gene dosage effect. Therefore, the level of RNA expression in mutant and normal mice was compared. An expression of 50% would be expected if the level of expression were dependent on gene dosage. However, the level of expression from the intact allele in heterozygous mutant mice was less than 10% of that in the normal littermates. Further analysis revealed that the gene is imprinted (De Chiara et al 1991), i.e. only one allele is expressed from the paternal or the maternal alleles. The paternal *IGF-II* allele is active, whereas the maternal allele is silent in most tissues. Consequently, mice homozygous for the *IGF-II* mutation exhibited the identical phenotype to the heterozygous animals, which were derived from male chimeric animals.

Interestingly, Barlow et al (1991) have recently demonstrated that the gene for the *IGF-II* receptor is also imprinted, but maternally.

Can Chimeras Have Informative Phenotypes?

In all examples described so far, a phenotype was detected in mutant offspring derived from chimeric animals. The contribution of ES cells to chimeric tissues can, however, be quite extensive. Thus, we might expect functional consequences of genetic changes in ES cells in chimeric animals.

Gain of function mutations could program the ES cells to differentiate inappropriately, thus disturbing the development of chimeras. Most mutations that have been introduced to date are gene disruptions and represent loss of function mutations. Heterozygous loss of function mutations can potentially result in phenotypic changes. If gene dosage is critical, highly chimeric animals that bear one mutant copy might be affected. Likewise, hemizygous mutations or mutations in imprinted genes could alter the phenotype of chimeras. Mutations might impair the ability of ES cells to differentiate into a particular cell type. If this defect can be compensated by host cells, the contribution of the ES cells to different tissues in the chimeras must be evaluated.

Pevny et al (1991) recently reported that targeting the *GATA-1* gene blocks the differentiation of hemizygous ES cells into erythrocytes in chimeric animals. *GATA-1* (previously termed *GF-1*, *Eryf-1*, *Nf-E1*; see Orkin 1990 for review) is a transcription factor with a presumed role in the activation of the erythroid cell lineage. It belongs to the family of zinc finger-containing genes and binds specifically to GATA consensus elements in the regulatory region of erythroid specific genes. *GATA-1* is located on the X chromosome; hence, it is only present in a single copy in male ES cells.

The PNS procedure was used to disrupt this gene, and chimeric animals were generated from two independently derived cell lines. These cells colonized all the tissues examined, including white blood cells, but failed to differentiate into mature red blood cells. Extensive chimeras were severely anemic and many died in utero. These results provide direct evidence that *GATA-1* is essential for erythrocyte differentiation, presumably at a late stage in the development of progenitor cells.

This mutation would obviously be lethal during the development of males if it were present in the germline. Many mutations probably lead to early embryonic lethality. The use of mutant ES cells in those cases might be of value to assess gene function in chimeric animals directly. Host cells in chimeras can potentially rescue these embryos, and a careful analysis might reveal clues of important functions that would not be identified otherwise.

Phenotypic changes from mutations in genes that are not X-linked might require the disruption of both gene copies. Te Riele et al (1990), who consecutively disrupted both alleles of the proto-oncogene *pim-1*, demonstrated that this is feasible. *Pim-1* is highly expressed in ES cells. Therefore, the authors first used a promotorless *neo*[r] cassette without a translational start signal to disrupt *pim-1* in exon 4. After homologous recombination a *pim-1/neo*[r] fusion, protein is produced to render the cells resistant to G418. The targeting frequency was remarkably high; 85% of all G418 resistant colonies carried the planned mutation. Subsequently, te Riele and colleagues disrupted the second allele by using a similar construct with a hygromycin (*hyg*) gene. Again, the frequency was very high, and the second gene was disrupted in 62% of all *hyg*[r] colonies. These results demonstrate that it is possible to mutate both alleles in ES cells with high efficiency, although the recombination frequency described above is certainly exceptional.

Surprisingly, the mutation in the *pim-1* gene had no effect on the growth properties of the embryonic stem cells, or their capability to differentiate in vitro. Mice homozygous for the *pim-1* mutation showed no gross abnormalities; they were all healthy and bred normally (A. Berns, personal communication).

CONCLUSION

The results of the gene targeting experiments, which I have summarized, are very exciting and surprising.

The phenotype of the *wnt-1*[−] homozygous and the *GATA-1*[−] chimeric mice strongly suggests that these genes have important functions in the development of the midbrain or in the differentiation of erythrocytes, respectively. This confirms and extends findings of in vitro and in vivo studies (Orkin 1990; Papkoff & Schreyver 1990; Shackleford & Varmus 1987; Wilkinson et al 1987).

However, the developmental and physiologic consequences of most mutations were modest. In the case of the *en-2* mutation, for example, most scientists have expected rather dramatic phenotypic consequences (Davis et al 1988). This expectation was based on the function of the *Drosophila* homolog as a segmentation gene and the fact that *en* genes are highly conserved during evolution. We currently do not understand the physiologic consequences of the altered cerebellum foliation in mutant mice.

Moreover, the *pim-1*, *β2-m*, *hprt*, and *IGF-II* disruptions did not result in the phenotypic changes that one might have predicted. None of these mutations yielded severe developmental abnormalities, and adult animals appeared healthy. Interestingly, no developmental defect is detectable in some tissues in which these genes are expressed. For example, the *β2-m* gene is already expressed at the two-cell stage in mouse embryos and is detectable in virtually all somatic cells. Yet, the phenotype is very restricted. *Pim-1* is expressed at high levels in hematopoietic tissues, testes, and ovaries, but the mutation does not result in a readily detectable defect. The *IGF-II* mutation apparently does not affect the growth of the brain, as has been suggested.

These results teach us the importance of performing gene targeting experiments, as proposed gene functions that are based on circumstantial evidence are sometimes incorrect. Because we have to accept the facts and assume that nature did not invest in conserving these genes just to put us on the wrong scent, the following questions arise: Is our understanding of the physiology and development of the mouse lacking, so that we fail to detect defects due to mutant genes, or is our understanding of the genetics insufficient?

It is tempting to speculate that the *en-2* mutation might have behavioral consequences in mice that are not readily apparent to human observers. The same could be said for the *hprt* mutation, a genetic disorder with devastating consequences in humans, but without any obvious impact on mice. Alternatively, the genetic control of the mammalian development and homeostasis may be more redundant than anticipated. Some mutations may also be epistatic (where a particular mutation is obscured by an other genetic locus).

The phenotypic consequences of mutations isolated by classical means may have given us false expectations. In most cases, we could easily correlate a mutation with a phenotype. But, the important difference is that the classical geneticists attempted to identify responsible genes based on a phenotype. However, the classical route has the important limitation that a defect may go undetected if the phenotype is weak. Thus, some of the mutations discussed here would have never been identified by classical means.

Certainly, genetics has entered a new age. Thirty-five years after the material basis of the genetic program was discovered, we are now able to introduce subtle changes into the program. Perhaps, this will eventually lead us to understand how the amazingly complex series of developmental events from the fertilized egg to the adult organism is orchestrated.

ACKNOWLEDGMENTS

I thank my colleagues for communicating their results before publication and Drs. Michael Brownstein, Robert Cohen, Dervla Mellerick-Dressler, Ulrich Eisel, and Moncef Jendoubi for critically reading the manuscript. This work was supported by a stipend of the Deutsche Forschungsgemeinschaft.

Literature Cited

Balling, R., Mutter, G., Gruss, P., Kessel, M. 1989. Craniofacial abnormalities induced by ectopic expression of the homeobox gene *Hox-1.1* in transgenic mice. *Cell* 55: 337–47

Barlow, D. P., Stöger, R., Herrmann, B. G., Saito, K., Schweifer, N. 1991. The mouse insulin-like growth factor type-2 receptor is imprinted and closely linked to the *Tme* locus. *Nature* 349: 84–87

Beck, F., Samani, N. J., Byrne, S., Morgan, K., Gebhard, R., Brammar, W. J. 1988. Histochemical localization of *IGF-I* and *IGF-II* mRNA in the rat between birth and adulthood. *Development* 104: 29–39

Beck, F., Samani, N. J., Penschow, J. D., Thorley, B., Tregear, G. W., Coghlan, J. P. 1987. Histochemical localization of *IGF-I* and *-II* mRNA in the developing rat embryo. *Development* 101: 175–84

Bollag, R. J., Waldman, A. S., Liskay, R. M. 1989. Homologous recombination in mammalian cells. *Annu. Rev. Genet.* 23: 199–225

Bradley, A. 1987. Production and analysis of chimaeric mice. In *Teratocarcinomas and Embryonic Stem Cells: A Practical Approach*, ed. E. J. Robertson, pp. 113–51. Oxford: IRL

Bradley, R. S., Brown, A. M. C. 1990. The proto-oncogene *int-1* encodes a secreted protein associated with the extracellular matrix. *EMBO J.* 9: 1569–75

Breese, G. R., Baumeister, A. A., McCown, T. J., Emerick, S. G., Frye, G. D., Mueller, R. A. 1984. Neonatal 6-hydroxidopamine treatment: model of susceptibility for self-mutilation in the Lesh-Nyhan syndrome. *Pharmacol. Biochem. Behav.* 21: 459–61

Brinster, R. L., Braun, R. E., Lo, D., Avarbock, M. R., Oram, F., Palmiter, R. D. 1989. Targeted correction of a major histocompatibility class II E_a gene by DNA microinjected into mouse eggs. *Proc. Natl. Acad. Sci. USA* 86: 7087–91

Caddy, K. W. T., Biscoe, T. J. 1979. Structural and quantitative studies on the normal C3H and lurcher mutant mouse. *Philos. Trans. R. Soc. London* 287: 167–201

Capecchi, M. R. 1980. High efficiency transformation by direct microinjection of DNA into cultured mammalian cells. *Cell* 22: 479–88

Capecchi, M. R. 1989a. Altering the genome by homologous recombination. *Science* 244: 1288–92

Capecchi, M. R. 1989b. The new mouse genetics: altering the genome by gene targeting. *Trends Genet.* 5: 70–76

Dargemont, C., Dunon, D., Deugnier, M.-A., Denoyelle, M., Girault, J.-M., et al. 1989. Thymotaxin, a chemotactic protein, is identical to β2-microglobulin. *Science* 246: 803–5

Davis, C. A., Joyner, A. L. 1988. Expression patterns of the homeo box-containing genes *En-1* and *En-2* and the proto-oncogene *int-1* diverge during mouse development. *Genes Dev.* 2: 1736–44

Davis, C. A., Noble-Topham, S. E., Rossant, J., Joyner, A. L. 1988. Expression of the homeo box-containing gene *En-2* delineates a specific region of the developing mouse brain. *Genes Dev.* 2: 361–71

DeChiara, T. M., Efstratiadis, A., Robertson, E. J. 1990. A growth-deficiency phenotype in heterozygous mice carrying an insulin-like growth factor II gene disrupted by targeting. *Nature* 345: 78–80

DeChiara, T. M., Robertson, E. J., Efstratiadis, A. 1991. Parental imprinting of mouse insulin-like growth factor II gene. *Cell* 64: 849–59

Di Cicco-Bloom, E., Black, I. B. 1988. Insulin growth factors regulate the mitotic cycle in cultured rat sympathetic neuroblasts. *Proc. Natl. Acad. Sci. USA* 85: 4066–70

Doetschman, T. C., Eistetter, H., Katz, M., Schmidt, W., Kemler, R. 1985. The in vitro development of blastocyst-derived embryonic stem cell lines: formation of visceral yolk sac, blood islands and myocardium. *J. Embryol. Exp. Morphol.* 87: 27–45

Doetschman, T., Gregg, R. G., Maeda, N., Hooper, M. L., Melton, D. W., et al. 1987. Targeted correction of a mutant *HPRT* gene in mouse embryonic stem cells. *Nature* 330: 576–78

Doetschman, T., Maeda, N., Smithies, O. 1988. Targeted mutation of the *Hprt* gene in mouse embryonic stem cells. *Proc. Natl. Acad. Sci. USA* 85: 8583–87

Dunnett, S. B., Sirinathsinghji, D. J., Heavens, R., Rogers, D. C., Kuehn, M. R. 1989. Monoamine deficiency in a transgenic (*Hprt⁻*) mouse model of Lesch-Nyhan syndrome. *Brain Res.* 501: 401–6

Evans, M. J., Kaufmann, M. H. 1981. Establishment in culture of pluripotential cells from mouse embryos. *Nature* 292: 154–56

Folger, K. R., Wong, E. A., Wahl, G., Capecchi, M. R. 1982. Patterns of integration of DNA microinjected into cultured mammalian cells: evidence for homologous recombination between injected plasmid DNA molecules. *Mol. Cell. Biol.* 2: 1372–87

Goffinet, A. M. 1984. Events governing organization of postmigratory neurons:

studies on brain development in normal and reeler mice. *Brain Res. Rev.* 7: 261–96

Herrup, K., Trencker, E. 1987. Regional differences in cytoarchitecture of the weaver cerebellum suggest a new model for weaver gene action. *Neuroscience* 23: 871–85

Hooper, M., Hardy, K., Handyside, A., Hunter, S., Monk, M. 1987. *HPRT*-deficient (Lesch-Nyhan) mouse embryos derived from germline colonization by cultured cells. *Nature* 326: 292–95

Inouye, M., Oda, S.-I. 1980. Strain-specific variations in the folial pattern of the mouse cerebellum. *J. Comp. Neurol.* 190: 357–62

Jasin, M., Berg, P. 1988. Homologous integration in mammalian cells without target gene selection. *Genes Dev.* 2: 1353–63

Johnson, R. S., Sheng, M., Greenberg, M. E., Kolodner, R. D., Papaioannou, V. E., Spiegelman, B. M. 1989. Targeting of nonexpressed genes in embryonic stem cells via homologous recombination. *Science* 245: 1234–36

Joyner, A. L., Herrup, K., Auerbach, B. A., Davis, C. A., Rossant, J. 1991. Subtle cerebellar phenotype in mice homozygous for a targeted deletion of the *En-2* homeobox. *Science* 251: 1239–43

Joyner, A. L., Skarnes, W. C., Rossant, J. 1989. Production of a mutation in mouse *En-2* gene by homologous recombination in embryonic stem cells. *Nature* 338: 153–56

Kelley, W. N., Wyngaarden, J. B. 1983. Clinical syndromes associated with hypoxanthine-guanine phosphorybosyltransferase deficiency. In *The Metabolic Basis of Inherited Disease*, ed. J. B. Stanbury, J. B. Wyngaarden, D. S., Frederickson, J. L., Goldstein, M. S. Brown, p. 2031. New York: McGraw-Hill

Kessel, M., Balling, R., Gruss, P. 1990. Variations of cervical vertebrae after expression of a *hox-1.1* transgene in mice. *Cell* 61: 301–8

Kim, H.-S., Smithies, O. 1988. Recombinant fragment assay for gene targetting based on the polymerase chain reaction. *Nucleic Acids Res.* 16: 8887–8903

Kittur, D., Shimizu, Y., DeMars, R., Edidin, M. 1987. Insulin binding to human B lymphoblasts is a function of HLA haplotype. *Proc. Natl. Acad. Sci. USA* 84: 1351–55

Koller, B. H., Marrack, P., Kappler, J. W., Smithies, O. 1990. Normal development of mice deficient in β2M, MHC class I proteins, and CD8+ T cells. *Science* 248: 1227–30

Koller, B. H., Smithies, O. 1989. Inactivating the β2-microglobulin locus in mouse embryonic stem cells by homologous recombination. *Proc. Natl. Acad. Sci. USA* 86: 8932–35

Kucherlapati, R. 1986. Homologous recombination in mammalian somatic cells. In *Gene Transfer*, ed. R. Kucherlapati, pp. 363–81. New York: Plenum

Kuehn, M. R., Bradley, A., Robertson, E. J., Evans, M. J. 1987. A potential animal model for Lesch-Nyhan syndrome through introduction of *HPRT* mutations into mice. *Nature* 326: 295–98

Lawrence, P. A., Johnston, P. 1984. On the role of the engrailed gene in the internal organs of Drosophila. *EMBO J.* 3: 2839–44

Lawrence, P. A., Struhl, G. 1982. Further studies of the engrailed phenotype in Drosophila. *EMBO J.* 1: 827–33

Lin, F.-L., Sperle, K., Sternberg, N. 1985. Recombination in mouse L cells between DNA introduced into cells and homologous chromosomal sequences. *Proc. Natl. Acad. Sci. USA* 82: 1391–95

Mansour, S. L., Thomas, K. R., Capecchi, M. R. 1988. Disruption of the proto-oncogene *int-2* in mouse embryo-derived stem cells: a general strategy for targeting mutations to nonselectable genes. *Nature* 336: 348–52

Mansour, S. L., Thomas, K. R., Deng, C., Capecchi, M. 1990. Introduction of a *lacZ* reporter gene into the mouse *int-2* locus by homologous recombination. *Proc. Natl. Acad. Sci. USA* 87: 7688–92

Martin, G. R. 1981. Isolation of a pluripotent cell line from early mouse embryos cultured in medium conditioned by teratocarcinoma stem cells. *Proc. Natl. Acad. Sci. USA* 78: 7634–38

Matis, L. A. 1990. The molecular basis of T-cell specificity. *Annu. Rev. Immunol.* 8: 65–82

McMahon, A. P., Bradley, A. 1990. The *Wnt-1* (*int-1*) proto-oncogene is required for development of a large region of the mouse brain. *Cell* 62: 1073–85

McMahon, J. A., McMahon, A. P. 1989. Nucleotide sequence, chromosomal localization and developmental expression of the mouse *int-1* related gene. *Development* 107: 643–51

Murnane, J. P., Yezzi, M. J., Young, B. R. 1989. Recombination events during integration of transfected DNA into normal human cells. *Nucleic Acids Res.* 18: 2733–38

Neumann, P. E., Mueller, G. G., Sidman, R. L. 1990. Identification and mapping of a mouse gene influencing cerebellar folial pattern. *Brain Res.* 524: 85–89

Nusse, R. 1988. The activation of cellular oncogenes by proviral insertions in murine mammary tumors. In *Breast Cancer:*

Cellular and Molecular Biology, ed. M. E. Lippman, R. Dikson, pp. 283–306. Boston: Kluwer Academic

Nusse, R., Brown, A., Papkoff, J., Scambler, P., Shackleford, G., et al. 1991. A new nomenclature for *int-1* and related genes: *Wnt* gene family. *Cell* 64: 231

Orkin, S. H. 1990. Globin gene regulation and switching: circa 1990. *Cell* 63: 665–72

Orr-Weaver, T. L., Szostak, J. W., Rothstein, R. J. 1981. Yeast transformation: A model system for study of recombination. *Proc. Natl. Acad. Sci. USA* 78: 6354–58

Papkoff, J., Schreyver, B. 1990. Secreted int-1 protein is associated with the cell surface. *Mol. Cell Biol.* 10: 2723–30

Pevny, L., Simon, M. C., Robertson, E., Klein, W. H., Tsai, S.-F., et al. 1991. Erythroid differentiation in chimeric mice blocked by a targeted mutation in the gene for transcription factor *GATA-1*. *Nature* 349: 257–60

Rathjen, P. D., Toth, S., Willis, A., Heath, J. K., Smith, A. G. 1990. Differentiation inhibiting activity is produced in matrix-associated and diffusible forms that are generated by alternate promotor usage. *Cell* 62: 1105–14

Robertson, E. J. 1987. Embryo-derived stem cell lines. In *Teratocarcinomas and Embryonic Stem Cells: A Practical Approach*, ed. E. J. Robertson, pp. 71–112. Oxford: IRL

Roelink, H., Wagenaar, E., Lopes da Silva, S., Nusse, R. 1990. *wnt-3*, a gene activated by proviral insertion in mouse mammary tumors, is homologous to *int-1/wnt-1* and is normally expressed in mouse embryos and adult brain. *Proc. Natl. Acad. Sci. USA* 87: 4519–23

Saiki, K. R., Gelfand, D. H., Stoffel, S., Scharf, S. J., Higuchi, R., et al. 1985. Enzymatic amplification of β-globin genomic sequences and restriction site analysis for diagnosis of sickle cell anemia. *Science* 230: 1350–54

Sara, V. R., Carlsson-Skwirut, C. 1988. The role of the insulin-like growth factors in the regulation of brain development. *Prog. Brain Res.* 73: 87–99

Schwartzberg, P. L., Robertson, E. J., Goff, S. P. 1990. Targeted gene disruption of the endogenous c-abl locus by homologous recombination with DNA encoding a selectable fusion protein. *Proc. Natl. Acad. Sci. USA* 87: 3210–14

Sedivy, J. M., Sharp, P. A. 1989. Positive genetic selection for gene disruption in mammalian cells by homologous recombination. *Proc. Natl. Acad. Sci. USA* 86: 227–31

Shackleford, G. M., Varmus, H. E. 1987. Expression of the proto-oncogene *int-1* is restricted to postmeiotic male germ cells

and the neural tube of mid-gastation embryos. *Cell* 50: 89–95

Siedmann, R. L. 1968. Development of interneuronal connections in brains of mutant mice. In *Physiological and Biochemical Aspects of Nervous Integration*, ed. F. D. Carlson, pp. 163–93. Englewood, NJ: Prentice-Hall

Singh, P. B., Brown, R. E., Roser, B. 1987. MHC antigens in urine as olfactory recognition cues. *Nature* 327: 161–64

Smith, A. G., Heath, J. K., Donaldson, D. D., Wong, G. G., Moreau, J., et al. 1988. Inhibition of pluripotential embryonic stem cell differentiation by purified polypeptides. *Nature* 336: 688–90

Smith, A. J. H., Berg, P. 1984. Homologous recombination between defective *neo* genes in mouse 3T6 cells. *Cold Spring Harbor Symp. Quant. Biol.* 49: 171–81

Smithies, O., Gregg, R. G., Boggs, S. S., Koralewski, M. A., Kucherlapati, R. S. 1985. Insertion of DNA sequences into the human chromosome *b-globin* locus by homologous recombination. *Nature* 317: 230–34

Smithies, O., Koralewski, M. A., Song, K.-Y., Kucherlapati, R. S. 1984. Homologous recombination with DNA introduced into mammalian cells. *Cold Spring Harbor Symp. Quant. Biol.* 49: 161–70

Song, K.-Y., Schwartz, F., Maeda, N., Smithies, O., Kucherlapati, R. 1987. Accurate modification of a chromosomal plasmid by homologous recombination in human cells. *Proc. Natl. Acad. Sci. USA* 84: 6820–24

Stanton, B. R., Reid, S. W., Parada, L. F. 1990. Germ line transmission of an inactive *N-myc* allele generated by homologous recombination in mouse embryonic stem cells. *Mol. Cell Biol.* 10: 6755–58

Stylianopoulou, F., Efstratiadis, A., Herbert, J., Pintar, J. 1988. Pattern of insulin-like growth factor II gene expression during rat embryogenesis. *Development* 103: 497–506

te Riele, H., Maandag, E. R., Clarke, A., Hooper, M., Berns, A. 1990. Consecutive inactivation of both alleles of the *pim-1* proto-oncogene by homologous recombination in embryonic stem cells. *Nature* 348: 649–51

Thomas, K. R., Capecchi, M. R. 1986. Introduction of homologous DNA sequences into mammalian cells induces mutations in the cognate gene. *Nature* 324: 34–38

Thomas, K. R., Capecchi, M. R. 1987. Site-directed mutagenesis by gene targeting in mouse embryo-derived stem cells. *Cell* 51: 503–12

Thomas, K. R., Capecchi, M. R. 1990. Targeted disruption of the murine *int-1* proto-oncogene resulting in severe abnormalities in midbrain and cerebellar development. *Nature* 346: 847–50

Thomas, K. R., Folger, K. R., Capecchi, M. R. 1986. High frequency targeting of genes to specific sites in the mammalian genome. *Cell* 44: 419–28

Thompson, S., Clarke, A. R., Pow, A. M., Hooper, M. L., Melton, D. W. 1989. Germ line transmission and expression of a corrected *HPRT* gene produced by gene targeting in embryonic stem cells. *Cell* 56: 313–21

Tsukamoto, A. S., Grosschedl, R., Guzman, R. C., Parslow, T., Varmus, H. E. 1988. Expression of the *int-1* gene in transgenic mice is associated with mammary gland hyperplasia and adenocarcinomas in male and female mice. *Cell* 55: 619–25

Ungerstedt, U. 1971. Postsynaptic super-sensitivity after 6-hydroxidopamine induced degeneration of the nigrostriatal dopamine system. *Acta Physiol. Scand.* 367: 69 ff

von Boehmer, H. 1988. The developmental biology of T lymphocytes. *Annu. Rev. Immunol.* 6: 309–26

Wilkinson, D. G., Bailes, J. A., McMahon, A. P. 1987. Expression of the proto-oncogene *int-1* is restricted to specific neural cells in the developing mouse embryo. *Cell* 50: 79–88

Williams, R. L., Hilton, D. J., Pease, S., Willson, T. A., Stewart, C. L., et al. 1988. Myeloid leukaemia inhibitory factor maintains the developmental potential of embryonic stem cells. *Nature* 336: 684–87

Wolgemuth, D. J., Behringer, R. R., Mostoller, M. P., Brinster, R. L.,

Palmiter, R. D. 1989. Transgenic mice overexpressing the mouse homeobox containing gene *Hox-1.4* exhibit abnormal gut development. *Nature* 337: 464–67

Yamazaki, K., Beauchamp, G. K., Kupniewski, D., Bard, J., Thomas, L., Boyse, E. A. 1988. Familial imprinting determines H-2 selective mating preference. *Science* 240: 1331–32

Zheng, H., Wilson, J. H. 1990. Gene targeting in normal and amplified cell lines. *Nature* 344: 170–73

Zijlstra, M., Bix, M., Simister, N. E., Loring, J. M., Raulet, D. H., Jaenisch, R. 1990. β2-Microglobulin deficient mice lack CD4$-$8$+$ cytolytic T cells. *Nature* 344: 742–46

Zijlstra, M., Li, E., Sajjadi, F., Subramani, S., Jaenisch, R. 1989. Germ-like transmission of a disrupted β2-microglobulin gene produced by homologous recombination in embryonic stem cells. *Nature* 342: 435–38

Zimmer, A., Gruss, P. 1989. Chimeric mice, produced with embryonal stem (ES) cells carrying a homeobox Hox-1.1 allele mutated by homologous recombination. *Nature* 338: 150–53

Zimmer, A., Wang, Z.-Q., Wagner, E. F., Gruss, P. 1990. Homologous recombination in ES cells as a means to generate mice with defined mutations. In *PNS Biomedical Symposia Series*, ed. H. Vogel. In press

Zinkernagel, R. M., Doherty, P. C. 1979. MHC-restricted cytotoxic T cells: Studies on the biological role of polymorphic major transplantation antigens determining T cell restriction-specificity and responsiveness. *Adv. Immunol.* 27: 51–117

Annu. Rev. Neurosci. 1992. 15:139–65

EXPRESSION OF A FAMILY OF POU-DOMAIN PROTEIN REGULATORY GENES DURING DEVELOPMENT OF THE CENTRAL NERVOUS SYSTEM

Maurice N. Treacy and Michael G. Rosenfeld

Eukaryotic Regulatory Biology Program, Howard Hughes Medical Institute, Cellular and Molecular Medicine, University of California, San Diego, California 92093-0648

KEY WORDS: gene regulation, transcription factors, homeo- and heterodimerization, development

INTRODUCTION

Understanding the molecular basis for the appearance of phenotypically distinct cell types with an organism is a central issue in development. This developmental program involves a precise spatial and temporal pattern of expression of certain genes. In *Drosophila*, the development of the body plan may be partly determined by homeotic genes, which encode defined homeodomains (Akam 1987; Anderson 1987; Ingham 1988). The importance of proteins that contain homeodomains in the normal body plan development of *Drosophila*, was identified on the basis of "homeotic" mutations in *Drosophila* (reviewed in Gehring 1987; Scott & Carroll 1987). The homeodomain alone seems to mediate sequence-specific binding to DNA, which allows homeodomain proteins to regulate or influence gene expression (Hoey & Levine 1988; Jaynes & O'Farrell 1989). Additionally, studies have demonstrated that some homeodomain proteins can influence

139

0147–006X/92/0301–0139$02.00

the expression of their own and other homeodomain genes (Beachy et al 1988), which implicates their involvement in a hierarchical development array of gene expression.

The recent identification of the POU-domain family, a subclass of homeodomain proteins based on cell-specific gene expression studies in mammals and analysis of a developmental mutant in nematodes, has provided new insights into the process of development of specialized cell phenotypes within an organism. Moreover, the discovery of multiple new POU-domain genes that coexpress in the developing brain in mammals (He et al 1989) raises the intriguing possibility that the phenotype maturation of groups of neurons may be partly regulated by combinatorial interactions of POU-domain proteins. This review summarizes the structural features of POU-domain proteins, the spatial profiles of both mammalian and *Drosophila* POU-domain protein expression, and the initial insights into the potential role of these novel proteins in neuronal development.

DISCOVERY OF A FAMILY OF POU-DOMAIN TRANSCRIPTION FACTORS

Analysis of an anterior pituitary-specific transcription factor that bound to related *cis*-active motifs in the rat prolactin and growth hormone genes, and the B cell-specific and ubiquitous octomer binding proteins, permitted the identification and cloning of cDNAs that encode the transcription factors Pit-1 (GHF-1), Oct-2 (OTF-2), and the universal octomer binding protein Oct-1 (OTF-1, NFIII) (Bodner et al 1988; Clerc et al 1988; Finney et al 1988; Ingraham et al 1988; Müller et al 1988; Scheidereit et al 1988; Sturm et al 1988). Pit-1 and Oct-2 were established to serve as transcription factors, which activate expression of appropriate fusion genes in heterologous cell types, respectively (Bodner et al 1988; Gerster et al 1990; Ingraham et al 1988; Müller et al 1988; Müller-Immerglück et al 1990). These three mammalian proteins and the gene encoding a regulator of cell fate in *C. elegans*, referred to as *unc*-86 (Finney et al 1988), share an extensively conserved domain of approximately 150 amino acids, referred to as the POU-domain (Herr et al 1988). Subsequently, numerous additional POU-domain genes in mammals, *Drosophila*, and *C. elegans* have been identified (Burglin et al 1989; He et al 1989; Johnson & Hirsh 1990; Monuki et al 1990; Okamoto et al 1990; Rosner et al 1990; Schöler et al 1990; Suzuki et al 1990) (see Figure 1), which indicates the existence of a large gene family. Post-transcriptional events generate further heterogeneity of POU-domain proteins, as exemplified by the 79 bp insertion into the 3' end of the Oct-2 transcript that results in an altered C-terminus,

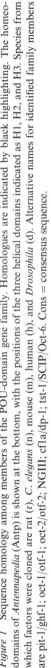

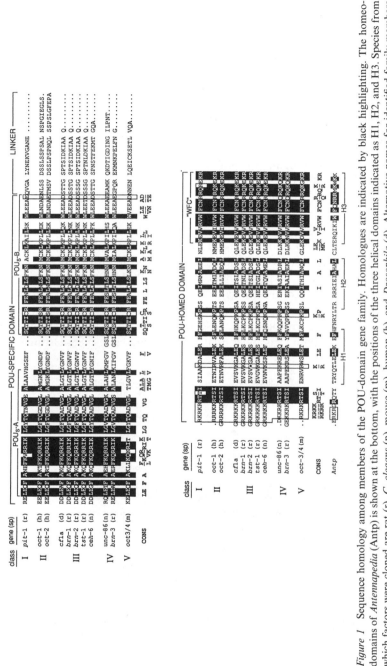

Figure 1 Sequence homology among members of the POU-domain gene family. Homologues are indicated by black highlighting. The homeo-domains of *Antennapedia* (Antp) is shown at the bottom, with the positions of the three helical domains indicated as H1, H2, and H3. Species from which factors were cloned are rat (r), *C. elegans* (n), mouse (m), human (h), and *Drosophila* (d). Alternative names for identified family members are pit-1/ghf-1; oct-1/otf-1; oct-2/otf-2; NGIII; cf1a/dp-1; tst-1/SCIP/Oct-6. Cons = consensus sequence.

in the *oct*-2a and *oct*-2b proteins (Hatzopoulos et al 1990; Schreiber et al 1988).

STRUCTURE OF THE POU-DOMAIN

Based on the sequences of the identified POU-domain proteins, the POU-domain can vary from 147–156 amino acids in length and contains two major regions of similarity: POU-specific (POU_S) and the POU-homeo-domain (POU_{HD}) (Figure 1). The POU_S domain (69–71 to 76–78 amino acids dependent on the placement of the N′-terminal boundary, which does not correspond to an intron-exon junction), which is the most similar region, contains 48 conserved residues. This region contains two distinct regions of particularly high similarity: the POU_S-A region (23 of 36 residues conserved) and the POU_S-B region (25 of 33 residues conserved) (see Figure 1). The POU_S domain is separated by a poorly conserved spacer region, 14–25 amino acids in length, from the variant POU_{HD}, which is clearly related to the classic homeodomains found in *Drosophila* (Scott et al 1989) The POU_{HD} is highly related among POU-domain proteins; it contains 32 conserved amino acids in the 60 amino acid homeodomain (see Figure 1). The POU_{HD} is predicted to contain three helices, diagramed in Figure 2, that correspond to those now established in the case of the classic homeodomain (Kissinger et al 1990; Otting et al 1988; Otting et al

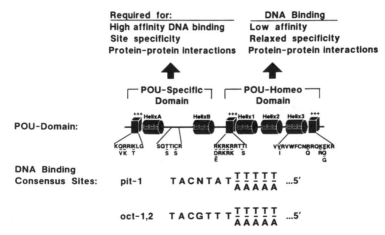

Figure 2 Schematic representation of the POU-domain, which indicates predicted helical domains, and the most highly conserved sequences. The known functions of the POU_S and POU_{HD} are listed above. These probes bind A/T-rich sequences, with consensus sequences for *pit*-1 and *oct*-1, *oct*-2 shown below.

1990). The region encompassing the third "recognition" helix is strikingly conserved; it contains the RVWFCN residues among all members of the POU-domain gene family. The N'- and C'-terminal boundaries of the POU_{HD} contain clusters of basic amino acid residues (5–6 of 8 and 4–5 of 7 conserved residues, respectively), which are critical for their function as DNA binding and transcription factors (see below). The POU_S domain is also bounded at its N'-terminus by a cluster of basic amino acids (4–5 of 7 conserved residues).

Based on the primary sequence throughout the POU domain, including the basic amino acid cluster at the N'-terminus of the POU_{HD} and patterns of highest conservation in the linker region that separates the POU_S and POU_{HD} domains, the POU-domain proteins may be classified into five groups (POU-I to POU-V) (Figure 1). Within each class, such as the large POU-III group of factors, there is significant conservation within the spacer region, even between species as diverged as *C. elegans* and mammals (Figure 1). The spacer region of Oct-1 is functionally capable of accommodating additional residues (Sturm & Herr 1988). Outside the POU-domain, there is enormous divergence between all known family members. However, many of the POU-domain gene family contain regions rich in specific amino acids, including serine-threonine-rich (Pit-1, Oct-1, Oct-2), glutamine-rich (Brn-2, Oct-1, Oct-2), and glycine-rich (Tst-1/SCIP/Oct-6) domains, which suggests a functional role for these regions.

EXPRESSION OF A LARGE FAMILY OF POU-DOMAIN REGULATORY GENES IN BRAIN DEVELOPMENT

The establishment of many neuronal phenotypes and the accompanying complex pattern of neuronal interconnections, so characteristic of brain development, involves an intricate program of gene expression. Although many of these nuclear events remain largely unknown, the precise temporal and spatial patterns characteristic of mammalian brain development probably reflect sequential activation of a complex network of regulatory factors that are similar to those presumed to establish structural patterns in *Drosophila* (Gehring 1987; Scott & Carroll 1987).

The possibility that novel POU-domain proteins are expressed with distinct spatial and temporal patterns during establishment of the nervous system, thus potentially exerting some roles in specifying neuronal phenotypes, was investigated by using a strategy based on the structure of the initial four members of the POU-domain gene family. Degenerate oligonucleotides, based on two highly conserved amino acid sequences in the A region of the POU-specific domain and in the C-terminal portion of the POU homeodomain, were used as primers for the polymerase chain reaction, which employs DNA complementary to mammalian brain and testes mRNAs and adult *Drosophila* mRNAs as templates. Six new members of the POU-domain gene family were identified: three from brain cDNA (Brn-1, Brn-2, Brn-3), one from rat testes cDNA (Tst-1/SCIP/Oct-6), and two from *Drosophila* cDNA (I-POU, DP-1). Predicted amino acid residues for each new member are shown in Figure 1.

Structural comparisons of the six new POU-domain proteins revealed that they are all highly related to Pit-1, Oct-1, Oct-2, and Unc-86. These 12 proteins constitute a distinct POU-domain protein family. Four of the new proteins (Brn-1, Brn-2, Tst-1, Cf1-a) are highly similar; >94% of the amino acid residues are identical among them throughout the entire POU-domain. Even the variable region between the POU-specific domain and the POU-homeodomain is well conserved among these proteins. I-POU and Brn-3 are highly related to *unc*-86, including three characteristic additional amino acid residues in the region between the A and B portions of the POU-specific domain. Pit-1 is distinct from the other 11 proteins in the POU-domain family at several amino acid positions that otherwise would be totally conserved; thus, Pit-1 is the most divergent member of the family. The known POU-domain proteins appear to segregate into five classes, which are arbitrarily referred to as POU-I (Pit-1), POU-II (Oct-1, Oct-2), POU-III (Brn-1, Brn-2, Tst-1, Cf1-a), POU-IV (Brn-3, Unc-86, I-POU), and POU-V (Oct-3/4). The N′ terminal basic part of the POU-homeodomain is identical for the first 11 or 12 amino acids, and 15–18 amino acids in the C-terminal "WFC" region (see Figure 1) are particularly well-conserved between all members of each class. A consensus sequence for the POU-domain emphasizes the high degree of primary amino acid sequence conservation among all 12 family members. Based upon the precedents of classical homeodomain proteins, many related genes will probably encode for proteins that involve residues that are substituted for the C′ residue in the WFC sequence.

TEMPORAL AND SPATIAL EXPRESSION PROFILES

Novel Mammalian POU-Domain Protein Genes

Hybridization histochemistry was employed to determine whether the transcripts of the four novel mammalian POU-domain protein genes exhibited either widespread or restricted patterns of expression during development of the central nervous system (CNS) and in the mature brain. Representative examples of this analysis are shown in Figure 3, and a summary of the complete adult survey is presented in Table 1. Brn-1 and

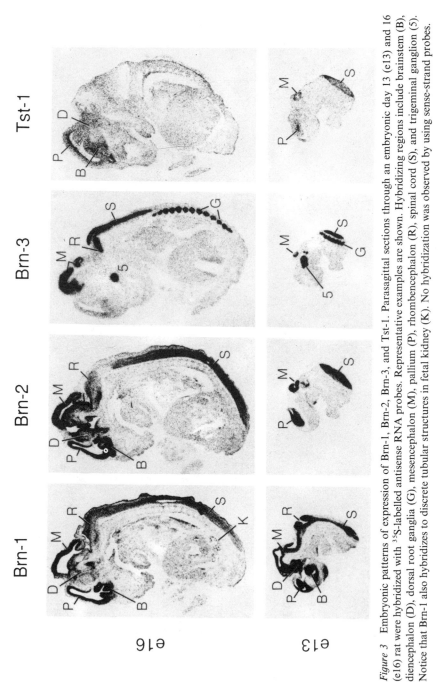

Figure 3 Embryonic patterns of expression of Brn-1, Brn-2, Brn-3, and Tst-1. Parasagittal sections through an embryonic day 13 (e13) and 16 (e16) rat were hybridized with ^{35}S-labelled antisense RNA probes. Representative examples are shown. Hybridizing regions include brainstem (B), diencephalon (D), dorsal root ganglia (G), mesencephalon (M), pallium (P), rhombencephalon (R), spinal cord (S), and trigeminal ganglion (5). Notice that Brn-1 also hybridizes to discrete tubular structures in fetal kidney (K). No hybridization was observed by using sense-strand probes.

Table 1 Distribution of POU-domain mRNA expression in the adult rat nervous system determined by in situ hybridization

	Pit-1	Brn-1	Brn-2	Brn-3	Tst-1	Oct-1	Oct-2
SENSORY GANGLIA	—	—	—	++++	—	—	—
CORTEX							
isocortex, layers 2–5	—	+	+	—	—	—	—
layers 5–6	—	—	—	—	+	—	—
olfactory bulb, mitral	—	+	+	—	+	—	—
periglomerular	—	+	+	—	—	—	—
islands of Calleja	—	+++	++	—	+	—	—
piriform	—	++	+	—	—	—	—
presubiculum, layer 2	—	++	++	—	—	—	—
CA1/subiculum	—	++	++	—	+++	—	—
dentate gyrus	—	+	+	—	—	—	—
subependymal zone	—	++	++	—	—	—	—
AMYGDALA							
n. lateral olfactory tract	—	+	+	—	++	—	—
SEPTUM							
medial n./n. diagonal band	—	+	+	—	—	—	—
bed n. stria terminalis	—	+	+	—	—	—	—
BASAL GANGLIA							
striatum	—	—	—	—	+	—	—
substantia innominata	—	++	++	—	—	—	—
s. nigra/ventral tegmental a.	—	+	++	—	+	—	—
THALAMUS							
medial habenula	—	+++	++	+++	++	++	—
lateral habenula	—	—	—	++	—	—	—
parafascicular n.	—	+	+	—	—	—	—
reticular n.	—	++	++	—	—	—	—
zona incerta	—	+	+	—	—	—	—

HYPOTHALAMUS								
suprachiasmatic n.	—	—	—	—	—	—	—	—
supraoptic/paraventricular n.	++	+++	—	+++	—	—	—	++
medial preoptic n.	++	+++	—	+	—	—	—	—
tuberomammillary n.	++	+++	—	+	—	—	—	—
lateral mammillary n.	—	+	—	—	—	—	—	—
medial mammillary n.	+	—	+	+	++	—	+	++
posterior hypothalamic a.	+	+	++	+	—	—	—	—
BRAINSTEM—SENSORY								
superior colliculus, inter. gray	++	+++	+	+	++	—	—	—
parabigeminal n.	++	+++	—	—	+++	—	—	—
superior olive	—	—	+	+	—	—	—	—
inferior colliculus, dorsal n.	++	+	++	++	—	—	—	—
area postrema	+	++	+	—	—	+	—	—
BRAINSTEM—MOTOR								
motor n. trigeminal	++	+++	+	—	—	—	—	—
facial n.	++	+++	+	—	—	—	—	—
hypoglossal n.	+	+	—	—	—	—	—	—
n. ambiguus	—	—	++	++	++	—	—	—
dorsal motor n. vagus	—	—	+	—	—	—	—	—
BRAINSTEM—CORE								
periaqueductal gray	++	+	+	+	+	—	—	—
n. incertus	++	++	—	—	+++	—	—	—
interpeduncular n.	—	—	+	+	—	—	—	—
CEREBELLUM & RELATED								
deep nuclei	++	+	—	—	+	—	—	—
Purkinje	++	++	++	++	+++	—	—	—
granule	—	—	+	—	+	+++	+++	++
red n.	—	—	—	—	+	+	—	+
inferior olive	+	+	+	—	++++	—	—	—
SPINAL CORD								
ventral horn	+	+	—	—	+	—	—	—

Relative density of in situ hybridization indicated as follows: —, not detected; +, low; ++, moderate; +++, high; ++++, very high. Sense-strand control hybridization was (—) for all regions indicated.

Brn-2 exhibited virtually identical patterns of expression in the CNS; however, Brn-1 was clearly expressed in the medullary zone of the kidney, whereas Brn-2 was not. Hybridization for Brn-1 and Brn-2, which are the most widely expressed members of the POU-domain family examined thus far, was found in at least some classes of neurons at all levels of the neuraxis (see below). Notably, almost all regions of the cerebral (layers II–V) and cerebellar (Purkinje cells) cortices specifically hybridized, as did the basal forebrain cholinergic system, ventral midbrain dopamine system, paraventricular and supraoptic neuroendocrine system, somatic moto-neurons in the cranial nerve nuclei and ventral horn, and tectum.

In contrast, the Brn-3 gene transcript exhibited a more restricted pattern; dense hybridization was limited to the habenula, posterior hypothalamic area, inferior olive, inferior colliculus, and nucleus amibiguus. Further-more, it was the only clone examined that hybridized to sensory ganglion cells (in the trigeminal and dorsal root ganglia, which are derived from the neural crest and in the spiral ganglion). The Tst-1 gene was expressed in the cerebral (layers V–VI) and cerebellar (granule cells) cortices, as well as in the medial habenula, superior colliculus and parabigeminal nucleus, and dorsal motor nucleus of the vagus. Tst-1, also referred to as SCIP or Oct-6, is transiently expressed in myelinating glia. Thus, each gene ex-hibited a unique, restricted pattern of expression.

The developmental patterns of expression of the four new mammalian POU-domain proteins were examined from rat embryonic day 8 through birth (see Figure 3). All four were widely expressed in all levels of the neural tube (including the retina) during at least some part of this period. Hybridization in the ventricular (proliferative) zone of the neuro-epithelium, which gives rise to neurons in CNS, was evident for all four probes at all levels. However, the intensity of hybridization signals and the time course of this anatomic restriction in the developing neural tube for Brn-1 and Brn-2 together, Brn-3, and Tst-1 were distinct. The patterns tended to reflect the adult loci of expression. Clear hybridization was also observed in the mantle layer, the early cortical plate (embryonic day 18 for Brn-1, Brn-2, Tst-1), and the external granular layer of the cerebellum (Tst-1). These data indicate that the genes are expressed during the migratory phase of at least some types of young neurons. RNase protection analyses confirmed the temporal expression pattern of these cloned genes.

Neuronal Expression of Pit-1, Oct-1, and Oct-2 Transcripts

In contrast to the four new mammalian POU-domain family members that are expressed in the neural tube during development and in the mature brain, it was assumed that Pit-1 and Oct-2 are expressed exclusively in nonneuronal tissues, the anterior pituitary, and lymphoid B cells, respectively. Unexpectedly, Pit-1 transcripts were detected in the neural plate and tube on embryonic days 10–13 and disappeared thereafter, as assessed by in situ hybridization and RNase protection assays (Figure 4A, and data not shown). On embryonic day 16, Pit-1 transcripts reappeared and were expressed exclusively in the developing anterior lobe of the pituitary gland; thereafter, they could not be detected in any brain region. Thus, Pit-1 exhibits a biphasic pattern of expression, which is reminiscent of several *Drosophila* segmentation and homeodomain genes, such as *fushi-tarazu* (Ftz) and *engrailed* (en). These genes are expressed early in development and play a critical role in pattern formation and later reappear in a limited number of neurons (Doe et al 1988; Patel et al 1989).

An unexpected result for the expression patterns of Oct-2 was also observed: Oct-2 transcripts were widely expressed in the developing neural tube, including the diencephalon, brainstem, spinal cord, and sensory ganglia (Figure 4B). Unlike Pit-1, Oct-2 transcripts were found in the brain of the adult animal; expression was limited to the medial mammillary nucleus, or suprachiasmatic nucleus, which is involved in generating circadian rhythms, cerebellar granule cells, and red nucleus (Table 1).

Sturm et al (1988) have suggested that Oct-1 is ubiquitously expressed. Thus, based on the expression of Pit-1 and Oct-2, it was interesting to determine whether Oct-1 was also expressed during neurogenesis. The Oct-1 gene was indeed widely expressed in the neural tube (embryonic day 13), but at much lower levels than any of the other POU-domain genes (Figure 4C). In adult brain, Oct-1 expression was highly restricted to cerebellar granule cells and to the medial habenula, medial mammillary nucleus, and area postrema (Table 1). Oct-1 transcripts were also detected in thymus, lymph nodes, spleen, thyroid, ovary, testes, and mammary tissue, but not in kidney, heart, liver, or salivary gland (data not shown). This pattern of expression is much more widespread than that observed for any other member of the family.

The potential early expression of the POU-domain gene family before induction of the neural plate was examined on embryonic day 8. Oct-1 and Tst-1 were highly expressed in the egg cylinder, whereas weaker hybridization signals were detected for Brn-1 and Brn-2. In contrast, Pit-1, Oct-2, and Brn-3 transcripts were not detected at this stage.

When the adult pattern of neuronal expression of the six mammalian POU-domain genes (all but Pit-1) are compared (Table 1), several intriguing features emerge. First, none of the patterns for individual members of the family corresponds entirely to any known topographical division or functional system, to the distribution of any known neurotransmitter or receptor, or to any one cell type. Second, some regions of the nervous system appear to express only one of the members examined

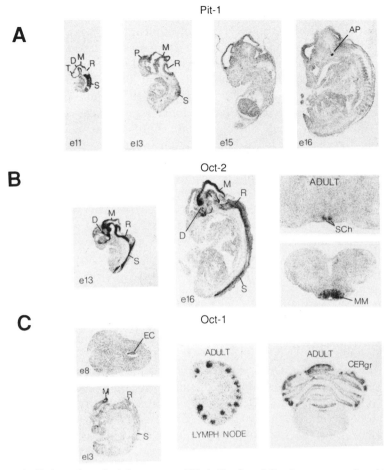

Figure 4   Embryonic and adult patterns of Pit-1, Oct-2, and Oct-1 gene expression. (*A*) In situ hybridization of Pit-1 transcripts. Pit-1 transcripts were present in the developing neural tube. In diencephalon (D), mesencephalon (M), Rhombencephalon (R), spinal cord (S), and telencephalon (T). Day e13 shows trace hybridization to indicated areas; no hybridization was observed on e15. On e16 (and e18, not shown), hybridization reappeared exclusively in the anterior pituitary gland (AP). (*B*) In situ hybridization of Oct-2 transcripts showed that on e13 and e16, Oct-2 transcripts are present in diencephalon (D), mesencephalon (M), rhombencephalon (R), and spinal cord (S). In adult rat brain, Oct-2 transcripts were restricted to a few regions, including the suprachiasmatic (SCh) and medial mammillary (MM) nuclei. (*C*) In situ hybridization of Oct-1 transcripts revealed that Oct-1 was present in the egg cylinder (EC, day e8). By e13, trace hybridization was observed in mesencephalon (M), rhombencephalon (R), and spinal cord (S). Hybridization was shown in adult cerebellum in the granule cell layer (CERgr) and in the lymph nodes.

here, whereas other regions clearly express multiple members of the family. In some instances, different cell types in a particular region express different members of the family. The clearest example of this is in the cerebellum, in which Brn-1 and Brn-2 are expressed in Purkinje cells, whereas Tst-1, Oct-1, and Oct-2 are expressed in granule cells. Only one of the known POU-domain proteins, Oct-2, is expressed in the suprachiasmatic nucleus; Brn-1, Brn-2, Brn-3, Tst-1, and Oct-1 are all expressed in the medial habenula. Therefore, neurons in a region like the medial habenula apparently express as many as five of the eight mammalian POU-domain genes identified thus far; however, there are many brain regions in which no expression of any known POU-domain protein is detected. An important aspect of the spatial and temporal expression profile presented for the mammalian POU-domain proteins is that, unlike classic Hox genes, they are widely expressed in the developing and mature forebrain and midbrain (see Figure 5). Thus, in addition to their wide expression in the neural tube, these proteins may play a crucial role in the development of neuronal phenotypes in these brain regions. Additionally, the transient expression of Pit-1 in the neural tube and continuous expression of Oct-1 in neural tube and adult brain, coupled with their well-characterized transcription activation function in the mature pituitary (Pit-1) and B lymphocytes (Oct-2), indicates that all members of the POU-domain gene family exert their effects by modulating specific patterns of gene transcription.

Drosophila POU-Domain Protein Genes

In situ hybridization analyses of the two *Drosophila* POU-domain proteins genes (I-POU, Cf1-a) indicated that I-POU transcripts were first detectable following germ-band shortening (stage 13; as defined by Campos-Ortega & Hartenstein 1985) and were localized specifically in a subset of neurons in the supraesophageal ganglia and the ventral nerve cord. No other tissue stained positive for I-POU, which indicates a strictly neuron-specific

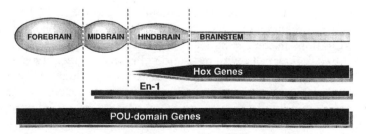

Figure 5 Spatial expression profiles of Hox and POU-domain genes. Notice that all Hox genes (except En-1) are restricted out of the fore and midbrain.

pattern of expression. This is reminiscent of the mammalian brain-specific POU-domain protein Brn-3, which shows high homology to I-POU.

The expression profile for Cf1-a was more complex. Initially, transcripts were detected in the ectoderm of stage 9 embryos (third h of embryogenesis) at regular repeating intervals, which is consistent with a segmentation pattern. However, this staining was later colocalized with periodic concavities in the laterodorsal region of the prospective dorsal epidermis in invaginations that are the origins of the 11 tracheal placodes. At stage 12/13, when the pits have lost their contact to the outside and begin to form lateral processes that eventually fuse to create the tracheal tree, expression of Cf1-a extended to the epidermis in positions between the pits. This intense epidermal staining, which did not involve gnathal segments, was not maintained and was completely absent in a stage 15/16 embryo. Cf1-a expression also appeared in a longitudinal strip of cells along the ventral midline of the embryo. These cells correspond to the mesectoderm that generates a mixed population of neurons and glia (Crews et al 1988; Jacobs et al 1984; Nambu et al 1990; Thomas et al 1988), which indicates that Cf1-a might exert an important early role in specifying a particular neuronal fate. Once the ventral nerve cord and supraesophageal ganglia were formed, Cf1-a was expressed within a subset of neuronal cells. Examination of these staining patterns revealed that Cf1-a and I-POU were coexpressed in a highly overlapping subset of cells within the CNS (Treacy et al 1991). This region-specific expression of many POU-domain proteins in the CNS indicates that conserved characteristics of structure and tissue distribution extend across several evolutionary barriers, which implies a fundamental requirement for these genes during neuronal development.

DEVELOPMENTAL FUNCTIONS OF POU-DOMAIN TRANSCRIPTION FACTORS

C. elegans *Development*

Genetic analysis of *C. elegans* development revealed that a POU-domain gene, *unc*-86, is required for cell-fate commitment in several neuroblast lineages. Mutations in the *unc*-86 locus prevent mother cells from differentiating into daughter cells, with restriction of maternal phenotype (Chalfie et al 1981; Finney et al 1988). In each case, the daughter cells continue to exhibit the phenotype of mother cells; therefore, specific types of neurons fail to appear. Failure to express *unc*-86 may lead to either inappropriate cell death or appearance of neurons that fail to serve specific functions. Recent analyses (Finney & Ruvkun 1990) have revealed that expression of *unc*-86 protein appears in the nuclei of cells affected by the *unc*-86 mutations several minutes after cell division, which is consistent with the proposed role of *unc*-86 in modulating the pattern of gene expression that phenotypically distinguishes daughter cells from mother cells. The molecular basis for this asymmetric activation of *unc*-86 remains unknown, although it is not dependent upon cell-cell interactions. Interestingly, the *unc*-86 protein is also expressed in several neuronal types that are not linked by any known common marker. These neuronal types do not disappear with genetic mutants of the *unc*-86 locus (Finney & Ruvkun 1990). It seems likely that *unc*-86 directs the patterns of gene expression in these cells by modulating their mature phenotype. The homeodomain gene *mec*-3, which is important in establishing specific neuronal phenotypes (Chalfie & An 1989), depends on *unc*-86 function for its expression. However, the distribution of the encoded proteins indicates that expression of *unc*-86 is itself not sufficient to activate *mec*-3. Thus, the products of *unc*-86 and *mec*-3 act within a regulatory hierarchy that specifies the same differentiation pathway for the same set of sensory neurons.

Mammalian Development

Although the precise roles of POU domain proteins in neuronal development remain to be established, the resolution of the developmental role of Pit-1 in mammalian anterior pituitary gland development provides insights into strategies that will likely be prototypic of those characterized in the CNS. Expression of Pit-1 is confined to three cell types in the anterior pituitary gland, which are defined on the basis of the trophic factor elaborated and referred to as thyrotrophs (express TSH-β), lactotrophs (express prolactin), and somatotrophs (express growth hormone) (Simmons et al 1990). However, Pit-1 gene expression in thyrotrophs occurs well after this cell type has initially expressed the TSH-β gene (Simmons et al 1990). Pit-1 transcripts and protein are initially expressed in the rat anterior pituitary gland immediately preceding the initial appearance of prolactin and growth hormone transcripts on e16-17 (Dollé et al 1990; Simmons et al 1990). Pit-1 has been independently confirmed to be capable of transactivating both the prolactin and growth hormone promoters (Fox et al 1990; Ingraham et al 1988; Larkin et al 1990; Mangalam et al 1989; Sharp & Cao 1990).

Genetically transmitted dwarfism in mice, characterized as a single autosomal recessive mutation on chromosome 16 (Eicher & Beamer 1980; Snell 1929) produces no detectable growth hormone, prolactin, or TSH; mature somatotroph, lactotroph, and thyrotroph cell types are depleted in these dwarfs (Roux et al 1982; Wilson & Wyatt 1986). Both allelic mutations involve disruptions in the Pit-1 gene: The Jackson dwarf results in a restriction fragment–length polymorphism (RFLP) in the Pit-1 gene; the

Snell dwarf mutation involves a G to T transversion that converts the tryptophan residue in the WFC homology in the POU_{HD} to a cysteine residue (WFC → CFC), which generates a mutant protein that is incapable of binding to its recognition elements (Li et al 1990). Thus, Pit-1 is required for the specification of three of the five cell types in the anterior pituitary gland, including the somatotroph and lactotroph phenotypes. The ultimate selective expression of the prolactin and growth hormone genes in lactotrophs and somatotrophs, respectively, and the results of transgenic analyses (Crenshaw et al 1989; Lira et al 1988) indicate that additional activating and restricting mechanisms must be required for the physiologic, quantitative expression of these genes in distinct cell types.

The hypoplastic nature of the genetic dwarf pituitary indicates that cell proliferation and/or survival are important components of the program specified by Pit-1 and that a POU-domain protein can exert direct or indirect roles in DNA replication. Such roles are consistent with the potential functions of Oct-1 in replication events in vitro.

Combinatorial codes for both *unc*-86 and Pit-1 appear to be required for the quantitative and qualitative pattern of target gene expression. Thus, differential expression of the Pit-1-dependent prolactin and growth hormone target genes in distinct cell types and *unc*-86-dependent expression of mec-3 in specific neurons in *C. elegans* must both reflect the actions of additional activating and restricting factors. In specifying the thyrotroph cell type initially present before detectable *pit*-1 expression, Pit-1 could potentially be required for survival and proliferation of a specific cell type and may exert functions comparable to the role of *unc*-86 in determination of neuronal lineages and programed cell death. These data link POU-domain transcription factors to the processes of proliferation of specific cell types and progression and commitment events in organogenesis.

The Role of POU-Domain Proteins: Tst-1, Brn-2, and I-POU

Specific functions of neuronally expressed POU-domain proteins have been explored in the case of Tst-1, Brn-2, and the *Drosophila* neural-specific POU-domain protein, I-POU. Tst-1 (SCIP/Oct-6), a member of the POU-III class family, is expressed in specific neurons and in myelinating glia in the mammalian nervous system (He et al 1991; Monuki et al 1990). Maximum expression levels are evident at the onset of myelination in glial cells, which is consistent with a Tst-1 involvement in certain aspects of the myelinating process. Consistent with known functions of POU-domain transcription factors (e.g. Pit-1, Oct-2), Tst-1 appears to be capable of functioning as a sequence-specific DNA-binding transcription regulator

(He et al 1991; Monuki et al 1990). The regulated expression of SCIP/Tst-1 by cAMP in myelinating glia (Monuki et al 1990) inferred potential gene-specific functions. It is of particular interest whether this protein could regulate a member of the immunoglobulin gene superfamily, the Po gene, which is a major component of the Schwann cell-myelin sheath and a cell-surface adhesion molecule (Filbin et al 1990). Tst-1/SCIP/Oct-6 protein binds specifically to five sites within the promoter of the Po gene and in cotransfection assays; Tst-1 specifically repressed the Po promoter activity (see Figure 6). These data indicate that, in concert with members of other

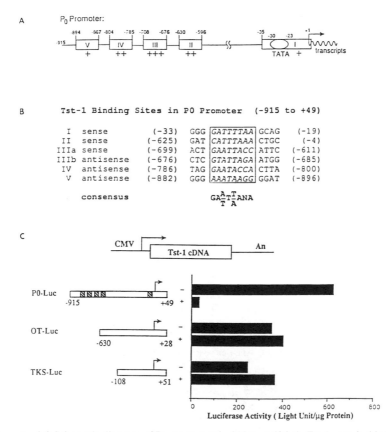

Figure 6   (*A*) Schematic diagram of Po promoter (−915 to +49 bp). Tst-1 protein-binding sites are boxed. The relative affinity of each site in the DNase I footprinting analysis is indicated by a + symbol; + + + is the highest. (*B*) Nucleotide sequence of Tst-1 binding sites and derived Tst-1-binding consensus sequence. (*C*) Cotransfection in CV-1 cells. Tst-1 cDNA was cloned in cytomegalovirus plasmids in both sense (+) and antisense (−) orientations. Reporter plasmids were the luciferase gene driven by the promoters shown in the figure. OY = oxytocin; TK = thymadine kinase.

families of transcriptional regulators, a single POU-domain protein can exert negative transcriptional effects dependent upon promoter and cellular contexts.

Brn-2, a highly conserved POU-domain protein, is expressed in the rat hypothalamus. Brn-2 transcripts are expressed in the dorsal medial paracellular part of the paraventricular hypothalamus, which synthesizes and secretes corticotrophin releasing hormone (CRH) (Swanson & Simmons 1989). Corticotrophin releasing hormone regulates release of adrenocorticotrophin and other products of the proopiomelanocortin gene from corticotrophs in the anterior pituitary and serves as a major mediator of the stress response (Vale et al 1983). By DNase I footprinting analysis, Brn-2 protein was shown to bind selectively to five sites within the CRH promoter. A consensus for these binding sites is described as (A/G)ATAAT(T/C); in cotransfection assays, Brn-2 activates transcription from the CRH promoter 40-fold (X. He and M. N. Treacy, unpublished data). These Brn-2 and Tst-1 data clearly establish that mammalian POU-domain genes expressed in the nervous system encode functional transcription factors that can selectively and alternatively regulate expression of distinct classes of genes found in neurons and glia.

A further, novel type of developmental regulation can be imposed by specific binding of the POU-domain gene family, as demonstrated by analysis of the *Drosophila* POU-domain protein, I-POU. This protein lacks two basic residues in the N′-terminal region of the POU$_{HD}$ (see Figure 1) and possesses no intrinsic DNA-binding activity. I-POU is speculated to form a heterodimeric complex with a second *Drosophila* POU-protein (Cf1-a) and inhibit its ability to bind to specific cognate DNA elements, thereby inhibiting expression of a subset of neuronally expressed genes (Treacy et al 1991) (see Figure 7).

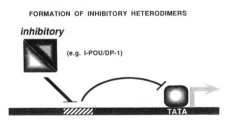

FORMATION OF INHIBITORY HETERODIMERS

inhibitory

(e.g. I-POU/DP-1)

TATA

Figure 7   Schematic of the proposed heterodimeric interaction of I-POI with DP-1, which results in the inability of DP-1 to bind site-specificity to DNA. The activating transcriptional activity of DP-1 is thus prevented.

FUNCTIONS OF THE POU SPECIFIC AND POU-HOMEODOMAIN

The large family of POU-domain genes expressed in the mammalian forebrain, to the exclusion of all Hox genes, raised the question of the important role of the POU-specific domain. Based on a series of mutagenic studies involving Pit-1 and Oct-1, both the POU_S and POU_{HD} are apparently combinatorially required to permit high affinity, site-specific DNA binding. In the case of Pit-1, the POU_{HD} is sufficient for low affinity binding to A/T-rich DNA sequences with a relaxed specificity; however, the POU_S domain is additionally required for high-affinity binding, thus increasing affinity for specific physiologic Pit-1 response elements up to 1000-fold (Ingraham et al 1990). The POU_S domain is similarly critical for high affinity binding of Oct-1 to its recognition elements (Sturm & Herr 1988; Verrijzer et al 1990a). Additionally, each of the three basic amino acid clusters at the N'-terminus of the POU_S domain and both N'- and C'-termini of the POU_{HD} serve critical functions in high affinity DNA binding by POU-domain proteins, as mutations in any of these regions abolish high affinity binding (Ingraham et al 1990; Sturm & Herr 1988; Treacy et al 1991). The Oct-3/4 gene product differs uniquely in a single amino acid in the basic region in the POU_S domain (KRITLG vs. RRIKLG), yet is capable of binding and stimulating promoters containing an octomer binding site (Rosner et al 1990; Schöler et al 1990; Suzuki et al 1990). Creating POU-domain chimeras between Oct-1 and Pit-1 POU_S and POU_{HD} revealed that both POU_S and POU_{HD} exert critical functions in discriminating octomer and Pit-1 binding sites (Ingraham et al 1990; Tanaka & Herr 1990). The invariant cysteine residue sequence in the POU_{HD} "recognition helix" is apparently not responsible for site-specificity of recognition, as has been suggested for several DNA sites in the case of the comparable sequence in classic homeodomain proteins (Hanes & Brent 1989; Treisman et al 1989). This suggestion was raised because POU-domain proteins recognize related, yet distinct, DNA sequences, and mutations in this residue fail to effect Pit-1 binding activity (Elsholtz et al 1990; Ingraham et al 1990). Disruption of the predicted first helix (helix A, Figure 2) in the POU_S domain or helix 3 (WFC) in the POU_{HD} abolishes DNA binding. However, disruption of the other putative helical structures in the POU_S and POU_{HD} does not effect binding (Figure 2), which contrasts with the proposed hydrophobic helical interactions between the engrailed homeodomain helix 1, helix 2 required to present the N'-terminal arm for minor groove contacts important for DNA binding (Kissinger et al 1990). This analysis of the α-helical domains and conserved structures, and the contacts of POU-domain on cognate sites, suggests that POU-domain proteins interact with their DNA recognition sites quite differently than homeodomain proteins, as both the POU_S and the POU_{HD} contact DNA. The isolated Pit-1 homeodomain binds an entirely distinct class of recognition sites compared with the full POU-domain, which is indicative of a POU_S involvement in binding. However, the contact of the third "recognition" helix in the POU_{HD} and the minor groove contacts involving residues in the amino terminus of the POU_{HD} will likely prove to be highly similar to those observed in classic homeodomain proteins (Kissinger et al 1990; Otting et al 1990). Although the DNA binding sites of classic homeodomain proteins are generally A/T-rich sequences, the best described sites for POU-domain proteins for Oct-2, Oct-1, and Pit-1 are variants of $(A/T)_{6-12}$TTTGCAT or $(A/T)_{6-12}$TATNCAT, respectively (Elsholtz et al 1990; Nelson et al 1988; Wirth et al 1987; and Figure 2). Although additional data are clearly required, the POU_S and POU_{HD} domains appear to interact with the TATNCA and A/T-rich sequences, respectively, in the binding site (Ingraham et al 1990; Kapiloff et al 1991; Kristie & Sharp 1990; Verrijzer et al 1990). Interestingly, the Pit-1 POU-homeodomain can itself bind with relatively high affinity to certain recognition sites for the *Drosophila* homeodomain regulators (Ingraham et al 1990). We speculate that the advantage of the large, apparently asymmetric Pit-1 recognition elements is that post-translation regulation and/or heterodimer formation can impose differential binding and conformation of different *cis*-active elements dependent upon sequences removed from the canonical TATNCAT core binding motif. Similarly, the isolated Oct-1 POU-homeodomain reportedly maintains high affinity for the TAATGRAT sequence, whereas binding on the classical octomer motif is quite compromised (Verrijzer et al 1990b). These data imply that a subset of POU domain binding sites is distinct from those that contain the TATNCAT motif. For example, Tst-1 and Brn-2 *cis*-active elements contain ATTA and ATAAT core motifs, respectively (He et al 1991).

POU-DOMAIN PROTEIN: PROTEIN INTERACTIONS

Oct-1, Oct-2, and Pit-1 can bind as monomers to their cognate DNA recognition elements; each has been described to behave as apparent monomers in solution (Ingraham et al 1990; Lebowitz et al 1989; Poellinger & Roeder 1989; Poellinger et al 1989). Cooperative binding interactions are exhibited by Oct-2 on adjacent octomer elements on a natural binding site, and many of the Pit-1 binding sites in the rat prolactin and growth hormone promoters permit DNA-dependent Pit-1 dimer formation via cooperative binding interactions (Ingraham et al 1990; Lebowitz et al 1989; Poellinger et al 1989). Cooperative interactions between Oct-2 proteins increase transcriptional activity (Poellinger et al 1989). Although many Pit-1 binding sites might contain direct repeats of the TATNCAT motif, Pit-1 apparently binds preferentially and with high affinity as a monomer on certain sites. Based on the precedents of Oct-2 and Pit-1, the cooperative binding, which is dependent on the POU_S domain, may be a feature of many, if not all, POU-domain proteins; however, dimers apparently occur only on a subset of binding sites. Because analysis of exon-intron boundaries of the Pit-1 gene (Li et al 1990) reveals that the highly conserved RRIKLG sequence in the POU_S-A region represents the 3' boundary of an exon, the POU_S-A and POU_S-B regions might subserve differential functions with respect to protein-protein interactions and for high affinity, sequence-specific DNA binding. In addition to the ability to confer DNA-dependent protein-protein interactions, the POU_S domain may be important in formation of heterodimers in solution between family members that are often coexpressed in a single cell type (He et al 1989; Voss et al 1991), comparable to heterodimers between members of other gene families (e.g. Glass et al 1989; Murré et al 1989a, 1989b; Rauscher et al 1988). Specific DNA-independent interactions between different POU-domain proteins have been observed in mammals and *Drosophila* (Treacy et al 1991; Voss et al 1991). One consequence of these DNA-independent interactions is apparently to prevent binding of specific, positive POU-domain regulators. Thus, a combinatorial interaction of various coexpressed POU-domain proteins may contribute to the mature phenotype of a group of developing cells within an organism.

Crucial functional protein-protein interactions are also conferred by specific sequences in the POU_{HD}, established by the intriguing ability of Oct-1 to bind to the Herpes virus VP16 gene product, dependent on specific residues in helix 2 of the POU_{HD} (Gerster & Roeder 1988; O'Hare & Godding 1988; Stern et al 1989). VP16 interacts with host Oct-1 protein in activation of viral early genes (O'Hare & Godding 1988; Preston et al 1988). This apparently involves the formation of a complex with a third protein (Gerster et al 1990; Kristie et al 1989). The mutation of the three critical residues in the POU_{HD} helix 2 to those present in Oct-2 significantly decreases the VP16 binding (Sturm et al 1988). Mammalian homologues or analogues of VP16 have yet to be established. Based on these data, both the POU_S and the POU_{HD} appear to permit specific, DNA-dependent and DNA-independent protein-protein interactions that may be functionally crucial in both binding and transactivation. Both regions apparently modify the specificity and the affinity of those proteins for their cognate recognition elements.

REGULATION OF POU-DOMAIN PROTEIN GENES

Based on precedents in *Drosophila* for other classes of developmental regulations (Akam 1987; Scott et al 1989), the molecular mechanisms of activation and maturation of POU-domain proteins are likely to involve the actions of numerous other classes of transcriptional regulators. *Oct-3/4*, expressed early in development and in embryonic stem cells, are markedly inhibited by retinoic acid, which induces phenotypic alteration along several pathways (Jones-Villeneuve et al 1983; Strickland & Mahdavi 1981). These events can be mimicked by introduction of the *c-jun* gene (deGroot et al 1990). The finding that *oct-3/4* and *tst-1*/SCIP/*oct-6* are strongly negatively regulated by retinoic acid provides a correlation of expression of these POU-domain proteins and early commitment events in development (He et al 1989, 1991; Monuki et al 1990; Okamoto et al 1990; Schöler et al 1989; Suzuki et al 1990). The cloning of the *pit-1* gene has permitted an initial assessment of the regulatory mechanisms responsible for its initial activation, maintenance of expression, and regulation (Chen et al 1990; McCormick et al 1990). Two Pit-1 binding and regulatory elements, which flank the CAP site, were identified in the Pit-1 gene. The 5′ sequence was a positive regulatory element, which could confer Pit-1-dependent gene expression; the 3′ element was an inhibitory element, which attenuated expression tenfold (Chen et al 1990) (see Figure 8). These data are consistent with a positive, attenuated autoregulatory loop that seems to function in maintaining *pit-1* gene expression; in a sense, this affects a memory of cell commitment. The very low levels of *pit-1*

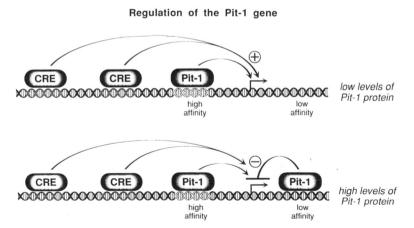

Regulation of the Pit-1 gene

Figure 8 Proposed mechanism for autoregulation of the Pit-1 gene. Both the high and low affinity binding sites are depicted, as are the two CRE binding sites.

transcript and protein in the *dw* genetic dwarfs are consistent with the model that *pit*-1 transcriptional autoregulation exerts an important function in the maintenance of *pit*-1 gene expression (Li et al 1990). Functional elements in the *pit*-1 promoter that bind cyclic AMP response element binding protein (CREB) (Chen et al 1990; McCormick et al 1990), could also serve developmental regulatory functions (see Figure 8). In this regard, it is intriguing that the *tst*-1/SCIP/*oct*-6 gene is transiently restricted during a phase of rapid cell division that precedes the myelinating phases of Schwann cell differentiation stimulated by cAMP (Monuki et al 1990).

CONCLUSIONS

Studies of cell-specific transcription factors and mutant loci in *C. elegans* have simultaneously led to the recognition of a novel class of transcription factors, referred to as POU-domain proteins, which can serve as developmental transactivators of genes that define specific cell phenotypes and of the proliferation of these cell types within an organ. An unexpected aspect of mammalian POU-domain gene expression is that the majority of known POU-domain genes are expressed, in distinct spatial and temporal patterns, in the developing forebrain, as well as in the rest of the CNS. Classic homeodomain gene expression has not yet been described in the forebrain, which plays such an important role in learning, memory, and the organization of behavior. POU-domain proteins may exert functions in developing forebrain neurons comparable to those exerted by *unc*-86 and Pit-1 in sensory neurons and anterior pituitary cells, respectively. These proteins also play a role in the earliest stages of pattern formation ("segmentation" of the primary brain vesicles) in the neural plate and tube as a whole. Although both the POU_S and POU_{HD} domains are involved in DNA binding by this class of transcription factors, the additional protein-protein interactions imprinted by the POU_S domain may provide additional, critical advantages to forebrain development via both homodimeric and heterodimeric protein-protein interactions, which could expand the specificity imposed by this class of protein. The binding sites for POU-domain factors are not palindromic; depending on their precise sequence, the sites can accommodate either stable binding dimer or monomer, which would perhaps permit differential types of interactions with members of other families of transcription factors and an even more complex pattern of positive and negative gene regulation. The POU_{HD} is itself capable of protein-protein interactions between members of the POU-domain family and with exogenous proteins, such as the Herpes VP16 gene product. It now becomes of significant interest to elucidate the precise nature of potential POU-domain protein interactions and the roles of

POU-domain proteins in development of the CNS. The role of a non-DNA-binding POU-domain protein in transcription regulation has also been discussed, which again implicates the importance of hetero-dimerization in this family of developmental regulators.

ACKNOWLEDGMENTS

The discussions, suggestions, and contributions of Xi He, Jeff Voss, Holly Ingraham, Bryan Crenshaw, Sen Li, Ruoping Chen, Renee Gerrero, Donna Simmons, and Larry Swanson are gratefully acknowledged. We are indebted to Susan Inglis for aid in preparing this review. Michael G. Rosenfeld is an investigator with the Howard Hughes Medical Institute.

Literature Cited

Akam, M. 1987. The molecular basis for metameric patterns in the *Drosophila* embryo. *Development* 101: 1–22

Anderson, K. A. 1987. Dorsal-ventral embryonic pattern genes of *Drosophila*. *Trends Genet.* 3: 91–97

Beachy, P. A., Krasnow, M. A., Gavis, E. R., Hogness, D. S. 1988. An ultrabithorax protein binds sequences near its own and the *Antennapedia* P1 promoter. *Cell* 55: 1069–81

Bodner, M., Castrillo, J.-L., Theill, L. E., Deerinck, T., Ellisman, M., Karin, M. 1988. The pituitary-specific transcription fator GHF-1 is a homeobox-containing protein. *Cell* 50: 267–75

Burglin, T., Finney, M., Coulson, A., Ruvkun, G. 1989. *Caenorhabditis elegans* has scores of homeobox-containing genes. *Nature* 341: 239–43

Campos-Ortega, J. A., Hartenstein, V. 1985. *The Embryonic Development of Drosophila Melanogasater.* Berlin: Springer-Verlag

Chalfie, M., An, M. 1989. Genetic control of differentiation of the *Caenorhabditis elegans* touch receptor neurons. *Science* 243: 1027–33

Chalfie, M., Horvitz, H. R., Sulston, J. E. 1981. Mutations that lead to reiterations in the cell lineages of *C. elegans. Cell* 24: 59–69

Chen, R., Ingraham, H. A., Treacy, M. N., Albert, V. R., Wilson, L., Rosenfeld, M. G. 1990. Autoregulation of Pit-1 gene expression is mediated by two cis-active promoter elements. *Nature* 346: 583–86

Clerc, R. G., Corcoran, L. M., LeBowitz, J. H., Baltimore, D., Sharp, P. A. 1988. The B-cell-specific Oct-2 protein contains POU box- and homeo-box-type domains. *Genes Dev.* 2: 1570–81

Crenshaw, E. B. III, Kalla, K., Simmons, D.

M., Swanson, L. W., Rosenfeld, M. G. 1989. Cell-specific expression of the prolactin gene in transgenic mice is controlled by synergistic interactions between promoter and enhancer elements. *Genes Dev.* 3: 959–72

Crews, S. T., Thomas, J. B., Goodman, C. S. 1988. The *Drosophila* single-minded gene encodes a nuclear protein with sequence similarity to the per gene product. *Cell* 52: 143–51

deGroot, R. P., Kruyt, F. A., vanderSaag, P. T., Kruiger, W. 1990. Ectopic expression of *c-jun* leads to differentiation of P19 embryonic carcinoma cells. *EMBO J.* 9: 1831–37

Doe, C. Q., Hiromi, Y., Gehring, W. J., Goodman, C. S. 1988. Expression and function of the segmentation gene *Fushitarazu* during Drosophila neurogenesis. *Science* 239: 170–75

Dollé, P., Castrillo, J.-L., Theill, L. E., Deerinck, T., Ellisman, M., Karin, M. 1990. Expression of GHF-1 protein in mouse pituitaries correlates both temporally and spatially with the onset of growth hormone gene activity. *Cell* 60: 809–20

Eicher, E. M., Beamer, W. G. 1980. New mouse *dw* allele: genetic location and effects on life span and growth hormone levels. *J. Hered.* 71: 187–90

Elsholtz, H. P., Albert, V. R., Treacy, M. N., Rosenfeld, M. G. 1990. A two-base change in a POU factor-binding site switches pituitary-specific to lymphoid-specific gene expression. *Genes Dev.* 4: 43–51

Filbin, M. T., Walsh, F. S., Trapp, B. D., Pizzey, J. A., Tennekoon, G. I. 1990. Role of myelin Po protein as a homophilic adhesion molecule. *Nature* 344: 871–72

Finney, M., Ruvkun, G. 1990. The *unc-86*

gene product couples cell lineage and cell identity in *C. elegans*. *Cell* 63: 895–905

Finney, M., Ruvkun, G., Horvitz, H. R. 1988. The *C. elegans* cell lineage and differentiation gene *unc-86* encodes a protein containing a homeodomain and extended sequence similarity to mammalian transcription factors. *Cell* 55: 757–69

Fox, S. R., Jong, M. T. C., Casanova, J., Ye, S. F., Stanley, F., Samuels, H. H. 1990. The homeodomain protein, Pit-1/Chf-1, is capable of binding to and activating cell-specific elements of both the growth hormone and prolactin gene promoters. *Mol. Endocrinol.* 4: 1069–80

Gehring, W. J. 1987. Homeoboxes in the study of development. *Science* 236: 1245–52

Gerster, T., Balmacedo, C. G., Roeder, R. G. 1990. The cell type-specific octomer transcription factor OTF-2 has two domains required for the activation of transcription. *EMBO J.* 9: 1635–43

Gerster, T., Roeder, R. G. 1988. A herpes virus transactivating protein interacts with transcription factor OTF-1 and other cellular proteins. *Proc. Natl. Acad. Sci. USA* 85: 6347–51

Glass, C. K., Lipkin, S. M., Devary, O. V., Rosenfeld, M. G. 1989. Positive and negative regulation of gene transcription by a retinoic acid-thyroid hormone receptor heterodimer. *Cell* 59: 697–708

Hanes, S., Brent, R. 1989. DNA specificity of the bicoid activator protein is determined by homeodomain recognition helix residue 9. *Cell* 57: 1275–83

Hatzopoulos, A. K., Stoykova, A. S., Erselius, J. R., Golding, M., Neuman, T., Gruss, P. 1990. Structure and expression of the mouse Oct2a and Oct2b, two differentially spliced products of the same gene. *Development* 109: 349–62

He, X., Gerrero, R., Simmons, D. M., Park, R. E., Lin, C. R., et al. 1991. Tst-1, a member of the POU-domain gene family, binds the promoter of the gene encoding the cell surface adhesion molecule Po. *Mol. Cell. Biol.* 3: 1739–44

He, X., Treacy, M. N., Simmons, D. M., Ingraham, H. A., Swanson, L. W., Rosenfeld, M. G. 1989. Expression of a large family of POU-domain regulatory genes in mammalian brain development. *Nature* 340: 35–42

Herr, W., Sturm, R. A., Clerc, R. G., Corcoran, L. M., Baltimore, D., et al. 1988. The POU domain: a large conserved region in the mammalian *pit-1*, *oct-1*, *oct-2*, and *Caenorhabditis elegans unc-86* gene products. *Genes Dev.* 2: 1513–16

Hoey, T., Levine, M. 1988. Divergent homeobox proteins recognize similar DNA-sequences in *Drosophila*. *Nature* 332: 858–61

Ingham, P. W. 1988. The molecular genetics of embryonic pattern formation in *Drosophila*. *Nature* 335: 25–34

Ingraham, H. A., Chen, R., Mangalam, H. J., Elsholtz, H. P., Flynn, S. E., et al. 1988. A tissue-specific transcription factor containing a homeodomain specifies a pituitary phenotype. *Cell* 55: 519–29

Ingraham, H. A., Flynn, S. E., Voss, J. W., Albert, V. R., Kapiloff, M. S., et al. 1990. The POU-specific domain of Pit-1 is essential for sequence-specific, high affinity DNA binding and DNA-dependent Pit-1-Pit-1 interactions. *Cell* 61: 1021–33

Jacobs, J. R., Hiromi, Y., Patel, N. H., Goodman, C. S. 1984. Lineage migration and morphogenesis of longitudinal glia in the *Drosophila* CNS as revealed by a molecular lineage marker. *Neuron* 2: 1625–31

Jaynes, J. B., O'Farrell, P. H. 1988. Activation and precision of transcription by homeodomain-containing proteins that bind a common site. *Nature* 336: 744–49

Johnson, W. A., Hirsh, J. 1990. Binding of a *Drosophila* POU-domain protein to a sequence element regulating gene expression in specific dopaminergic neurons. *Nature* 343: 467–70

Jones-Villeneuve, E. M. V., Rudnicki, M. A., Harris, J. K., McQuerney, M. W. 1983. Retinoic acid-induced neural differentiation of embryonal carcinoma cells. *Mol. Cell. Biol.* 3: 2271–79

Kapiloff, M. S., Farkash, Y., Wegner, M., Rosenfeld, M. E. 1991. Variable effects of phosphorylation of Pit-1 dictated by the DNA response elements. *Science*. In press

Kissinger, C. R., Liu, B., Martin-Blanco, E., Kornberg, T. B., Pabo, C. O. 1990. Crystal structure of an engrailed homeodomain-DNA complex at 2.8 Å resolution: a framework for understanding homeodomain-DNA interactions. *Cell* 63: 579–90

Kristie, T. M., LeBowitz, J. H., Sharp, P. A. 1989. The octomer binding proteins form multi-protein DNA complexes with the HSV αTIF regulatory protein. *EMBO J.* 8: 4229–38

Kristie, T. M., Sharp, P. A. 1990. Interactions of the Oct-1 POU subdomains with specific DNA sequences and with the HSV α-transactivation protein. *Genes Dev.* 4: 2383–96

Larkin, S., Tait, S., Treacy, M. N., Martin, F. 1990. Characterization of tissue-specific *trans*-acting factor binding to a proximal element in the rat growth hormone gene promoter. *Eur. J. Biochem.* 191: 605–15

Lebowitz, J. H., Clerc, R. G., Brenowitz, M., Sharp, P. 1989. The Oct-2 protein binds cooperatively to adjacent octomer sites. *Genes Dev.* 3: 1625–38

Li, S., Crenshaw, E. B., Rawson, E. J., Simmons, D. M., Swanson, L. W., Rosenfeld, M. G. 1990. Dwarf locus mutants lacking three pituitary cell types result form mutations in the POU-domain gene Pit-1. *Nature* 347: 528–33

Lira, S. A., Crenshaw, E. B. III, Glass, C. K., Swanson, L. W., Rosenfeld, M. G. 1988. Identification of rat growth hormone genomic sequences targeting pituitary expression in transgenic mice. *Proc. Natl. Acad. Sci. USA* 85: 4755–59

Mangalam, H. J., Albert, V. R., Ingraham, H. A., Kapiloff, M., Wilson, L., et al. 1989. A pituitary POU domain protein, Pit-1, activates both growth hormone and prolactin promoters transcriptionally. *Genes Dev.* 3: 946–58

McCormick, A., Brady, H., Theill, L. E., Karin, M. 1990. Regulation of the pituitary-specific homeobox gene GHF-1 by cell autonomous and environmental cues. *Nature* 345: 829–32

Monuki, E. S., Kuhn, R., Weinmaster, G., Trapp, B., Lemke, G. 1990. Expression and activity of the POU transcription factor SCIP. *Science* 249: 1300–3

Müller, M., Ruppert, S., Schaffner, W., Mathias, P. 1988. A cloned octomer transcription factor stimulates transcription from lymphoid-specific promoters in non-B cells. *Nature* 336: 544–51

Müller-Immerglück, M. M., Schaffner, W., Matthias, P. 1990. Transcription factor Oct-2A contains functionally redundant activating domains and works selectively from a promoter but not from a remote enhancer position in non-lymphoid (HeLa) cells. *EMBO J.* 9: 1625–34

Murré, C., McCaw, P. S., Baltimore, D. 1989a. A new DNA binding and dimerization motif in immunoglobulin enhancer binding, daughterless, MyoD, and myc proteins. *Cell* 56: 777–83

Murré, C., McCaw, P. S., Vaessin, H., Caudy, M., Jan, L. Y., et al. 1989b. Interactions between heterologous helix-loop-helix proteins generate complexes that bind specifically to a common DNA sequence. *Cell* 58: 537–44

Nambu, J. R., Franks, R. G., Hu, S., Crews, S. T. 1990. The single-minded gene of *Drosophila* is required for the expression of genes important for the development of CNS midline cells. *Cell* 63: 63–75

Nelson, C., Albert, V. R., Elsholtz, H. P., Lu, L. E.-W., Rosenfeld, M. G. 1988. Activation of cell-specific expression of rat growth hormone and prolactin genes by a

common transcription factor. *Science* 239: 1400–5

O'Hare, P., Godding, C. R. 1988. Herpes simplex virus regulatory elements and the immunoglobulin octomer domain bind a common factor and are both targets for virion transactivation. *Cell* 52: 435–45

Okamoto, K., Okazawa, H., Okuda, A., Sakai, M., Muramatsu, M., Hamada, H. 1990. A novel octamer transcription factor is differentially expressed in mouse embryonic cells. *Cell* 60: 461–72

Otting, G., Qian, Y.-Q., Müller, M., Affolter, M., Gehring, W., Wüthrich, L. 1988. Secondary structure determination for the *Antennapedia* homeodomain by nuclear magnetic resonance and evidence for a helix-turn-helix motif. *EMBO J.* 7: 4305–9

Otting, G., Qian, Y.-Q., Billeter, M., Müller, M., Affolter, M., et al. 1990. Protein-DNA contacts in the structure of a homeodimer-DNA complex determined by nuclear magnetic resonance spectroscopy in solution. *EMBO J.* 9: 3085–92

Patel, N. H., Schafer, B., Goodman, C. S., Holmgren, R. 1989. The role of segment polarity genes during *Drosophila* neurogenesis. *Genes Dev.* 3: 890–904

Poellinger, L., Roeder, R. G. 1989. Octamer transcription factors 1 and 2 each bind to two different functional elements in the immunoglobulin heavy-chain promoter. *Mol. Cell. Biol.* 9: 747–56

Poellinger, L., Yoza, B. K., Roeder, R. G. 1989. Functional cooperativity between protein molecules bound at two distinct sequence elements of the immunoglobulin heavy-chain promoter. *Nature* 337: 573–76

Preston, C. M., Frame, M. C., Campbell, M. E. M. 1988. A complex formed between cell components and an HSV structural polypeptide binds to a viral immediate early gene regulatory DNA sequences. *Cell* 52: 425–34

Rauscher, F. J., Cohen, D. R., Curran, T., Bos, T. J., Vogt, P. R., et al. 1988. Fos-associated protein P39 is the product of the *jun* proto-oncogene. *Science* 240: 1010–16

Rosner, M. H., Vigano, M. A., Ozato, K., Timmons, P. M., Poirier, F., et al. 1990. A POU-domain transcription factor in early stem cells and germ cells of the mammalian embryo. *Nature* 345: 686–91

Roux, M., Bartke, A., Dumont, F., Dubois, M. P. 1982. Immunohistological study of the anterior pituitary gland—pars distalis—and pars intermedia—in dwarf mice. *Cell Tissue Res.* 223: 415–20

Scheidereit, C., Cromlish, J. A., Gerster, T., Kawakami, K., Balmaceda, C.-G., et al.

1988. A human lymphoid-specific transcription factor that activates immunoglobulin genes in a homeobox protein. *Nature* 336: 552–57

Schöler, H. R., Hatzopoulos, A. K., Balling, R., Suzuki, N., Gruss, P. 1989. A family of octomer-specific proteins present during mouse embryogenesis: evidence for germline-specific expression of an oct factor. *EMBO J.* 8: 2543–50

Schöler, H. R., Ruppert, S., Balling, R., Suzuki, N., Chowdhury, K., Gruss, P. 1990. New type of POU domain in germ line-specific protein Oct-4. *Nature* 344: 435–39

Schreiber, E., Mathias, P., Müller, M. N., Schaffner, W. 1988. Identification of a novel lymphoid specific octomer binding protein (OTF-2B) by proteolytic clipping bandshift assay (PCBA). *EMBO J.* 7: 4221–29

Scott, M. P., Carroll, S. B. 1987. The segmentation and homeotic gene network in early *Drosophila* development. *Cell* 51: 689–98

Scott, M. P., Tamkun, J. W., Hartzell, G. W. 1989. The structure and function of the homeodomain. *Biochem. Biophys. Acta* 989: 25–48

Sharp, Z. D., Cao, Z. 1990. Regulation of cell-type-specific transcription and differentiation of the pituitary. *Bioessays* 12: 80–85

Simmons, D. M., Voss, J. W., Ingraham, H. A., Holloway, J. M., Broide, R. S., et al. 1990. Pituitary cell phenotypes involve cell-specific Pit-1 mRNA translation and synergistic interactions with other classes of transcription factors. *Genes Dev.* 4: 695–711

Snell, G. D. 1929. Dwarf, a new Mendelian recessive character of the house mouse. *Proc. Natl. Acad. Sci. USA* 15: 733–34

Stern, S. A., Tanaka, M., Herr, W. 1989. The Oct-1 homeodomain direct formation of a multiprotein-DNA complex with the HSV transactivator VP16. *Nature* 341: 624

Strickland, S., Mahdavi, V. 1981. The induction of differentiation in teratocarcinoma stem cells by retinoic acid. *Cell* 15: 393–403

Sturm, R. A., Das, G., Herr, W. 1988. The ubiquitous octomer-binding protein Oct-1 contains a POU domain with a homeo box subdomain. *Genes Dev.* 2: 1582–99

Sturm, R. A., Herr, W. 1988. The POU-domain is a bipartite DNA-binding structure. *Nature* 336: 601–4

Suzuki, N., Rohdenohld, H., Neuman, T., Gruss, P., Schöler, H. R. 1990. Oct-6: a POU domain transcription factor expressed in embryonal and stem cells and in the developing brain. *EMBO J.* 9: 3723–32

Swanson, L. W., Simmons, D. M. 1989. Differential steroid hormone and neural influences on peptide mRNA levels in CRH cells of the paraventricular nucleus: A hybridization histochemical study in the rat. *J. Comp. Neurol.* 285: 413–35

Tanaka, M., Herr, W. 1990. Differential transcriptional activation by Oct-1 and Oct-2: Interdependent activation domains induce Oct-2 phosphorylation. *Cell* 60: 375–86

Thomas, J. B., Crews, S. T., Goodman, C. S. 1988. Molecular genetics of the single-minded locus: a gene involved in the development of the *Drosophila* embryonic nervous system. *Cell* 58: 133–41

Treacy, M. N., He, X., Rosenfeld, M. G. 1991. I-POU: A novel POU-domain protein that inhibits neuron-specific gene activation. *Nature* 350: 577–84

Treisman, J., Gonczy, P., Vashishtha, M., Harris, E., Desplan, C. 1989. A single amino acid can determine the DNA binding specificity of homeodomain proteins. *Cell* 59: 553–62

Vale, W., Rivier, C., Brown, M. R., Spiess, J., Koob, G., et al. 1983. Chemical and biological characterization of corticotropin releasing factor. *Recent Prog. Horm. Res.* 39: 245–70

Verrijzer, C. P., Kal, A. J., van der Vliet, P. C. 1990a. The DNA binding domain (POU domain) of transcription factor Oct-1 suffices for stimulation of DNA replication. *EMBO J.* 9: 1883–88

Verrijzer, C. P., Kal, A. J., van der Vliet, P. C. 1990b. The oct-1 homeo domain contacts only part of the octamer sequence and full oct-1 DNA-binding activity requires the POU-specific domain. *Genes Dev.* 4: 1964–74

Voss, J. W., Wilson, L., Rosenfeld, M. G. 1991. The POV-domain proteins, Pit-1 and Oct-1, interact to form a heterodimeric complex and can cooperate to induce regulation of the protactin promoter. *Genes Dev.* 5: 1309–20

Wilson, D. B., Wyatt, D. P. 1986. Ultrastructural immunocytochemistry of somatotrophs and mammotrophs in embryos of the dwarf mutant mouse. *Anat. Rec.* 215: 282–87

Wirth, T., Staudt, L., Baltimore, D. 1987. An octamer oligonucleotide upstream of a TATA motif is sufficient for lymphoid-specific promoter activity. *Nature* 329: 174–78

Annu. Rev. Neurosci. 1992. 15:167–91

MOVING IN THREE-DIMENSIONAL SPACE: FRAMES OF REFERENCE, VECTORS, AND COORDINATE SYSTEMS

J. F. Soechting and M. Flanders

Department of Physiology, University of Minnesota, Minneapolis, Minnesota 55455

KEY WORDS: sensorimotor transformations, arm movements, eye movements, head movements

INTRODUCTION

We move in a three-dimensional world. What are the motor commands that generate movements to a target in space, and how is sensory information used to control and coordinate such movements? To answer these questions, one must determine how spatial parameters are encoded by the activity of neurons. Within the last decade, experimenters have begun to study a variety of movements in three-dimensional space. Among these are "reflexive" (or postural) eye, head, and body movements elicited by vestibular and visual stimuli; orienting movements of the eyes, head, and body subserved by the superior colliculus (or in lower vertebrates, the optic tectum); and arm movements with their neural correlates in motor cortex.

The neural systems that are involved in the production of each of these movements must deal with aspects that are particular to that task, and specialized reviews are available on each of these topics (Georgopoulos 1986; Knudsen et al 1987; Simpson 1984; Sparks 1986). Nevertheless, the question of spatial representation is a theme common to each of these

167

0147–006X/92/0301–0167$02.00

areas, and in this review we focus on that question. We show that multi-dimensional information can be, and is, represented in a variety of ways such as topographically, vectorially, or in coordinate systems. Underlying each of these representations is the notion of a frame of reference. We begin by defining these terms. Then, we summarize experimental data for each of the above-mentioned tasks and attempt to identify how spatial parameters are represented. We conclude by examining some common concepts that have begun to emerge from the study of this variety of motor tasks.

DEFINITIONS

Frames of Reference

Central to any spatial description is the concept of a frame of reference. As a textbook example of a frame of reference, consider a passenger standing on a moving train and an observer watching the train go by. We can imagine two frames of reference: one fixed to the train, the other fixed to the earth (Figure 1*A*). The passenger is moving in the earth's frame of reference, but is stationary in the train's frame of reference. If the passenger

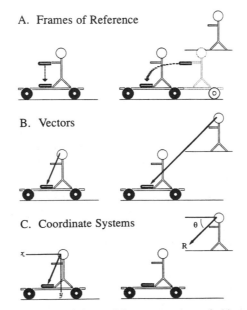

A. Frames of Reference

B. Vectors

C. Coordinate Systems

Figure 1   Schematic illustration of the spatial representations of objects in frames of reference (*A*), vectorially (*B*) and by coordinate systems (*C*). On the left, the frame of reference moves with the passenger; on the right, the observer's frame of reference is fixed to the earth.

drops a book, it will fall straight down in the train's frame of reference. However, from the perspective of the observer in the earth's frame of reference, the book will drop along a curved path.

Closer to the problem at hand, we can imagine a retinocentric frame of reference, i.e. one fixed to the eye. We can also imagine other frames of reference fixed to the head, to the trunk, and to the earth. As was demonstrated by the simple example above, our (or a neuron's) description of events depends on the frame of reference that is adopted. The criterion for identifying a frame of reference is straightforward. For example, if a neuron encodes the location of an object in a retinocentric frame of reference, then the neuron's activity should remain constant as long as the object's image falls on the same locus on the retina, irrespective of the eye's position in the head, or the head's position on the trunk.

Vectors

Once we have defined a frame of reference, one way to define the location of any point in this frame of reference (e.g. the book in Figure 1B) is by means of a vector, with a direction and an amplitude. To do so, we must first define an origin for the frame of reference. In the illustrated example, the origin is the eye of the passenger (*left*) or of the observer (*right*). The amplitude of the vector is its distance from the origin, and its direction is given by the line that connects the origin with the point.

Coordinate Systems

Sometimes, it is convenient to define a coordinate system within the frame of reference by choosing a set of base vectors. Any point in the reference frame is now defined in terms of an amplitude along each of the base vectors (coordinate axes). In Figure 1C, a coordinate system in the passenger's frame of reference might be given by the horizontal (x) and vertical (y) axes, i.e. a Cartesian coordinate system. In the observer's frame of reference, a coordinate system could be defined by the distance from the observer to the book (in the radial direction, R), the angle between the radial direction and the horizontal (elevation, θ), and a second angle that defines the deviation of the radial direction from the sagittal plane (azimuth), i.e. in a spherical coordinate system.

COORDINATE SYSTEMS DEFINED BY NEURAL ACTIVITY

Are coordinate systems defined by neural activity? If so, how can one recognize them? These questions are more easily answered at the periphery of the nervous system, where coordinate systems (sensory and motor) are clearly defined by the geometry of the sensory receptors or the musculo-skeletal system. The base vectors of the motor coordinate system are provided by the direction in which each of the muscles exerts force (Pellionisz & Llinás 1980). Sensory coordinate systems are defined by the direction of the stimulus that most effectively activates peripheral receptors. For muscle stretch receptors, the coordinate axes would also coincide with the direction in which each muscle exerts force. For semicircular canal afferents, the coordinate axes would be defined by the axes of head rotation that provide the most effective stimuli (Robinson 1982).

As Pellionisz & Llinás (1980, 1982) first pointed out, motor and sensory coordinate systems usually have nonorthogonal axes. In such a case, it becomes necessary to distinguish between the two types of coordinate representations, which are illustrated in Figure 2. Although sensory (receptor) representations are formed by projections onto coordinate axes (Figure 2A), and motor (effector) actions follow the rules of vector summation (Figure 2B), both cases predict a cosine tuning of neural activity around a "best" direction. In the bottom half of Figure 2, the amplitudes of the x and y components of point P are plotted as a function of the angle between the x axis and a vector from the origin to the point. The best direction is the angle for which the amplitude is the largest, and one might expect this best direction to correspond to the maximal neural activity. For vector summation (Figure 2B), the best directions do not coincide with the coordinate axes.

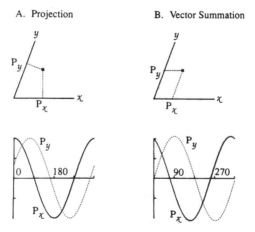

Figure 2   In coordinate systems with nonorthogonal axes, the coordinates of a point can be defined by projection onto the coordinate axes (A) or by vector summation (B). In both types of representation, the amplitudes of the x and y components vary sinusoidally with the angle between the x-axis and the vector from the origin to the point.

Thus, independently of whether a coordinate system is defined by projection or by vector summation, a neural representation in such a coordinate system should generally define a best direction along which activity is maximal. Neural activity should decrease by an amount proportional to the cosine of the angle, for inputs or outputs oriented along directions other than best directions. A vectorial code should exhibit tuning characteristics that are similar to one encoding a coordinate system, with one major difference: A coordinate system is defined by a limited number of base vectors; therefore, the number of best directions in a population of neurons should be similarly limited. In a simple vectorial code, one might expect the best directions to be more numerous and widely distributed.

In summary, to understand central processing of information in sensorimotor systems, it might be useful to begin by first identifying the frame of reference in which the information is encoded. The next steps would be to determine whether parameters in that frame of reference are encoded vectorially, and to ascertain whether the vectorial code also implies a coordinate system. If the criteria can be satisfied, it then becomes possible to describe neural processing in geometric terms, i.e. transformations from one frame of reference to another and transformations between coordinate systems within a single frame of reference. In the following sections we examine several examples in which this approach has been useful for understanding the neural representations involved in sensorimotor transformations.

VESTIBULO-OCULAR COORDINATE SYSTEMS

The semicircular canals and the extraocular muscles provide the clearest example of coordinate systems imposed by the geometric arrangement of the sensors and the motor apparatus. The afferents are linked to the efferents by a three-neuron arc (the vestibulo-ocular reflex), which acts to rotate the eyes in the direction opposite to the head rotation sensed by the semicircular canals.

Each of the three canals defines a plane; head rotation about an axis perpendicular to this plane is the most effective stimulus, whereas rotations about axes lying in this plane are ineffective (Blanks et al 1972; Estes et al 1975). Canal planes have been determined anatomically for several species (Ezure & Graf 1984a; Reisine et al 1988).

There are six extraocular muscles for each eye, and the pulling directions of these muscles have been computed from anatomic measurements (Ezure & Graf 1984a). The neural innervation of muscle pairs is organized in a push-pull fashion (Baker et al 1988a); thus, one can combine the antagonistic action of muscle pairs to define three axes of eye rotation, each evoked by activation of one of the three pairs (Robinson 1982). These three axes are not perpendicular to each other and they do not align exactly with the axes of the semicircular canals. In this nonorthogonal motor coordinate system, the axis of eye movement for which each muscle pair is most active does not coincide with the axis defined by the muscles' pulling directions (Baker et al 1988a), as predicted in Figure 2*B*. Also in accord with the prediction, the amplitude of the modulation in eye muscle activity in response to sinusoidal head rotation decreases as a cosine function of the angle between the best direction and the direction of rotation (Baker et al 1988b).

Thus, both the semicircular canals and the extraocular muscles define three-dimensional coordinate systems in a reference frame fixed to the head. Furthermore, because the axes of the two coordinate systems do not coincide, a coordinate transformation is implied. As there are only three neurons in the reflex arc, the coordinate transformation can occur in only two places: by convergence of vestibular afferents from different canals onto vestibulo-ocular relay neurons in the vestibular nuclei, or by convergence of these relay neurons in the oculomotor nuclei. This problem has received considerable attention, both theoretically (Pellionisz 1985; Pellionisz & Graf 1987; Robinson 1982) and experimentally (Ezure & Graf 1984b; Peterson & Baker 1991). Experimental evidence indicates that part of the coordinate transformation occurs at both sites.

The function of the vestibulo-ocular reflex is to stabilize gaze in an earth-fixed frame of reference. Visual input also contributes to stabilizing gaze, and there is substantial convergence of vestibular and visual inputs in the vestibular nuclei (Dichgans & Brandt 1978). Although the geometry of the semicircular canals and the eye muscles virtually imposes a coordinate system on the vestibulo-ocular pathway, retinal receptors do not define a coordinate system. How, then, is motion of the visual image encoded centrally? Is it also defined by a coordinate system? If so, what are the coordinate transformations on this visual input?

Simpson (1984) and colleagues have addressed these questions by studying the rabbit's accessory optic system, which consists of three target nuclei that receive input from retinal ganglion cells and make efferent projections to the inferior olive and, hence, to the cerebellum (Maekawa & Simpson 1973). Neurons in this system respond preferentially to movements of large visual stimuli at slow speeds (Simpson 1984), i.e. to stimuli that would arise naturally during slow speed head motion in a stationary environment. Visual input to the accessory optic system could help compensate for the semicircular canal afferents' low gain at such speeds (Fernandez & Goldberg 1971).

In the accessory optic system, image motion is also represented in coordinates whose axes are aligned with the axes of the semicircular canals and the extraocular muscles (Simpson et al 1988; Sodak & Simpson 1988). Neural activity in the dorsal cap of the inferior olive and in the climbing fiber and mossy fiber inputs to the flocculo-nodular lobe of the cerebellum clearly defines a coordinate system (Graf et al 1988; Leonard et al 1988). One class of neurons in the dorsal cap responds best to rotation of the visual field about a vertical axis, i.e. to rotation in the plane of the horizontal canals. Two other types of neurons respond best to rotation about horizontal axes aligned with the axes of the anterior and posterior semicircular canals. One axis is located anterior at 45° to the sagittal plane, the other is oriented in the opposite direction (posterior, 135° to the sagittal plane). Climbing fiber activity in Purkinje cells shows the same preferential orientations (Figure 3A), as does simple spike activity.

Visual input to vestibular nuclei neurons (which, in turn, project to extraocular muscles) also defines a coordinate system aligned with the semicircular canals (Graf 1988). As shown in Figure 3B, one type of neuron, which also receives input from the posterior semicircular canal, shows a polarization in line with that of the posterior canal. (A second type responds best to rotations of the visual surround about the axis of the anterior canal.) The visual receptive field of these neurons is bipartite in nature, as indicated by the hatching in the right part of Figure 3B. Upward movement in one part of the receptive field is excitatory, as is downward movement in the other part. Rotation of the visual surround about the axis of the posterior canals (as indicated schematically in Figure 3B) would lead to upward motion on one side of the axis and downward motion on the other.

Activity of retinal ganglion cells is not in a vestibulo-oculomotor coordinate system; therefore, a coordinate transformation is required to go from retinal ganglion cell activity to the activity of neurons in the accessory

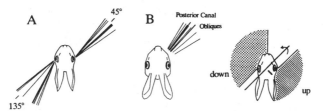

Figure 3 Coordinate axes defined by neural activity in cerebellum (*A*) and vestibular nuclei (*B*). Each line defines the best direction of one neuron for rotation of the visual surround. (*A*) is redrawn from Graf et al (1988), (*B*) from Graf (1988).

optic system or the vestibular nuclei. Simpson and coworkers have also worked out some of the details of this coordinate transformation. In the rabbit, which is a lateral-eyed animal, there are retinal ganglion cells that exhibit tuning for movement in one of three directions (Oyster et al 1972). One axis of this coordinate system is oriented anteriorly, i.e. it is aligned with the plane of the horizontal canals. This horizontal coordinate axis is maintained at subsequent stages in the terminal nuclei of the accessory optic system and beyond. The other two coordinate axes of retinal ganglion cells are oriented superiorly and slightly posteriorly, and inferiorly and slightly posteriorly. These vertically oriented axes undergo a transformation. The tuning of neurons in the accessory optic system nuclei is similar, but their orientation selectivity suggests a monocular combination of excitatory input from superior retinal ganglion cells with inhibitory input from inferior retinal ganglion cells, and vice versa (Sodak & Simpson 1988). More interestingly, a few neurons in the medial terminal nucleus exhibited bipartite monocular receptive fields (Simpson et al 1988), which would be stimulated by rotation of the visual surround about a horizontal axis between the two receptive fields (see Figure 3B). Thus, several distinct coordinate systems can be associated with the accessory optic system, providing for a gradual transformation of information about linear image motion to information about image rotation in a coordinate system aligned approximately with that of the semicircular canals.

COORDINATE SYSTEMS FOR POSTURAL RESPONSES

Afferent activity from the semicircular canals also contributes to stabilizing the head in an earth-fixed frame of reference by means of the vestibulo-collic reflex. This reflex exhibits a considerable increase in complexity over the vestibulo-ocular reflex: There are many more muscles involved (about 30 in the cat, see Pellionisz & Peterson 1988); there is apparently more extensive convergence from other sensors (muscle stretch receptors and vestibular macular afferents), and the neural circuitry underlying this reflex is more complex.

Are there sensorimotor transformations to align the signals from the different sensors in a common frame of reference? How are these signals transformed to activate the neck muscles? Investigators have begun to address these questions experimentally and theoretically. The pulling directions of the neck muscles exhibit a wide range of orientations (Pellionisz & Peterson 1988). There is no unique solution for the manner in which the activation of neck muscles should vary with the axis of head torque, as there are more muscles than degrees of freedom. Theoretical activation vectors (best directions) for the muscles have been predicted (Pellionisz & Peterson 1988), based on the idea of coordinate transformations from canal coordinates to neck muscle coordinates to minimize the extent of muscle coactivation. As one would expect (Figure 2*B*), these vectors are not colinear with the muscles' pulling directions. When patterns of neck muscle activation in response to whole body rotation (activating vestibular receptors) were measured by Baker et al (1985), and compared with the theoretical predictions (Peterson et al 1988, 1989), they were found to be in good qualitative agreement.

Less is known about the intermediate stages in this sensorimotor transformation and the extent to which signals from other afferents are aligned with those from the semicircular canal afferents. Wilson and colleagues (Kasper et al 1988a,b; Wilson et al 1990) have begun to record activity in vestibulospinal neurons during head rotation about horizontal axes. The activity of most of these neurons defined a vector orientation for rotation, i.e. neural activity fell off as a cosine function of the angle between the axis of rotation and a best axis (see also Baker et al 1984). The orientations of these vectors do not appear to cluster about a few directions (i.e. to define coordinate axes), but they are also not distributed uniformly. Most appear to be oriented close to the roll (antero-posterior) axis or at a 45° angle to either side of this axis.

From the frequency response of the units, these investigators deduced contributions of otolith afferent input to some of the neurons. In most cases, the spatial orientation of the otolith and canal inputs was in alignment. Because the orientation of otolith response vectors to tilt shows a wide range of distributions (Fernandez & Goldberg 1976), such an alignment would not be expected by chance. About 50% of vestibulospinal neurons also responded to passive neck rotation; in most of them, the vestibular and neck response vectors were also in alignment, differing by close to 180°. These neurons do not respond to head rotation about a stationary trunk, as the vestibular and neck inputs would tend to cancel. They would respond to trunk rotation about a stationary head or to whole body rotation, i.e. movement of the trunk in an earth-fixed frame of reference. The tuning of the other 50% would be appropriate to signal head rotation in the earth-fixed frame of reference.

In summary, vestibulospinal neurons appear to provide a vectorial code of rotation in an earth-fixed frame of reference, of either the head or the trunk. In most instances, the vectors of each of the afferent inputs (semicircular canals, otoliths, and neck afferents) are in approximate alignment.

Responses in limb muscles evoked by perturbations to the surface of support during posture also involve concurrent input from a variety of sensors: vestibular, visual, and proprioceptive (Nashner & McCollum 1985). How postural information from limb proprioceptors is transformed into a common reference frame with visual and vestibular information remains to be determined (Droulez & Darlot 1990).

The control of limb musculature is apparently not effected muscle by muscle; instead, it has been suggested that global variables are controlled (Nashner & McCollum 1985; Lacquaniti et al 1990). Can these global variables be associated with a coordinate system? Nashner & McCollum (1985) have found it convenient to describe bipedal posture in terms of both the distance from the center of gravity to the base of support, and the ankle and hip angles. Maioli and coworkers (1988, 1989) have also suggested limb length to be one controlled variable in quadrupeds, along with the orientation of the limb relative to the vertical in the sagittal plane (see Figure 4). They found that these two variables remained constant when the base of support was tilted (around the pitch axis) or the location of the animals' center of gravity was shifted by adding weights. Subsequent work (Maioli & Poppele 1989) suggested limb length and orientation were controlled independently of each other. Thus, these parameters may provide two of the axes of a postural coordinate system in an earth-fixed frame of reference. At least a third axis would be needed to regulate the sideways tilt of the animal.

Ground reaction forces in posture also appear to define a coordinate system. Macpherson (1988) measured the tangential reaction forces on cat fore- and hindlimbs when the cats were subjected to translation of the support surface in different directions. During quiet stance, these forces were directed at angles of 45° or 135° relative to the anterior direction. Following perturbation, actively evoked reaction forces were also oriented along these two directions, irrespective of the direction of the perturbation, whereas passive forces were always aligned with the direction of pertur-

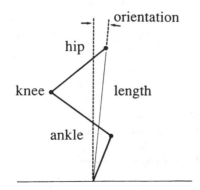

Figure 4 Limb length and orientation are two coordinates that can describe quadrupedal limb posture. A cat hindlimb is shown schematically; length is the distance from the base of support to the hip, and orientation is the angle of the vector from the base of support to the hip from the vertical axis.

bation. Thus, both limb kinematics (movements) and actively produced limb kinetics (forces) define coordinate systems in reference frames fixed in space. Whether these coordinate systems are independent of one another, or one is a consequence of the other, remains to be determined.

FRAMES OF REFERENCE AND COORDINATE REPRESENTATIONS FOR ORIENTING MOVEMENTS

Orienting movements of the eyes, head, and whole body can be evoked by visual, acoustic, and somesthetic stimuli. Information from each of these sensors is represented in a different frame of reference: visual in one fixed to the eyes, acoustic in one fixed to the head, and somesthetic in one fixed to the body. Because the eyes can move in the head, and the head on the body, the question arises: is information from these sensors transformed into a common frame of reference, and if so, what is it? How is information represented in each frame of reference? How are the transformations achieved?

The superior colliculus (or its analogue in lower vertebrates, the optic tectum) is a key structure for orienting movements. There is a topographic map of target location in the layers of the superior colliculus or the tectum (Knudsen et al 1987; Sparks 1986). Each neuron is preferentially activated by a stimulus located in one region of space. In the deeper layers, neurons respond to stimuli from more than one sensory modality, and the receptive fields defined by each sensory modality are approximately in register (Knudsen 1982; Meredith & Stein 1986; Middlebrooks & Knudsen 1984) when eyes, head, and body are in alignment. Visual and acoustic stimuli that are in spatial and temporal congruence enhance the response, whereas two stimuli that are spatially or temporally disparate can lead to a depression of the neuron's activity (Meredith et al 1987; Newman & Hartline 1981).

The auditory map of space is synthesized from interaural time and intensity differences. In the barn owl, maps of interaural time difference (Carr & Konishi 1990; Sullivan & Konishi 1986) and maps of interaural intensity differences (Manley et al 1988) are formed in separate nuclei. Azimuth of target location is primarily related to interaural time difference, and target elevation to interaural intensity difference. However, the separation of the mapping between the two acoustic parameters and the two spatial parameters is not complete (Moiseff 1989). The elevation and azimuth of the location to which a barn owl turns its head depends in a linear fashion on both acoustic parameters. In any case, intensity and time information (or equivalently, elevation and azimuth) is combined in the superior colliculus.

In the barn owl, the range of eye movements is limited. Therefore, the problem of misalignment between the head-fixed auditory map and the eye-fixed visual map does not arise. Nevertheless, the auditory map is apparently in a visually defined frame of reference in this species. The auditory map remains aligned with the visual map when auditory input is altered by ear plugs (Knudsen 1985), or when the visual map is shifted by the use of displacing prisms (Knudsen & Knudsen 1989); the map is degraded when owls are raised with eyelids sutured (Knudsen 1988).

In cats and monkeys, the range of eye movement is much greater; thus, the potential for misalignment is also greater. Jay & Sparks (1984, 1987) have shown that the auditory map of space shifts with eye position. They trained monkeys to gaze at a fixation point and to make saccades (with the head fixed) to auditory and visual stimuli. They varied the fixation point and found that the receptive fields of neurons that responded to auditory stimuli shifted with the fixation point, i.e. with eye position. On average, the shift was by an amount smaller than the shift in eye position from one fixation point to another (Figure 5). Strictly speaking, the frame of reference for auditory space for these neurons is between a head-fixed and an eye-fixed one.

In the experiments of Jay & Sparks, the monkey, whose head was fixed, made only saccadic eye movements. What is the frame of reference of collicular maps when the head is also free to move? Is the frame of reference

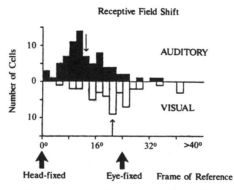

Figure 5   The reference frame of neurons in superior colliculus for representing the location of auditory and visual stimuli. The histogram describes the shift in neurons' receptive field after eye position (gaze) has shifted by 24°. The heavy arrows point to the amount of shift predicted if information were encoded in head-fixed (0°) or eye-fixed (24°) frames of reference. The median receptive field shift is indicated by the light arrows. Redrawn from Jay & Sparks (1987).

for eye and head movements the same? What is the frame of reference for the somesthetic map of body surface? These questions remain to be answered. Orienting movements of the eyes and the head only require information about the direction of target location (azimuth and elevation), but whole body orienting movements may also require information about the distance of the target (see below). Whether distance information is also encoded in the collicular map is not known.

Electrical stimulation of a site in the deeper layers of the superior colliculus evokes saccadic eye movements in the direction defined by the visual topographic map (Robinson 1972; Sparks 1986). The activity of neurons in the deeper layers is also correlated temporally with saccade onset (Sparks 1986). For these reasons, Sparks (1988) has suggested that the deeper layers represent a "motor map" for goal-directed movements (see also Grobstein 1988 for a discussion of this point).

The movement signal in superior colliculus, however, is not in the coordinate system of the muscles. For eye movements, the axes of the eye muscles' coordinate system are oriented vertically and horizontally (see above), and a separation of horizontal and vertical saccadic components in brain stem nuclei has been noted (Büttner & Büttner-Ennever 1988; Cohen et al 1985). There must be a transformation from the (coordinate-free) topographic map in superior colliculus to the different coordinate systems of eye, neck, and limb muscles. There is evidence (primarily from lower vertebrates) that this transformation involves an intermediate coordinate system whose axes are the spatial azimuth, elevation (and distance) of the movement (Grobstein 1988); this intermediate coordinate system is common to all effectors; and the transformation involves a population vector coding by collicular neurons (van Gisbergen et al 1987; Lee et al 1988).

Lee et al (1988) have demonstrated vector coding by reversibly inactivating small regions of the deep layers of superior colliculus and measuring saccadic error for eye movements in different directions. Saccades to targets lying within the center of the receptive field of the inactivated area were not in error, but those to targets at directions to either side were. These results imply that each collicular neuron provides a vectorial contribution to the code for movement; this contribution is in the neuron's best direction, and the movement is predicted by the vectorial average of the activity of all active neurons, i.e. a population vector code.

Evidence in favor of intermediate coordinate systems comes from two sources. Masino & Knudsen (1990) took advantage of the fact that there is refractoriness to electrical stimulation of the tectum—there is no movement evoked by the second of two stimuli presented in brief succession at the same locus (Robinson 1972). In the barn owl, they stimulated two different tectal sites in brief succession. The direction of the head movement evoked by the first stimulus was arbitrary; the direction of movement in response to the second stimulus was either horizontal or vertical, but never oblique (Figure 6A). For example, if stimulation of the first site evoked upward, leftward head movement, and stimulation of the second site in isolation evoked downward, leftward head movement, then the response to the second of the two stimuli presented in quick succession would be restricted to the downward direction, i.e. the direction that was not in common with the first movement. There was a refractoriness to the left-ward component of the movement, as that was a coordinate axis common to the two tectal sites. The pulling directions of the neck muscles are widely distributed; thus, the horizontal and vertical axes of this intermediate coordinate system are not aligned with the coordinate axes of the neck muscles.

Experiments on whole body orienting movements in the frog suggest that the same spatial intermediate coordinate system may also be used to encode body movements. Presented with a worm, a frog orients its body to the target by turning (dependent on the azimuthal location of the target) and by hopping or snapping (dependent on the distance of the target from the frog). Large lesions in the optic tectum abolish this response, but hemisection of the caudal mesencephalon leads to a very different deficit (Kostyk & Grobstein 1987). Frogs still respond by hopping or snapping, but fail to turn if the stimulus is located to one side of the sagittal plane.

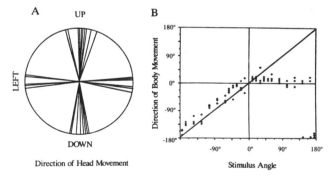

Figure 6 Intermediate coordinate systems for head and body orienting movements. (*A*) The directions of head movements evoked by the second of two electric stimuli to a region of the optic tectum in the owl are restricted to the horizontal or vertical directions. (*B*) Brain stem lesions in the frog abolish the horizontal (azimuthal) component of body orienting responses to one side. For stimulus angles greater than 0°, the direction of body movement was straight ahead. (*A*) is redrawn from Masino & Knudsen (1990), (*B*) from Masino & Grobstein (1989a).

Normally, there is a transition from snap to hop at a characteristic distance that depends on the azimuthal location. Lesioned frogs also exhibit this transition, but always at the distance characteristic of targets located straight ahead. That is, the frogs produce a behavior that would have been appropriate had the worm been located straight ahead. A similar deficit can be evoked by localized lesions at the junction of the medulla and the spinal cord (Masino & Grobstein 1989a,b) as shown in Figure 6*B*. An intact tecto-tegmento-spinal pathway is necessary to produce normal behavior.

ARM MOVEMENTS TO A SPATIAL TARGET

Arm movements to a spatial target also utilize sensory information that is initially represented in different frames of reference, and the sensory signals that specify target location need to be transformed into motor commands to arm muscles. Thus, the same questions concerning frames of reference and coordinate transformations that we have dealt with for eye, head, and body movements also arise in the study of arm movements. However, arm movements also illustrate an additional aspect of sensorimotor transformations: the distinction between forces and the movements that the forces produce.

For eye movements, a torque applied to the eye produces rotation about the torque axis. Therefore, forces and movements are colinear, and the coordinate system for forces and movements can be assumed to be the same. This is not usually the case for the arm, as illustrated in Figure 7. Consider a force directed downward (F_2) that is resisted by muscle activation. If the force is suddenly released, the arm does not begin to move

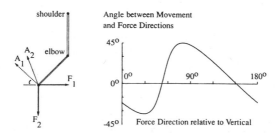

Figure 7   The directions of force and movement are not colinear for the arm. On the left, the dashed lines indicate the initial direction of hand acceleration (A) when a force (F) is suddenly released. On the right is shown how the difference between force direction and movement direction varies with the force direction. These results were computed from the equations of motion of the arm (Hollerbach & Flash 1982) by using typical values for the moments of inertia of the arm.

in the direction opposite to the force (i.e. straight up). Instead, the arm moves upward and forward (A_2). Similarly, release of a posteriorly directed force (F_1) also leads to forward and upward movement (A_1). The difference between the direction in which muscles exert their force and the direction in which the arm moves depends on the orientation of the force vector (Figure 7, *right*), on the posture of the arm, and on the arm's angular motion.

Thus, the transformation between kinematics (movement) and kinetics (forces) is nontrivial in the case of arm motion. Not much is known about how this transformation might be implemented by neural circuits. Mathematical formulations of the problem have been provided by several investigators (Hollerbach & Flash 1982; Hoy & Zernicke 1986; Zajac & Gordon 1989). Other investigators have quantified biomechanical factors, such as muscle stiffness (Mussa-Ivaldi et al 1985) and the changes in the muscles' lever arms with posture (Wood et al 1989), which also affect the relationship between force and movement.

Arm muscle activation vectors for isometric forces have been empirically determined (Buchanan et al 1986, 1989; Flanders & Soechting 1990b). In contrast to the patterns for neck muscle activation, static arm muscle activation sometimes deviates substantially from single cosine tuning functions, which suggests a complex vector code. Arm muscle activation onsets (Hasan & Karst 1989) and activation waveforms (Flanders 1991) have been empirically related to the direction of movement.

There is evidence (described below) that neurons in motor cortex, like those in the superior colliculus, encode the direction of movement by a population vector code. We now focus on three questions: What is the sensory information required to compute movement direction? In which frame(s) of reference is it represented? What is known about sensorimotor transformations for arm movements?

To move to a target accurately in the absence of visual guidance, the starting point of the movement, as well as the desired final point, must be sensed (Bizzi et al 1984; Hogan 1985), as is the case also for eye movements (Mays & Sparks 1980). Information about target location is provided by the visual system, whereas proprioceptors are adequate to signal initial arm posture. Because proprioceptors sense muscle length and joint angles (McCloskey 1978), the initial frames of reference for kinesthesis are fixed to the limb segments, i.e. elbow joint angles are initially sensed in the frame of reference fixed to the upper arm. There is psychophysical evidence that this representation of joint angles is transformed to a frame of reference fixed in space (Soechting 1982). Soechting & Ross (1984) found that subjects were best able to match joint angles of their right and left arms when they were measured relative to the vertical axes and the sagittal plane (see also Worringham & Stelmach 1985; Worringham et al 1987). In particular, these experiments identified yaw and elevation angles as a spatial coordinate system for arm orientation.

Target location is initially defined in a reference frame centered at the eyes. Other psychophysical experiments indicate that the origin of this reference frame is shifted toward the shoulder during the neural processing for targeted arm movements (Soechting et al 1990). In this shoulder-centered frame of reference, target location is defined by three parameters: distance, elevation, and azimuth, i.e. a spherical coordinate system (Soechting & Flanders 1989a).

The direction of hand movement is the difference between initial hand location and the location of the target. An analysis of human pointing errors suggests that there is a coordinate transformation from target coordinates to hand (arm) coordinates. The intended, final arm position is computed from target location by a linear transformation that is only approximately accurate (Soechting & Flanders 1989b). This transformation involves two separate channels: Arm elevation is computed from target distance and elevation, and arm yaw is computed from target azimuth (Flanders & Soechting 1990a). Thus, visually derived target coordinates are transformed into a common frame of reference with kinesthetically derived arm coordinates (Helms Tillery et al 1991).

A model that synthesizes these observations (Flanders et al 1992; Soechting & Flanders 1991) ends at the point at which a movement vector is defined by the difference between the intended arm orientation and the initial arm orientation. Thus, the model provides a description of the kinematic coordinate transformations required for goal-directed arm movements, and the transformation to kinetics is beyond its scope.

Because these transformations involve cortical processing, it is interesting to consider which parameters the cortical activity encodes. Since the pioneering work of Evarts (1968), who studied one-dimensional movements, researchers have recognized that discharge of motor cortical neurons is strongly correlated with force (see also Humphrey et al 1970). This, plus the strong monosynaptic connections of pyramidal tract neurons to motoneurons of distal muscles (cf. Kuypers 1981), leads to the interpretation that kinetic parameters are encoded by motor cortical activity.

A different perspective has been provided by Georgopoulos and co-workers (reviewed by Georgopoulos 1986, 1990), who studied the neural correlates of two- or three-dimensional reaching movements. Activity in motor cortex and in area 5 was best correlated with the direction of the movement (i.e. the difference between the initial and final hand positions in space) in a vectorial code (Georgopoulos et al 1982, 1984; Kalaska et al 1983; Schwartz et al 1988). Each neuron's activity defined a direction in space (the "best direction"); for other directions, activity was proportional to the cosine of the angle between that direction and the best direction. The best directions were distributed uniformly in space.

From these observations, Georgopoulos et al (1984) deduced that the motor command for movement direction is determined by the discharge of the entire population (the population vector), and that each cell provides a vectorial contribution to this command. This vector is in the cell's best direction and has an amplitude proportional to the cell's discharge (see Figure 8). The neuronal population vector agrees well with the observed hand trajectories (Figure 8), even when it is computed every 20 ms (Georgopoulos & Massey 1988; Georgopoulos et al 1984, 1988).

Taken at face value, the results of Georgopoulos and coworkers imply that motor cortical activity encodes movement direction, i.e. a kinematic parameter. Kalaska (1991) has attempted to reconcile these findings with earlier observations that neural activity was correlated with force. He suggested that the population vector encodes a kinetic parameter, such as

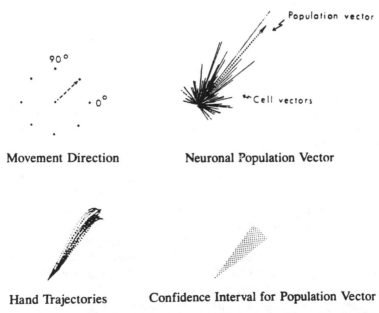

Movement Direction   Neuronal Population Vector

Hand Trajectories   Confidence Interval for Population Vector

Figure 8   Movement direction is encoded vectorially by the activity of a population of motor cortical neurons. For hand movements in the 45° direction, each cortical neuron makes a vectorial contribution in its best direction (*top right*). The vector sum of the cell vectors is the population vector. The 95% confidence interval of the population vector (*bottom right*) approximates the variability in the hand trajectories (*bottom left*). Redrawn from Georgopoulos et al (1984).

the direction of force. He has interpreted his experimental evidence in favor of this suggestion. Kalaska et al (1989) applied static loads to the monkey's arm and found that the neural discharge was tuned to both the direction of the static load and to the direction of a planar arm movement. Although the best directions for movement and for static load were, on average, 180° apart, there was a broad distribution in the angular difference between the two directions.

One would not expect such a broad divergence if the activity of each cell encoded a single parameter measured under two conditions. However, this divergence might be expected if the tonic and phasic activities of the cell were related to two different parmeters (i.e. static load direction and movement kinematics). Also, as shown in Figure 7, a cosine tuning to a kinematic parameter (such as movement direction) would not generally correspond with a cosine tuning of a kinetic parameter, such as force direction, because the difference between force and movement is a non-linear function of force direction. Without a more precise kinematic and dynamic analysis of the movements, the results of Kalaska et al (1989) are inconclusive. Finally, the population vector does not reverse direction as it evolves over time (Georgopoulos et al 1984), but force does reverse direction as the movement is decelerated.

For these reasons it appears that a kinematic representation of movement direction in motor cortical neurons is compatible with experimental evidence, at least for proximal muscles. Connections between motor cortical neurons and proximal motoneurons are primarily via interneurons (Kuypers 1981; Preston et al 1967), such as the propriospinal neurons described by Lundberg (1979) and Alstermark et al (1981, 1986). These interneuronal circuits could provide the substrate for the transformation from movement kinematics to movement kinetics.

CONCLUDING REMARKS

We have discussed how spatial parameters may be represented by the activities of neurons involved in several different motor tasks. We applied geometric constructs borrowed from classical physics and outlined a step-wise procedure to answer this question. Central to the procedure is the concept of a frame of reference. We have given this term its traditional meaning, even though activity in the central nervous system may never conform exactly to the criteria outlined at the beginning of the review. For example, in the superior colliculus, the frame of reference for auditory stimuli is not exactly eye-fixed, and the direction vector of motor cortical neurons is not exactly in an earth-fixed frame of reference (Caminiti et al 1990). Thus, the concept of an eye-fixed frame of reference in the former case, and of one fixed in space in the latter, is only an approximation.

Nevertheless, reference frames provide a useful point of departure for understanding information processing by neural structures. This is not a given. For example, connectionist models can lead to a very different perspective. In such models, activity in both input and output layers is defined in specific frames of reference, but activity in intervening (hidden) layers need not be in any frame of reference. These hidden layers receive and send highly divergent and convergent projections from other layers. The synaptic weights of these connections are initially random and are then modified iteratively to produce the desired behavior (Sejnowski et al 1988). Because the initial pattern of connectivity is random, the receptive fields of elements in the network would be different from one implementation to the next. Each neuron would have its own idiosyncratic frame of reference. Such a model has been useful in interpreting the visual receptive fields of neurons in parietal cortex (Andersen & Zipser 1988). These receptive fields cannot be defined in any specific frame of reference; instead, they behave as if these neurons were part of an intermediate layer in the transformation from eye-fixed to head-fixed frames of reference.

However, in the examples reviewed here, approximate frames of reference do appear definable. Once a frame of reference has been identified, we can ask how information is encoded in that frame of reference. A variety of neural codes exist, such as topographic (place) codes, vectorial coding, and coding along coordinate axes. In any given system, these different codes may coexist. For example, the spatial coordinates (azimuth and elevation) of sound location appear to be segregated initially (i.e. as time and intensity differences), then combined in the optic tectum in a place code, only to be segregated again in the brainstem. Similarly, the representations of the target location for arm movements appear to be encoded topographically in the retina, in a coordinate system in the intermediate representation, and vectorially in motor cortex.

Coordinate systems have been identified for the three motor tasks we have discussed, either electrophysiologically (Peterson & Baker 1991; Simpson 1984) or behaviorally (Flanders et al 1992; Maioli & Poppele 1989; Masino & Grobstein 1989a,b). It may not be coincidental that in all three motor tasks, one of the coordinate axes was defined by the gravitational vertical. Another coordinate was defined by a sagittal horizontal axis. Thus, one may suggest that, ultimately, there is a common, earth-fixed frame of reference utilized for all motor tasks.

We move in a three-dimensional world dominated by the force of gravity and by the visual horizon. Although one may not be consciously aware of gravitational force (Lackner & Graybiel 1984), its influence on movement is readily appreciated when one observes the movements of astronauts under conditions of microgravity. The vestibular system provides a primary, but not sole (Berthoz et al 1979), indicator of the vertical direction, and one can suggest that other coordinate systems may be aligned with the one defined by the vestibular afferents. In this context, it is noteworthy that the head is usually stabilized in space (Pozzo et al 1990), thus providing an inertial platform for sensing the vertical.

One advantage of representing information in different parts of the brain in a common, spatial frame of reference might be that the exchange of information is facilitated. This would be especially true if the same parameters (e.g. the same coordinate system) were represented in each part. Electrophysiological data on superior colliculus and motor cortex (two major command centers for movement) suggest that this is the case. Neural activity in both structures appears to encode movement kinematics, specifically the movement direction (vector difference between initial and final position). A transformation from kinematic to kinetic parameters occurs much later, perhaps in spinal cord (Georgopoulos 1990).

Representations of kinematics can be effector-independent, whereas codes of kinetics (or muscle activation) are not. Thus, the same kinematic signal could be used to encode an orienting movement if it was effected by the eyes, the head, the body, or a combination of all three. The structure provided by kinematic codes in common coordinate systems can provide the ability for a system to process information from a variety of stimuli concurrently and to respond to one stimulus by a variety of movements.

ACKNOWLEDGMENTS

The authors thank Drs. A. P. Georgopoulos, P. Grobstein, R. E. Poppele, and J. I. Simpson for helpful discussions on topics discussed in this review. The authors' work was supported by National Institutes of Health Grants NS-15018 and NS-27484.

Literature Cited

Alstermark, B., Gorska, T., Johannisson, T., Lundberg, A. 1986. Hypermetria in forelimb target-reaching after interruption of the inhibitory pathway from forelimb afferents to C3–C4 propriospinal neurones. *Neurosci. Res.* 3: 457–61

Alstermark, B., Lundberg, A., Norsell, U., Sybirska, E. 1981. Integration in descending motor pathways controlling the forelimb in the cat. 9. Differential behavioral defects after spinal cord lesions interrupting defined pathways from higher centers to motoneurones. *Exp. Brain Res.* 42: 299–318

Andersen, R. A., Zipser, D. 1988. The role of posterior parietal cortex in coordinate transformations for visual-motor integration. *Can. J. Physiol. Pharmacol.* 66: 488–501

Baker, J. F., Banovetz, J. M., Wickland, C. R. 1988a. Models of sensorimotor transformations and vestibular reflexes. *Can. J. Physiol. Pharmacol.* 66: 532–39

Baker, J., Goldberg, J., Herrmann, G., Peterson, B. 1984. Optimal response planes and canal convergence in secondary neurons in vestibular nuclei of alert cats. *Brain Res.* 294: 133–37

Baker, J., Goldberg, J., Peterson, B. 1985. Spatial and temporal responses of the vestibulocollic reflex in decerebrate cats. *J. Neurophysiol.* 54: 735–56

Baker, J., Wickland, C., Goldberg, J., Peterson, B. 1988b. Motor output to lateral rectus in cats during the vestibulo-ocular reflex in three-dimensional space. *Neuroscience* 25: 1–12

Berthoz, A., Lacour, M., Soechting, J. F., Vidal, P. P. 1979. The role of vision in the control of posture during linear motion. *Progr. Brain Res.* 50: 197–210

Bizzi, E., Accornero, N., Chapple, W., Hogan, N. 1984. Posture control and trajectory formation during arm movement. *J. Neurosci.* 4: 2738–44

Blanks, R. H. I., Curthoys, I. S., Markham, C. H. 1972. Planar relationships of semicircular canals in the cat. *Am. J. Physiol.* 223: 55–62

Buchanan, T. S., Almdale, D. P. J., Lewis, J. L., Rymer, W. Z. 1986. Characteristics of synergic relations during isometric contractions of human elbow muscles. *J. Neurophysiol.* 56: 1225–41

Buchanan, T. S., Rovai, G. P., Rymer, W. Z. 1989. Strategies for muscle activation during isometric torque generation at the human elbow. *J. Neurophysiol.* 62: 1201–12

Büttner, U., Büttner-Ennever, J. A. 1988. Present concepts of oculomotor organization. *Rev. Oculomot. Res.* 2: 3–32

Caminiti, R., Johnson, P. B., Urbano, A. 1990. Making arm movements within different parts of space: dynamic aspects in the primate motor cortex. *J. Neurosci.* 10: 2039–58

Carr, C. E., Konishi, M. 1990. A circuit for the detection of interaural time differences in the brain stem of the barn owl. *J. Neurosci.* 10: 3227–46

Cohen, B., Matsuo, V., Raphan, T., Waitzman, D., Fradin, J. 1985. Horizontal saccades induced by stimulation of the central mesencephalic reticular formation. *Exp. Brain Res.* 57: 605–16

Dichgans, J., Brandt, T. 1978. Visual-vestibular interactions: Effects on self-motion perception and postural control. In *Handbook of Sensory Physiology*, ed. R. Held, H. Leibowitz, H. L. Teuber, 8: 756–804. Berlin: Springer

Droulez, J., Darlot, C. 1990. The geometric and dynamic implications of the coherence constraints in three-dimensional sensorimotor interactions. In *Attention and Performance XIII. Motor Representation and Control*, ed. M. Jeannerod, pp. 495–526. Hillsdale, NJ: Erlbaum

Estes, M., Blanks, R., Markham, C. 1975. Physiological characteristics of vestibular first order canal neurons in the cat. I. Response plane determination and resting discharge characteristics. *J. Neurophysiol.* 38: 1232–49

Evarts, E. V. 1968. Relation of pyramidal tract activity to force exerted during voluntary movement. *J. Neurophysiol.* 31: 14–27

Ezure, K., Graf, W. 1984a. A quantitative analysis of the spatial organization of the vestibulo-ocular reflexes in lateral- and frontal-eyed animals. I. Orientation of semicircular canals and extraocular muscles. *Neuroscience* 12: 85–94

Ezure, K., Graf, W. 1984b. A quantitative analysis of the spatial organization of the vestibulo-ocular reflexes in lateral- and frontal-eyed animals. II. Neuronal networks underlying vestibulo-oculomotor coordination. *Neuroscience* 12: 95–110

Fernandez, C., Goldberg, J. M. 1971. Physiology of peripheral neurones innervating semicircular canals of the squirrel monkey. II. Response to sinusoidal stimulation and dynamics of peripheral vestibular system. *J. Neurophysiol.* 34: 661–75

Fernandez, C., Goldberg, J. M. 1976. Physiology of peripheral neurones innervating otolith organs of the squirrel monkey. II. Directional sensitivity and force response relations. *J. Neurophysiol.* 39: 985–95

Flanders, M. 1991. Temporal patterns of muscle activation for arm movements in three-dimensional space. *J. Neurosci.* 11: 2680–93

Flanders, M., Soechting, J. F. 1990a. Parcellation of sensorimotor transformations for arm movements. *J. Neurosci.* 10: 2420–27

Flanders, M., Soechting, J. F. 1990b. Arm muscle activation for static forces in three-dimensional space. *J. Neurophysiol.* 64: 1818–37

Flanders, M., Helms-Tillery, S. I., Soechting, J. F. 1992. Early stages in a sensorimotor transformation. *Behav. Brain Sci.* In press

Georgopoulos, A. P. 1986. On reaching. *Annu. Rev. Neurosci.* 9: 147–70

Georgopoulos, A. P. 1990. Neurophysiology of reaching. In *Attention and Performance XIII. Motor Representation and Control*, ed. M. Jeannerod, pp. 227–64. Hillsdale: Erlbaum

Georgopoulos, A. P., Kalaska, J. F., Caminiti, R., Massey, J. T. 1982. On the relations between the direction of two-dimensional arm movements and cell discharge in primate motor cortex. *J. Neurosci.* 2: 1527–37

Georgopoulos, A. P., Kalaska, J. F., Crutcher, M. D., Caminiti, R., Massey, J. T. 1984. The representation of movement direction in the motor cortex: single cell and population studies. In *Dynamical Aspects of Cortical Function*, ed. G. M.

Edelman, W. E. Gall, W. M. Cowan, pp. 453–73. New York: Wiley

Georgopoulos, A. P., Kettner, R. E., Schwartz, A. B. 1988. Primate motor cortex and free arm movements to visual targets in three-dimensional space. II. Coding of the direction by a neuronal population. *J. Neurosci.* 8: 2928–37

Georgopoulos, A. P., Massey, J. T. 1988. Cognitive spatial-motor processes. 2. Information transmitted by the direction of two-dimensional arm movements and by neuronal populations in primate motor cortex and area 5. *Exp. Brain Res.* 69: 315–26

Graf, W. 1988. Motion detection in physical space and its peripheral and central representation. In *Representation of Three-dimensional Space in the Vestibular, Oculomotor and Visual Systems, Ann. NY Acad. Sci.*, eds. B. Cohen, V. Henn, 545: 154–69. New York: NY Acad. Sci.

Graf, W., Simpson, J. I., Leonard, C. S. 1988. Spatial organization of visual messages of the rabbit's cerebellar flocculus. II. Complex and simple spike responses of Purkinje cells. *J. Neurophysiol.* 60: 2091–2121

Grobstein, P. 1988. Between the retinotectal projection and directed movement: topography of a sensorimotor interface. *Brain Behav. Evol.* 31: 34–48

Hasan, Z., Karst, G. M. 1989. Muscle activity for initiation of planar, two-joint arm movements in different directions. *Exp. Brain Res.* 76: 651–55

Helms Tillery, S. I., Flanders, M., Soechting, J. F. 1991. A coordinate system for the synthesis of visual and kinesthetic information. *J. Neurosci.* 11: 770–78

Hogan, N. 1985. The mechanics of multijoint posture and movement control. *Biol. Cybern.* 52: 315–32

Hollerbach, J. M., Flash, T. 1982. Dynamic interactions between limb segments during planar arm movement. *Biol. Cybern.* 44: 67–77

Hoy, M. G., Zernicke, R. F. 1986. The role of intersegmental dynamics during rapid limb oscillations. *J. Biomech.* 19: 867–77

Humphrey, D. R., Schmidt, E. M., Thompson, W. D. 1970. Predicting measures of motor performance from multiple cortical spike trains. *Science* 170: 758–61

Jay, M. F., Sparks, D. L. 1984. Auditory receptive fields in primate superior colliculus shift with changes in eye position. *Nature* 309: 345–47

Jay, M. F., Sparks, D. L. 1987. Sensorimotor integration in the primate superior colliculus. II. Coordinates of auditory signals. *J. Neurophysiol.* 57: 35–55

Kalaska, J. F. 1991. What parameters of reaching are encoded by the discharge of cortical cells? In *Motor Control: Concepts and Issues. Dahlem Konferenzen*, ed. D. R. Humphrey, H.-J. Freund, pp. 307–30. Chichester: Wiley

Kalaska, J. F., Caminiti, R., Georgopoulos, A. P. 1983. Cortical mechanisms related to the direction of two-dimensional arm movements: Relations in parietal Area 5 and comparison with motor cortex. *Exp. Brain Res.* 51: 247–60

Kalaska, J. F., Cohen, D. A. D., Hyde, M. L., Prud'homme, M. 1989. A comparison of movement direction-related versus load direction-related activity in primate motor cortex, using a two-dimensional reaching task. *J. Neurosci.* 9: 2080–2102

Kasper, J., Schor, R. H., Wilson, V. J. 1988a. Response of vestibular neurons to head rotations in the vertical plane. I. Response to vestibular stimulation. *J. Neurophysiol.* 60: 1753–64

Kasper, J., Schor, R. H., Wilson, V. J. 1988b. Response of vestibular neurons to head rotations in the vertical plane. II. Response to neck stimulation and vestibular-neck interaction. *J. Neurophysiol.* 60: 1765–78

Knudsen, E. I. 1982. Auditory and visual maps of space in the optic tectum of the owl. *J. Neurosci.* 2: 1177–94

Knudsen, E. I. 1985. Experience alters the spatial tuning of auditory units in the optic tectum during a sensitive period in the barn owl. *J. Neurosci.* 5: 3094–3109

Knudsen, E. I. 1988. Early blindness results in degraded auditory map of space in the optic tectum of the barn owl. *Proc. Natl. Acad. Sci. USA* 85: 6211–15

Knudsen, E. I., du Lac, S., Esterly, S. 1987. Computational maps in the brain. *Annu. Rev. Neurosci.* 10: 41–65

Knudsen, E. I., Knudsen, P. F. 1989. Visuomotor adaptation to displacing prisms by adult and baby barn owls. *J. Neurosci.* 9: 3297–3305

Kostyk, S. K., Grobstein, P. 1987. Neuronal organization underlying visually elicited prey orienting in the frog. I. Effects of various unilateral lesions. *Neuroscience* 21: 41–55

Kuypers, H. G. J. M. 1981. Anatomy of the descending pathways. In *Handbook of Physiology. The Nervous System*, ed. J. M. Brookhart, V. B. Mountcastle, 2: 597–666. Bethesda: Am. Physiol. Soc.

Lackner, J. R., Graybiel, A. 1984. Perception of body weight and body mass at twice earth-gravity acceleration levels. *Brain* 107: 133–44

Lacquaniti, F., LeTaillanter, M., Lopiàno, L., Maioli, C. 1990. The control of limb geometry in cat posture. *J. Physiol. (London)* 426: 177–92

Lee, C., Rohrer, W. H., Sparks, D. L. 1988. Population coding of saccadic eye movements by neurons in the superior colliculus. *Nature* 332: 357–59

Leonard, C. S., Simpson, J. I., Graf, W. 1988. Spatial organization of visual messages of the rabbit's cerebellar flocculus. I. Typology of inferior olive neurons of the dorsal cap of Kooy. *J. Neurophysiol.* 60: 2073–90

Lundberg, A. 1979. Integration in a propriospinal centre controlling the forelimb in the cat. In *Integration in the Nervous System*, ed. H. Asanuma, V. J. Wilson, pp. 47–69. Tokyo: Igaku-Shoin

Macpherson, J. M. 1988. Strategies that simplify the control of quadrupedal stance. I. Forces at the ground. *J. Neurophysiol.* 60: 204–17

Maekawa, K., Simpson, J. J. 1973. Climbing fiber responses evoked in vestibulocerebellum of rabbit from visual system. *J. Neurophysiol.* 36: 649–66

Maioli, C., Lacquaniti, F. 1988. Determinants of postural control in cats: a biomechanical study. In *Posture and Gait: Development, Adaptation and Modulation*, ed. B. Amblard, A. Bethoz, F. Clarac, pp. 371–79. Amsterdam: Elsevier

Maioli, C., Poppele, R. E. 1989. Dynamic postural responses in the cat involve independent control of limb length and orientation. *Soc. Neurosci. Abstr.* 15: 392

Manley, G. A., Koppl, C., Konishi, M. 1988. A neural map of interaural intensity differences in the brain stem of the barn owl. *J. Neurosci.* 8: 2665–76

Masino, T., Grobstein, P. 1989a. The organization of descending tectofugal pathways underlying orienting in the frog, Rana pipiens. I. Lateralization, parcellation, and an intermediate representation. *Exp. Brain Res.* 75: 227–44

Masino, T., Grobstein, P. 1989b. The organization of descending tectofugal pathways underlying orienting in the frog, Rana pipiens. II. Evidence for the involvement of a tecto-spinal pathway. *Exp. Brain Res.* 75: 245–64

Masino, T., Knudsen, E. I. 1990. Horizontal and vertical components of head movement are controlled by distinct neural circuits in the barn owl. *Nature* 345: 434–37

Mays, L. E., Sparks, D. L. 1980. Saccades are spatially, not retinocentrically, coded. *Science* 208: 1163–65

McCloskey, D. I. 1978. Kinesthetic sensibility. *Physiol. Rev.* 58: 763–820

Meredith, M. A., Nemitz, J. W., Stein, B. E. 1987. Determination of multisensory integration in superior colliculus neurons.

I. Temporal factors. *J. Neurosci.* 7: 3215–29

Meredith, M. A., Stein, B. E. 1986. Spatial factors determine the activity of multisensory neurons in cat superior colliculus. *Brain Res.* 365: 350–54

Middlebrooks, J. C., Knudsen, E. I. 1984. A neural code for auditory space in the cat's superior colliculus. *J. Neurosci.* 4: 2621–34

Moiseff, A. 1989. Bi-coordinate sound localization by the barn owl. *J. Comp. Physiol. A* 164: 637–44

Mussa-Ivaldi, F. A., Hogan, N., Bizzi, E. 1985. Neural, mechanical and geometric factors subserving arm posture in humans. *J. Neurosci.* 5: 2732–43

Nashner, L. M., McCollum, G. 1985. The organization of human postural movements: a formal basis and experimental synthesis. *Behav. Brain Sci.* 8: 135–72

Newman, E. A., Hartline, P. H. 1981. Integration of visual and infrared information in bimodal neurons of the rattlesnake optic tectum. *Science* 213: 789–91

Oyster, C. W., Takahashi, E., Collewijn, H. 1972. Direction selective retinal ganglion cells and control of optokinetic nystagmus in the rabbit. *Vision Res.* 12: 183–93

Pellionisz, A. 1985. Tensorial aspects of the multi-dimensional approach to the vestibulo-oculomotor reflex and gaze. In *Reviews of Oculomotor Control. I. Adaptive Mechanisms in Gaze Control*, ed. A. Berthoz, G. Melvill Jones, pp. 281–96. Amsterdam: Elsevier

Pellionisz, A., Graf, W. 1987. Tensor network model of the "three-neuron vestibulo-ocular reflex-arc" in cat. *J. Theor. Neurobiol.* 5: 127–51

Pellionisz, A., Llinás, R. 1980. Tensorial approach to the geometry of brain function: cerebellar coordination via a metric tensor. *Neuroscience* 5:1125–36

Pellionisz, A., Llinás, R. 1982. Space-time representation of the brain. The cerebellum as a predictive space-time metric tensor. *Neuroscience* 7: 2949–70

Pellionisz, A., Peterson, B. W. 1988. A tensorial model of neck motor activation. In *Control of Head Movement*, ed. B. W. Peterson, F. Richmond, pp. 178–86. Oxford: Oxford Univ. Press

Peterson, B. W., Baker, J. F. 1991. Spatial transformations in vestibular reflex systems. In *Motor Control: Concepts and Issues. Dahlem Konferenzen*, ed. D. R. Humphrey, H.-J. Freund, pp. 121–36. Chichester: Wiley

Peterson, B. W., Baker, J. F., Goldberg, J., Banovetz, J. 1988. Dynamic and kinematic properties of the vestibulocollic and cervicocollic reflexes in the cat. *Progr. Brain Res.* 76: 163–72

Peterson, B. W., Pellionisz, A. J., Baker, J. F., Keshner, E. A. 1989. Functional morphology and neural control of neck muscles in mammals. *Am. Zool.* 29: 139–49

Pozzo, T., Berthoz, A., Lefort, L. 1990. Head stabilization during various locomotor tasks in humans. *Exp. Brain Res.* 82: 97–106

Preston, J. B., Shende, M. C., Uemura, K. 1967. The motor cortex-pyramidal system: patterns of facilitation and inhibition on motoneurons innervating the limb musculature of cat and baboon. In *Neurophysiological Basis of Normal and Abnormal Motor Activities*, ed. M. D. Yahr, D. P. Purpura, pp. 61–72. New York: Raven

Reisine, H., Simpson, J. I., Henn, V. 1988. A geometric analysis of semicircular canals and induced activity in their peripheral afferents in the rhesus monkey. In *Representation of Three-dimensional Space in the Vestibular, Oculomotor and Visual Systems*, Ann. NY Acad. Sci., ed. B. Cohen, V. Henn, 545: 10–20. New York: NY Acad. Sci.

Robinson, D. A. 1972. Eye movements evoked by collicular stimulation in the alert monkey. *Vision Res.* 12: 1795–1808

Robinson, D. A. 1982. The use of matrices in analyzing the three-dimensional behavior of the vestibulo-ocular reflex. *Biol. Cybern.* 46: 53–66

Schwartz, A. B., Kettner, R. E., Georgopoulos, A. P. 1988. Primate motor cortex and free arm movements to visual targets in three-dimensional space. I. Relations between single cell discharge and direction of movement. *J. Neurosci.* 8: 2913–27

Sejnowski, T. J., Koch, K., Churchland, P. S. 1988. Computational neuroscience. *Science* 241: 1299–1306

Simpson, J. I. 1984. The accessory optic system. *Annu. Rev. Neurosci.* 7: 13–41

Simpson, J. I., Leonard, C. S., Sodak, R. E. 1988. The accessory optic system of rabbit. II. Spatial organization of direction selectivity. *J. Neurophysiol.* 60: 2055–72

Sodak, R. E., Simpson, J. I. 1988. The accessory optic system of rabbit. I. Basic visual response properties. *J. Neurophysiol.* 60: 2037–54

Soechting, J. F. 1982. Does position sense at the elbow reflect a sense of elbow joint angle or one of limb orientation? *Brain Res.* 248: 392–95

Soechting, J. F., Flanders, M. 1989a. Sensorimotor representations for pointing to targets in three-dimensional space. *J. Neurophysiol.* 62: 582–94

Soechting, J. F., Flanders, M. 1989b. Errors in pointing are due to approximations in sensorimotor transformations. *J. Neurophysiol.* 62: 595–608

Soechting, J. F., Flanders, M. 1991. Deducing central algorithms of arm movement control from kinematics. In *Motor Control: Concepts and Issues. Dahlem Konferenzen*, ed. D. R. Humphrey, H.-J. Freund, pp. 293–306. Chichester: Wiley

Soechting, J. F., Ross, B. 1984. Psychophysical determination of coordinate representation of human arm orientation. *Neuroscience* 13: 595–604

Soechting, J. F., Tillery, S. I. H., Flanders, M. 1990. Transformation from head- to shoulder-centered representation of target direction in arm movements. *J. Cogn. Neurosci.* 2: 32–43

Sparks, D. L. 1986. Translation of sensory signals into commands for control of saccadic eye movements: role of primate superior colliculus. *Physiol. Rev.* 66: 118–71

Sparks, D. L. 1988. Neural cartography: sensory and motor maps in the superior colliculus. *Brain Behav. Evol.* 31: 49–56

Sullivan, W. E., Konishi, M. 1986. Neural map of interaural phase difference in the owl's brainstem. *Proc. Natl. Acad. Sci. USA* 83: 8400–4

van Gisbergen, J. A. M., van Opstal, A. J., Tax, A. A. M. 1987. Collicular ensemble coding of saccades based on vector summation. *Neuroscience* 21: 541–55

Wilson, V. J., Yamagata, Y., Yates, B. J., Schor, R. H., Nonaka, S. 1990. Response of vestibular neurones to head rotations in vertical planes. III. Response of vestibulocollic neurons to vestibular and neck stimulation. *J. Neurophysiol.* 64: 1695–1703

Wood, J. E., Meek, S. G., Jacobsen, S. C. 1989. Quantitation of human shoulder anatomy for prosthetic arm control. II. Anatomy matrices. *J. Biomech.* 22: 309–26

Worringham, C. J., Stelmach, G. E. 1985. The contribution of gravitational torques to limb position sense. *Exp. Brain Res.* 61: 38–42

Worringham, C. J., Stelmach, G. E., Martin, Z. E. 1987. Limb segment inclination sense in proprioception. *Exp. Brain Res.* 66: 653–58

Zajac, F. E., Gordon, M. E. 1989. Determining muscle's force and action in multiarticular movement. *Exercise Sport Sci. Rev.* 17: 187–230

Annu. Rev. Neurosci. 1992. 15:193–225

GUANYLYL CYCLASE-LINKED RECEPTORS

Peter S. T. Yuen and David L. Garbers

Howard Hughes Medical Institute and Department of Pharmacology, University of Texas Southwestern Medical Center, Dallas, Texas 75235-9050

KEY WORDS: spermatozoa, natriuretic peptides, enterotoxin, nitric oxide

PERSPECTIVES AND SUMMARY

Cyclic GMP was identified as a component of urine in 1963 (Ashman et al) shortly after the classical works of Sutherland & Rall had proposed the second messenger hypothesis (Rall et al 1957, Rall & Sutherland 1958, Sutherland & Rall 1957, 1958, 1960). Guanylyl cyclase and cyclic nucleotide phosphodiesterase activities were subsequently reported to exist in virtually all tissues (Appleman et al 1973, Hardman et al 1971, Hardman & Sutherland 1969, Ishikawa et al 1969, Schultz et al 1969, White & Aurbach 1969), and cyclic GMP concentrations were shown to change in response to a wide variety of different agents, including hormones and neurotransmitters (Goldberg & Haddox 1977). Progress in understanding the mechanisms by which cellular cyclic GMP concentrations were elevated in response to these various agents moved rather slowly, however, in part, because guanylyl cyclase (enzyme activity is found in both soluble and particulate fractions of most tissue homogenates) could not be activated in broken cell preparations (Murad et al 1979).

That guanylyl cyclases in both particulate and soluble fractions of tissue homogenates serve as receptors for a variety of different and specific ligands has become realized over the last ten years. This review concentrates on the guanylyl cyclase/receptor family and the nature of the ligands that specifically bind and activate these receptors. Other recent reviews that address this topic include Brenner et al (1990), Chinkers & Garbers (1991), Garbers (1989a,b), and Inagami (1989).

193

0147–006X/92/0301–0193$02.00

PURIFICATION AND SUBCELLULAR LOCALIZATION OF GUANYLYL CYCLASES

In the mid 1970s, two groups established that guanylyl cyclase activities present in the particulate and soluble parts of mammalian tissue homogenates were due to different proteins as opposed to the translocation of a single enzyme between soluble and particulate fractions (Kimura & Murad 1974, Chrisman et al 1975). During the same period, Garbers et al (1974) reported that a particulate form of guanylyl cyclase found in sea urchin spermatozoa closely resembled, with respect to kinetic properties, the particulate form found in various mammalian tissues; in later studies, polyclonal antibody to the sea urchin sperm enzyme was shown to immunoprecipitate detergent-solubilized guanylyl cyclase from rat lung and other rat tissues, thus establishing the likely presence of common amino acid sequences between the sea urchin and mammalian enzymes (Garbers 1979). Of some interest was the observation that not all of the rat lung detergent-solubilized enzyme activity was precipitated by the antibody, thereby suggesting heterogeneity among the membrane forms. Eventually, the particulate activity found in various mammalian tissues was subdivided into detergent-soluble and detergent-insoluble fractions. Relatively detergent-insoluble activity is especially prevalent in retina (Fleischman et al 1980, Fleischman & Denisevich 1979) and gastrointestinal mucosa (deJonge 1975a,b); these forms have not yet been purified to homogeneity.

A detergent-soluble plasma membrane form was first purified to apparent homogeneity from sea urchin spermatozoa (Garbers 1976, Radany et al 1983). It was shown to be a glycoprotein with an M_r of 135,000. Subsequently, guanylyl cyclase from a second sea urchin species was purified and shown to have an M_r of 150,000/160,000, depending on its state of phosphorylation (Suzuki et al 1984, Ward et al 1985). Not only could the guanylyl cyclase from the latter species of sea urchin be specifically activated by a peptide (resact; see Table 1) obtained from the egg-conditioned media of the same species (Bentley et al 1986, Suzuki et al 1984), but resact also could be specifically crosslinked to guanylyl cyclase in the presence of disuccinimidyl suberate (Shimomura et al 1986). These experiments suggested that guanylyl cyclase was a sperm surface receptor for resact.

It should be noted that speract (see Table 1), a peptide obtained from a different species of sea urchin, was crosslinked, in similar experiments, to a protein whose M_r was 77,000 (Dangott et al 1989, Dangott & Garbers 1984). Speract and resact do not crossreact in any detectable manner across those two species. The mRNA for the crosslinked protein (M_r 77,000) was

Table 1   The structures of peptides that activate guanylyl cyclase[a]

Peptide	Species	Structure
ANP	Rat, mouse, rabbit	SLRRSSCFGGRIDRIGAQSGLGCNSFR-Y
	Human, bovine, dog	SLRRSSCFGGRMDRIGAQSGLGCNSFR-Y
	Consensus	SLRRSSCFGGR DRIGAQSGLGCNSFR-Y
BNP	Rat	NS-K-MAHSSSCFGQKIDRIGAVSRLGCDGLRLF
	Human	SPK-MVQGSGCFGRKMDRISSSSGLGCKVLRRH
	Pig	SPKTM-RDSGCFGRRLDRIGSLSGLGCNVLRRY
	Consensus	S K M S CFG DRI S LGC LR
CNP	Pig	GLSKGCFGLKLDRIGSMSGLGC
	Frog	GYSRGCFGVKLDRIGAFSGLGC
	Chicken	GLSRSCFGVKLDRIGSMSGLGC
	Eel	GWNRGCFGLKLDRIGSLSGLGC
	Consensus	G CFG KLDRIG SGLGC
Speract	S. purpuratus	GFDLNGGGVG
Resact	A. punctulata	CVTGAPGCVGGGRL-NH$_2$
Mosact	C. japonicus	DSDSAQNLIG
ST	E. coli	NSSNYCCELCCNPACTGCY
ST	E. coli	NTFYCCELCCNPACAGCY
	Y. enterocolitica	QACDPPSPPAEVSSDWDCCDVCCNPACAGC
	V. cholerae	IDCCEICCNPACFGCLN
	Consensus	CC CCNPAC GC

[a] The eel CNP was originally thought to be BNP (Takei et al 1990).

subsequently cloned and shown not to resemble guanylyl cyclase; in fact, the M_r 77,000 protein turned out to be homologous to the macrophage scavenger receptor (Freeman et al 1990). Although the function of this domain in the scavenger receptor is not known (it does not appear to function in the endocytosis of modified LDL), it is also found in other proteins such as CD5 and complement factor I (Freeman et al 1990). Since speract and resact cause the same general biochemical responses in spermatozoa (although in a species-specific manner), including the elevations of cyclic GMP, it has been suggested that distinctly different receptor molecules do not mediate the actions of the peptides across the species. Possible explanations for the crosslinking results, then, are that the functional peptide receptor consists of at least two proteins, the M_r 77,000 protein and guanylyl cyclase, or that the M_r 77,000 protein and/or guanylyl cyclase are closely associated with the actual receptor, thus resulting in the crosslinking of either protein.

The specific activity of detergent-solubilized guanylyl cyclase from various mammalian tissues is about 1000-fold lower than that found in invertebrate spermatozoa (Gray et al 1976), but four laboratories have purified a mammalian form of the detergent-soluble particulate enzyme from various tissues to specific activities similar to those reported for the purified sea urchin enzyme (Kuno et al 1986a, Meloche et al 1988, Paul et al 1987, Takayanagi et al 1987). The small amounts of recovered protein precluded rigorous examination of purity, but the four groups found that radio-labeled atrial natriuretic peptide (ANP) specifically bound to a component of the preparation, possibly guanylyl cyclase itself. Added ANP failed to stimulate the activity of the purified enzyme. The M_r of the major-staining protein in the various preparations ranged between 130,000–180,000, dependent upon the tissue from which the enzyme was purified and the laboratory reporting the purification.

The soluble form of guanylyl cyclase was first purified to apparent homogeneity from rat lung (Garbers 1979). The protein was suggested to exist as a dimer (M_r 151,000) with subunit M_r of 79,400 and 74,000. The specific activity of the purified enzyme, however, was about 20 times lower than that found with the purified particulate enzyme of sea urchin spermatozoa (Garbers 1976, Radany et al 1983). The enzyme was subsequently purified from rat lung by Lewicki et al (1980), who also reported low specific activities relative to those of the purified sea urchin enzyme. An M_r of 150,000 agreed with that of Garbers (1979), but the presence of a single subunit of M_r 72,000 suggested that the enzyme could exist as a homodimer. In 1981, the soluble enzyme was purified in milligram quantities from bovine lung (Gerzer et al 1981a,b); the significance of these studies was the discovery of heme as a component of the purified enzyme. This represented one of the most important observations made at the time, since its discovery provided a model for the mechanisms by which nitric oxide could activate the enzyme. As shown by Lewicki et al (1980), a single subunit (M_r 72,000) was observed after electrophoresis on SDS gels. Whether the soluble form of guanylyl cyclase existed as a heterodimer or homodimer was finally resolved when Kamisaki et al (1986) separated the rat lung enzyme into two subunits of M_r 82,000 (α-subunit) and 70,000 (β-subunit) and demonstrated by the use of antibodies that the two subunits were unique.

Many activators of the soluble form of guanylyl cyclase were described during the 1970–1980 era (Murad et al 1979), but it was mainly research from Craven & DeRubertis and coworkers and from Murad's group that led to the discovery of nitric oxide as a potent regulator of the soluble enzyme (Braughler et al 1979, Craven et al 1979, Craven & DeRubertis 1978, DeRubertis & Craven 1976, Kimura et al 1975a,b). There also were significant reports at about the same time by others (Miki et al 1976, 1977).

PRIMARY STRUCTURES AND LIGANDS OF PLASMA MEMBRANE FORMS OF GUANYLYL CYCLASE

The mRNA encoding an invertebrate sperm plasma membrane form of guanylyl cyclase was first cloned in 1988 (Singh et al 1988). The predicted amino acid sequence suggested the presence of a single transmembrane domain that divided the protein in about one half. Only one region of the sea urchin enzyme was similar to primary amino acid sequences already present in protein databases, this being a region just within the trans-membrane domain that was homologous to the catalytic domains of protein kinases (Hanks et al 1988). Initially it was thought that this region could represent the cyclase catalytic domain, but later studies showed this not to be the case (Chinkers & Garbers 1989, Thorpe & Morkin 1990). The general topological model for the sea urchin plasma membrane guanylyl cyclase, as well as the mammalian plasma membrane forms to be discussed below, is shown in Figure 1. Since guanylyl cyclase could be crosslinked to resact, the cDNA was expressed in COS-7 cells to determine whether or not it would bind this egg peptide. Neither binding of resact nor guanylyl cyclase activity were observed in the transfected COS-7 cell extracts. The M_r of the overproduced protein, however, was considerably less than the M_r of the native protein (possibly due to glycosylation differences), and unlike the native enzyme, it was not phosphorylated (Singh et al 1988). Subsequently, the cDNA for a plasma membrane guanylyl cyclase from a different species of sea urchin was cloned (Thorpe & Garbers 1989). This cDNA predicted an extended carboxyl region relative to the first sea urchin enzyme, but expression studies again failed to demonstrate either egg peptide binding or guanylyl cyclase activity.

Since a long-term goal had been to determine whether or not any new cell surface receptor found in sea urchin spermatozoa would also be found in mammalian tissues, a 3'-cDNA probe (coding region) from the sea

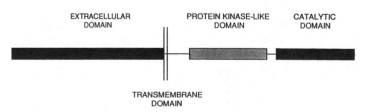

Figure 1   Schematic model for plasma membrane forms of guanylyl cyclase receptors.

urchin was used to isolate positive-hybridizing clones from human cDNA libraries (Lowe et al 1989). Full-length clones from rat (GC-A) (Chinkers et al 1989) and from human (ANP-A receptor) (Lowe et al 1989) were subsequently isolated. The predicted proteins from human and rat were greater than 90% identical. In later studies, Pandey & Singh (1990) isolated a cDNA clone encoding the same apparent receptor from mouse Leydig cells.

By the time the GC-A clone was obtained, various mammalian peptides had been shown to activate a particulate form of guanylyl cyclase in broken cell preparations (Table 1). These included sea urchin egg peptides, peptides with natriuretic properties, and heat-stable enterotoxins. Since four groups had independently reported that ANP continued to bind to a component of a highly purified preparation of guanylyl cyclase (Kuno et al 1986a, Meloche et al 1988, Paul et al 1987, Takayanagi et al 1987), this peptide was first tested for effects on expressed GC-A. ANP (nanomolar concentrations) caused marked elevations of cyclic GMP in cells transfected with GC-A and bound specifically to such cells (Chinkers et al 1989, Lowe et al 1989). In addition, crosslinking with ^{125}I-ANP resulted in a single radiolabeled protein of M_r 130,000, consistent with previous estimates of receptor size (Kuno et al 1986a, Leitman et al 1986, 1988, Meloche et al 1986, Misono et al 1985, Takayanagi et al 1987, Yip et al 1985). From these and subsequent experiments it became clear that GC-A was the ANP receptor.

The existence of multiple peptides with natriuretic properties and the presence of plasma membrane guanylyl cyclase activity on most cells in the body raised an obvious question of whether or not there were other members of this new receptor family. Low-stringency hybridization resulted in the isolation of full-length cDNA clones for another guanylyl cyclase labeled as GC-B (Schulz et al 1989) or ANP-B receptor (Chang et al 1989). GC-B and GC-A were about 79% identical at the amino acid level within the intracellular region; however, the identity fell off to 43% within the extracellular, putative ligand-binding domain (Chang et al 1989, Schulz et al 1989). In the initial studies on ligand specificity, two important observations were made: First, brain natriuretic peptide (BNP) was slightly more potent than ANP in stimulating GC-B, and, second and possibly more importantly, neither ANP nor BNP were active at concentrations lower than 10^{-7} M (Schulz et al 1989). It was suggested, therefore, that a ligand specific for GC-B had yet to be discovered.

In 1990, a third peptide with natriuretic properties was discovered and named C-type natriuretic peptide (CNP) (Table 1). CNP is a carboxyl-truncated member of the ANP family that appears to be highly conserved across species (Arimura et al 1991, Sudoh et al 1990, Yoshihara et al 1990). Recently, Mizuno et al (1990) reported that intracellular cyclic GMP concentrations were elevated by ANP or BNP in SW-13 cells, endothelial cells, or in lung membranes; CNP, however, was relatively ineffective. Based on recent work of Koller et al (1991), these responses are consistent with the properties of GC-A. Koller et al (1991) expressed human GC-A in cells and showed that CNP was essentially ineffective at stimulating enzyme activity. Human GC-B, in contrast, when expressed in the same type of cell, was stimulated at considerably lower CNP than BNP or ANP concentrations, thus suggesting that CNP may be the natural ligand for GC-B (Koller et al 1991). These authors also showed that BNP from different species had profoundly different effects on GC-B activation, thereby demonstrating a need to perform studies on receptor/ligand specificity within the same species. Whereas ANP and CNP appear to be highly conserved across the species, BNP shows considerable diversity (Table 1).

With respect to sites of action, ANP and BNP are found principally in heart but also in other tissues (Brenner et al 1990). CNP, however, appears to be brain- or nervous system-specific (Arimura et al 1991). Although CNP has been suggested to represent a neurotransmitter (Arimura et al 1991), its cellular localization has not yet been reported.

Whether or not the distribution of ANP, BNP, or CNP coincides with the tissue and cellular localization of GC-A and GC-B is now under study. Wilcox et al (1991) defined tissues in which GC-A and GC-B mRNA are expressed in the monkey by in situ hybridization. GC-B mRNA was detected in brain and adrenal medulla possibly consistent with the proposed specific neural localization of CNP. However, work in our laboratory has shown the presence of GC-B in 3T3 fibroblasts (M. Takada, T. D. Chrisman & D. L. Garbers, unpublished observations), bovine trachea (S. Schulz, T. D. Chrisman & D. L. Garbers, unpublished observations), and many other tissues, suggesting that the protein is expressed in a variety of cell types. Thus, it seems likely that either CNP (the proposed ligand) is found in parts of the body other than the brain, or another physiological ligand for GC-B has yet to be discovered.

Following the successful cloning of the various guanylyl cyclases described above, conserved amino acid sequences within the various guanylyl cyclases were identified for the purpose of producing degenerate oligonucleotide probes that would be expected to anneal to mRNA encoding other guanylyl cyclases. Yuen et al (1990) identified two highly conserved regions of both soluble and plasma membrane forms of guanylyl cyclase that were not conserved in adenylyl cyclase (Krupinski et al 1989), and these were then used to design degenerate oligonucleotides to search for new guanylyl cyclases by use of the polymerase chain reaction (PCR).

Rat small intestine cDNA was one of the first to be screened for new members of the cyclase family with these PCR primers. The small intestine was initially chosen because it had been known since the late 1970s that small bacterial peptides (heat-stable enterotoxins; ST) could stimulate an intestinal guanylyl cyclase in broken cell preparations (Field et al 1978, Guerrant et al 1980, Hughes et al 1978, Rao et al 1980). These small peptides (Table 1) cause acute secretory diarrhea in humans and domestic animals, presumably due to their effects on cyclic GMP concentrations (Field et al 1978, Guerrant et al 1980). Various reports had suggested that guanylyl cyclase was not an enterotoxin receptor (Kuno et al 1986b, Waldman et al 1986), but this conclusion was not definitive (Schulz et al 1990).

PCR-amplified DNA fragments obtained from small intestine cDNA were sequenced and a unique fragment was subsequently used to screen for full-length clones. A full-length cDNA clone, labeled GC-C, was obtained (Schulz et al 1990). The general features of GC-C were the same as those of the other plasma membrane guanylyl cyclases. At the primary sequence level, however, GC-C and GC-A appeared to be no more closely related than GC-A was related to the sea urchin cyclase. GC-C, therefore, appeared to fall outside the natriuretic peptide receptor family. Subsequent experiments demonstrated that cells transfected with GC-C specifically bound a heat-stable enterotoxin; cyclic GMP concentrations also were markedly elevated by added enterotoxin, thus showing that GC-C is a cell surface receptor for these bacterial peptides. The physiological relevance of the receptor remains unknown, but a major question is whether or not an endogenous ligand for GC-C exists (see Actions of Cyclic GMP, below).

HOMOLOGOUS CELL SURFACE RECEPTORS THAT DO NOT CONTAIN GUANYLYL CYCLASE ACTIVITY

In addition to the two natriuretic peptide receptors (GC-A and GC-B) that are linked to guanylyl cyclase, a natriuretic peptide receptor containing only 37 amino acids within the proposed intracellular region has been cloned (Fuller et al 1988). This truncated receptor (ANP-CR) has been postulated by some to be involved in G-protein-coupled signal transduction (Anand-Srivastava et al 1985, 1987, 1990, Anand-Srivastava & Cantin 1984, 1986, Drewett et al 1990, Hirata et al 1989), and by others to regulate the concentration of ANP by virtue of its binding capacity and/or subsequent internalization, thereby serving as a clearance receptor (Maack et al 1987). A similar nontransducing role for a member of a signal transduction family of receptors has been postulated for a member of the *trk* oncogene family (Klein et al 1990, Middlemas at al 1991). Among several apparent splice variants of the *trk* receptor/protein tyrosine kinase, one that lacks the tyrosine kinase catalytic domain is suggested by Klein et al (1990) to serve as a transporter at the blood brain barrier.

A number of different proteins are crosslinked to ^{125}I-ST in the presence of a variety of crosslinking reagents (Ivens et al 1990, Kuno et al 1986b, Thompson & Giannella 1990, Waldman et al 1986); one of these has an M_r of that predicted for GC-C. Whether lower M_r crosslinked proteins represent ST receptors, possibly analogous to the ANP-C receptor (Fuller et al 1988), proteolytic fragments of GC-C, or components of the functional receptor, is not yet known. The native ST receptor could exist as a complex of proteins, similar to the interleukin-2 receptor (Dukovich et al 1987, Robb et al 1987, Sharon et al 1986, Teshigawara et al 1987, Tsudo et al 1986), one member being GC-C. Antibodies to GC-C may define whether or not tightly associated molecules are immunoprecipitated with the cyclase, and whether they coincide with the previously defined ^{125}I-ST crosslinked proteins. Some support for the existence of a multicomponent receptor comes from the sea urchin research, in which ^{125}I-resact is covalently bound in the presence of disuccinimidyl suberate to guanylyl cyclase (Shimomura et al 1986), whereas ^{125}I-speract crosslinks to a 77,000 M_r protein (Dangott et al 1989). Yet, both peptides cause marked elevations of cyclic GMP in intact cells as well as in detergent-solubilized membrane preparations (Bentley et al 1988).

The other major family of proteins homologous to the plasma membrane guanylyl cyclases are the protein kinase cell surface receptors (see below).

IDENTIFICATION AND FUNCTIONS OF DOMAINS WITHIN THE PLASMA MEMBRANE FORMS OF GUANYLYL CYCLASE

Ligand-binding Domain

Evidence from various studies shows that the amino terminal half of guanylyl cyclase represents the ligand-binding domain. The lines of evidence are two-fold. First, comparison with the ANP-C receptor demonstrates that the putative extracellular domain of ANP-CR is about 34% identical with GC-A (Chinkers et al 1989, Fuller et al 1988, Lowe et al 1989). When a termination signal is introduced in the ANP-C receptor cDNA at a position 5′ to the putative transmembrane domain, the protein is secreted from transfected cells and continues to bind ANP (Porter et al 1989). Second, when either the protein kinase-like or carboxyl terminal domains of GC-A are deleted, the expressed, mutated protein continues to bind ANP (Chinkers & Garbers 1989). The conservation of cysteine residues within the extracellular domain of GC-A and GC-B suggests the presence of critical disulfide bonds for formation of the natriuretic peptide binding site, although detailed studies on the binding site have not yet appeared.

Cyclase Catalytic Domain

The guanylyl cyclase catalytic domain has been identified in two studies. In the first, deletion of the protein kinase-like domain of GC-A did not diminish guanylyl cyclase activity (Chinkers & Garbers 1989). In the second, Thorpe & Morkin (1990) deleted the entire sequence of GC-A, save the carboxyl region, and continued to find guanylyl cyclase activity after expression of the mutant protein in bacteria. This region is also similar to two internally homologous domains of bovine brain adenylyl cyclase (Krupinski et al 1989), which have been suggested to exist within the cytoplasm. There are also some regions of weak identity between the guanylyl cyclase catalytic domains and corresponding regions of adenylyl cyclase from *Saccharomyces cerevisiae* (Kataoka et al 1985) and *Schizosaccharomyces pombe* (Yamawaki-Kataoka et al 1989). The predicted primary amino acid sequence from the catalytic domain of the adenylyl cyclase from *Rhizobium meliloti*, however, is homologous to the guanylyl cyclase catalytic region (Beuve et al 1990). Although the carboxyl terminal sequence represents the cyclase catalytic region, the α- or β-subunits of heterodimeric guanylyl cyclase, when expressed separately, do not have demonstrable enzyme activity (Harteneck et al 1990, Nakane et al 1988, 1990, Yuen et al 1990). Explanations for these results are not yet clear, although the possibility of an improper conformation in the absence of a second subunit is one simple explanation. It is conceivable that dimerization is required for cyclase activity in all cases but that the plasma membrane forms can associate as homodimers whereas the soluble enzyme subunits cannot.

Protein Kinase-like Domain

The protein kinase-like domains of the sea urchin guanylyl cyclase and of GC-A, GC-B, and GC-C contain many of the highly conserved or invariant amino acids found in protein kinases (Hanks et al 1988). Of the 33 invariant or highly conserved residues identified in all protein kinases by Hanks et al (1988) (identified by an asterisk in Table 2), 22 are consensus amino acids in the guanylyl cyclases. The primary sequence of the protein kinase-like region resembles that of the protein tyrosine kinases more than the Ser/Thr protein kinases; GC-A, for example, is about 31% identical at the amino acid level with the platelet-derived growth factor receptor across the entire protein kinase domain (Chinkers et al 1989); it is about 22% identical with cyclic AMP-dependent protein kinase (A-kinase) across the same region (Uhler et al 1986). A notable deviation from established protein kinase consensus sequences occurs in the H/YRDL/I sequence, where Asp is invariant in established protein kinases but is an N, S, R, or H in the various cyclases. Conversion of the corresponding Asp to Asn in c-*kit* (murine white spotting locus) results in a loss of protein kinase activity

Table 2 Comparison of the protein kinase domains of GC-A, GC-B, GC-C, and a sea urchin sperm guanylyl cyclase[a]

```
GC-A       --GRGSNYGSLLTTEGQFQNFAKTAYYKGNLVAVKRVNRKRIELT---
GC-B       --LRGSSYGSLMTAHGKYQIFANTGHFKGNVVAIKHVNKKRIELT---
GC-C       --TNHVSLKIDDDRRRDTIQRVRQCKYDKKKVILKDLKHCDGNFS---
S.U.       --NMVMSAISVISNAEKQQIFATIGTYRGTVCALHAVHKNHIDLT---
CONSENSUS          S   S          Q FA    Y G  V*  *    I LT
FES        QIGRGNFGEVFSGRLRA---------DNTLVAVKSCRET-LPPDIK-
AKIN       TLGTGSFGRVMLVKHME----------TGNHYAMKILDKQKVVKLKQI
SRC        KLGQGCFGEVWMGTWNG----------TTRVAIKTLKPGTMSPEA--
LTK        ALGHGAFGEVYEGLVTG-----LPGDSSPLPVAIKTLPELCSHQDE--
CONSENSUS  ** *  * *                           *  *

GC-A       RKVLFELKHMRDVQNEHLTRFVGACTDPPNICILTEYCPRGSLQDI--
GC-B       RQVLFELKHMRDVQFNHLTRFIGACIDPPNICIVTEYCPRGSLQDI--
GC-C       EKQKIELNKLLQSDYYNLTKFYGTVKLDTRIFGVVEYCERGSLREV--
S.U.       RAVRTELKIMRDMRHDNICPFIGACIDRPHISILMHYCAKGSLQDI--
CONSENSUS  R V  *LK MRD    *T F GAC D P I I   EYC RGSLQDI
FES        AKFLQEAKILKQYSHPNIVRLIGVCTQKQPIYIVMELVQGGDFLTF--
AKIN       EHTLNEKRILQAVNFPFLVKLEFSFKDNSNLYMVMEYVPGGEMFSH--
SRC        --FLQEAQVMKKLRHEKLVQLYAVVS-EEPIYIVTEYMSKGSLLDF--
LTK        LDFLMEALIISKFSHQNIVRCVGLSFRSAPRLILLELMSGGDMKSF--
CONSENSUS      *          *

GC-A       LENE-------SITLDWMFRYSLTNDIVKGMLFLHNGAICSHGNLKSS
GC-B       LEND-------SINLDWMFRYSLINDLVKGMAFLHNSIISSHGSLKSS
GC-C       LNDT--ISYPDGTFMDWEFKISVLNDIAKGMSYLHSSKIEVHGRLKST
S.U.       LEND--DIKLDSM-----FLSSLIADLVKGIVYLHSSEIKSHGHLKSS
CONSENSUS  LEN        S   DW F  S  ND VK*M  **H S I S*G  *KSS
FES        LRTEGARLRVKTLLQ-MVGDAAA------GMEYLE-SKCCIHRDLAAR
AKIN       LRRIGR--FSEPHARFYAAQIVL------TFEYLH-SLDLIYRDLKPE
SRC        LKGETGKYLRLPQLVDMAAQIAS------GMAYVE-RMNYVHRDLRAA
LTK        LRHSRPHPGQLAPLT-MQDLLQLAQDIAQGCHYLEENHF-IHRDIAAR
CONSENSUS                             *  **       *  **

GC-A       NCVVDGRF---VLKITDYGLESFRD-PEPEQGHTLFAK---KLWTAPE
GC-B       NCVVDSRF---VLKITDYGLASFRSTAEPDDSHALYAK---KLWTAPE
GC-C       NCVVDSRM---VVKITDFG----------CNSILPPKK---DLWTAPE
S.U.       NCVVDNRW---VLQITDYGLNEFKKGQKQDVDLGDHAKLARQLWTSPE
CONSENSUS  **VVD R    VLK*T*Y*L  F          AK    L*T*P*
FES        NCLVTEKN---VLKISDFGMSREEADGVYAASGGLRQ--VPVKWTAPE
AKIN       NLLIDQQG---YIQVTDFGFAKRVKGRT----WTLCG--TP-EYLAPE
SRC        NILVGENL---VCKVADFGLARLIEDNEY-TARQGAK--FPIKWTAPE
LTK        NCLLSCSGASRVAKIGDFGMARDIYQASYYRKGGRTL--LPVKWMPPE
CONSENSUS  **           *  ***       *        * * * *
```

Table 2 (*continued*)

```
GC-A       LL-RMASPPARGSQAGDVYSFGIILQEIALRSG-VFYVEGLDLSPKEI
GC-B       LL-SGNPLPTTGMQKADVYSFAIILQEIALRSG-PFYLEGLDLSPKEI
GC-C       HL-RQ----ATISQKGELYSFSIIAQEIILRKE-TFYTLSCRDQNEKI
S.U.       HLRQEGSMPTAGSPQGDIYSFAIILTELYSRQE-PFHENE--MDLADI
CONSENSUS  L       P  GSQ G* Y*F *ILQEI LR     FY          I
FES        AL-----NYGRYSSESDVWSFGILLWETFSLGASPYPNLSNQQTREFV
AKIN       II----LSKGYNKAV-DWWALGVLIYEMAA-GYPPFFADQPIQIYEKI
SRC        AA-----LYGRFTIKSDVWSFGILLTELTTKGRVPYPGMVNREVLDQV
LTK        AL-----LEGLFTSKTDSWSFGVLLWEIFSLGYMPYPGHTNQEVLDFI
CONSENSUS                  * ** **

GC-A       IERVTRGEQPPFRPSMDLQSHLE-ELG--QLMQRCWAEDPQERPPFVQ
GC-B       VQKVRNGQRPYFRPS-IDRTQLNEELV--LLMERCWAQDPTERPDFIG
GC-C       FRVENSYGTKPFRPDLFLETADEKELEVYLLVKSCWEEDPEKRPDFVK
S.U.       IGRVKSGEVPPYRPILNAVNAAAPDCV-LSAIRACWPEDPADRPNIIM
CONSENSUS  V   G   PPFRP          EL      L   CW EDP  *P F
FES        EKGG-RLPCPELCP-DAVFRLME----------QCWAYEPGQRPSFST
AKIN       VSGKVRFPSHFSSDLKDL--LRN-----------LLQVDLTKRFGNLK
SRC        ERGY-RMPCPPECPESLHDLMC-----------QCWRKEPEERPTFEY
LTK        ATGN-RMDPPRNCPGPVYRIMT----------QCWQHQPELRPDFGS
CONSENSUS                      *                         *

GC-A       QIRLALRK----FNKENSSNILDNLLSR
GC-B       QIKG----FIRRFNKEGGTSILDNLLLR
GC-C       KIESTLAKIFGLFHDQKNESYMDTLIRR
S.U.       AVRTMLAPLQK--GLKP--NILDNMIAI
CONSENSUS  I   L       F         ILDNL  R
FES        IYQELQSIRKR
AKIN       DGVNDIKNHKW
SRC        LQAFLEDYFTS
LTK        ILERIQYCTQD
CONSENSUS
```

[a] Consensus amino acid sequences are those that appear in at least three of the listed guanylyl cyclases. Four protein kinases are chosen arbitrarily to represent example amino acid sequences throughout the protein kinase catalytic domain. Positions marked with * represent the invariant or conserved residues found in all protein kinases. The guanylyl cyclases have * in the same position if it is also conserved.

(Tan et al 1990) and the same mutation in *fps* has also been reported to abolish protein kinase activity (Moran et al 1988). In addition, the GXGXXG consensus sequence of protein kinases is absent in GC-C and only partially retained in the other cyclases. The protein kinase-like domain, therefore, may not possess actual protein kinase activity, although studies directly addressing this possibility have not yet been reported.

The high conservation of protein kinase-like sequences suggests that this domain binds ATP, although no published studies address this issue either. ATP either augments the stimulation of guanylyl cyclase activity observed with natriuretic peptides (Chang et al 1990, Chinkers & Garbers

1989, Kurose et al 1987, Song et al 1988) or is absolutely required for such stimulation (Chinkers et al 1991). Nonhydrolyzable analogues of ATP such as ATPγS or AMP-PNP are as effective as ATP, thus demonstrating that phosphoryl transfer is not required for cyclase activation. A simple model to explain these observations would be that the binding of natriuretic peptide promotes the binding of ATP to the protein kinase-like domain, which leads to activation of the cyclase catalytic domain. Further support for an involvement of the protein kinase-like domain in signal transduction comes from deletion mutagenesis studies. When the protein kinase-like domain is deleted from GC-A, the enzyme retains guanylyl cyclase and ANP-binding activity, but the addition of ANP to either intact cells or to isolated plasma membranes carrying the mutant receptor no longer activates the cyclase (Chinkers & Garbers 1989). Chinkers & Garbers (1989) have suggested that the protein kinase-like domain serves as a negative regulatory element and that the binding of ATP relieves the cyclase inhibition.

There are potential parallels between hormone receptor/adenylyl cyclase regulation and that of guanylyl cyclase. Cyclic AMP concentrations are increased in cells by virtue of hormones binding to specific receptors that subsequently activate nucleotide-binding regulatory proteins (G-proteins); these proteins specifically bind GTP or GDP. The G-proteins appear to be "activated" by an exchange of GTP for GDP, which subsequently leads to an interaction of a G-protein subunit with an effector molecule such as adenylyl cyclase (Bourne et al 1990, Gilman 1987). The G-proteins also possess intrinsic GTPase activity, and hydrolysis of GTP subsequently results in apparent inactivation of the protein. With guanylyl cyclase, the receptor (extracellular binding domain) may be directly coupled to the proposed nucleotide-binding domain (protein kinase-like domain); ATP rather than GTP could serve as the specific regulatory nucleotide. The binding of ATP would "activate" the protein kinase-like domain, leading, by mechanisms not understood, to marked stimulations of guanylyl cyclase activity. Experiments to test this model will not only need to determine whether or not the protein kinase-like domain binds ATP, but also whether the domain possesses ATPase activity (phosphoryl transfer also could be to another protein); if so, such activity could represent a mechansim of desensitization.

PRIMARY STRUCTURES AND LIGANDS OF HETERODIMERIC FORMS OF GUANYLYL CYCLASE

The mRNA encoding the two subunits of guanylyl cyclase from bovine (Koesling et al 1988, 1990) and rat (Nakane et al 1988, 1990) lung have been cloned. Both subunits contain putative catalytic regions near their carboxyl termini that are homologous to the catalytic domain of plasma membrane forms of guanylyl cyclase. Highly conserved sequences within these catalytic domains were used to design degenerate oligonucleotide primers, which have been used to amplify cDNA of other forms of guanylyl cyclase. One additional clone has been obtained from rat kidney; it is similar to both subunits of the soluble guanylyl cyclase (Yuen et al 1990).

The diversity of the soluble guanylyl cyclase subfamily creates a need for consistent nomenclature; we propose that the subunits be referred to as α (larger) and β (smaller). GC-S is the acronym used for the heterodimeric form of guanylyl cyclase in lung, since the enzyme is known to be soluble; this may not be the case for all heterodimeric forms. The 76,300 Da form found preferentially in kidney is most similar to β_1, the subunit from lung, and is therefore referred to as β_2. Whether or not this form is found in the cytoplasm is not known. The final four amino acids (-C-V-V-L) of β_2 correspond to the carboxyl terminal consensus sequence -C-A-A-X (where A is an aliphatic amino acid and X is any amino acid) for multistep post-translational modification of several membrane-associated proteins, including *ras* (Glomset et al 1990). An isoprenyl group is covalently attached to the cysteine through a thioether bond, and proteolysis of the final three amino acids is followed by carboxymethylation of the new terminal cysteine (Gutierrez et al 1989, Hancock et al 1989). Mutation of these four amino acids in *ras* results in redistribution from the membrane to the cytosol and a loss of transforming activity (Gibbs et al 1989, Willumsen et al 1984). Addition of these four amino acids to an unrelated soluble protein conversely causes redistribution to the membrane fraction (Hancock et al 1989).

The expression of the individual α_1, β_1, or β_2 subunits does not result in detectable guanylyl cyclase activity, even though each subunit contains a putative catalytic domain (Harteneck et al 1990, Nakane et al 1988, Yuen et al 1990). When α_1- and β_1-subunits are co-expressed, however, not only is guanylyl cyclase activity observed, but it is stimulated by sodium nitroprusside (Buechler et al 1991, Harteneck et al 1990, Nakane et al 1990). The mixing of α- and β-subunits after expression does not result in guanylyl cyclase activity (Buechler et al 1991, Harteneck et al 1990). Similarly, antisense RNA to either of the co-expressed subunits also blocks expression of enzyme activity (Buechler et al 1991).

Based on primary amino acid sequence, three domains appear to be conserved among the heterodimeric forms of guanylyl cyclase (Figure 2). As with the plasma membrane forms of guanylyl cyclase, the putative catalytic domain is near the carboxyl terminus. Adjacent to this domain

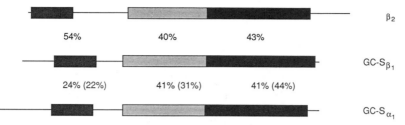

Figure 2 Consensus regions of the α and β subunits of heterodimeric guanylyl cyclase. Numbers represent identity between adjacent subunits (β_2 vs β_1 or β_1 vs α_1) and numbers in parentheses represent identity between α_1 and β_2.

is a region that is conserved between the α- and β-subunits; this region is not found in the plasma membrane forms of guanylyl cyclase. The domain nearest the amino terminus is also conserved among heterodimeric forms, but the similarity between the β_1- and β_2-subunits is much greater than between the β- and α-subunits. No apparent similarity between the α- and β-subunits and heme-binding proteins has been identified, even though heme seems to be associated with the enzyme (Gerzer et al 1981a). The heme-binding site of most other heme proteins is considered to be a part of the catalytic site (e.g. cytochrome oxidase) or as a binding site for oxygen in molecules such as hemoglobin and myoglobin; this may explain, in part, the failure to detect similarities at the primary sequence level. Reconstitution of a heme-deficient, nitric-oxide-unresponsive form of soluble guanylyl cyclase with heme-related molecules is known to restore its sensitivity to nitric oxide or similar agents, thereby establishing that the heme substituent is necessary for regulation, most probably as the recognition site (Craven & DeRubertis 1978, 1983, Gerzer et al 1982, Ignarro et al 1982, Ohlstein et al 1982). Assuming that the apparent catalytic sites do not bind heme, one can speculate that the amino terminal regions bind this regulatory group. The enzyme, then, may exist in an inhibited state, possibly analogous to the plasma membrane forms; the binding of nitric oxide to the heme would then relieve the inhibition.

The carboxyl terminal region (the final 86 amino acids) of the β_2-subunit extends beyond the conserved regions of guanylyl cyclase catalytic domains. This domain contains 50% hydrophilic residues. Similarly, the ST-sensitive plasma membrane guanylyl cyclase (GC-C) has a 63 amino acid carboxyl extension that is also relatively hydrophilic (Schulz et al 1990), although there is no significant similarity with the extended region of the β_2-subunit. Since ST-sensitive guanylyl cyclase activity in the intestine is relatively detergent-insoluble, this form of guanylyl cyclase has been postulated to be associated with cytoskeletal elements (Kuno et al 1986b, Waldman et al 1986).

The failure to observe enzyme activity when α- or β-subunits are expressed alone suggests that both subunits are required for guanylyl cyclase activity and for its regulation by nitric oxide. The formation of the catalytic site of guanylyl cyclase may require dimerization, in which case the plasma membrane forms of guanylyl cyclase would also be active only in the homodimeric state; this possibility has not yet been carefully tested. Another alternative is that other proteins inhibit the activity of the α- or β-subunits when expressed separately, whereas the expression of both subunits protects against such inhibition. Also, the proper folding of the holoenzyme could require the coincident expression of both subunits, since mixing of the two subunits following separate expression does not reconstitute enzyme activity (Buechler et al 1991, Harteneck et al 1990). If each subunit contains a separate catalytic domain, apparently they are not kinetically distinct (Garbers 1979, Ignarro et al 1982, 1986, Lewicki et al 1980, Ohlstein et al 1982).

Since the α_1- and β_1-subunits are located on different genes (Nakane et al 1990), differential regulation of their expression could well occur, including combinations with other heterodimeric subunits, for example, an $\alpha_1\beta_2$-heterodimer.

Several compounds that can serve as precursors to nitric oxide, as well as nitric oxide itself, stimulate soluble guanylyl cyclases from several sources. Included among other agents that stimulate guanylyl cyclase are sodium nitroprusside and nitroglycerin, which are used clinically in the treatment of angina (Murad 1990). The correlation of vascular smooth muscle relaxation with a stimulation of guanylyl cyclase by these compounds has suggested that the enzyme is a principal site of action for these drugs (Ignarro & Kadowitz 1985, Rapoport & Murad 1983). For many years these compounds were thought only to be of pharmacologic importance; many endogenous oxidants could stimulate soluble guanylyl cyclase, including prostaglandin endoperoxides (Graff et al 1978), fatty acid hydroperoxides (Graff et al 1978, Hidaka & Asano 1977), and dehydroascorbate (Haddox et al 1978), but none were clearly associated with cyclic GMP-related functions, including vasodilation.

Nitric oxide is now considered an important mediator of several physiological processes. The newly appreciated general role for nitric oxide is a result of the convergence of several separate lines of investigation, including endothelium-dependent vasorelaxation (Furchgott & Vanhoutte 1989, Ignarro 1990, Moncada 1990), macrophage-mediated cytotoxicity (Hibbs et al 1990), and the endogenous production of nitrosamines (Marletta 1989).

In 1980, Furchgott and Zawadzki demonstrated that acetylcholine-induced relaxation of vascular smooth muscle required the presence of the endothelium, and that this relaxation was mediated by a diffusible substance that they named endothelium-derived relaxing factor (EDRF). Several other vasoactive substances also have been demonstrated to generate EDRF, including bradykinin, histamine, substance P, serotonin, and vasopressin (Furchgott & Vanhoutte 1989). In 1986, comparisons between nitric oxide and EDRF led Furchgott (1988) and Ignarro et al (1988) to postulate that nitric oxide itself was EDRF. Subsequently, Moncada and colleagues were able to quantitate the nitric oxide released from cultured endothelial cells stimulated by bradykinin (Palmer et al 1987). Because nitric oxide is extremely reactive and its half-life within a biological system is very short, it is difficult to measure directly. Therefore, although it is generally regarded as a major component of EDRF, other factors released from the endothelium may account for subtle differences between the effects and properties of nitric oxide and EDRF (Shikano et al 1988, Tracey et al 1990). An endothelium-derived hyperpolarizing factor that is released upon stimulation by acetylcholine has been distinguished from EDRF (Taylor & Weston 1988). Alternatively, metabolites of nitric oxide such as nitrosothiols may also have nitric oxide-like effects (Myers et al 1990).

Nitric oxide also appears to be a cytotoxic agent generated by macrophages. Stimulation of the immune system by lipopolysaccharide or *Bacille Calmette-Guerin* increases the endogenous production of nitrate, which is mediated by macrophages (Stuehr & Marletta 1985). The guanido nitrogens of L-arginine are the source of macrophage-derived nitrite and nitrate, and nitric oxide is an intermediate in this oxidation reaction (Marletta et al 1988). Nitric oxide apparently exerts its cytotoxic effect on target tumor cells by nitrosylating iron-sulfur-containing proteins (Pellat et al 1990, Lancaster & Hibbs 1990) and may have additional targets distinct from guanylyl cyclase (Garg & Hassid 1991). These seemingly unrelated systems may have common regulatory elements. Lipopolysaccharides are potent stimulators of tumor necrosis factor secretion (Beutler & Cerami 1988). Both agents cause a marked hypotension that is thought to be responsible for septic shock. Tumor necrosis factor or lipopolysaccharide appear to stimulate nitric oxide generation in cultured endothelial cells (Kilbourn & Belloni 1990, Salvemini et al 1990). Intravenous administration of tumor necrosis factor causes a hypotension that can be reversed with *N*-(guanidino)-methyl-L-arginine, a specific inhibitor of nitric oxide synthase (Kilbourn et al 1990). Therefore, in addition to any effect on the local vasculature that macrophage-derived nitric oxide may have, macrophages may also regulate blood pressure systemically through release of tumor necrosis factor and subsequent endothelium-dependent relaxation mediated by nitric oxide.

Nitric oxide also has been suggested as a mediator of cyclic GMP elevations induced by N-methyl-D-aspartate receptors in the central nervous system (Bredt & Snyder 1989, Garthwaite et al 1988). Glutamate neurotoxicity has been shown to be mediated by nitric oxide in vitro (Dawson et al 1991), and this neurotoxicity is thought to be associated with Huntington's and Alzheimer's diseases. A population of neurons identified by their NADPH diaphorase activity is resistant in these degenerative diseases. NADPH diaphorase has now been identified as nitric oxide synthase (Hope et al 1991). The mechanisms underlying these observations may further define the relative inter- and intracellular roles for nitric oxide. Possible functions include long-term potentiation (Gally et al 1990, Garthwaite et al 1988, Shibuki & Okada 1991) or dendrite/synapse localization that is regulated by neuronal activity during development (Gally et al 1990).

Differential effects of inhibitors (Moncada et al 1989, Nathan & Stuehr 1990) and cofactors (Bredt & Snyder 1990, Kwon et al 1989, Tayeh & Marletta 1989) have distinguished at least three general types of nitric oxide synthase. One form, an activity found in cell-soluble fractions, has been purified to apparent homogeneity and is known to be regulated by calcium/calmodulin (Bredt & Snyder 1990); its mRNA has been recently cloned from rat brain (Bredt et al 1991). The enzyme is homologous to cytochrome P-450 reductase. Also, recently Forstermann et al (1991) have purified a form of nitric oxide synthase found in the particulate fraction of bovine aortic endothelial cells. This enzyme also appears to be regulated by Ca^{2+}/calmodulin. Another form that is Ca^{2+}/calmodulin independent has been purified from the cytosol of rat neutrophils (Yui et al 1991); its degree of similarity to the enzyme found in macrophages is currently under examination. It should be noted that Deguchi & Yoshioka (1982) reported L-arginine as an endogenous activator of the soluble enzyme before the current studies on nitric oxide synthase. Cellular sources of nitric oxide synthase activity are summarized in Table 3. Indirect measurement of nitric oxide synthase activity, such as generation of EDRF-like activity or generation of nitrite, nitrate, or citrulline, are not included but have been measured in several other tissues and cell types. Many regions of the central nervous system also appear to contain nitric oxide synthase: the olfactory bulb, the posterior pituitary, the supraoptic and paraventricular nuclei of the hypothalamus, the superior and inferior colliculi, the dentate gyrus of the hippocampus, the stria terminalis, the islands of Callejae, the diagonal band of Broca, the cerebral cortex, and the choroid and pigment epithelium layers of the retina (Bredt et al 1990).

Table 3   Cellular localization of nitric oxide synthase

Cell	References
Vascular endothelial cells	Palmer et al 1988
Central nervous system	
Neurons	
Basket cells (cerebellum, molecular layer)	Bredt et al 1990
Golgi II cells (cerebellum, granule layer)	Bredt et al 1990
Myenteric plexus (innervating inner circular muscle)	Bredt et al 1990
Adrenal medulla (innervating chromaffin cells)	Bredt et al 1990
Astroglial cells	Murphy et al 1990
Immune system	
Macrophages	Marletta et al 1988, Kwon et al 1990
Neutrophils	Schmidt et al 1989, Wright et al 1989

The heterodimeric subunits (α_1, β_1) are expressed in various tissues as determined by Northern hybridization analysis (Nakane et al 1988, 1990) and by immunosorbent assays (Lewicki et al 1983). Immunoreactivity, presumably due to GC-S$_{\alpha_1\beta_1}$, has been reported in the soma and dendrites of the large pyramidal neurons in the neocortex, the medium spiny neurons of the caudate-putamen, the Purkinje cells in the molecular layer of the cerebellum, and the stellate and basket cells of the granular layer of the cerebellum (Ariano et al 1982).

ACTIONS OF CYCLIC GMP

The guanylyl cyclase-linked receptors presumably signal principally by virtue of their increased production of cyclic GMP, although signaling through pathways other than cyclic GMP (e.g. protein kinase activity; ligand-induced association with other molecules) also may occur. With respect to receptors for cyclic GMP, reviews on cyclic GMP-regulated phosphodiesterases (Beavo 1988), cyclic GMP-regulated protein kinases (G-kinases; Corbin et al 1990), and cyclic GMP-regulated ion-channels (Yau & Baylor 1989) address recent research.

Goldberg et al (1983) have proposed in an alternative hypothesis that cyclic GMP concentrations need not increase for signaling. Cyclic nucleotides are suggested to act as a signal-directed energy source upon hydrolysis by phosphodiesterases to transduce a work-dependent function analogous to ATPases. When phosphodiesterase activity of intact cells is stimulated in retina with illumination (Dawis et al 1988, Goldberg et al 1983) and in the parotid gland by β-adrenergic agonists (Deeg et al 1988), under conditions in which changes in cyclic nucleotide content are apparently not detectable, marked increases in cyclic nucleotide fluxes can be seen. (These experiments do not rule out possible changes in cyclic nucleotide concentrations within a microenvironment of the cell.) Although several potential mechanisms for the utilization of hydrolysis energy have been proposed (Goldberg et al 1983, Goldberg & Walseth 1985), none has been associated with the regulation of a known function. Furthermore, the currently established targets for cyclic GMP action appear to involve binding proteins that detect changes in cyclic GMP concentration.

The number of different guanylyl cyclase receptors is not yet known, nor is the cellular and tissue distribution defined for the known guanylyl cyclases. From the cloning studies and mRNA hybridization analysis, however, GC-A is known to be expressed in rat brain, human kidney, adrenal gland, adipose tissue, and terminal ileum (Chinkers et al 1989, Lowe et al 1989). The ANP-C receptor is expressed in aortic smooth muscle and endothelium, adrenal cortex, and kidney, based on mRNA hybridization analysis (Fuller et al 1988) and pulmonary parenchyma, bronchiolar smooth muscle, and renal glomeruli by immunohistochemistry (Kawaguchi et al 1989). GC-B mRNA has been detected in human and rat brain (Chang et al 1989, Schulz et al 1989) and in human placenta, pituitary gland, and terminal ileum (Chang et al 1989) and porcine atrium (Chang et al 1989). GC-B is also apparently present in 3T3 fibroblasts (M. Takada, T. D. Chrisman & D. L. Garbers, unpublished observations). RNA hybridization studies suggest that GC-C is found only in rat intestine (Schulz et al 1990), but PCR and cDNA cloning have shown the presence of GC-C mRNA in tissues such as olfactory tissue, adrenal gland, brain, and retina (S. Schulz, T. D. Chrisman, S. Singh & D. L. Garbers, unpublished data).

Binding studies with ^{125}I-ANP have suggested that GC-A or GC-B (or perhaps another member of the guanylyl cyclase natriuretic peptide receptor family) are present in Leydig tumor cells (Pandey et al 1988), brain (Wildey & Glembotski 1986), kidney (Yip et al 1985), adrenal cortex (Meloche et al 1986), smooth muscle cells (Vandlen et al 1985), and other cells and tissues.

The sites of synthesis of the specific ligands for the various cyclases is not yet well defined. Nitric oxide synthase, which produces nitric oxide, a putative ligand for heterodimeric forms of guanylyl cyclase, has been identified in many different tissues, including brain (Bredt et al 1990; see Table 3). Peptides with natriuretic properties also have been found in various tissues, including heart and brain (Brenner et al 1990), but whereas

BNP and ANP appear to predominate in heart, CNP may exist principally or only in the brain (Arimura et al 1991).

Whether an endogenous ligand for GC-C exists is not known, but its existence seems likely now that a combination of in situ hybridization, cDNA cloning, and PCR analysis has shown mRNA encoding the receptor in various tissues other than the small intestine. A caveat is that whether the GC-C protein, in fact, is expressed in the varous tissues is not yet known. Previously, Krause et al (1990) had shown ^{125}I-ST binding in epithelial cells from various tissues of the North American opossum, but this has not been observed in other species (Guerrant et al 1980). Figure 3 envisions a physiological function of an endogenous ligand for GC-C, if it exists, basically suggesting the presence of another signaling system that regulates secretion. If such a ligand exists, its site of synthesis and regulation will become an important new area of research. Regulation of the cystic fibrosis transmembrane conductance regulator (CFTR) by cyclic AMP and A-kinase has been studied in some depth (Gregory et al 1990, Rogers et al 1990, Welsh 1990), and overproduction of CFTR in various cells results in the expression of cyclic AMP-regulated chloride secretion (Anderson et al 1991). Whether CFTR is the channel itself or a regulator of the channel is not yet certain, but inactivating mutations appear to fall within a putative nucleotide-binding region (Riordan et al 1989). CFTR appears to be found in considerably higher amounts in a colonic cell line

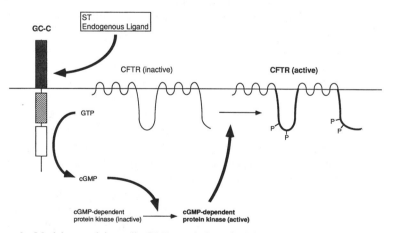

Figure 3 Model to explain cyclic GMP regulation of epithelial cell chloride channels. The proposed endogenous ligand for GC-C has not yet been identified. Small peptides (ST) from various bacteria can bind to GC-C, causing increases in cyclic GMP concentrations. CFTR is the cystic fibrosis transmembrane conductance regulator, which either regulates chloride channels or is the channel itself.

(T_{84}) than in airway epithelium (Gregory et al 1990), where the basic mechanisms of regulation of secretion through cyclic AMP and cyclic GMP have been suggested to be similar (Huott et al 1988). It also has been proposed that G-kinase mediates the cyclic GMP effects on chloride ion channels possibly by phosphorylating the same sites as the A-kinase (Lin et al 1991).

With respect to receptors for cyclic GMP, we discuss only a few current articles here. Cyclic GMP-regulated phosphodiesterases exist in many different tissues and may demonstrate different structure/binding profiles than the G-kinases or cyclic GMP-activated ion channels due to distinct differences in primary amino acid sequence within the cyclic GMP-binding site (Kaupp et al 1989, Takio et al 1984). Although an ANP effect to inhibit aldosterone synthesis had been thought not to involve cyclic GMP (Ganguly et al 1989, Matsuoka et al 1987), thereby leading to speculation of other signaling pathways, a primary role for cyclic GMP has now been proposed for the ANP-mediated inhibition of aldosterone synthesis in the adrenal cortex glomerulosa (MacFarland et al 1991). One form of cyclic nucleotide phosphodiesterase, known as cyclic GMP-stimulable phospho-diesterase, contains an allosteric site for cyclic GMP; when occupied, a catalytic site for cyclic AMP is activated (Martins et al 1982). This phosphodiesterase is abundant in adrenal cortical glomerulosa cells, the site of aldosterone production. ACTH-stimulated aldosterone synthesis is mediated by cyclic AMP, which appears to be inhibited by ANP distal to adenylyl cyclase. The action of ANP is mimicked by the addition of cell-permeant cyclic GMP analogs to intact cells, and their efficacy correlates with their ability to stimulate the cyclic GMP-stimulable phospho-diesterase rather than G-kinase (MacFarland et al 1991). Thus, cyclic GMP may induce aldosterone inhibition by virtue of its effects on cyclic AMP. This does not preclude a possible signaling function of the ANP-C receptor as well. A similar mechanism has been proposed for the action of ANP on fibroblasts (Lee et al 1988), although the physiological role of ANP has not been established in these cells. There also are cyclic GMP-inhibited phosphodiesterases (Harrison et al 1986), whose occurrence raises the possibility that cyclic GMP could cause increases or decreases in cyclic AMP depending upon the cell type.

Sodium channels from two tissues are cloned and expressed that contain cyclic GMP-binding sites whose primary sequence appears similar to that of the cyclic GMP-dependent protein kinases (Dhallan et al 1990, Kaupp et al 1989, Ludwig et al 1990). The channels from the olfactory bulb bind cyclic AMP, but about ten-fold less effectively than cyclic GMP (Dhallan et al 1990, Nakamura & Gold 1987). The finding that the olfactory channel is more sensitive to cyclic GMP is especially intriguing since there are no reports demonstrating a role for cyclic GMP in odorant signaling. Particulate guanylyl cyclase activity, however, is present in olfactory cilia at moderately high specific activities (Steinlen et al 1990), thus raising the possibility that a family of guanylyl cyclase-linked receptors could exist that recognize certain classes of odorants. In the kidney, cyclic GMP-regulated sodium transport also has been studied; cyclic GMP, both directly and indirectly through G-kinase, regulates channel activity (Light et al 1990). The amiloride-sensitive sodium channel from kidney has been cloned recently, but expression of the protein did not result in sodium transport, most likely because the channel is a multicomponent unit (Barbry et al 1990). Therefore, an analysis for cyclic GMP regulatory sites or G-kinase consensus phosphorylation sites awaits the cloning of the mRNA for the other putative subunits.

An additional recent report of considerable interest with respect to the action of cyclic GMP on ion transport showed that the specificity of a channel could be altered (Sorbera & Morad 1990). In these studies, ANP was shown to alter the selectivity of the sodium channel of atrial cells toward calcium ions, thereby repressing sodium current. Presumably the change in selectivity was mediated by cyclic GMP.

OTHER GUANYLYL CYCLASE-LINKED RECEPTORS

Based on the diversity of other cell surface receptor families, other plasma membrane forms of guanylyl cyclase probably exist. This likelihood also suggests that important physiological regulators of cyclic GMP concentrations will yet be discovered. Since the sea urchin enzyme is distinctly different from the mammalian forms, we can also expect to find different ligands across the species. Preliminary studies have shown that *Drosophila* contain members of the guanylyl cyclase cell surface receptor family, one of which is similar to GC-A (D. K. Thompson & D. L. Garbers, personal communication); the potential ligands for the *Drosophila* receptors and the function of such receptors remain an intriguing avenue to pursue. No reports of mutations in these insects have attributed a phenotype to a cyclic GMP defect. Cyclic GMP also has been reported to exist in yeast (Eckstein 1988), but the nature of the guanylyl cyclases and the possible function of this putative second messenger in yeast have not been pursued.

A guanylyl cyclase involved in phototransduction that appears to be regulated by Ca^{2+} (Koch & Stryer 1988) has not been purified or cloned. There are also Ca^{2+}-regulated guanylyl cyclases in *Paramecium* (Klumpp et al 1983) and *Tetrahymena* (Schultz et al 1983). Although the effects of Ca^{2+} on GC-A, GC-B, or GC-C have not been specifically evaluated, the above Ca^{2+}-regulated forms are probably distinct. Ca^{2+} has been suggested, however, to modulate the effects of ANP on guanylyl cyclase activity (Mukopadhyay et al 1989). Based on enzyme purification (Paul et al 1987), the adrenal gland has also been suggested to contain a unique form of guanylyl cyclase.

With respect to heterodimeric forms, no apparent physiological data in the literature suggest the existence of different forms; yet at least one unique β-subunit is now known to exist (Yuen et al 1990). Therefore, discovery of differential regulation of heterodimeric forms may be forthcoming.

We will probably soon see the discovery of many other members of this family, which in turn will lead to a search for potential ligands, much the same as we now see occurring in research on the receptor/protein tyrosine kinases (Flanagan & Leder 1990, Ullrich & Schlessinger 1990, Williams et al 1990) or protein phosphatases (Fischer et al 1990). Then will come the need to understand the integrated physiological role of these ligands and receptors, all of which may lead to new and important avenues for the treatment of specific animal diseases.

Literature Cited

Anand-Srivastava, M. B., Cantin, M. 1984. Atrial natriuretic factor inhibits adenylate cyclase activity. *Biochem. Biophys. Res. Commun.* 121: 855–62

Anand-Srivastava, M. B., Cantin, M. 1986. Atrial natriuretic factor receptors are negatively coupled to adenylate cyclase in cultured atrial and ventricular cardiocytes. *Biochem. Biophys. Res. Commun.* 138: 427–36

Anand-Srivastava, M. B., Cantin, M., Genest, J. 1985. Inhibition of pituitary adenylate cyclase by atrial natriuretic factor. *Life Sci.* 36: 1873–79

Anand-Srivastava, M. B., Sairam, M. R., Cantin, M. 1990. Ring-deleted analogs of atrial natriuretic factor inhibit adenylate cyclase/cAMP system: Possible coupling of clearance atrial natriuretic receptors to adenylate cyclase/cAMP signal transduction system. *J. Biol. Chem.* 265: 8566–72

Anand-Srivastava, M. B., Srivastava, A. K., Cantin, M. 1987. Pertussis toxin attenuates atrial natriuretic factor-mediated inhibition of adenylate cyclase: Involvement of inhibitory guanine nucleotide regulatory protein. *J. Biol. Chem.* 262: 4931–34

Anderson, M. P., Rich, D. P., Gregory, R. J., Smith, A. E., Welsh, M. J. 1991. Generation of cAMP-activated chloride currents by expression of CFTR. *Science* 251: 679–82

Appleman, M. M., Thompson, W. J., Russell, T. R. 1973. Cyclic nucleotide phosphodiesterases. In *Advances in Cyclic Nucleotide Research*, ed. P. Greengard, G. A. Robison, 3: 65–98. New York: Raven

Ariano, M. A., Lewicki, J. A., Brandwein, H. J., Murad, F. 1982. Immunohistochemical localization of guanylate cyclase within neurons of rat brain. *Proc. Natl. Acad. Sci. USA* 79: 1316–20

Arimura, J. J., Minamino, N., Kangawa, K., Matsuo, H. 1991. Isolation and identification of C-type natriuretic peptide in chicken brain. *Biochem. Biophys. Res. Commun.* 174: 142–48

Ashman, D. F., Lipton, R., Melicow, M. M., Price, T. D. 1963. Isolation of adenosine 3′,5′-monophosphate and guanosine 3′,5′-monophosphate from rat urine. *Biochem. Biophys. Res. Commun.* 11: 330–34

Barbry, P., Champe, M., Chassande, O., Munemitsu, S., Champigny, G., et al. 1990. Human kidney amiloride-binding protein: cDNA structure and functional expression. *Proc. Natl. Acad. Sci. USA* 87: 7347–51

Beavo, J. A. 1988. Multiple isozymes of cyclic nucleotide phosphodiesterase. In *Ad-*

vances in Second Messenger and Phospho-protein Research, ed. P. Greengard, G. A. Robison, 22: 1–38. New York: Raven

Bentley, J. K., Khatra, A. S., Garbers, D. L. 1988. Receptor-mediated activation of detergent-solubilized guanylate cyclase. *Biol. Reprod.* 39: 639–47

Bentley, J. K., Tubb, D. J., Garbers, D. L. 1986. Receptor-mediated activation of spermatozoan guanylate cyclase. *J. Biol. Chem.* 261: 14859–62

Beutler, B., Cerami, A. 1988. Cachectin (tumor necrosis factor): A macrophage hormone governing cellular metabolism and inflammatory response. *Endocr. Rev.* 9: 57–66

Beuve, A., Boesten, B., Crasnier, M., Danchin, A., O'Gara, F. 1990. *Rhizobium meliloti* adenylate cyclase is related to eucaryotic adenylate and guanylate cyclases. *J. Bacteriol.* 172: 2614–21

Bourne, H. R., Sanders, D. A., McCormick, F. 1990. The GTPase superfamily: A conserved switch for diverse cell functions. *Nature* 348: 125–32

Braughler, J. M., Mittal, C. K., Murad, F. 1979. Effects of thiols, sugars, and proteins on nitric oxide activation of guanylate cyclase. *J. Biol. Chem.* 254: 12450–54

Bredt, D. S., Hwang, P. M., Glatt, C. E., Lowenstein, C., Reed, R. R., Snyder, S. H. 1991. Cloned and expressed nitric oxide synthase structurally resembles cytochrome P-450 reductase. *Nature* 351: 714–18

Bredt, D. S., Hwang, P. M., Snyder, S. H. 1990. Localization of nitric oxide synthase indicating a neural role for nitric oxide. *Nature* 347: 768–70

Bredt, D. S., Snyder, S. H. 1989. Nitric oxide mediates glutamate-linked enhancement of cGMP levels in the cerebellum. *Proc. Natl. Acad. Sci. USA* 86: 9030–33

Bredt, D. S., Snyder, S. H. 1990. Isolation of nitric oxide synthetase, a calmodulin-requiring enzyme. *Proc. Natl. Acad. Sci. USA* 87: 682–85

Brenner, B. M., Ballermann, B. J., Gunning, M. E., Zeidel, M. L. 1990. Diverse biological actions of atrial natriuretic peptide. *Physiol. Rev.* 70: 665–99

Buechler, W. A., Nakane, M., Murad, F. 1991. Expression of soluble guanylate cyclase activity requires both enzyme subunits. *Biochem. Biophys. Res. Commun.* 174: 351–57

Chang, C. H., Kohse, K. P., Chang, B., Hirata, M., Jiang, B., et al. 1990. Characterization of ATP-stimulated guanylate cyclase activation in rat lung membranes. *Biochim. Biophys. Acta* 1052: 159–65

Chang, M. S., Lowe, D. G., Lewis, M., Hellmiss, R., Chen, E., Goeddel, D. V.

1989. Differential activation by atrial and brain natriuretic peptides of two receptor guanylate cyclases. *Nature* 341: 68–72

Chinkers, M., Garbers, D. L. 1989. The protein kinase domain of the ANP receptor is required for signaling. *Science* 245: 1392–94

Chinkers, M., Garbers, D. L. 1991. Signal transduction by guanylyl cyclases. *Annu. Rev. Biochem.* 60: 553–75

Chinkers, M., Garbers, D. L., Chang, M. S., Lowe, D. G., Chin, H., et al. 1989. Molecular cloning of a new type of cell surface receptor: A membrane form of guanylate cyclase is an atrial natriuretic peptide receptor. *Nature* 338: 78–83

Chinkers, M., Singh, S., Garbers, D. L. 1991. Adenine nucleotides are required for activation of rat atrial natriuretic peptide receptor/guanylyl cyclase expressed in a baculovirus system. *J. Biol. Chem.* 266: 4088–93

Chrisman, T. D., Garbers, D. L., Parks, M. A., Hardman, J. G. 1975. Characterization of particulate and soluble guanylate cyclases from rat lung. *J. Biol. Chem.* 250: 374–81

Corbin, J. D., Thomas, M. K., Wolfe, L., Shabb, J. B., Woodford, T. A., Francis, S. H. 1990. New insights into cGMP action. In *Advances in Second Messenger and Phosphoprotein Research: The Biology and Medicine of Signal Transduction*, ed. Y. Nishizuka, M. Endo, C. Tanaka, 24: 411–18. New York: Raven

Craven, P. A., DeRubertis, F. R. 1978. Restoration of the responsiveness of purified guanylate cyclase to nitrosoguanidine, nitric oxide, and related activators by heme and hemeproteins: Evidence for involvement of the paramagnetic nitrosyl-heme complex in enzyme activation. *J. Biol. Chem.* 253: 8433–43

Craven, P. A., DeRubertis, F. R. 1983. Requirement for heme in the activation of purified guanylate cyclase by nitric oxide. *Biochim. Biophys. Acta* 745: 310–21

Craven, P. A., DeRubertis, F. R., Pratt, D. W. 1979. Electron spin resonance study of the role of NO · catalase in the activation of guanylate cyclase by NaN_3 and NH_2OH. *J. Biol. Chem.* 254: 8213–22

Dangott, L. J., Garbers, D. L. 1984. Identification and partial characterization of the receptor for speract. *J. Biol. Chem.* 259: 13712–16

Dangott, L. J., Jordan, J. E., Bellet, R. A., Garbers, D. L. 1989. Cloning of the mRNA for the protein that crosslinks to the egg peptide speract. *Proc. Natl. Acad. Sci. USA* 86: 2128–32

Dawis, S. M., Graeff, R. M., Heyman, R. A., Walseth, T. F., Goldberg, N. D. 1988.

Regulation of cyclic GMP metabolism in toad photoreceptors: Definition of the metabolic events subserving photoexcited and attenuated states. *J. Biol. Chem.* 263: 8771–85

Dawson, V. L., Dawson, T. M., London, E. D., Bredt, D. S., Snyder, S. H. 1991. Nitric oxide mediates glutamate neurotoxicity in primary cortical cultures. *Proc. Natl. Acad. Sci. USA* 88: 6368–71

Deeg, M. D., Graeff, R. M., Walseth, T. F., Goldberg, N. D. 1988. A Ca^{2+}-linked increase in coupled cAMP synthesis and hydrolysis is an early event in cholinergic and β-adrenergic stimulation of parotid secretion. *Proc. Natl. Acad. Sci. USA* 85: 7867–71

Deguchi, T., Yoshioka, M. 1982. L-arginine identified as an endogenous activator for soluble guanylate cyclase from neuroblastoma cells. *J. Biol. Chem.* 257: 10147–51

DeJonge, H. R. 1975a. The localization of guanylate cyclase in rat small intestinal epithelium. *FEBS Lett.* 53: 237–42

DeJonge, H. R. 1975b. Properties of guanylate cyclase and levels of cyclic GMP in rat small intestinal villous and crypt cells. *FEBS Lett.* 55: 143–52

DeRubertis, F. R., Craven, P. A. 1976. Properties of the guanylate cyclase-guanosine 3′:5′-monophosphate system of rat renal cortex: Activation of guanylate cyclase and calcium-independent modulation of tissue guanosine 3′:5′-monophosphate by sodium azide. *J. Biol. Chem.* 251: 4651–58

Dhallan, R. S., Yau, K. W., Schrader, K. A., Reed, R. R. 1990. Primary structure and functional expression of a cyclic nucleotide-activated channel from olfactory neurons. *Nature* 347: 184–87

Drewett, J. G., Ziegler, R. J., Trachte, G. J. 1990. Neuromodulatory effects of atrial natriuretic factor are independent of guanylate cyclase in adrenergic neuronal pheochromocytoma cells. *J. Pharmacol. Exp. Ther.* 255: 497–503

Dukovich, M., Wano, Y., Thuy, L. B., Katz, P., Cullen, B. R., et al. 1987. A second human interleukin-2 binding protein that may be a component of high-affinity interleukin-2 receptors. *Nature* 327: 518–22

Eckstein, H. 1988. Evidence for cyclic GMP in the yeast *Saccharomyces cerevisiae*, and studies on its possible role in growth. *Z. Naturforsch.* 43: 386–96

Field, M., Graf, L. H. Jr., Laird, W. J., Smith, P. L. 1978. Heat-stable enterotoxin of *Escherichia coli*: In vitro effects on guanylate cyclase activity, cyclic GMP concentration, and ion transport in small intestine. *Proc. Natl. Acad. Sci. USA* 75: 2800–4

Fischer, E. H., Tonks, N. K., Charbonneau, H., Cicirelli, M. F., Cool, D. E., et al. 1990. Protein tyrosine phosphatases: A novel family of enzymes involved in transmembrane signalling. In *Advances in Second Messenger and Phosphoprotein Research: The Biology and Medicine of Signal Transduction*, ed. Y. Nishizuka, M. Endo, C. Tanaka, 24: 273–79. New York: Raven

Flanagan, J. G., Leder, P. 1990. The *kit* ligand: A cell surface molecule altered in Steel mutant fibroblasts. *Cell* 63: 185–94

Fleischman, D., Denisevich, M. 1979. Guanylate cyclase of isolated bovine retinal rod axonemes. *Biochemistry* 18: 5060–66

Fleischman, D., Denisevich, M., Raueed, D., Pannbacker, R. G. 1980. Association of guanylate cyclase with the axoneme of retinal rods. *Biochim. Biophys. Acta* 630: 176–86

Fostermann, U., Pollock, J. S., Schmidt, H. H. H. W., Heller, M., Murad, F. 1991. Calmodulin-dependent endothelium-derived relaxing factor/nitric oxide synthase activity is present in the particulate and cytosolic fractions of bovine aortic endothelial cells. *Proc. Natl. Acad. Sci. USA* 88: 1788–92

Freeman, M., Ashkenas, J., Rees, D. J. G., Kingsley, D. M., Copeland, N. G., et al. 1990. An ancient, highly conserved family of cysteine-rich domains revealed by cloning type I and type II murine macrophage scavenger receptors. *Proc. Natl. Acad. Sci. USA* 87: 8810–14

Fuller, F., Porter, J. G., Arfsten, A. E., Miller, J., Schilling, J. W., et al. 1988. Atrial natriuretic peptide clearance receptor: Complete sequence and functional expression of cDNA clones. *J. Biol. Chem.* 263: 9395–9401

Furchgott, R. F. 1988. Studies on relaxation of rabbit aorta by sodium nitrite: The basis for the proposal that the acid activatable inhibitory factor from retractor penis is inorganic nitrite and the endothelium-derived relaxing factor is nitric oxide. In *Vasodilation: Vascular Smooth Muscle, Peptides, Autonomic Nerves and Endothelium*, ed. P. M. Vanhoutte, pp. 401–14. New York: Raven

Furchgott, R. F., Vanhoutte, P. M. 1989. Endothelium-derived relaxing and contracting factors. *FASEB J.* 3: 2007–18

Furchgott, R. F., Zawadzki, J. V. 1980. The obligatory role of endothelial cells in the relaxation of arterial smooth muscle by acetylcholine. *Nature* 288: 373–76

Gally, J. A., Montague, P. R., Reeke, G. N. Jr., Edelman, G. M. 1990. The NO hypothesis: Possible effects of a short-lived, rapidly diffusible signal in the development and function of the nervous system. *Proc. Natl. Acad. Sci. USA* 87: 3547–51

Ganguly, A., Chiou, S., West, L. A., Davis, J. S. 1989. Atrial natriuretic factor inhibits angiotensin-induced aldosterone secretion: Not through cGMP or interference with phospholipase C. *Biochem. Biophys. Res. Commun.* 159: 148–54

Garbers, D. L. 1976. Sea urchin sperm guanylate cyclase: Purification and loss of cooperativity. *J. Biol. Chem.* 251: 4071–77

Garbers, D. L. 1979. Purification of soluble guanylate cyclase from rat lung. *J. Biol. Chem.* 254: 240–43

Garbers, D. L. 1989a. Guanylate cyclase, a cell surface receptor. *J. Biol. Chem.* 264: 9103–6

Garbers, D. L. 1989b. Molecular basis of fertilization. *Annu. Rev. Biochem.* 58: 719–42

Garbers, D. L., Hardman, J. G., Rudolph, F. B. 1974. Kinetic analysis of sea urchin sperm guanylate cyclase. *Biochemistry* 13: 4166–71

Garg, U. C., Hassid, A. 1991. Nitric oxide decreases cytosolic free calcium in Balb/c 3T3 fibroblasts by a cyclic GMP-independent mechanism. *J. Biol. Chem.* 266: 9–12

Garthwaite, J., Charles, S. L., Chess-Williams, R. 1988. Endothelium-derived relaxing factor release on activation of NMDA receptors suggests role as intercellular messenger in the brain. *Nature* 336: 385–88

Gerzer, R., Bohme, E., Hoffmann, F., Schultz, G. 1981a. Soluble guanylate cyclase purified from bovine lung contains heme and copper. *FEBS Lett.* 132: 71–74

Gerzer, R., Hoffmann, F., Schultz, G. 1981b. Purification of a soluble, sodium-nitroprusside-stimulated guanylate cyclase from bovine lung. *Eur. J. Biochem.* 116: 479–86

Gerzer, R., Radany, E. W., Garbers, D. L. 1982. The separation of the heme and apoheme forms of soluble guanylate cyclase. *Biochem. Biophys. Res. Commun.* 108: 678–86

Gibbs, J. B., Schaber, M. D., Schofield, T. L., Scolnick, E. M., Sigal, I. S. 1989. *Xenopus* oocyte germinal-vesicle breakdown induced by [val^{12}] ras is inhibited by a cytosol-localized ras mutant. *Proc. Natl. Acad. Sci. USA* 86: 6630–34

Gilman, A. G. 1987. G-proteins: Transducers of receptor-generated signals. *Annu. Rev. Biochem.* 56: 615–49

Glomset, J. A., Gelb, M. H., Farnsworth, C. C. 1990. Prenyl proteins in eukaryotic cells: A new type of membrane anchor. *Trends Biochem. Sci.* 15: 139–42

Goldberg, N. D., Ames, A. III, Gander, J. E., Walseth, T. F. 1983. Magnitude of increase in retinal cGMP metabolic flux determined by ^{18}O incorporation into nucleotide α-phosphoryls corresponds with intensity of photic stimulation. *J. Biol. Chem.* 258: 9213–19

Goldberg, N. D., Haddox, M. K. 1977. Cyclic GMP metabolism and involvement in biological regulation. *Annu. Rev. Biochem.* 46: 823–96

Goldberg, N. D., Walseth, T. F. 1985. A second role for second messengers: Uncovering the utility of cyclic nucleotide hydrolysis. *Bio/technology* 3: 235–39

Graff, G., Stephenson, J. H., Glass, D. B., Haddox, M. K., Goldberg, N. D. 1978. Activation of soluble splenic cell guanylate cyclase by prostaglandin endoperoxides and fatty acid hydroperoxides. *J. Biol. Chem.* 253: 7662–76

Gray, J. P., Drummond, G. I., Luk, D. W., Hardman, J. G., Sutherland, E. W. 1976. Enzymes of cyclic nucleotide metabolism in invertebrate and vertebrate sperm. *Arch. Biochem. Biophys.* 172: 20–30

Gregory, R. J., Cheng, S. H., Rich, D. P., Marshall, J., Paul, S., et al. 1990. Expression and characterization of cystic fibrosis transmembrane conductance regulator. *Nature* 347: 382–86

Guerrant, R. L., Hughes, J. M., Chang, B., Robertson, D. C., Murad, F. 1980. Activation of intestinal guanylate cyclase by heat-stable enterotoxin of *Escherichia coli*: Studies of tissue specificity, potential receptors, and intermediates. *J. Infect. Dis.* 142: 220–28

Gutierrez, L., Magee, A. I., Marshall, C. J., Hancock, J. F. 1989. Post-translational processing of p21^{ras} is two-step and involves carboxyl-methylation and carboxyl-terminal proteolysis. *EMBO J.* 8: 1093–98

Haddox, M. K., Stephenson, J. H., Moser, M. E., Goldberg, N. D. 1978. Oxidative-reductive modulation of guinea pig splenic cell guanylate cyclase activity. *J. Biol. Chem.* 253: 3143–52

Hancock, J. F., Magee, A. I., Childs, J. E., Marshall, C. J. 1989. All ras proteins are polyisoprenylated but only some are palmitoylated. *Cell* 57: 1167–77

Hanks, S. K., Quinn, A. M., Hunter, T. 1988. The protein kinase family: Conserved features and deduced phylogeny of the catalytic domains. *Science* 241: 42–52

Hardman, J. G., Beavo, J. A., Gray, J. P., Chrisman, T. D., Patterson, W. D., Sutherland, E. W. 1971. The formation and metabolism of cyclic GMP. *Ann. NY Acad. Sci.* 185: 27–35

Hardman, J. G., Sutherland, E. W. 1969. Guanyl cyclase, an enzyme catalyzing the formation of guanosine 3′,5′-monophosphate from guanosine triphosphate. *J. Biol. Chem.* 244: 6363–70

Harrison, S. A., Reifsnyder, D. H., Gallis, B., Cadd, G. G., Beavo, J. A. 1986. Isolation and characterization of bovine cardiac muscle cGMP-inhibited phosphodiesterase: A receptor for new cardiotonic drugs. *Mol. Pharmacol.* 29: 506–14

Harteneck, C., Koesling, D., Soling, A., Schultz, G., Bohme, E. 1990. Expression of soluble guanylate cyclase: Catalytic activity requires two enzyme subunits. *FEBS Lett.* 272: 221–23

Hibbs, J. B. Jr., Taintor, R. R., Vavrin, Z., Granger, D. L., Drapier, J. C., et al. 1990. Synthesis of nitric oxide from a terminal guanidino nitrogen atom of L-arginine: A molecular mechanism regulating cellular proliferation that targets intracellular iron. In *Nitric Oxide from L-Arginine: A Bioregulatory System*, ed. S. Moncada, E. A. Higgs, pp. 189–223. Amsterdam: Elsevier

Hidaka, H., Asano, T. 1977. Stimulation of human platelet guanylate cyclase by unsaturated fatty acid peroxides. *Proc. Natl. Acad. Sci. USA* 74: 3657–61

Hirata, M., Chang, C. H., Murad, F. 1989. Stimulatory effects of atrial natriuretic factor on phosphoinositide hydrolysis in cultured bovine aortic smooth muscle cells. *Biochim. Biophys. Acta* 1010: 346–51

Hope, B. T., Michael, G. J., Knigge, K. M., Vincent, S. R. 1991. Neuronal NADPH diaphorase is a nitric oxide synthase. *Proc. Natl. Acad. Sci. USA* 88: 2811–14

Hughes, J. M., Murad, F., Chang, B., Guerrant, R. 1978. Role of cyclic GMP in the action of heat-stable enterotoxin of *Escherichia coli. Nature* 271: 755–56

Huott, P. A., Liu, W., McRoberts, J. A., Gianella, R. A., Dharmsathaphorn, K. 1988. Mechanism of action of *Escherichia coli* heat stable enterotoxin in a human colonic cell line. *J. Clin. Invest.* 82: 514–23

Ignarro, L. J. 1990. Biosynthesis and metabolism of endothelium-derived nitric oxide. *Annu. Rev. Pharmacol. Toxicol.* 30: 535–60

Ignarro, L. J., Adams, J. B., Horwitz, P. M., Wood, K. S. 1986. Activation of soluble guanylate cyclase by NO-hemoproteins involves NO-heme exchange: Comparison of heme-containing and heme-deficient enzyme forms. *J. Biol. Chem.* 261: 4997–5002

Ignarro, L. J., Byrns, R. E., Wood, K. S. 1988. Biochemical and pharmacological properties of endothelium-derived relaxing factor and its similarity to nitric oxide radical. In *Vasodilation: Vascular Smooth Muscle, Peptides, Autonomic Nerves and Endothelium*, ed. P. M. Vanhoutte, pp. 427–35. New York: Raven

Ignarro, L. J., Degnan, J. N., Baricos, W. H., Kadowitz, P. J., Wolin, M. S. 1982. Activation of purified guanylate cyclase by nitric oxide requires heme: Comparison of heme-deficient, heme-reconstituted and heme-containing forms of soluble enzyme from bovine lung. *Biochim. Biophys. Acta* 718: 49–59

Ignarro, L. J., Kadowitz, P. J. 1985. The pharmacological and physiological role of cyclic GMP in vascular smooth muscle relaxation. *Annu. Rev. Pharmacol. Toxicol.* 25: 171–91

Inagami, T. 1989. Atrial natriuretic factor. *J. Biol. Chem.* 264: 3043–46

Ishikawa, E., Ishikawa, S., Davis, J. W., Sutherland, E. W. 1969. Determination of guanosine 3′,5′-monophosphate in tissues and of guanyl cyclase in rat intestine. *J. Biol. Chem.* 244: 6371–76

Ivens, K., Gazzano, H., O'Hanley, P., Waldman, S. A. 1990. Heterogeneity of intestinal receptors for *Escherichia coli* heat-stable enterotoxin. *Infect. Immun.* 58: 1817–20

Kamisaki, Y., Saheki, S., Nakane, M., Palmieri, J. A., Kuno, T., et al. 1986. Soluble guanylate cyclase from rat lung exists as a heterodimer. *J. Biol. Chem.* 261: 7236–41

Kataoka, T., Broek, D., Wigler, M. 1985. DNA sequence and characterization of the *S. cerevisiae* gene encoding adenylate cyclase. *Cell* 43: 493–505

Kaupp, U. B., Niidome, T., Tanabe, T., Terada, S., Bonigk, W., et al. 1989. Primary structure and functional expression from complementary DNA of the rod photoreceptor cyclic GMP-gated channel. *Nature* 342: 762–66

Kawaguchi, S., Uchida, K., Ito, T., Kozuka, M., Shimonaka, M., et al. 1989. Immunohistochemical localization of atrial natriuretic peptide receptor in bovine kidney and lung. *J. Histochem. Cytochem.* 37: 1739–42

Kilbourn, R., Belloni, P. 1990. Endothelial cells produce nitrogen oxides in response to interferon-γ, tumour necrosis factor and endotoxin. In *Nitric Oxide from L-Arginine: A Bioregulatory System*, ed. S. Moncada, E. A. Higgs, pp. 61–67. Amsterdam: Elsevier

Kilbourn, R. G., Gross, S. S., Jubran, A., Adams, J., Griffith, O. W., Levi, R., Lodato, R. F. 1990. N^G-methyl-L-arginine inhibits tumor necrosis factor-induced hypotension: Implications for the involvement of nitric oxide. *Proc. Natl. Acad. Sci. USA* 87: 3629–32

Kimura, H., Mittal, C. K., Murad, F. 1975a. Activation of guanylate cyclase from rat liver and other tissues by sodium azide. *J. Biol. Chem.* 250: 8016–22

Kimura, H., Mittal, C. K., Murad, F. 1975b. Increases in cyclic GMP levels in brain and liver with sodium azide an activator of guanylate cyclase. *Nature* 257: 700–2

Kimura, H., Murad, F. 1974. Evidence for two different forms of guanylate cyclase in rat heart. *J. Biol. Chem.* 249: 6910–16

Klein, R., Conway, D., Parada, L. F., Barbacid, M. 1990. The *trk*B tyrosine protein kinase gene codes for a second neurogenic receptor that lacks the catalytic kinase domain. *Cell* 61: 647–56

Klumpp, S., Kleefeld, G., Schultz, J. E. 1983. Calcium/calmodulin-regulated guanylate cyclase of the excitable ciliary membrane from *Paramecium*. *J. Biol. Chem.* 258: 12455–59

Koch, K. W., Stryer, L. 1988. Highly cooperative feedback control of retinal rod guanylate cyclase by calcium ions. *Nature* 334: 64–66

Koesling, D., Harteneck, C., Humbert, P., Bosserhoff, A., Frank, R., et al. 1990. The primary structure of the larger subunit of soluble guanylyl cyclase from bovine lung: Homology between the two subunits of the enzyme. *FEBS Lett.* 266: 128–32

Koesling, D., Herz, J., Gausepohl, H., Niroomand, F., Hinsch, K. D., et al. 1988. The primary structure of the 70 kDa subunit of bovine soluble guanylate cyclase. *FEBS Lett.* 239: 29–34

Koller, K. J., Lowe, D. G., Bennett, G. L., Minamino, N., Kangawa, K., Matsuo, H., Goeddel, D. V. 1991. C-type natriuretic peptide (CNP) selectively activates the B natriuretic peptide receptor (ANPR-B). *Science* 252: 120–23

Krause, W. J., Freeman, R. A., Forte, L. R. 1990. Autoradiographic demonstration of specific binding sites for *E. coli* enterotoxin in various epithelia of the North American opossum. *Cell Tissue Res.* 260: 387–94

Krupinski, J., Coussen, F., Bakalyar, H. A., Tang, W. J., Feinstein, P. G., et al. 1989. Adenylyl cyclase amino acid sequence: Possible channel- or transporter-like structure. *Science* 244: 1558–64

Kuno, T., Andresen, J. W., Kamisaki, Y., Waldman, S. A., Chang, L. Y., et al. 1986a. Co-purification of an atrial natriuretic factor receptor and particulate guanylate cyclase from rat lung. *J. Biol. Chem.* 261: 5817–23

Kuno, T., Kamisaki, Y., Waldman, S. A., Gariepy, J., Schoolnik, G., Murad, F. 1986b. Characterization of the receptor for heat-stable enterotoxin from *E. coli* in rat intestine. *J. Biol. Chem.* 261: 1470–76

Kurose, H., Inagami, T., Ui, M. 1987. Participation of adenosine 5′-triphosphate in the activation of membrane-bound guanylate cyclase by the atrial natriuretic factor. *FEBS Lett.* 219: 375–79

Kwon, N. S., Nathan, C. F., Gilker, C., Griffith, O. W., Matthews, D. E., Stuehr, D. J. 1990. L-citrulline production from L-arginine by macrophage nitric oxide synthase. *J. Biol. Chem.* 265: 13442–45

Kwon, N. S., Nathan, C. F., Stuehr, D. J. 1989. Reduced biopterin as a cofactor in the generation of nitrogen oxides by murine macrophages. *J. Biol. Chem.* 264: 20496–20501

Lancaster, J. R., Hibbs, J. B. 1990. EPR demonstration of iron-nitrosyl complex formation by cytotoxic activated macrophages. *Proc. Natl. Acad. Sci. USA* 87: 1223–27

Lee, M. A., West, R. E. Jr., Moss, J. 1988. Atrial natriuretic factor reduces cyclic adenosine monophosphate content of human fibroblasts by enhancing phosphodiesterase activity. *J. Clin. Invest.* 82: 388–93

Leitman, D. C., Andresen, J. W., Catalano, R. M., Waldman, S. A., Tuan, J. J., Murad, F. 1988. Atrial natriuretic peptide binding, cross-linking, and stimulation of cyclic GMP accumulation and particulate guanylate cyclase activity in cultured cells. *J. Biol. Chem.* 263: 3720–28

Leitman, D. C., Andresen, J. W., Kuno, T., Kamisaki, Y., Chang, J. K., Murad, F. 1986. Identification of multiple binding sites for atrial natriuretic factor by affinity cross-linking in cultured endothelial cells. *J. Biol. Chem.* 261: 11650–55

Lewicki, J. A., Brandwein, H. J., Waldman, S. A., Murad, F. 1980. Purified guanylate cyclase: Characterization, iodination and preparation of monoclonal antibodies. *J. Cyc. Nuc. Res.* 6: 283–96

Lewicki, J. A., Chang, B., Murad, F. 1983. Quantification of guanylate cyclase concentrations by a direct double determinant tandem immunoradiometric assay. *J. Biol. Chem.* 258: 3509–15

Light, D. B., Corbin, J. D., Stanton, B. A. 1990. Dual ion-channel regulation by cyclic GMP and cyclic GMP-dependent protein kinase. *Nature* 344: 336–39

Lin, M., Nairn, A. C., Guggino, S. E. 1991. Cyclic GMP-dependent protein kinase regulation of chloride channels in T_{84} cells. *Biophys. J.* 59: 254a (Abstr.)

Lowe, D. G., Chang, M. S., Hellmiss,

R., Chen, E., Singh, S., Garbers, D. L., Goeddel, D. V. 1989. Human atrial natriuretic peptide receptor defines a new paradigm for second messenger signal transduction. *EMBO J.* 8: 1377–84

Ludwig, J., Margalit, T., Eismann, E., Lancet, D., Kaupp, U. B. 1990. Primary structure of cAMP-gated channel from bovine olfactory epithelium. *FEBS Lett.* 270: 24–29

Maack, T., Suzuki, M., Almeida, F. A., Nussenzveig, D., Scarborough, R. M., McEnroe, G. A., Lewicki, J. A. 1987. Physiological role of silent receptors of atrial natriuretic factor. *Science* 238: 675–78

MacFarland, R. T., Zelus, B. D., Beavo, J. A. 1991. High concentrations of a cGMP-stimulated phosphodiesterase mediate ANP-induced decreases in cAMP and steroidogenesis in adrenal glomerulosa cells. *J. Biol. Chem.* 266: 136–42

Marletta, M. A. 1989. Nitric oxide: Biosynthesis and biological significance. *Trends Biochem. Sci.* 14: 488–92

Marletta, M. A., Yoon, P. S., Iyengar, R., Leaf, C. D., Wishnok, J. S. 1988. Macrophage oxidation of L-arginine to nitrite and nitrate: Nitric oxide is an intermediate. *Biochemistry* 27: 8706–11

Martins, T. J., Mumby, M. C., Beavo, J. A. 1982. Purification and characterization of a cyclic GMP-stimulated cyclic nucleotide phosphodiesterase from bovine tissues. *J. Biol. Chem.* 257: 1973–79

Matsuoka, H., Ishii, M., Hirata, Y., Atarashi, K., Sugimoto, T., Kangawa, K., Matsuo, H. 1987. Evidence for lack of a role of cGMP in effect of α-hANP on aldosterone inhibition. *Am. J. Physiol.* 252: E643–47

Meloche, S., McNicoll, N., Liu, B., Ong, H., DeLean, A. 1988. Atrial natriuretic factor R1 receptor from bovine adrenal glomerulosa: Purification, characterization, and modulation by amiloride. *Biochemistry* 27: 8151–58

Meloche, S., Ong, H., Cantin, M., DeLean, A. 1986. Affinity cross-linking of atrial natriuretic factor to its receptor in bovine adrenal glomerulosa. *J. Biol. Chem.* 261: 1525–28

Middlemas, D. S., Lindberg, R. A., Hunter, T. 1991. *trk*B, a neural receptor protein-tyrosine kinase: Evidence for a full-length and two truncated receptors. *Mol. Cell. Biol.* 11: 143–53

Miki, N., Kawabe, Y., Kuriyama, K. 1977. Activation of cerebral guanylate cyclase by nitric oxide. *Biochem. Biophys. Res. Commun.* 75: 851–56

Miki, N., Nagano, M., Kuriyama, K. 1976. Catalase activates cerebral guanylate cyclase in the presence of sodium azide. *Biochem. Biophys. Res. Commun.* 72: 952–59

Misono, K. S., Grammer, R. T., Rigby, J. W., Inagami, T. 1985. Photoaffinity labeling of atrial natriuretic factor receptor in bovine and rat adrenal cortical membranes. *Biochem. Biophys. Res. Commun.* 130: 994–1001

Mizuno, T., Katafuchi, T., Hagiwara, H., Ito, T., Kangawa, K., Matsuo, H., Hirose, S. 1990. Human adrenal tumor cell line SW-13 contains a natriuretic peptide receptor system that responds preferentially to ANP among various natriuretic peptides. *Biochem. Biophys. Res. Commun.* 173: 886–93

Moncada, S. 1990. Introduction. In *Nitric Oxide from L-Arginine: A Bioregulatory System,* ed. S. Moncada, E. A. Higgs, pp. 1–4. Amsterdam: Elsevier

Moncada, S., Palmer, R. M. J., Higgs, E. A. 1989. Biosynthesis of nitric oxide from L-arginine: A pathway for the regulation of cell function and communication. *Biochem. Pharmacol.* 38: 1709–15

Moran, M. F., Koch, C. A., Sadowski, I., Pawson, T. 1988. Mutational analysis of a phosphotransfer motif essential for v-*fps* tyrosine kinase activity. *Oncogene* 3: 665–72

Mukhopadhyay, A. Y., Helbing, J., Leidenherger, F. A. 1989. The role of Ca^{2+} and calmodulin in the regulation of atrial natriuretic peptide-stimulated guanosine 3′,5′-cyclic monophosphate accumulation by isolated mouse Leydig cells. *Endocrinology* 125: 686–92

Murad, F. 1990. Drugs used for the treatment of angina: Organic nitrates, calcium channel blockers, and β-adrenergic antagonists. In *The Pharmacological Basis of Therapeutics,* ed. A. G. Gilman, T. W. Rall, A. S. Nies, P. Taylor, pp. 764–83. New York: Pergamon

Murad, F., Arnold, W. P., Mittal, C. K., Braughler, J. M. 1979. Properties and regulation of guanylate cyclase and some proposed functions for cyclic GMP. In *Advances in Cyclic Nucleotide Research,* ed. P. Greengard, G. A. Robison, 24: 175–204. New York: Raven

Murphy, S., Minor, R. L. Jr., Welk, G., Harrison, D. G. 1990. Evidence for an astrocyte-derived vasorelaxing factor with properties similar to nitric oxide. *J. Neurochem.* 55: 349–51

Myers, P. R., Minor, R. L. Jr., Guerra, R., Bates, J. N., Harrison, D. G. 1990. Vasorelaxant properties of the endothelium-derived relaxing factor more closely resemble S-nitrosocysteine than nitric oxide. *Nature* 345: 161–63

Nakamura, T., Gold, G. H. 1987. A cyclic

nucleotide-gated conductance in olfactory receptor cilia. *Nature* 325: 442–44

Nakane, M., Arai, K., Saheki, S., Kuno, T., Buechler, W., Murad, F. 1990. Molecular cloning and expression of cDNAs coding for soluble guanylate cyclase from rat lung. *J. Biol. Chem.* 265: 16841–45

Nakane, M., Saheki, S., Kuno, T., Ishii, K., Murad, F. 1988. Molecular cloning of a cDNA coding for 70 kilodalton subunit of soluble guanylate cyclase from rat lung. *Biochem. Biophys. Res. Commun.* 156: 1000–6

Nathan, C. F., Stuehr, D. J. 1990. Does endothelium-derived nitric oxide have a role in cytokine-induced hypotension? *J. Natl. Cancer Inst.* 82: 726–28

Ohlstein, E. H., Wood, K. S., Ignarro, L. J. 1982. Purification and properties of heme-deficient hepatic soluble guanylate cyclase: Effects of heme and other factors on enzyme activation by NO, NO-heme, and protoporphyrin IX. *Arch. Biochem. Biophys.* 218: 187–98

Palmer, R. M. J., Ashton, D. S., Moncada, S. 1988. Vascular endothelial cells synthesize nitric oxide from L-arginine. *Nature* 333: 664–66

Palmer, R. M. J., Ferrige, A. G., Moncada, S. 1987. Nitric oxide release accounts for the biological activity of endothelium-derived relaxing factor. *Nature* 327: 524–26

Pandey, K. N., Pavlou, S. N., Inagami, T. 1988. Identification and characterization of three distinct atrial natriuretic factor receptors: Evidence for tissue-specific heterogeneity of receptor subtypes in vascular smooth muscle, kidney tubular epithelium, and leydig tumor cells by ligand binding, photoaffinity labeling, and tryptic proteolysis. *J. Biol. Chem.* 263: 13406–13

Pandey, K. N., Singh, S. 1990. Molecular cloning and expression of murine guanylate cyclase/atrial natriuretic factor receptor cDNA. *J. Biol. Chem.* 265: 12342–48

Paul, A. K., Marala, R. B., Jaiswal, R. K., Sharma, R. K. 1987. Coexistence of guanylate cyclase and atrial natriuretic factor receptor in a 180 kD protein. *Science* 235: 1224–26

Pellat, C., Henry, Y., Drapier, J. C. 1990. IFN-γ-activated macrophages: Detection by electron paramagnetic resonance of complexes between L-arginine-derived nitric oxide and non-heme iron proteins. *Biochem. Biophys. Res. Commun.* 166: 119–25

Porter, J. G., Scarborough, R. M., Wang, Y., Schenk, D., McEnroe, G. A., et al. 1989. Recombinant expression of a

secreted form of the atrial natriuretic peptide clearance receptor. *J. Biol. Chem.* 264: 14179–84

Radany, E. W., Gerzer, R., Garbers, D. L. 1983. Purification and characterization of particulate guanylate cyclase from sea urchin spermatozoa. *J. Biol. Chem.* 258: 8346–51

Rall, T. W., Sutherland, E. W. 1958. Formation of a cyclic adenine ribonucleotide by tissue particles. *J. Biol. Chem.* 232: 1065–76

Rall, T. W., Sutherland, E. W., Berthet, J. 1957. The relationship of epinephrine and glucagon to liver phosphorylase: Effect of epinephrine and glucagon on the reactivation of phosphorylase in liver homogenates. *J. Biol. Chem.* 224: 463–75

Rao, M. C., Guandalini, S., Smith, P. L., Field, M. 1980. Mode of action of heat-stable *Escherichia coli* enterotoxin: Tissue and subcellular specificities and role of cyclic GMP. *Biochim. Biophys. Acta* 632: 35–46

Rapoport, R. M., Murad, F. 1983. Endothelium-dependent and nitrovasodilator-induced relaxation of vascular smooth muscle: Role of cyclic GMP. *J. Cyc. Nuc. Prot. Phosph. Res.* 9: 281–96

Riordan, J. R., Rommens, J. M., Kerem, B., Alon, N., Rozmahel, R., et al. 1989. Identification of the cystic fibrosis gene: Cloning and characterization of complementary DNA. *Science* 245: 1066–72

Robb, R. J., Rusk, C. M., Yodoi, J., Greene, W. C. 1987. Interleukin 2 binding molecule distinct from the Tac protein: Analysis of its role in formation of high-affinity receptors. *Proc. Natl. Acad. Sci. USA* 84: 2002–6

Rogers, K. V., Goldman, P. S., Frizzell, R. A., McKnight, G. S. 1990. Regulation of Cl⁻ transport in T84 cell clones expressing a mutant regulatory subunit of cAMP-dependent protein kinase. *Proc. Natl. Acad. Sci. USA* 87: 8975–79

Salvemini, D., Korbut, R., Anggard, E., Vane, J. 1990. Immediate release of a nitric oxide-like factor from bovine aortic endothelial cells by *Escherichia coli* lipopolysaccharide. *Proc. Natl. Acad. Sci. USA* 87: 2593–97

Schmidt, H. H. H. W., Seifert, R., Bohme, E. 1989. Formation and release of nitric oxide from human neutrophils and HL-60 cells induced by chemotactic peptide, platelet activating factor and leukotriene B₄. *FEBS Lett.* 244: 357–60

Schultz, G., Bohme, E., Munske, K. 1969. Guanyl cyclase: Determination of enzyme activity. *Life Sci.* 8: 1323–32

Schultz, J. E., Schonefeld, U., Klumpp, S. 1983. Calcium/calmodulin-regulated guanylate cyclase and calcium permeability in the ciliary membrane from *Paramecium*. *Eur. J. Biochem.* 137: 89–94

Schulz, S., Green, C. K., Yuen, P. S. T., Garbers, D. L. 1990. Guanylyl cyclase is a heat-stable enterotoxin receptor. *Cell* 63: 941–48

Schulz, S., Singh, S., Bellet, R. A., Singh, G., Tubb, D. J., Chin, H., Garbers, D. L. 1989. The primary structure of a plasma membrane guanylate cyclase demonstrates diversity within this new receptor family. *Cell* 58: 1155–62

Sharon, M., Klausner, R. D., Cullen, B. R., Chizzonite, R., Leonard, W. J. 1986. Novel interleukin-2 receptor subunit detected by cross-linking under high-affinity conditions. *Science* 234: 859–63

Shibuki, K., Okada, D. 1991. Endogenous nitric oxide release required for long-term synaptic depression in the cerebellum. *Nature* 349: 326–28

Shikano, K., Long, C. J., Ohlstein, E. H., Berkowitz, B. A. 1988. Comparative pharmacology of endothelium-derived relaxing factor and nitric oxide. *J. Pharmacol. Exp. Ther.* 247: 873–81

Shimomura, H., Dangott, L. J., Garbers, D. L. 1986. Covalent coupling of a resact analogue to guanylate cyclase. *J. Biol. Chem.* 261: 15778–82

Singh, S., Lowe, D. G., Thorpe, D. S., Rodriguez, H., Kuang, W. J., Dangott, L. J., Chinkers, M., Goeddel, D. V., Garbers, D. L. 1988. Membrane guanylate cyclase is a cell surface receptor with homology to protein kinases. *Nature* 334: 708–12

Song, D. L., Kohse, K. P., Murad, F. 1988. Brain natriuretic factor: Augmentation of cellular cyclic GMP, activation of particulate guanylate cyclase and receptor binding. *FEBS Lett.* 232: 125–29

Sorbera, L. A., Morad, M. 1990. Atrionatriuretic peptide transforms cardiac sodium channels into calcium-conducting channels. *Science* 247: 969–73

Steinlen, S., Klumpp, S., Schultz, J. E. 1990. Guanylate cyclase in olfactory cilia from rat and pig. *Biochim. Biophys. Acta* 1054: 69–72

Stuehr, D. J., Marletta, M. A. 1985. Induction of nitrite/nitrate biosynthesis: Mouse macrophages produce nitrite and nitrate in response to *Escherichia coli* lipopolysaccharide. *Proc. Natl. Acad. Sci. USA* 82: 7738–42

Sudoh, T., Minamino, N., Kangawa, K., Matsuo, H. 1990. C-type natriuretic peptide (CNP): A new member of natriuretic peptide family identified in porcine brain.

Biochem. Biophys. Res. Commun. 168: 863–70

Sutherland, E. W., Rall, T. W. 1957. The properties of an adenine ribonucleotide produced with cellular particles, ATP, Mg^{++}, and epinephrine or glucagon. *J. Am. Chem. Soc.* 79: 3608

Sutherland, E. W., Rall, T. W. 1958. Fractionation and characterization of a cyclic adenine ribonucleotide formed by tissue particles. *J. Biol. Chem.* 232: 1077–91

Sutherland, E. W., Rall, T. W. 1960. The relation of adenosine-3′,5′-phosphate and phosphorylase to the actions of catecholamines and other hormones. *Pharmacol. Rev.* 12: 265–99

Suzuki, N., Shimomura, H., Radany, E. W., Ramarao, C. S., Ward, G. E., Bentley, J. K., Garbers, D. L. 1984. A peptide associated with eggs causes a mobility shift in a major plasma membrane protein of spermatozoa. *J. Biol. Chem.* 259: 14874–79

Takayanagi, R., Inagami, T., Snajdar, R. M., Imada, T., Tamura, M., Misono, K. S. 1987. Two distinct forms of receptors for atrial natriuretic factor in bovine adrenocortical cells: Purification, ligand binding, and peptide mapping. *J. Biol. Chem.* 262: 12104–13

Takei, Y., Takahashi, A., Watanabe, T. X., Nakajima, K., Sakakibara, S., et al. 1990. Amino acid sequence and relative biological activity of a natriuretic peptide isolated from eel brain. *Biochem. Biophys. Res. Commun.* 170: 883–91

Takio, K., Wade, R. D., Smith, S. B., Krebs, E. G., Walsh, K. A., Titani, K. 1984. Guanosine cyclic 3′,5′-phosphate dependent protein kinase, a chimeric protein homologous with two separate protein families. *Biochemistry* 23: 4207–18

Tan, J. C., Nocka, K., Ray, P., Traktman, P., Besmer, P. 1990. The dominant W^{42} *spotting* phenotype results from a missense mutation in the c-*kit* receptor kinase. *Science* 247: 209–12

Tayeh, M. A., Marletta, M. A. 1989. Macrophage oxidation of L-arginine to nitric oxide, nitrite, and nitrate: Tetrahydrobiopterin is required as a cofactor. *J. Biol. Chem.* 264: 19654–58

Taylor, S. G., Weston, A. H. 1988. Endothelium-derived hyperpolarizing factor: A new endogenous inhibitor from the vascular endothelium. *Trends Pharmacol. Sci.* 9: 272–74

Teshigawara, K., Wang, H. M., Kato, K., Smith, K. A. 1987. Interleukin high-affinity receptor expression requires two distinct binding proteins. *J. Exp. Med.* 165: 223–38

Thompson, M. R., Gianella, R. A. 1990.

Different crosslinking agents identify distinctly different putative *Escherichia coli* heat-stable enterotoxin rat intestinal cell receptor proteins. *J. Receptor Res.* 10: 97–117

Thorpe, D. S., Garbers, D. L. 1989. The membrane form of guanylate cyclase: Homology with a subunit of the cytoplasmic form of the enzyme. *J. Biol. Chem.* 264: 6545–49

Thorpe, D. S., Morkin, E. 1990. The carboxyl region contains the catalytic domain of the membrane form of guanylate cyclase. *J. Biol. Chem.* 265: 14717–20

Tracey, W. R., Linden, J., Peach, M. J., Johns, R. A. 1990. Comparison of spectrophotometric and biological assays for nitric oxide (NO) and endothelium-derived relaxing factor (EDRF): Nonspecificity of the diazotization reaction for NO and failure to detect EDRF. *J. Pharmacol. Exp. Ther.* 252: 922–28

Tsudo, M., Kozak, R. W., Goldman, C. K., Waldmann, T. A. 1986. Demonstration of a non-Tac peptide that binds interleukin 2: A potential participant in a multichain interleukin 2 receptor complex. *Proc. Natl. Acad. Sci. USA* 83: 9694–98

Uhler, M. D., Carmichael, D. F., Lee, D. C., Chrivia, J. C., Krebs, E. G., McKnight, G. S. 1986. Isolation of cDNA clones coding for the catalytic subunit of mouse cyclic AMP-dependent protein kinase. *Proc. Natl. Acad. Sci. USA* 83: 1300–4

Ullrich, A., Schlessinger, J. 1990. Signal transduction by receptors with tyrosine kinase activity. *Cell* 61: 203–12

Vandlen, R. L., Arcuri, K. E., Napier, M. A. 1985. Identification of a receptor for atrial natriuretic factor in rabbit aorta membranes by affinity cross-linking. *J. Biol. Chem.* 260: 10889–92

Waldman, S. A., Kuno, T., Kamisaki, Y., Chang, L. Y., Gariepy, J., et al. 1986. Intestinal receptor for heat-stable enterotoxin of *Escherichia coli* is tightly coupled to a novel form of particulate guanylate cyclase. *Infect. Immun.* 51: 320–26

Ward, G. E., Garbers, D. L., Vacquier, V. D. 1985. Effects of extracellular egg factors on sperm guanylate cyclase. *Science* 227: 768–70

Welsh, M. J. 1990. Abnormal regulation of ion channels in cystic fibrosis epithelia. *FASEB J.* 4: 2718–25

White, A. A., Aurbach, G. D. 1969. Detection of guanyl cyclase in mammalian tissues. *Biochim. Biophys. Acta* 191: 686–97

Wilcox, J. N., Augustine, A., Goeddel, D. V., Lowe, D. G. 1991. Differential regional expression of three natriuretic peptide receptor genes within primate tissues. *Mol. Cell. Biol.* 11: 3454–62

Wildey, G. M., Glembotski, C. C. 1986. Crosslinking of atrial natriuretic peptide to binding sites in the rat olfactory bulb. *J. Neurosci.* 6: 3767–76

Williams, D. E., Eisenman, J., Baird, A., Rauch, C., Van Ness, K., et al. 1990. Identification of a ligand for the *c-kit* protooncogene. *Cell* 63: 167–74

Willumsen, B. M., Christensen, A., Hubbert, N. L., Papageorge, A. G., Lowy, D. R. 1984. The p21 *ras* C-terminus is required for transformation and membrane association. *Nature* 310: 583–86

Wright, C. D., Mulsch, A., Busse, R., Osswald, H. 1989. Generation of nitric oxide by human neutrophils. *Biochem. Biophys. Res. Commun.* 160: 813–19

Yamawaki-Kataoka, Y., Tamaoki, T., Choe, H. R., Tanaka, H., Kataoka, T. 1989. Adenylate cyclases in yeast: A comparison of the genes from *Schizosaccharomyces pombe* and *Saccharomyces cerevisiae. Proc. Natl. Acad. Sci. USA* 86: 5693–97

Yau, K. W., Baylor, D. A. 1989. Cyclic GMP-activated conductance of retinal photoreceptor cells. *Annu. Rev. Neurosci.* 12: 289–327

Yip, C. C., Laing, L. P., Flynn, T. G. 1985. Photoaffinity labeling of atrial natriuretic factor receptors of rat kidney cortex plasma membranes. *J. Biol. Chem.* 260: 8229–32

Yoshihara, A., Kozawa, H., Minamino, N., Kangawa, K., Matsuo, H. 1990. Isolation and sequence determination of frog C-type natriuretic peptide. *Biochem. Biophys. Res. Commun.* 173: 591–98

Yuen, P. S. T., Potter, L. R., Garbers, D. L. 1990. A new form of guanylyl cyclase is preferentially expressed in rat kidney. *Biochemistry* 29: 10872–78

Yui, Y., Hattori, R., Kosuga, K., Eizawa, H., Hiki, K., et al. 1991. Calmodulin-independent nitric oxide synthase from rat polymorphonuclear neutrophils. *J. Biol. Chem.* 266: 3369–71

Annu. Rev. Neurosci. 1992. 15:227–50

NEURAL MECHANISMS OF TACTUAL FORM AND TEXTURE PERCEPTION

K. O. Johnson and S. S. Hsiao

The Phillip Bard Laboratories of Neurophysiology, Department of
Neuroscience, The Johns Hopkins University School of Medicine,
Baltimore, Maryland 21205

KEY WORDS: pattern recognition, tactile, psychophysics, roughness, neurophysiology

INTRODUCTION

Form and texture perception at the fingertips is a critical component
of sensory and motor function in the hand. Recent psychophysical and
neurophysiological studies have provided evidence linking form and texture perception with specific neural coding mechanisms. We review the
results of these studies and propose a general hypothesis concerning the
role of each of the primary afferent fiber types in tactual perception.

The experimental design employed in most of the neurophysiological
studies reviewed here was pioneered by Mountcastle in his studies of the
neural mechanisms underlying vibratory sensation (Mountcastle et al 1972,
Talbot et al 1968). In that design, psychophysical studies of a particular
aspect of perceptual behavior are followed by neurophysiological studies
in which exactly the same stimuli and stimulus conditions are used. The
object of the neurophysiological studies is to reconstruct the population
responses to all of the stimuli in order "to inquire which quantitative
aspects of the neural response tally with psychophysical measurements"
(Mountcastle et al 1963). Studies using this approach have shown that
it is reasonable to expect extensive quantitative agreement between the
predictions of a good hypothesis and observed psychophysical behavior.

Hypotheses linking observed somatosensory peripheral neural activity

227

0147–006X/92/0301–0227$02.00

to psychophysical behavior have two parts: (*a*) Which one or more afferent fiber populations is used by the higher neural mechanisms as the basis for the observed behavior? (*b*) How does the observed behavior depend on the neural activity in that population? There are four classes of primary afferents: SAI, RA, SAII, and PC (for a recent review of their properties see Darian-Smith 1984). Because the hypotheses considered in this review concern central mechanisms, we use the terms SAI, RA, SAII, and PC systems throughout the review. By SAI system, we mean the primary slowly adapting type I afferent population and all of the central mechanisms that convey its signals to memory and perception. Similar definitions hold for the RA, cutaneous rapidly adapting, SAII, slowly adapting type II, and PC, Pacinian, systems. We do not mean to imply that there is no central convergence, i.e. no overlap between the systems. A neural representation of a stimulus is defined here as being all of the neural activity evoked by the stimulus at a specific level of processing (in pathways leading to perception).

FORM PERCEPTION

Form perception at the finger pads consists of the appreciation of the spatial features of an object or surface through direct contact with the skin. A prime example of form perception is the human ability to read Braille. Braille is a digital code based on six cells in a 3 × 2 pattern in which the cells are spaced at 2.3 mm center to center, and each cell may or may not contain an embossed dot (Nolan & Kederis 1969). Braille codes for the letters A through R are illustrated at the top of Figure 1. Adults read grade 2 (contracted) Braille at mean rates of 70–90 words per minute and many read at much higher rates (Nolan & Kederis 1969). This performance must be accounted for by a detailed representation of form at all levels of processing leading to recognition and perception.

The responses of typical human primary afferent fibers to Braille patterns scanned at 60 mm/sec are illustrated in Figure 1 (Phillips et al 1990). Comparable monkey data are similar (Johnson & Lamb 1981). SAI afferents respond most acutely, providing a clear, robust representation of the Braille dot patterns, thereby suggesting that the SAI system provides the information on which recognition is based. RA afferent responses yield less distinct representations of individual Braille characters but cannot be ruled out; the RA response patterns are visually discriminable and therefore discriminable by some neural mechanism. The SAII and PC afferents are not viable candidates for form perception both because of their low innervation densities (Johansson & Vallbo 1979) and their inability to resolve spatial detail, which is evident in Figure 1.

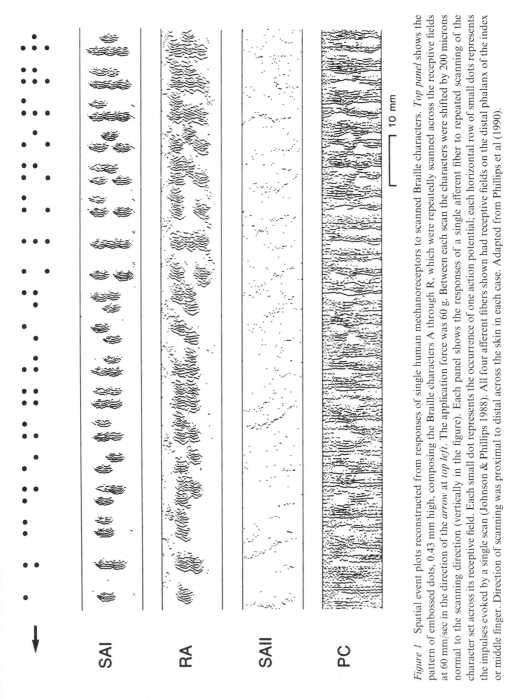

Figure 1   Spatial event plots reconstructed from responses of single human mechanoreceptors to scanned Braille characters. *Top panel* shows the pattern of embossed dots, 0.43 mm high, composing the Braille characters A through R, which were repeatedly scanned across the receptive fields at 60 mm/sec in the direction of the *arrow* at *top left*. The application force was 60 g. Between each scan the characters were shifted by 200 microns normal to the scanning direction (vertically in the figure). Each panel shows the responses of a single afferent fiber to repeated scanning of the character set across its receptive field. Each small dot represents the occurrence of one action potential; each horizontal row of small dots represents the impulses evoked by a single scan (Johnson & Phillips 1988). All four afferent fibers shown had receptive fields on the distal phalanx of the index or middle finger. Direction of scanning was proximal to distal across the skin in each case. Adapted from Phillips et al (1990).

The SAI and RA population responses to scanned Braille letters can be inferred from Figure 1. The small receptive fields and high innervation densities of SAI and RA afferents (Darian-Smith & Kenins 1980, Johansson & Vallbo 1979) dictate that the basic principle of spatial representation in those two populations is isomorphism, i.e. the neural representation of a stimulus is in roughly the same form as the stimulus. The exact form of representation depends on the response properties of individual afferents. SAI and RA afferents respond differently but the responses of individual afferent fibers are relatively homogeneous within classes; thus, neural images like the patterns illustrated in Figure 1 scroll across the SAI and RA populations as a finger scans across Braille dot patterns.

Two questions concerning tactile form processing are addressed below. First, which afferent population (SAI, RA, or both?) is used in form perception? The SAI afferent population provides a more acute representation of form, as can be seen in Figure 1, but that does not prove that the SAI system is the critical spatial system, i.e. the system actually used by the higher mechanisms for the finest spatial discriminations. Second, how is form recognition related to the primary neural representation of form?

Afferent Population Responsible for Form Perception

The available evidence provides a strong argument that the SAI system is the critical spatial system and that the RA system acuity may be as much as three times poorer than the SAI acuity. The evidence is derived from studies in which three modes of stimulation are used that engage the SAI and RA systems differently: static touch, scanned touch, and the Optacon, which is a 6 by 24 array of piezoelectric probes.

STATIC TOUCH   Psychophysical grating orientation studies show that subjects discriminate gratings with 0.5 mm gaps and bars at a level slightly greater than chance and that their performance rises to threshold, $d' = 1.35$, for gratings with 0.9 mm gaps and bars (Johnson & Phillips 1981). Studies with complex patterns such as Braille and embossed letters produce similar estimates of spatial acuity (Johnson & Phillips 1981, Loomis 1981, 1985, Phillips et al 1983, Vega-Bermudez et al 1991). When gratings are used as stimuli in neurophysiological experiments, only the SAI afferents resolve the spatial structure of all gratings that produce better than chance discrimination behavior. The RA afferents, in contrast, only begin to resolve the structure of gratings with 1.5 mm gaps and bars, and some fail even to resolve 3 mm gaps and bars. Thus, in static touch the SAI system accounts for the limits of acuity.

SCANNED TOUCH   When the finger pad is scanned across Braille and embossed letters, a 10 to 20% improvement in acuity relative to static touch is observed (Loomis 1985, Phillips et al 1983). The argument that the SAI system is the critical system in scanned as well as static touch seems strong for two reasons. First, the responses of SAI afferents to scanned stimuli with dimensions near the limits of acuity are sufficiently acute to account for recognition behavior (Johnson et al 1991, Johnson & Lamb 1981, Vega-Bermudez et al 1991). The modest increase in acuity with scanning is accounted for by a tenfold increase in SAI impulse rate with no loss of spatial resolution (Johnson & Lamb 1981). In contrast, the RA afferents provide a marginal representation even of patterns whose size is well above threshold, as in Figure 1, thus making it unlikely that they can account for the limits of acuity. Second, the responses of neurons in area 3b of the alert monkey show that neurons with slowly adapting properties preserve spatiotemporal information more effectively than do neurons with rapidly adapting properties (Phillips et al 1988).

OPTACON   The Optacon is a tactual stimulator comprising 144 cantilevered piezoelectric pins in a 6 × 24 array spanning an area of 11 × 28 mm, i.e. 2.2 × 1.2 mm spacings between pins. The pins move normal to the skin, can be activated individually or in combination, and deliver single or repetitive transient taps to the skin. The limit of spatial acuity in using the Optacon appears to be three times poorer than in static or scanned touch. In three extensive studies of letter recognition with the Optacon, the letter sizes required for 50% correct judgments ranged from 12 to 20 mm (Bliss 1969, Craig 1979, Loomis 1980). The comparable range in studies with embossed letters and scanned or static touch is 4 to 6 mm (Johnson & Phillips 1981, Loomis 1981, 1985, Phillips et al 1983, Vega-Bermudez et al 1991). A recent study by Gardner & Palmer (1990) provides some clues to the basis for this difference. Gardner & Palmer (1990) have shown that the Optacon activates only RA and PC afferents and have provided a picture of the RA population response to grating patterns, which is illustrated in Figure 2. The separation between bars in an Optacon pattern required to produce a 50% drop in mean rate between the bars is 3.6 mm. When the responses of SAI afferents to mechanical gratings with narrow bars are examined, the comparable figure is 1 to 2 mm (Phillips & Johnson 1981a), which explains the three-to-one difference in sizes required for comparable performance.

Linking SAI Responses to Pattern Recognition

The challenge in relation to form perception is to understand the transformations leading from the peripheral, isomorphic representation of form to the representations that flow into memory and perception. In this section

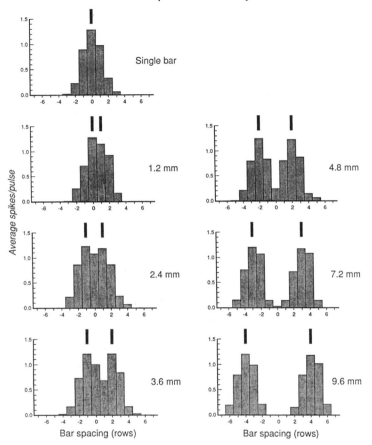

Figure 2   RA population response to bar patterns spaced 1.2 to 9.6 mm apart on the Optacon. The ordinate represents the mean number of impulses per indentation pulse. The abscissa represents the locations of receptive field centers relative to a point midway between the bar patterns. Modified from Gardner & Palmer (1990).

we discuss psychophysical studies of tactual pattern recognition and show that there appears to be a causal link between the spatial structure of peripheral neural responses to patterned stimuli and pattern recognition behavior.

The patterns used most frequently in studies of tactual pattern recognition are Roman letters, although other patterns, including Braille, have been used. The advantage of Roman letters is that they require no training and they yield quantitative data that are reliable and repeatable within and between laboratories (Loomis 1982, Vega-Bermudez et al 1991). Psychophysical studies of tactile letter recognition have shown that (*a*) there are large differences in performance (percentage correct) between untrained subjects (Johnson & Phillips 1981, Loomis 1981, 1982, Phillips et al 1983, Vega-Bermudez et al 1991), (*b*) subjects' performance improve rapidly even without training before testing or feedback during testing (Epstein et al 1989, Vega-Bermudez et al 1991), (*c*) there is no statistically detectable difference between the responses of subjects in active and passive tactual pattern recognition (Vega-Bermudez et al 1991), (*d*) variations in scanning velocity have a small effect between 20 and 80 mm/sec (Vega-Bermudez et al 1991), and (*e*) variations in contact force and relief height have little or no significant effect (Loomis 1985).

The more interesting data in relation to the underlying neural mechanisms derive from subjects' classification and misclassification behavior (Vega-Bermudez et al 1991). The pattern recognition task is a 26-alternative, forced-choice design in which letters of the alphabet are presented in random order and subjects are required to respond with one of the 26 possibilities. The result is a stimulus-response matrix P, called a confusion matrix, in which the element p_{ij} represents the fraction of presentations of S_i that yield the response R_j. The matrix illustrated in Figure 3 was derived from experiments in which the embossed forms and scanning methods were identical to those used in neurophysiological experiments. There are large variations in hit, false, and total response rates that require explanation. The hit rates vary by 6:1, from 15% for the letter N to 98% for the letter I. The false response rates vary by 25:1, from 4% for the letter I to 103% for the letter O, i.e. O was reported falsely slightly more often than it was presented. Fifty percent of all false responses are concentrated in 7% of all possible confusion pairs (i.e. 22 out of 325 confusion pairs), which are enclosed in boxes in Figure 3. In 17 of those 22 confusion pairs, the response asymmetry is significant at the 0.001 level, e.g. C is called G frequently but G is rarely called C.

The first question to be addressed is whether it is reasonable to attempt to account for the complex behavior exhibited in the confusion matrix displayed in Figure 3 in terms of primary sensory mechanisms. The patterns of confusion and correct responses might be the result of cognitive rather than sensory effects, i.e. the result of expectation, familiarity, preference, and so forth. Analysis of the confusion matrix illustrated in Figure 3 shows that true and false response rates for individual letters are not correlated with the frequency of letters in American English, and that the pattern of responses is not accounted for by any general bias that can be detected through an analysis of true and false response rates for individual letters (Loomis 1982, Vega-Bermudez et al 1991). Excepting the highly

Response

Stimulus	A	B	C	D	E	F	G	H	I	J	K	L	M	N	O	P	Q	R	S	T	U	V	W	X	Y	Z	n
A	71	1	.	.	.	1	.	.	.	1	3	1	2	8	.	.	.	5	.	.	.	3	1	2	.	1	324
B	2	18	2	27	1	.	5	2	.	.	1	.	2	6	6	4	3	6	3	.	6	.	1	.	.	.	324
C	.	1	46	3	.	.	15	.	.	.	.	.	.	.	15	.	15	2	1	.	2	.	.	.	.	.	324
D	1	2	2	49	1	.	2	1	.	.	.	.	1	.	22	5	4	1	1	.	7	.	1	.	.	.	324
E	2	6	1	1	60	2	1	2	.	.	8	.	1	2	.	.	1	8	1	.	1	1	1	1	.	1	324
F	1	1	.	.	7	57	1	1	1	.	4	.	.	2	.	11	.	3	3	2	.	2	.	1	2	.	324
G	1	11	2	3	1	.	41	.	.	.	.	.	4	1	6	1	18	6	5	.	.	.	.	.	.	.	324
H	1	1	.	1	1	1	1	49	.	.	4	.	9	15	.	1	1	5	.	.	6	1	4	1	.	.	324
I	.	.	.	.	.	.	.	.	98	1	.	.	.	.	.	.	.	.	.	.	.	.	.	.	.	.	324
J	.	1	.	1	1	1	1	.	1	84	1	.	.	.	.	.	.	.	3	.	1	1	.	1	1	2	324
K	1	1	2	.	10	1	.	.	.	.	57	.	3	1	.	1	.	9	.	.	1	.	2	7	.	4	324
L	1	.	1	.	.	.	.	.	.	1	1	93	.	.	.	.	.	1	.	.	.	.	.	.	.	1	324
M	1	4	.	.	2	1	2	2	.	.	8	.	34	4	.	.	1	5	1	.	6	1	30	.	.	.	324
N	.	5	.	1	1	1	.	12	.	.	7	.	20	15	.	1	.	9	2	.	4	.	19	2	.	.	324
O	.	1	2	6	.	.	4	.	.	.	.	.	.	.	83	1	3	.	.	.	.	.	.	.	.	.	324
P	.	2	1	1	1	23	2	1	.	.	3	.	.	2	.	44	1	2	4	5	.	3	1	.	4	1	324
Q	.	2	2	8	.	.	7	.	.	.	.	.	.	1	49	1	28	1	1	.	.	.	.	.	.	.	324
R	6	4	1	4	1	.	2	17	.	.	3	.	8	14	1	2	.	31	.	.	4	.	1	.	.	.	324
S	.	21	2	6	2	.	14	1	.	.	1	1	2	1	5	3	6	8	25	.	1	.	2	.	.	1	177
T	1	1	1	.	.	2	.	.	1	1	1	1	1	.	.	1	.	.	3	61	.	2	.	4	15	3	324
U	.	1	.	2	.	.	1	1	.	.	.	.	1	1	2	1	1	1	1	.	80	4	2	.	.	1	324
V	.	.	.	1	.	.	.	.	.	.	3	.	.	2	.	2	.	.	1	.	3	62	3	1	21	1	324
W	1	1	.	.	1	1	1	1	.	.	3	.	11	2	.	1	.	1	2	.	1	1	71	.	.	1	324
X	1	1	1	.	5	1	.	.	.	1	11	.	2	1	.	.	1	3	1	1	.	.	2	57	1	12	324
Y	.	.	.	.	.	2	1	.	.	2	5	.	1	2	.	1	.	1	1	6	.	15	.	3	58	3	324
Z	1	.	.	.	3	3	1	1	1	2	7	.	.	1	.	.	.	2	1	2	.	.	1	31	2	41	324
	92	86	66	114	98	97	102	91	102	93	131	96	102	81	189	81	83	110	60	77	123	96	142	111	104	73	8,278

Figure 3 Pooled confusion matrix for 64 subjects. Each entry, e.g. the entry in row I and column J, represents the percentage of trials on which the stimulus letter S_i evoked the response R_j. The stimuli were embossed sans-serif, Helvetica capital letters 6.0 mm in height. The Helvetica letters identifying the rows and columns are identical in form to the letters used in the psychophysical tasks. *Boxes* represent letter pairs whose mean confusion rates exceed 8%. For example, the mean confusion rate for B and G is 8% based on the fact that G is called B on 11% of trials in which G is presented and B is called G on 5% of trials in which B is presented. From Vega-Bermudez et al (1991).

distinctive letters I, J, and L, the frequencies of true and false response rates are uncorrelated, a result that suggests that the responses are entirely stimulus dependent.

Some aspects of the confusion matrix can be explained by the response properties of first-order afferent fibers. The response of a single, typical monkey SAI primary afferent fiber to scanned embossed letters is illustrated in Figure 4. A feature analysis of the stimulus letters explains some of the structure of the confusion matrix. For example, it shows, at least qualitatively, why N is confused with other letters more frequently than is J and why B and D are confused but not B and X. However, neither a feature analysis of the letters themselves nor decision theory explains the strong asymmetry in confusion between letters like B and D or C and G, etc. The basis for many of these asymmetries is explained by examination of the SAI response illustrated in Figure 4. For example, the internal horizontal bar and the trailing cusp that distinguish a B from a D are only weakly represented in the responses of primary afferent fibers; that is, the primary neural image of a B looks like a D but not vice versa. The expected result, that B is called D but not vice versa, is exactly the result displayed in the confusion matrix in Figure 3. In the pattern recognition studies, B was called D even more often than it was called B. In this case, the loss of internal and trailing features biases the psychophysical responses toward the simpler member of a pair. That is not the general rule, however, as C is called G and Q frequently, but G and Q are rarely called C. An explanation based on the neural responses illustrated in Figure 4 is that the protruding elements of G and Q evoke neural images that cannot be misclassified as a C. On the other hand, C is missing the distinguishing features of G and Q but that does not inhibit the subject from responding G or Q, perhaps because missing features are common in the neural representations. Similar neurophysiological explanations account for most of the asymmetries in the confusion matrix. The basis for learning without training or feedback, which was observed in the study by Vega-Bermudez et al (1991), may also be explained by such considerations: Once the subject learns that the tactual impression of a G has a distinct intensity in the center then it would seem to be ruled out as an acceptable false response to the stimulus letter C.

A comparison of this kind shows that there is a strong causal link between the first-order neural representations and subjects' behavior. It does not provide, however, the kind of quantitative result that is required for closer linkage of the psychophysical and neurophysiological data. What is needed is an adequate theoretical framework for analyzing the relationship between the structure of the stimuli, physical or neural, and the structure of the confusion matrix. A recent model of

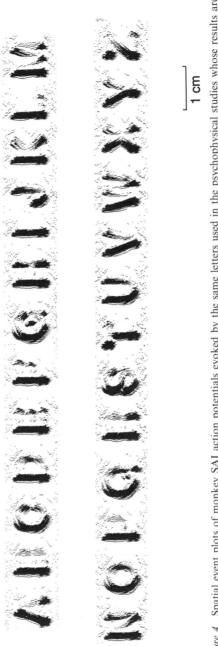

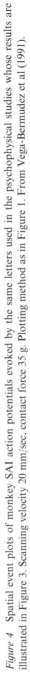

Figure 4 Spatial event plots of monkey SAI action potentials evoked by the same letters used in the psychophysical studies whose results are illustrated in Figure 3. Scanning velocity 20 mm/sec, contact force 35 g. Plotting method as in Figure 1. From Vega-Bermudez et al (1991).

character recognition and legibility by Loomis (1990) is one step in that direction.

TEXTURE PERCEPTION

Texture is a property of sensation and perception in audition (Guirao & Garavilla 1976) and vision (Gibson 1950) as well as touch. In each of the spatiotemporal senses, the perception of texture depends on repetitive spatial or temporal structure, which may be random, e.g. sandpaper or the sound of rain, or completely regular, as in a tightly woven fabric (Culbert & Stellwagen 1963). The dimensions of tactual texture are not well understood but the percept is at least two dimensional, including rough-smooth and hard-soft as independent unidimensional continua that allow rating judgments and greater-than, less-than judgments. The neural mechanisms of hard-soft judgments are beginning to be studied, but there were no published reports when this review was written. The neural mechanisms of roughness have, on the other hand, been studied in several laboratories.

The sense of roughness is a robust, intensive continuum that has been studied extensively (Connor et al 1990, Culbert & Stellwagen 1963, Cussler et al 1977, Ekman et al 1965, Green 1981, Inukai et al 1980, Katz 1925, Lederman 1982, Meenes & Zigler 1923, Sathian et al 1989, Stevens & Harris 1962, Stone 1967). Roughness allows ratio scaling (Stevens & Harris 1962), but unlike other intensive continua that have been studied within a neural coding framework, e.g. vibratory intensity (LaMotte 1987, Mountcastle 1975), its physical determinants are not understood. Many of the investigators cited above have studied the relationship between roughness magnitude and physical measures of structure, but none has been able to identify the physical determinant of roughness magnitude.

In this section we discuss the results from a range of psychophysical and neurophysiological studies aimed at roughness perception and its neural mechanisms. The hypothesis that emerges is that roughness perception is based on an SAI spatial mechanism when the fingers contact a surface directly.

Psychophysical Studies

Experimental studies of texture perception have occurred in two phases. The first consisted of the qualitative work of Meenes & Zigler (1923) and Katz (1925), which established many of the basic facts of texture perception. The current, quantitative phase, which yields psychophysical data that can be linked to neurophysiological observations, began with the application of magnitude estimation to roughness judgments (Stevens

& Harris 1962) and decision theory to the discrimination of textured surfaces (Lamb 1983a, Morley et al 1983).

Both qualitative and quantitative studies have shown that it makes no difference whether the finger is moved actively across the surface or the surface is moved across the passive finger (Katz 1925, Lamb 1983a, Lederman 1981, Meenes & Zigler 1923). Although movement is required, subjective roughness perception is affected little across a wide range of scanning velocities (10–100 mm/sec, Katz 1925; 10–250 mm/sec, Lederman 1974). Lubricants have no effect on perceived roughness (Taylor & Lederman 1975), a finding that rules out surface friction as a significant determinant of roughness (Ekman et al 1965). Contact force and skin temperature have small to moderate effects on roughness. Lederman (1974) has shown that the natural, unrestrained force used by a subject in a roughness judgment task is about 1 N, and that increasing contact force causes increased roughness judgments. Decreasing skin temperature causes a decline in perceived roughness (Green et al 1979). Lederman and her colleagues have studied roughness sensations by using metal gratings with gap widths, ridge widths, and spatial periods ranging from 0.4 to 1.6 mm. Roughness magnitude for these surfaces increases as a near linear function of gap width over the range from 0.125 to 1.0 mm and declines with increasing ridge width over the same range (Lederman 1974, 1981, Lederman & Taylor 1972, Taylor & Lederman 1975). Sathian et al (1989) confirmed these findings, using a wider range of gaps and bars, with periods ranging up to 3.1 mm. Connor et al (1990) obtained similar results with embossed dot patterns with center-to-center dot spacings ranging from 1.3 to 3.2 mm. They also described a decline in perceived roughness as the dot spacings increased from 3.2 to 6.3 mm, illustrated in the lower part of Figure 6. The near linear relationship between subjective roughness for spatial periods up to 3.0 mm is consistent with earlier studies with more complex surfaces such as sandpaper (Ekman et al 1965, Stevens & Harris 1962, Stone 1967).

The objective capacity to discriminate surfaces that differ in minute detail, which is evident in daily experience, was explored in detail by Meenes & Zigler (1923) and Katz (1925). The most fundamental fact of texture discrimination is its dependence on movement. Meenes & Zigler (1923) and Katz (1925) showed, using papers ranging from smooth to coarse, that it is difficult to discriminate any but the coarsest surfaces without lateral movement between the skin and the surface. With lateral movement, features raised 0.1 microns above a smooth background can be detected (LaMotte & Srinivasan 1991). Surfaces with feature dimensions well above the threshold for detection can be discriminated when the dimensions vary by as little as 1–2% (Lamb 1983a, Morley et al 1983).

Neural Representation of Textured Surfaces

Mechanical gratings and embossed dot patterns have been used widely in neurophysiological experiments (Connor et al 1990, Darian-Smith et al 1980, Goodwin & John 1991, Johnson 1983, Lamb 1983b, LaMotte 1977, Sathian et al 1989). Figure 5 illustrates the responses of SAI, RA, and PC afferents to surfaces with embossed dots. SAI afferents respond most acutely, and they display a form of surround suppression that is due to skin mechanics (Phillips & Johnson 1981b), i.e. the response to individual dots is diminished greatly at dot spacings of less than 3 mm. The RA afferents respond with lower acuity and display less surround suppression. The spatial structure of the finest surface studied by Connor et al (1990),

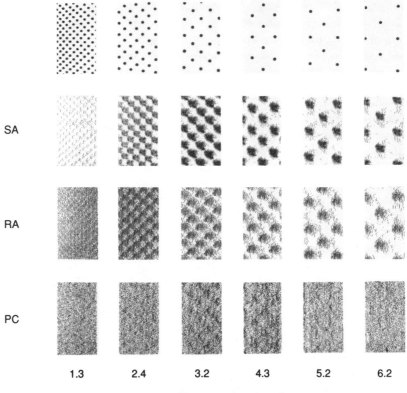

Figure 5 Spatial event plots of monkey SAI, RA, and PC action potentials evoked by the embossed dot patterns shown in the top row. Dot diameter 0.5 mm. Plotting method as in Figure 1. From Connor et al (1990).

1.3 mm dot spacing, is clearly represented in the responses of the SAI and RA afferents illustrated in Figure 5. The SAI and RA afferents form relatively homogeneous populations, and the responses illustrated in Figure 5 are typical. PC responses are less homogeneous, presumably because they are so sensitive and vary in mechanical location relative to the stimulus surface. In general, PC responses are only weakly sensitive to the surface structure; for example, mean impulse rate is almost independent of dot spacing. SAII afferents innervating the finger pad have not been reported in monkey studies and are not included in Figure 5 for that reason. Human SAII afferents respond weakly to raised dot patterns, as can be seen in Figure 1, and are not viable candidates for roughness coding.

Neural Codes for Roughness

It is evident that the neural activity displayed in Figure 5 provides a rich basis for perception. After running a finger over a surface of embossed dots or having the surface applied to the passive finger, one can describe and discriminate hardness, roughness, dot size, dot spacing, dot pattern, and scanning velocity. The problem is not to find some possible basis for each of these sensory capacities but rather to narrow the range of possibilities.

Neural codes of four types have been advanced as possible bases for roughness perception: intensive, modal, temporal, and spatial (Johnson & Lamb 1981). Intensive codes are measures of neural activity such as mean impulse rate that are independent of the spatial and temporal structure of the population response. Since roughness magnitude varies on an intensive continuum, intensive neural codes are obvious candidate codes and have been considered in every study. A modal code is based on the relative magnitudes of response intensity between neuronal populations with different transducer properties; the trichromatic theory of color vision is one example. Temporal codes depend on the temporal structure of firing in afferent fibers but not the spatial organization of the firing patterns. Spatial codes depend on the spatial structure of impulse rate across a population.

MEAN IMPULSE RATE   Mean rate codes have been examined by a number of investigators (Connor et al 1990, Goodwin & John 1991, LaMotte 1977, Sathian et al 1989), but no satisfactory general principles based on rate coding have emerged. LaMotte (1977) used four nylon fabrics in a pilot study, which showed that SAI afferents responded poorly and in no apparent relationship to the roughness of the surfaces. In contrast, RA afferents responded vigorously and the rank order of RA impulse rate and subjective roughness were similar, implying that RA rate might provide a suitable basis for roughness judgments. Goodwin and his colleagues (Goodwin & John 1991, Sathian et al 1989) used the same gratings that they had used previously in psychophysical experiments. The increase in psychophysical roughness magnitude with increasing gap width and the decline with increasing ridge width that they and Lederman & Taylor (1972) observed was matched only by the PC mean rate. Sathian et al (1989), however, found none of the correlations between impulse rate and subjective magnitude sufficiently convincing to attempt a quantitative comparison. Lamb (1983b) advocated mean rate codes to account for the human ability to discriminate textured surfaces but had to nominate different afferent populations for different surfaces. In fact, his data show that the spatial structure of the SAI population response provides a simple basis for the discrimination behavior that he observed.

Connor et al (1990) rejected all rate codes based on their inability to account for roughness magnitude across a wide range of surfaces. The SAI mean impulse rate and roughness magnitude evoked by each of the 18 surfaces used in their study is shown in the upper left corner of Figure 6. Surfaces that evoke similar SAI firing rates evoke very different reports of roughness, i.e. there is no consistent relationship between mean rate and roughness magnitude. RA and PC mean rate produced similar results.

MODAL CODES   Although modal codes have been suggested as a form of coding for texture perception (Richards 1979) and some evidence showing that the different afferent populations respond differentially to textured surfaces has been presented (Johnson 1983), there has thus far been no extensive study of modal coding. The only quantitative consideration of combinations of mean rates is by Connor et al (1990), who argued that no simple linear combination of rates across afferent classes could account for roughness magnitude because they all err in the same manner.

TEMPORAL AND SPATIAL CODES   A temporal or spatial code implies a mechanism that extracts some measure of temporal or spatial variation in the population response. Connor et al (1990) tested general measures of impulse rate variation such as the variance, standard deviation, and first absolute central moment of impulse rate variation in single fibers and obtained strong consistency between those measures and roughness judgments (see top middle plot in Figure 6). They argued, however, that statistical measures of impulse rate variation that have no spatial or temporal structure are not physiologically realistic and went on to test measures of variation based on specific spatial and temporal mechanisms. Analyses based on both spatial and temporal coding mechanisms produced results that were highly correlated with roughness magnitude. Only spatial variation, however, defined as the mean absolute difference in firing rate

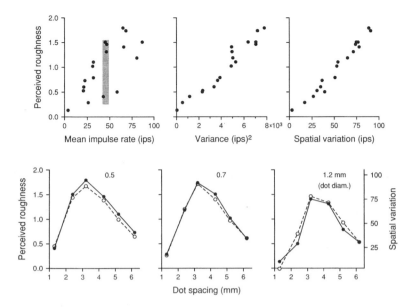

Figure 6 Subjective roughness magnitude compared with candidate neural codes for 18 surfaces composed of embossed dots with varying dot diameter and spacing. The left-hand ordinates in all six graphs represent mean roughness magnitude judgments for 21 subjects. Each abscissa in the *upper row* represents candidate code values derived from the SAI discharge evoked by each of the surfaces: *top left*, mean SAI impulse rate; *top middle*, SAI impulse rate variance; *top right*, spatial variation (see text). Each *filled circle* in the upper three graphs represents the mean roughness magnitude for one of the 18 surfaces and its associated neural code value. Graphs in the *bottom row* represent mean roughness judgments (*filled circles*) and SAI spatial variation values (*open circles*) versus center-to-center dot spacing. The left, center, and right graphs show results for surfaces with dot diameters of 0.5, 0.7, and 1.2 mm, respectively. The right-hand ordinate in the bottom row represents SAI spatial variation for each of the 18 surfaces. Modified from Connor et al (1990).

between afferent fibers separated by 2 mm, was independent of scanning velocity. The plot of roughness magnitude versus spatial variation illustrated in the top right-hand corner of Figure 6 suggests a linear relationship. The predicted psychophysical roughness judgments based on a linear causal relationship, i.e. a linear regression of roughness on spatial variation, is illustrated in the bottom row of Figure 6.

The spatial variation measure used by Connor et al (1990) has a simple physiological analogue. The mechanism requires higher-level neurons with receptive fields that contain excitatory and inhibitory subregions separated by about 2 mm, i.e. two afferent fiber spacings. If the effects from the subregions sum linearly, the synaptic effects are equal, and the impulse rate is proportional to the net excitation, then the neuron acts as a half-wave rectifier of the difference in afferent firing rates from the subregions; that is, the neuron fires at a rate proportional to the difference when the excitatory input exceeds the inhibitory input and zero otherwise. If a neighboring neuron is reciprocally connected to the same two subregions, then the summed output of the two neurons acts as a full-wave rectifier, i.e. the summed impulse rate is proportional to the absolute difference in firing rate between the two subregions. The mean impulse rate over a population of such neurons could serve as the neural basis for the sensation of roughness, as illustrated in Figure 6.

A mechanism like the one proposed by Connor et al (1990) accounts for roughness perception when the skin contacts a surface and the surface evokes significant SAI activity. There are stimuli, however, whose texture is perceived clearly that do not evoke a spatially structured SAI response. Under those circumstances, other mechanisms must account for texture perception.

Limitations of an SAI Spatial Coding Hypothesis

The hypothesis that subjective roughness depends on an SAI spatial mechanism is not universally applicable. There are two limitations, as expressed well by Katz (1925): "I do not attribute all roughness judgments to vibration sensations, but call upon the latter only for the levels of roughness that are no longer apparent to the spatial sense of the skin, as well as for cases where the presence of intermediaries precludes deformation of the skin."

The first is that surfaces with microscopic variations in surface level are not likely to be discriminated on the basis of SAI spatial mechanisms. Katz (1925) demonstrated that finely etched glass is easily distinguished from smooth glass and surmised that the irregularities were too fine to be detected by the "spatial sense." LaMotte and his colleagues have shown that finely machined asperities, 2 microns high, which are detected reliably in a psychophysical experiment, activate only RA afferent fibers (Johansson & LaMotte 1983, LaMotte & Whitehouse 1986). In other experiments, they showed that subjects could detect etched gratings and dot patterns raised 0.1 microns above the background. Under these conditions, only the PC afferents signaled the presence of microstructure (LaMotte & Srinivasan 1991). They have also shown that the detection of slip between skin and a surface requires surface irregularities that are at least large enough to cause activity in the RA or PC afferents (Srinivasan et al 1990). When a surface that activates only RA afferents is used, e.g. a surface with a small asperity, subjects report movement of a dot across the skin. When a surface that activates only PC afferents is used, e.g. a surface with a fine grating, subjects report vibration at the interface between the finger and the surface. LaMotte & Srinivasan (1991) surmised that the perception of slip was based on perceived movement and local vibration in those two cases. Johansson & Westling (1987) provide evidence that subjects detect the coefficient of friction of a surface without overt horizontal movement between the skin and the surface and that the RA afferents provide a signal inversely related to the coefficient of friction. The probable mechanism advanced by Johansson & Westling (1987) is that surfaces with less friction allow more microscopic movement between skin and surface as the convex skin surface conforms to the stimulus surface. The failure of the SAI system to respond in all of these cases is probably based on its lower sensitivity to vibratory stimuli. Thus, it appears that the finest levels of surface structure are transduced by the RA and PC afferents but not the SAI afferents.

The second limitation of a spatial coding hypothesis for roughness magnitude judgments is that textured surfaces can be differentiated effectively through a rigid probe by scanning the probe across the surfaces. The basis for this discrimination is vibration induced in the probe; Katz (1925) demonstrated that this ability is diminished by any mechanism that attenuates vibrations transmitted through the probe or from the probe to skin. The primary candidate for the transduction of this vibratory information is the PC system because of its sensitivity to transmitted vibration. A diagnostic procedure for the preliminary classification of a neuron as a PC afferent during neurophysiological experimentation is to tap the table holding the experimental animal. Whereas PC afferents often respond vigorously to such stimuli, SAI and RA afferents never do. When an object is lifted from a platform and returned, there are abrupt mechanical transients that activate PCs vigorously but affect SAIs and RAs minimally (Johansson & Westling 1991). The original, mistaken belief that Pacinian corpuscles are slowly adapting pressure receptors (Adrian & Umrath 1929) derives from their extreme sensitivity to transmitted environmental vibrations (Hunt 1961). Thus, it is likely that roughness perception through a probe is based on the temporal response of the PC system.

The perception of texture through a rigid medium is a special case of a major category of tactual perception whose neural mechanisms have not been studied; that is, the perception of external events via transmitted forces and vibrations. Examples include the perception of events at the working surfaces of a tool grasped in the hand and the perception of the dynamic state of a machine through palpation. The Tadoma method, used by deaf and blind individuals to perceive speech in real time, demonstrates the capacity of this form of perception (Reed et al 1989). The PC system is, as discussed above, well suited to the transduction of the minute, high frequency vibrations that occur when rigid surfaces interact. The PC system is poorly suited, however, to the transduction of the low-frequency forces and movements that are part of perception through intermediaries; e.g. the perception that the blade of a screwdriver is well seated because of its resistance to rotation. When the transmitted forces are static, the relevant information must be transduced by the slowly adapting afferents. When the forces are dynamic, the RA afferents might also be activated.

ROLES OF AFFERENT SYSTEMS: WORKING HYPOTHESES

The data reviewed here suggest complementary roles for the SAI, RA, and PC afferent systems, which we present as working hypotheses: The *SAI system* is the primary spatial system and is responsible for tactual form and roughness perception when the fingers contact a surface directly and for the perception of external events through the distribution of forces across the skin surface. The *PC system* is responsible for the perception of external events that are manifested through transmitted high-frequency vibrations. The *RA system* has a lower spatial acuity than the SAI system but a higher sensitivity to local vibration and is responsible for the detection and representation of localized movement between skin and a surface as well as for surface form and texture when surface variation is too small to engage the SAI system. These hypotheses are supported by the psychophysical and neurophysiological studies reviewed above. Another line of evidence supporting these hypotheses derives from the transduction and fundamental response properties of the primary afferents.

The SAI afferent fibers exhibit surround suppression (Phillips & Johnson 1981a), which serves to diminish their responses to spatially uniform stimuli and enhance their responses to isolated elements on a surface (Phillips & Johnson 1981a) and edges (Johansson et al 1982, Phillips & Johnson 1981a, Vierck 1979). Phillips & Johnson (1981b) have provided evidence that the SAI receptors are responsive to a particular component of the subcutaneous strain field, which provides this surround suppression effect. LaMotte & Srinivasan have advanced the compatible theory that SAI afferents respond to curvature at the skin surface (LaMotte & Srinivasan 1987, Srinivasan & LaMotte 1991).

The RA afferent fibers terminate in fluid-filled Meissner corpuscles (Andres & von During 1973), which isolate the sensitive endings from large, low-frequency deformations of the kind that occur when holding an object. Their sensitivity to micron-level vibrations at intermediate frequencies (Mountcastle et al 1972) makes them ideally suited to the representation of slip (Srinivasan et al 1990), the frictional characteristics of a surface (Johansson & Westling 1987), and fine spatial detail (LaMotte & Srinivasan 1991).

The PC afferents terminate in receptors that are highly sensitive to transmitted vibration (Hunt 1961, Talbot et al 1968) and they respond in a manner that preserves the temporal structure of the vibratory pattern (Talbot et al 1968). The extensive encapsulation of the Pacinian corpuscle acts as a high pass mechanical filter with a time constant of about 4 msec (Loewenstein & Skalak 1965), which isolates the sensitive receptor from large, low-frequency strain. Another consequence of the short time constant is that PC responses to vibratory stimuli are fully described as a rate-modulated Poisson point process for vibratory frequencies up to about 100 Hz (Freeman & Johnson 1982a,b). At frequencies above 100 Hz, PCs respond in a more regular manner, consistent with the 4 msec time constant. The importance of this observation is that a population of Poisson processes exhibits an overall time-varying mean rate whose waveform is an undistorted representation of the complex driving function, except the inability to represent negative rate (Brugge et al 1969). The consequent fidelity may account for our ability to discriminate pure sinusoidal vibrations (Mountcastle et al 1990) or complex surfaces with the aid of a probe.

A puzzle that has existed since it was first hypothesized that the SAI system might be the critical spatial system even during scanned touch (Johnson & Lamb 1981) is the lower innervation density of SAI afferents relative to RA afferents. The answer, as in vision, where the foveal cone spacing matches the optical limits of resolution (DeValois & DeValois 1988), might be that the SAI innervation density matches the resolution limits imposed by skin mechanics (Phillips & Johnson 1981b). The higher RA innervation may function to provide increased sensitivity to micro-events at the skin surface. Based on the sizes of their receptive fields, 10–20 RA afferents innervate a single point on the skin (Johansson & Vallbo 1980, Talbot et al 1968). This redundancy may amplify the afferent signal associated with single microevents or provide a graded signal of the intensity of local events based on differences in RA thresholds (Johnson 1974).

The synergy provided by these hypothesized functions can be seen by considering the act of grasping and using an instrument of any kind. The SAI system provides an acute, continuing representation of the instrument itself and also the low-frequency forces transmitted from the working surfaces of the instrument. The PC receptors are well protected from low-frequency subcutaneous forces by their encapsulation and therefore remain sensitive to vibrations resulting from transient mechanical events that occur at the working surfaces of the instrument. Because these vibrations are transmitted rapidly to many PC afferents, the PC population response provides a high-fidelity temporal representation of those events. The on-going discharge of the SAI and PC afferents to low- and high-frequency events renders them insensitive to small dynamical events at the skin surface such as impending slip or overt movement between object and skin. The encapsulation of the RA receptors protects them from low-frequency forces and their relative insensitivity to high frequencies protects them from transmitted vibrations. As a result, they are ideally suited to detect and localize movement on the skin, even during periods of high force and high levels of transmitted vibration.

SUMMARY

In the last decade or so, there has been rapid movement toward the use of more complex stimuli in the study of perceptual function related to the hand. This review has focused on the neural mechanisms of form and texture perception. Evidence from neurophysiological and psychophysical studies in which static touch, scanning touch, and the Optacon were used indicate that the spatial acuity of the RA system may be as much as three times poorer than the SAI system, evidence that suggests that form perception is dominated by the SAI system. Pattern recognition behavior in a tactual letter recognition task appears to be directly related to the response properties of SAI afferent fibers. Psychophysical studies of rough-ness perception show that roughness magnitude is related to surface struc-ture in an orderly manner. Because roughness varies along an intensive continuum, mean impulse rate in one or more of the afferent systems is the most obvious coding possibility. No satisfactory relationship between mean impulse rate and roughness has been observed, however. The strong-est hypothesis is that tactual roughness perception is based on spatial variation in the SAI population response. The combined evidence from studies reviewed here suggests complementary roles for each of the afferent systems, which are presented as working hypotheses: The SAI system is the primary spatial system and is responsible for tactual form and roughness perception when the fingers contact a surface directly and for the per-ception of external events through the distribution of forces across the skin surface. The PC system is responsible for the perception of external events that are manifested through transmitted high-frequency vibrations of the kind that are critical in the use of objects as tools. The RA system is responsible for the detection and representation of localized movement between skin and a surface as well as for surface form and texture when surface variation is too small to activate the SAI afferents effectively.

Literature Cited

Adrian, E. D., Umrath, K. 1929. The impulse discharge from the Pacinian corpuscle. *J. Physiol.* 68: 139–54

Andres, K. H., von During, M. 1973. Morphology of cutaneous receptors. In *The Somatosensory System, Handbook of Sensory Physiology*, ed. A. Iggo, 2: 3–28. New York: Springer-Verlag

Bliss, J. C. 1969. A relatively high resolution reading aid for the blind. *IEEE Trans. Man-Machine Sys.* 10: 1–9

Brugge, J. F., Anderson, D. J., Hind, J. E., Rose, J. E. 1969. Time structure of discharge in single auditory nerve fibers of the squirrel monkey in response to complex periodic sounds. *J. Neurophysiol.* 32: 386–401

Connor, C. E., Hsiao, S. S., Phillips, J. R., Johnson, K. O. 1990. Tactile roughness: Neural codes that account for psychophysical magnitude estimates. *J. Neurosci.* 10: 3823–36

Craig, J. C. 1979. A confusion matrix for tactually presented letters. *Percept. Psychophys.* 26: 409–11

Culbert, S. S., Stellwagen, W. T. 1963. Tactual discrimination of textures. *Percept. Motor Skills* 16: 545–52

Cussler, E. L., Zlotnick, S. J., Shaw, M. C. 1977. Texture perceived with the fingers. *Percept. Psychophys.* 21: 504–12

Darian-Smith, I. 1984. The sense of touch: Performance and peripheral neural processes. In *Handbook of Physiology—The Nervous System III*, pp. 739–88. Bethesda, MD: Am. Physiol. Soc.

Darian-Smith, I., Davidson, I., Johnson, K. O. 1980. Peripheral neural representation of spatial dimensions of a textured surface moving across the monkey's finger pad. *J. Physiol.* 309: 135–46

Darian-Smith, I., Kenins, P. 1980. Innervation density of mechanoreceptive fibers supplying glabrous skin of the monkey's index finger. *J. Physiol.* 309: 147–55

DeValois, R. L., DeValois, K. K. 1988. *Spatial Vision*. Oxford: Oxford Univ. Press

Ekman, G., Hosman, J., Lindstrom, B. 1965. Roughness, smoothness and preference: A study of quantitative relation in individual subjects. *J. Exp. Psychol.* 70: 18–26

Epstein, W., Hughes, B., Schneider, S. L., Bach-y-Rita, P. 1989. Perceptual learning of spatiotemporal events: Evidence from an unfamiliar modality. *J. Exp. Psychol. Hum. Perc. Perf.* 15: 28–44

Franzen, O., Westman, J. 1991. *Information Processing in the Somatosensory System*. London: Macmillan. 467 pp.

Freeman, A. W., Johnson, K. O. 1982a. Cutaneous mechanoreceptors in macaque monkey: Temporal discharge patterns evoked by vibration, and a receptor model. *J. Physiol.* 323: 21–41

Freeman, A. W., Johnson, K. O. 1982b. A model accounting for effects of vibratory amplitude on responses of cutaneous mechanoreceptors in macaque monkey. *J. Physiol.* 323: 43–64

Gardner, E. P., Palmer, C. I. 1990. Simulation of motion on the skin. III. Mechanisms used by rapidly adapting cutaneous mechanoreceptors in the primate hand for spatiotemporal resolution and two-point discrimination. *J. Neurophysiol.* 63: 841–59

Gibson, J. J. 1950. *The Perception of the Visual World*. Boston: Houghton-Mifflin

Goodwin, A. W., John, K. T. 1991. Peripheral neural basis for the tactile perception of texture. See Franzen & Westman 1991, pp. 81–91

Green, B. G. 1981. Tactile roughness and the "paper effect." *Bull. Psychon. Soc.* 18: 155–58

Green, B. G., Lederman, S. J., Stevens, J. C. 1979. The effect of skin temperature on the perception of roughness. *Sensory Proc.* 3: 327–33

Guirao, M., Garavilla, J. M. 1976. Perceived roughness of amplitude-modulated tones and noise. *J. Acoust. Soc. Am.* 60: 1335–38

Hunt, C. C. 1961. On the nature of vibration receptors in the hind limbs of cats. *J. Physiol.* 155: 175–86

Inukai, Y., Saito, S., Mishima, I. 1980. A vector model analysis of individual differences in sensory measurement of surface roughness. *Human Factors* 22: 25–36

Johansson, R. S., LaMotte, R. H. 1983. Tactile detection thresholds for a single asperity on an otherwise smooth surface. *Somat. Res.* 1: 21–31

Johansson, R. S., Landstrom, U., Lundstrom, R. 1982. Sensitivity of edges of mechanoreceptive afferents units innervating the glabrous skin of the human hand. *Brain Res.* 244: 27–32

Johansson, R. S., Vallbo, A. 1979. Tactile sensitivity in the human hand: relative and absolute densities of four types of mechanoreceptive units in glabrous skin. *J. Physiol.* 286: 283–300

Johansson, R. S., Vallbo, A. B. 1980. Spatial properties of the population of mechanoreceptive units in the glabrous skin of the human hand. *Brain Res.* 184: 353–66

Johansson, R. S., Westling, G. 1987. Signals in tactile afferents from the fingers eliciting adaptive motor responses during precision grip. *Exp. Brain Res.* 66: 141–54

Johansson, R. S., Westling, G. 1991. Afferent signals during manipulative tasks in humans. See Franzen & Westman 1991, pp. 25–48

Johnson, K. O. 1974. Reconstruction of population response to a vibratory stimulus in quickly adapting mechanoreceptive afferent fiber populations innervating glabrous skin of the monkey. *J. Neurophysiol.* 37: 48–72

Johnson, K. O. 1983. Neural mechanisms of tactual form and texture discrimination. *Fed. Proc.* 42: 2542–47

Johnson, K. O., Lamb, G. D. 1981. Neural mechanisms of spatial tactile discrimination: Neural patterns evoked by Braille-like dot patterns in the monkey. *J. Physiol.* 310: 117–44

Johnson, K. O., Phillips, J. R. 1981. Tactile spatial resolution: I. Two-point discrimination, gap detection, grating resolution, and letter recognition. *J. Neurophysiol.* 46: 1177–91

Johnson, K. O., Phillips, J. R. 1988. A rotating drum stimulator for scanning embossed patterns and textures across the skin. *J. Neurosci. Methods* 22: 221–31

Johnson, K. O., Phillips, J. R., Hsiao, S. S., Bankman, I. N. 1991. Tactile pattern recognition. See Franzen & Westman 1991, pp. 305–18

Katz, D. 1925. *Der Aufbau der Tastwelt. Z. Psychol.* Erganzungsband 11. Transl. L. E. Krueger, 1989 as *The World of Touch.* Hillsdale, NJ: Erlbaum. 260 pp.

Lamb, G. 1983a. Tactile discrimination of textured surfaces: Psychophysical performance measurements in humans. *J. Physiol.* 338: 551–65

Lamb, G. 1983b. Tactile discrimination of textured surfaces: Peripheral neural coding in the monkey. *J. Physiol.* 338: 567–87

LaMotte, R. H. 1977. Psychophysical and neurophysiological studies of tactile sensibility. In *Clothing Comfort: Interaction of Thermal, Ventilation Construction and Assessment Factors,* ed. N. Hollies, R. Goldman, pp. 83–105. Ann Arbor: Ann Arbor Science

LaMotte, R. H. 1987. The sense of flutter-vibration in monkeys and man. In *Comparative Perception,* ed. M. Berkely, W. Stebbins, pp. 215–44. New York: Wiley

LaMotte, R. H., Srinivasan, M. A. 1987. Tactile discrimination of shape: Responses of slowly adapting mechanoreceptive afferents to a step stroked across the monkey fingerpad. *J. Neurosci.* 7: 1655–71

LaMotte, R. H., Srinivasan, M. A. 1991. Surface microgeometry: Neural encoding and perception. See Franzen & Westman 1991, pp. 49–58

LaMotte, R. H., Whitehouse, J. 1986. Tactile detection of a dot on a smooth surface: Peripheral neural events. *J. Neurosci.* 56: 1109–28

Lederman, S. J. 1974. Tactile roughness of grooved surfaces: The touching process and effects of macro- and microsurface structure. *Percept. Psychophys.* 16: 385–95

Lederman, S. J. 1981. The perception of surface roughness by active and passive touch. *Bull. Psychon. Soc.* 18: 253–55

Lederman, S. J. 1982. The perception of texture by touch. In *Tactual Perception: A Sourcebook,* ed. W. Schiff, E. Foulke, pp. 130–67. Cambridge: Cambridge Univ. Press

Lederman, S. J., Taylor, M. M. 1972. Fingertip force, surface geometry, and the perception of roughness by active force. *Percept. Psychophys.* 12: 401–8

Loewenstein, W. R., Skalak, R. 1965. Mechanical transmission in a Pacinian corpuscle. An analysis and a theory. *J. Physiol.* 182: 346–78

Loomis, J. M. 1980. Interaction of display mode and character size in vibrotactile letter recognition. *Bull. Psychon. Soc.* 16: 385–87

Loomis, J. M. 1981. On the tangibility of letters and Braille. *Percept. Psychophys.* 29: 37–46

Loomis, J. M. 1982. Analysis of tactile and visual confusion matrices. *Percept. Psychophys.* 31: 41–52

Loomis, J. M. 1985. Tactile recognition of raised characters: A parametric study. *Bull. Psychon. Soc.* 23: 18–20

Loomis, J. M. 1990. A model of character recognition and legibility. *J. Exp. Psychol. Hum. Percept.* 16: 106–20

Meenes, M., Zigler, M. J. 1923. An experimental study of the perceptions of roughness and smoothness. *Am. J. Psychol.* 34: 542–49

Morley, J. W., Goodwin, A. W., Darian-Smith, I. 1983. Tactile discrimination of gratings. *Exp. Brain Res.* 49: 291–99

Mountcastle, V. B. 1975. The view from within: Pathways to the study of perception. *Johns Hopkins Med. J.* 136: 109–31

Mountcastle, V. B., LaMotte, R. H., Carli, G. 1972. Detection thresholds for stimuli in humans and monkeys: Comparison with threshold events in mechanoreceptive afferent nerve fibers innervating the monkey hand. *J. Neurophysiol.* 35: 122–36

Mountcastle, V. B., Poggio, G., Werner, G. 1963. The relation of thalamic cell responses to peripheral stimuli varied over an intensive continuum. *J. Neurophysiol.* 26: 807–34

Mountcastle, V. B., Steinmetz, M. A.,

Romo, R. 1990. Frequency discrimination in the sense of flutter: Psychophysical measurements correlated with postcentral events in behaving monkeys. *J. Neurosci.* 10: 3032–44

Nolan, C. Y., Kederis, C. J. 1969. Perceptual factors in Braille word recognition. *Am. Found. Blind, Res. Ser.* no. 20

Phillips, J. R., Johansson, R. S., Johnson, K. O. 1990. Representation of Braille characters in human nerve fibers. *Exp. Brain Res.* 81: 589–92

Phillips, J. R., Johnson, K. O. 1981a. Tactile spatial resolution: II. Neural representation of bars, edges, and gratings in monkey afferents. *J. Neurophysiol.* 46: 1192–1203

Phillips, J. R., Johnson, K. O. 1981b. Tactile spatial resolution: III. A continuum mechanics model of skin predicting mechanoreceptor responses to bars, edges, and gratings. *J. Neurophysiol.* 46: 1204–25

Phillips, J. R., Johnson, K. O., Browne, H. 1983. A comparison of visual and two modes of tactual letter recognition. *Percept. Psychophys.* 34: 243–49

Phillips, J. R., Johnson, K. O., Hsiao, S. S. 1988. Spatial pattern representation and transformation in monkey somatosensory cortex. *Proc. Natl. Acad. Sci. USA* 85: 1317–21

Reed, C. M., Durlach, N. I., Braida, L. D., Schultz, M. C. 1989. Analytic study of the Tadoma method: Effects of hand position on segmental speech perception. *J. Speech Hearing Res.* 32: 921–29

Richards, W. 1979. Quantifying sensory channels: Generalizing colorimetry to orientation and texture, touch, and tones. *Sensory Proc.* 3: 207–29

Sathian, K., Goodwin, A. W., John, K. T.,

Darian-Smith, I. 1989. Perceived roughness of a grating: Correlation with responses of mechanoreceptive afferents innervating the monkey's fingerpad. *J. Neurosci.* 9: 1273–79

Srinivasan, M. A., LaMotte, R. H. 1991. Encoding of shape in the responses of cutaneous mechanoreceptors. See Franzen & Westman 1991, pp. 59–70

Srinivasan, M. A., Whitehouse, J. M., LaMotte, R. H. 1990. Tactile detection of slip: Surface microgeometry and peripheral neural codes. *J. Neurophysiol.* 63: 1312–32

Stevens, S. S., Harris, J. 1962. Scaling of roughness and smoothness. *J. Exp. Psychol.* 64: 489–94

Stone, L. A. 1967. Subjective roughness and smoothness for individual judges. *Psychon. Sci.* 9: 347–48

Talbot, W. H., Darian-Smith, I., Kornhuber, H. H., Mountcastle, V. B. 1968. The sense of flutter-vibration: Comparison of the human capacity with response patterns of mechanoreceptive afferents from the monkey hand. *J. Neurophysiol.* 31: 301–34

Taylor, M. M., Lederman, S. J. 1975. Tactual perception of grooved surfaces. A model and the effect of friction. *Percept. Psychophys.* 17: 23–36

Vega-Bermudez, F., Johnson, K. O., Hsiao, S. S. 1991. Human tactile pattern recognition: Active versus passive touch, velocity effects and patterns of confusion. *J. Neurophysiol.* 65: 531–46

Vierck, C. J. 1979. Comparisons of punctate, edge and surface stimulation of peripheral slowly adapting, cutaneous, afferent units of cats. *Brain Res.* 175: 155–59

Annu. Rev. Neurosci. 1992. 15:251–84

MOLECULAR BASES OF EARLY NEURAL DEVELOPMENT IN *XENOPUS* EMBRYOS

Chris Kintner

Salk Institute of Biological Studies, San Diego, California 92138

INTRODUCTION

The development of the vertebrate nervous system begins when a region called the neural plate forms on the dorsal surface of the embryo following the completion of gastrulation (see Figure 1). The embryonic tissue that forms the neural plate can be traced to a region of the blastula called the ectoderm. The ectoderm before gastrulation is a homogeneous epithelium, but by the end of gastrulation it assumes two developmental fates. The ectoderm that comes to lie on the dorsal surface of the embryo forms the neural plate and subsequently gives rise to the nervous system by folding into a neural tube and then generating neural cell types. In contrast, the ectoderm on the lateral and ventral sides of the embryo differentiates into nonneural tissue, primarily skin. Thus, the division of ectoderm into neural and nonneural fates is first marked by the formation of the neural plate. Moreover, the neural plate also marks a point in development at which many regional features of the vertebrate body axis are likely to be first established. The different regions of the vertebrate body plan can be traced to different points on the blastula fate map, but many features of both the dorsal/ventral (D/V) and anterior/posterior (A/P) axes are not apparent until the completion of gastrulation. These features are also evident within the neural plate, suggesting that the formation of the plate signifies the first step in neural regionalization.

The formation of the neural plate depends on a specific tissue interaction called *neural induction*. In amphibians, where this interaction has been studied extensively, the ectoderm is induced to form the neural plate by a region in the dorsal marginal zone of the blastula called *Spemann's orga-*

251

0147–006X/92/0301–0251$02.00

A   B

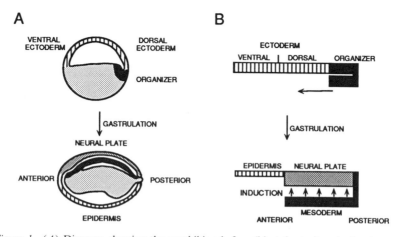

Figure 1 (*A*) Diagram showing the amphibian before (blastula, *top*) and after (neurula, *bottom*) gastrulation. The tissues involved in neural induction are highlighted. (*B*) The embryo diagram shown in (*A*) is highly schematized to show more clearly the tissue involved in neural induction.

nizer. The experimental evidence for this interaction is the famous organizer transplantation experiment of Mangold & Spemann in which the dorsal marginal zone is grafted from one blastula to the ventral side of another (Spemann 1938). The transplanted organizer undergoes gastrulation and contributes to a second dorsal axis that now forms on the ventral side of the host embryo (Figure 1). The striking outcome of this manipulation is that only part of the second axis, primarily the notochord, is derived from the donor tissue, whereas the neural tissue in the dorsal axis is derived from host ectoderm. The host ectoderm that forms neural tissue in response to the transplanted organizer would have formed ventral epidermis in the absence of a graft. This experiment and other analyses of this type have led to the idea that the ectoderm is destined to form epidermis in the blastula, but it can be directed to undergo neural development upon the appropriate inductive interaction with the organizer.

The analysis of embryonic induction initiated by Spemann and continued by his students in the early part of this century left an important legacy in the field of developmental neurobiology. In a recent book, Victor Hamburger recounts the experimental observations made during this era that established neural induction as the basis of early neural development in vertebrate embryos (Hamburger 1988). The importance of induction in establishing a neural anlage suggests that the molecular analysis of early neural development should focus on two questions. What are the molecules produced by the organizer that induce the neural plate and then regionalize it? What is the response of the ectoderm to these inducers, and how does this response lead to the formation of the neural tissue, or different regions of neural tissue, rather than epidermal differentiation?

Progress in several areas has recently revitalized the molecular analysis of early neural development in amphibian embryos. In this respect, one of the more influential areas of research is the analysis of the developmental processes that underlie pattern formation and cell determination in *Drosophila* embryos (Akam 1987). One insight to emerge from these studies is that the segmented insect body plan arises by an increasingly fine subdivision of embryonic tissue into regions with different development fates. These different fates are generated by the differential expression of one or several new gene products, usually transcription factors involved in gene regulation. By analogy, the initial separation of the ectoderm into epidermal and neural fates in vertebrate embryos may also depend on molecules whose expression distinguishes neural from nonneural ectoderm. Similarly, the regionalization of the neural system may depend on molecules whose expression distinguishes different parts of the neural plate from each other. Identifying molecules with both types of expression patterns has been the basis for the recent progress in analyzing early neural development.

Recent studies on the molecular basis of early neural development in *Xenopus* embryos is reviewed here in three parts. The first part describes the isolation of cDNA and antibody probes for molecules that mark the formation of the neural plate in *Xenopus* embryos. In the second section, I discuss experiments in which these probes have been used to analyze the interaction between the ectoderm and the organizer that underlies neural induction. The goal in these experiments is to determine how the inducing signals from the organizer affect the expression of these molecules in the ectoderm. Finally, the third section describes experiments in which the expression of neural plate molecules has been altered in embryos during early neural development. The goal in these experiments is to determine the function of these molecules when they are expressed in the ectoderm in response to induction.

MOLECULES EXPRESSED DURING NEURAL INDUCTION IN *XENOPUS* EMBRYOS

Table 1 contains a current listing of cDNA and antibody probes that have been isolated for molecules expressed in the ectoderm during neural induction in *Xenopus* embryos. For simplicity, these probes have been divided into two categories depending on whether they (*a*) distinguish

Table 1 Probes for molecules expressed in ectoderm of *Xenopus* embryos during neural induction

Markers	Reagent	Reference
Epidermal markers		
Epi-1	Mouse monoclonal antibody	Akers et al (1986)
2C7C	Mouse monoclonal antibody	Jones & Woodland (1986)
XEPI1		Itoh et al (1988)
XEPI2	Mouse monoclonal antibody	Itoh et al (1988)
XEPI3	Mouse monoclonal antibody	Itoh et al (1988)
DG 81	cDNA clone	Jonas et al (1985)
E-cadherin	Rabbit antibodies	Choi & Gumbiner (1989)
	Mouse monoclonal antibody	Angres et al (1991)
	cDNA clone	C. R. Kintner, unpublished
Neural markers		
2G9	Mouse monoclonal antibody	Jones & Woodland (1989)
24-10 (β-tubulin)	cDNA clones	Dworkin-Rastl et al (1986),
17-5		Richter et al (1988)
13-6		
NCAM	Antibodies	Jacobson & Rutishauser (1986)
	cDNA	Kintner & Melton (1987)
XIF-6	cDNA clones	Sharpe (1988)
XIF-3		
NF-3	cDNA clone	Charnas et al (1987)
Regional markers		
EN-2	Monoclonal antibody	Patel et al (1989)
	cDNA	Hemmati Brivanlou et al (1991)
Xhox-3	cDNA	Ruiz i Altaba & Melton (1989)
XlHBox6 (Hox 2.5)	cDNA	Sharpe et al (1987),
	Rabbit antibodies	Wright et al (1990)
XlHbox1 (Hox 3.3)	cDNA	Cho et al (1989), Oliver et al (1988)

neural tissue from other tissue in the embryo and bear on the question of neural specification, or whether they (*b*) distinguish neural tissue at different points along the body plan and bear on the question of neural regionalization. In addition, these probes can be divided into other categories based on the approach used in their isolation. For example, some of the probes in Table 1 were isolated by blindly screening monoclonal antibodies raised against embryonic tissues or cDNA libraries with differential or subtracted hybridization procedures. Many of the probes identified by this approach are directed against molecules expressed at abundant levels because of limitations inherent to these screening procedures. Other probes listed in Table 1 were isolated by directly screening for the members of certain gene families such as those encoding intermediate filaments proteins

(Steinert & Roop 1988) or cell adhesion molecules (Edelman 1988). Members of these gene families tend to be expressed during the formation of new tissue anlage. Finally, some probes listed in Table 1 were isolated through sequence similarity with genes that play a role in the development of the *Drosophila* embryo. Since many of these genes encode putative transcription factors, they contain a conserved sequence encoding an amino acid motif involved in protein-DNA interaction (McGinnis et al 1984b). By exploiting the ability of these sequences to cross hybridize, it has been possible to isolate vertebrate genes that encode molecules with similar structure and presumably function (Gehring 1987).

Markers that Recognize Epidermal Tissue

The ectoderm is either induced to form neural tissue or else it differentiates into epidermis as a default pathway. The first reagents that were isolated as markers of the induction process were ones that mark the default pathway of epidermal differentiation. Epidermal markers that have been isolated include two monoclonal antibodies, called Epi-1 and 2C7C (Akers et al 1986, Jones & Woodland 1986). These antibodies stain the epidermal cell layer but not the neural plate or any of the mesodermal or endodermal derivatives present in early *Xenopus* embryos. The 2C7C and Epi-1 antigen are first expressed in embryos soon after the formation of the neural plate. The antigens recognized by the two antibodies appear to be distinct because 2C7C, but not Epi-1, stains the cement gland. In addition, the 2C7C and Epi-1 antigens have distinct molecular weights of 250,000 and 300,000, respectively, upon electrophoresis in polyacrylamide gels and Western analysis (Itoh et al 1988). Since the molecules that react with these antibodies have yet to be characterized in any detail, the 2C7C and Epi-1 antibodies have been used primarily as markers of epidermal differentiation rather than as tools for functional studies. In this regard, these antibodies have been extensively used to monitor epidermal differentiation when the ectoderm divides and takes on a neural or epidermal fate (see below). Other antibody markers of epidermal differentiation were isolated as described by Itoh et al (1988) in an extensive screen for epidermal-specific antigens. This monoclonal antibody screen yielded five distinct antibodies that specifically recognize epidermal tissue. One of these antibodies, XEPI-1, appears to be directed against the same antigen recognized by 2C7C. Of the remaining antibodies, XEPI-2 and 3 are likely to be the most useful as epidermal markers, because they first stain embryos around stage 13 (neural plate formation), when staining occurs on the surrounding epidermis but not on the neural plate. On Western analysis, the XEPI-2 antibody recognizes a collection of proteins perhaps related to intermediate filaments (see below) whereas XEPI-3 fails to react with any protein.

The other class of epidermal markers that has been isolated are cDNA clones encoding epidermal-specific transcripts. The best characterized of the epidermal cDNAs encode a family of *Xenopus* epidermal keratins referred to as XK-81 (Miyatani et al 1986). The first member of this family, a cDNA clone called DG81, was isolated in a screening strategy designed to identify transcripts expressed in the embryo when gastrulation begins (Jonas et al 1985). Surprisingly, DG81 RNA is one of the most abundant RNAs expressed in embryos at gastrulation, although it is totally absent from maternal RNA pools. Despite its abundance, DG81 RNA is restricted to the epidermis, as described further below. Sequence analysis of the DG81 cDNA revealed that it encodes a type 1 keratin. It was subsequently used as a probe to isolate additional cDNAs that encode other keratins expressed in the epidermis of early embryos.

Markers that Recognize Neural Tissue

Molecules with the reciprocal expression in neural but not epidermal tissue have also been identified in general screens carried out with monoclonal antibodies or cDNA libraries. Jones & Woodland (1989) isolated and characterized a monoclonal antibody, 2G9, that reacts with an antigen found only in neural tissue and is first expressed at early neural tube stages. Although the 2G9 antibody is a useful marker of neural tissue, the molecular identity of the 2G9 epitope has not been reported. cDNA clones that encode transcripts found only in neural tissue have been isolated by Richter et al (1988), who employed a cDNA probe enriched for brain transcripts by extensive subtractive hybridization to RNA from nonneural tissues. cDNA clones isolated from early neurulae libraries with this probe were then selected for ones encoding RNA transcripts expressed just in brain tissue at late tadpole stages. Of the six distinct cDNA clones isolated in this approach, expression of the clone 24-10 is the best elucidated. 24-10 encodes a neural-specific form of β-tubulin that had been isolated previously by the Dorwkins, who studied its expression in late tadpole embryos by in situ hybridization (Dworkin-Rastl et al 1986). This analysis, along with the studies of Richter et al (1988), shows that 24–10 transcripts are expressed in the pattern expected for a marker of neural tissue. 24-10 RNA is not present at significant levels in maternal RNA but begins to appear in embryos at stage 11 just as the neural plate begins to form (early gastrulae). In addition, 24-10 RNA appears to be localized in stage 17 (late neural plate) embryos to neural plate tissue and at later stages to the neural tube. Three of the other cDNA clones identified by Richter et al (1988) as neural-specific at late tadpole stages were found to be expressed in the egg or in nonneural tissues in early embryos. In contrast to 24–10, these clones cannot be used as neural markers in early embryos. Finally, the remaining two cDNAs described by Richter et al, 17-5 and 13-6, appear to encode transcripts that are first expressed in embryos during gastrulation, and this expression appears to be restricted to the neural plate. These cDNAs are likely to be useful markers of neural induction and are presumably under further investigation.

Genes encoding intermediate filament proteins (IF) comprise a very large gene family, and different members of this family are known to exhibit tissue-specific patterns of expression during embryogenesis (Osborn & Weber 1982, Steinert & Roop 1988). For these reasons, IF proteins were considered to be good candidates for molecules that mark ectoderm as it forms new tissue anlagen. The first attempt to characterize IF protein expression in *Xenopus* embryos was undertaken by using a panel of antibodies raised against IFs that had been biochemically isolated from different tissues (Godsave et al 1986). Each antibody in this panel produced a different pattern of staining of embryonic tissues, but these particular antibodies were not useful in marking either the early formation of epidermal or neural tissue during neural induction. Using an IF gene probe, however, Sharpe (1988) was able to isolate cDNAs that encode members of the IF family that are differentially expressed in ectoderm during neural induction. One of these, called XIF-6, encodes the *Xenopus* homolog for the medium form of neurofilament (NF-M). From work in several species, it is known that the expression of neurofilament proteins is restricted to neurons, and that during neuronal differentiation the three forms are expressed sequentially, with NF-M appearing first (Tapscott et al 1981a,b). Over the time course of neural development, however, even NF-M transcripts appear relatively late. For example, the birthdates of primary neurons occur during neural plate formation, whereas XIF-6 transcripts do not appear until approximately stage 26 (early tadpole) or 12 hr later (Hartenstein 1989, Sharpe 1988). Another neurofilament-like protein, NF-3, also appears to be expressed at significant levels in early tadpole embryos (Charnas et al 1987, Dixon & Kintner 1989). Although these filament RNAs are expressed relatively late compared to other neural markers, they are useful general markers of neuronal maturation because they are present uniformly in all regions along the neural axis.

Sharpe has also reported the isolation of another *Xenopus* IF cDNA, XIF-3 (Sharpe et al 1989). XIF-3 is probably the *Xenopus* homolog of the peripherin gene, an intermediate filament first identified in peripheral neurons of higher vertebrates (Greene 1989). The expression of XIF-3 transcripts in early *Xenopus* embryos can be divided into two distinct phases (Sharpe et al 1989). In the first phase, XIF-3 transcripts are expressed at low levels at the beginning of gastrulation; this expression appears to occur uniformly in the ectoderm. In the second phase, the expression of XIF-3 transcripts increases in stage 14 embryos (neural plate) to levels that are tenfold higher than transcripts expressed during the first phase. Importantly, the abundant XIF-3 transcripts are expressed primarily by the neuroectoderm. In addition, the neural expression of XIF-3 does not occur uniformly along the neural axis but is primarily confined to anterior regions of the spinal cord and the hindbrain. Thus, XIF-3 transcripts found in abundant levels are a marker of neural tissue and an anterior marker of the neural plate.

Cell adhesion molecules represent another class of proteins associated with histogenesis in early embryos (Edelman 1988, Takeichi 1988). Cell adhesion molecules were first identified via antibodies that blocked aggregation of dissociated embryonic cells in vitro. Edelman and colleagues employed this approach to characterize the neural cell adhesion molecule, NCAM, by using an antibody that blocked the calcium independent aggregation of retinal ganglion cells (Cunningham et al 1987). The subsequent characterization of NCAM by molecular cloning, along with the molecular characterization of similar molecules, particularly in the immune system, has defined what is now a large gene family whose members share structural characteristics with the constant domain of immunoglobulins (Ig) (Hunkapiller & Hood 1986). Many of the immunoglobin-like cell adhesion molecules are expressed in the nervous system and have been implicated in cell-cell interactions necessary for tissue morphogenesis as well as axon fasciculation, outgrowth, and targeting (Jessell 1988).

At an early stage in the characterization of NCAM, Edelman and colleagues showed that NCAM antibodies stain the neural plate of a chick embryo but not the surrounding ectoderm that differentiates into nonneural tissue (Edelman et al 1983). Among the immunoglobin-like cell adhesion molecules, NCAM appears to be expressed earliest in neural development. Early expression of NCAM in neural development has also been found in *Xenopus* embryos (Jacobson & Rutishauser 1986, Kintner & Melton 1987, Levi et al 1987). NCAM expression outside the nervous system in *Xenopus* embryos, however, appears to be different from that observed in higher vertebrates. Although NCAM is expressed at significant levels in the mesodermal derivatives of early chick embryos, its expression in the frog is almost exclusively neural, at least until metamorphosis (Kay et al 1988). In addition, although NCAM RNA and proteins are present in *Xenopus* eggs, the amount of this maternal NCAM is extremely low relative to that in neural tissue (Kintner & Melton 1987, Levi et al 1987). Thus, for all practical purposes, the expression of NCAM RNA and proteins in early *Xenopus* embryo is first detected when the neural plate forms, and this expression is completely dependent on the induction of neural tissue. Since the expression of NCAM does not appear to be restricted to any particular neural cell type or to be localized to any point along the neural axis, it serves as an early general marker of neural tissue formation.

The cadherins are proteins that comprise another major class of cell adhesion molecules associated with histogenesis in vertebrate embryos (for reviews of cadherins see Jessell 1988, Takeichi 1988, 1991). The cadherins can be distinguished from the immunoglobin-like cell adhesion molecules on the basis of their structure and their requirement of calcium for activity. The first three members of the cadherin family to be described by molecular cloning were N-cadherin (also known as A-CAM), first identified in the chick neural tissue; E-cadherin (also known as uvomorulin), first identified in the epithelium of early mouse embryos; and P-cadherin, first identified in mouse placenta. The sequence of these molecules shows that the cadherins are closely related in primary sequence and contain extensive similarity in various functional domains that are now being extensively characterized (Nagafuchi & Takeichi 1988, 1989; Ozawa et al 1989, Ozawa & Kemler 1990). Although other cadherins are now known to be expressed in the nervous system, N-cadherin appears to be expressed earliest in neural development. Takeichi and colleagues showed that antibodies to N-cadherin stain the neural plate but not the surrounding ectoderm in chick embryos (Hatta & Takeichi 1986). Similarly, N-cadherin is expressed during the formation of the neural plate in *Xenopus* embryos (Detrick et al 1990). Expression of N-cadherin only occurs in the ectoderm of *Xenopus* embryos following neural induction, and thus N-cadherin can be used to mark the separation of the ectoderm into neural and epidermal tissue. The use of N-cadherin as a neural marker is limited to special cases, however, because N-cadherin is also expressed on nonneural tissues such as the dorsal mesoderm.

Of the other cadherins expressed in *Xenopus*, the one most likely to be useful as a tissue marker during neural induction is E-cadherin (Angres et al 1991, Choi & Gumbiner 1989, Levi et al 1991). E-cadherin expression in early *Xenopus* embryos is restricted primarily to differentiated epithelium such as the epidermis, much like E-cadherin in the mouse or LCAM in the chick. In higher vertebrates, however, E-cadherin is also expressed during blastula stages in early epithelial cell layers before tissue differentiation occurs. In contrast, E-cadherin is not expressed in *Xenopus* embryos until the beginning of gastrulation (Choi & Gumbiner 1989), so E-cadherin expression follows the differentiation of the epidermis much the same way as epidermal markers such as DG81, described further below (Angres et al 1991). E-cadherin, therefore, can be used as a marker to distinguish nonneural ectoderm from neural ectoderm in early embryos, although at later stages E-cadherin expression is more ubiquitous and occurs in several epithelial tissues, such as the kidney, liver, and lung. Although *Xenopus* differs from other vertebrates by not expressing E-cadherin in early embryos, it apparently compensates for this by expressing another cadherin related to E-cadherin called E/P cadherin, C-cadherin, or U-cadherin (Angres et al 1991, Choi et al 1990, Ginsberg et al 1991).

Molecules with Regionally Restricted Expression in the Xenopus Nervous System

The isolation of probes for molecules that show region-specific expression in the nervous system has generated invaluable tools for studying the processes underlying regionalization of the neural plate. In *Xenopus*, markers have not yet been described that are useful for marking the medial/lateral axis of the neural plate, which then becomes the dorsal/ventral axis of the neural tube. Instead, the markers currently in use are first expressed in embryos during gastrulation when their expression is localized to different points of the neural plate along the A/P axis. In many cases, the A/P markers are expressed in the derivatives of several germ layers, so they mark axial position in the embryo rather than tissue type. The most important development in isolating these markers is the ability to exploit sequence similarity found within a structural domain of putative transcription factors called the homeobox.

The homeobox was first identified in genes that play a role in establishing the proper number of body segments, segment polarity, and identity during the development of the *Drosophila* embryo (Akam 1987). The majority of these genes encode putative transcription factors and many contain a 60-amino-acid structural motif, the homeobox, involved in protein–DNA interactions (McGinnis et al 1984b). The homeobox genes establish the segments in the *Drosophila* body plan by generating restricted domains of gene expression through a hierarchical network of regulatory interactions. These observations, along with the finding that homeobox genes are present in many different species, including vertebrates, initially suggested that the process of segmentation may be conserved throughout evolution (McGinnis et al 1984a). This suggestion is not likely to be literally true, in that even organisms with no apparent segmentation contain homeobox genes (Holland & Hogan 1986); however, transcription factors that act as master regulators of gene expression may be a general mechanism for establishing new cell fates in development. Because the neural plate appears to be divided early on into regions with different fates, the homeobox genes may mediate and mark these early stages of neural regionalization.

Some of the first homeobox genes described in *Drosophila* are contained within a locus called the UBX-ANTP complex. These genes are clustered together on the chromosome and presumably arose by gene duplication, since their homeoboxes share significant sequence similarity. The vertebrate counterparts of the *Drosophila* UBX-ANTP complex, called antp-like homeobox genes, are remarkably conserved both in sequence and in chromosomal organization (Akam 1989). Even more remarkable, however, is that the conservation of this genetic locus also extends to the pattern by which individual genes in the cluster are expressed along the A/P axis of developing embryos (Duboule & Dolle 1989, Graham et al 1989). This expression pattern has been studied mainly in the mouse embryo, where the antp-like homeobox genes, referred to as Hox genes, are contained in at least four different chromosomal clusters. Although the analysis of the *Xenopus* Hox genes has not advanced as far as the mouse genes, it has been very informative. The Hox genes that have been isolated are listed in Table 1 along with their mouse Hox counterparts.

Although the embryonic expression of the *Xenopus* Hox genes is expected to be similar to the mouse genes, establishing the spatial distribution of the Hox gene transcripts in *Xenopus* embryos has been difficult for two reasons. First, the Hox RNA transcripts are not expressed at particularly abundant levels in early embryos (Harvey et al 1986). Second, in situ hybridization, which has been so successful for localizing Hox gene expression in the mouse, is notoriously insensitive when applied to early *Xenopus* embryos, whose tissue consists of large, yolky cells. Thus, with a few exceptions (Carrasco & Malacinski 1987), the localization of Hox RNA transcripts in *Xenopus* embryos has been achieved by subjecting dissected tissue to RNAase protection assays (Harvey et al 1986, Sharpe et al 1987). These assays are sufficiently sensitive to detect Hox transcripts in small amounts of tissue, but the tissue dissection results in very poor spatial resolution. The expression of Hox genes in *Xenopus* was not known in any detail until antibodies were raised against the *Xenopus* Hox genes X1HBox1 and X1HBox6 gene products expressed in bacteria (Oliver et al 1988, Wright et al 1990).

X1HBox1 and X1HBox6 cDNA encode Hox gene products that correspond to the mouse Hox 3.3 and Hox 2.5 genes, respectively. *Xenopus* embryos first express X1HBox1 and X1HBox6 RNA transcripts in early gastrulae stages (Carrasco et al 1984, Sharpe et al 1987). In contrast, antibodies raised against the X1HBox1 and X1HBox6 gene products do not give detectable staining of embryos until they reach early neurulae stages (Oliver et al 1988, Wright et al 1990). The reason for this discrepancy could be a difference in the ability to detect RNA versus protein, or a delay in translation of the X1HBox1 and X1HBox6 transcripts. Resolution of this tissue is important, since we would like to know when these genes first act during early neural development. Embryos show staining with the antibodies against X1HBox1 and X1HBox6 that is localized along the

A/P axis of both the nervous system and other embryonic tissues such as the mesoderm (Oliver et al 1988, Wright et al 1990); in the nervous system, X1HBox6 staining is localized to the posterior portion of the embryo, with an anterior limit of expression corresponding to the hindbrain/spinal cord boundary. X1HBox6 expression therefore marks the spinal cord portion of the central nervous system. Embryos reacted with the X1HBox1 antibody also show posterior neural staining, but this staining is restricted to a band of tissue located in cervical regions of the central nervous system. The expression pattern of the X1HBox1 protein is complicated because there are two forms of the X1HBox1 protein that differ slightly in where they are expressed in the nervous system (Cho et al 1989). The two forms are generated from the same open-reading frame but the longer form contains an additional 82 amino acids absent from the amino terminus of the shorter form. The expression patterns for each form overlap extensively but differ in the anterior limits of expression. The short form of X1HBox1 has an anterior limit that falls within the posterior hindbrain whereas the anterior limit of the long form lies within the spinal cord. Thus, both the X1HBox1 and the X1HBox6 antibodies are useful reagents to mark the formation of posterior neural tissue. In addition, antibodies to the short form of X1HBox1 can be used to distinguish the posterior hindbrain from the spinal cord within cervical regions of the central nervous system.

Drosophila genes with a homeobox sequence that is distinct from antp-like homeobox genes have also been useful in identifying regional markers for the *Xenopus* nervous system (Doe & Scott 1988). A monoclonal antibody, 4D9, was raised against a peptide sequence present in the homeobox of the *engrailed* gene, a gene involved in the establishment of segment polarity in *Drosophila* embryos (Patel et al 1989). The 4D9 antibody cross-reacts with an engrailed-like protein found in embryos representing a wide range of different species. In arthropod embryos such as *Drosophila*, the 4D9 antibody stains the posterior-half of each parasegment in accordance with the role of the *engrailed* gene in segment polarity. At later stages of development, however, the antibody also stains defined cell types in the larval nervous system, and the neural component of the 4D9 staining is remarkably conserved in different species (Patel et al 1989). Embryos from leech to higher vertebrates reacted with the 4D9 antibody show predominant staining within a small region in the anterior portions of the nervous system. In vertebrate embryos, this staining is confined to a band of tissue within the portion of the nervous system corresponding to the midbrain/hindbrain boundary. In *Xenopus* embryos, the 4D9 antigen and its expression has been analyzed in detail by Harland and his colleagues (Hemmati Brivanlou et al 1991, Hemmati Brivanlou & Harland 1989). They have shown that the 4D9 antibody first stains embryos during neural plate formation, when staining is observed in a stripe of tissue across the anterior portion of the neural plate. In late-stage tadpole embryos, 4D9 intensely stains the neural tissue around the midbrain/hindbrain boundary, but also stains tissue lightly in the mandibular arch, the optic tectum and the anterior pituitary. The 4D9 antigen has also been characterized by isolating cDNAs through bacterial expression cloning with the 4D9 antibody, and through screening libraries with a probe consisting of engrailed-homeobox sequences (Hemmati Brivanlou et al 1991). Two distinct but related cDNAs were isolated in these screens, both of which encode proteins very similar in sequence to the engrailed-2 (en-2) gene in the mouse (Joyner & Martin 1987). The expression of *Xenopus* en-2 RNA is first detected in neural plate embryos and, like the 4D9 antibody staining, appears to be restricted to anterior regions of the nervous system. These observations indicate that both the 4D9 antibody and en-2 cDNAs are useful probes for marking an anterior region of the *Xenopus* nervous system.

The *Xenopus* gene, Xhox-3, contains a homeobox whose sequence is similar to that found in the *Drosophila eve* gene (Ruiz i Altaba & Melton 1989). The expression of Xhox-3 transcripts in *Xenopus* embryos has been determined by both RNase protection analysis of dissected tissue and in situ hybridization on tissue sections. During an early phase of expression, Xhox-3 transcripts appear in embryos during gastrulation, where they can be found in a graded distribution within the developing mesoderm. This gradient of Xhox-3 expression is high in the posterior mesoderm and low in anterior regions. By late neurulae stages, expression of Xhox-3 enters a second phase wherein expression is found only in the nervous system in anterior regions and within the growing tailbud posteriorly. The neural expression of Xhox-3 is particularly relevant to this discussion because it is localized along the anterior-posterior axis (Ruiz i Altaba 1990). The neural expression of Xhox-3 appears to arise relatively late, in that transcripts are first detected at early tadpole stages. Xhox-3 neural expression, however, could occur as early as during the formation of the neural plate but at levels below the sensitivity of RNAase protection analysis and in situ hybridization. The neural expression of Xhox-3 in late tadpoles is localized to postmitotic cells in the hindbrain and anterior spinal cord and to neural crest cells derived from the same axial location. Xhox-3 transcripts therefore mark a region of the central nervous system similar to the region marked by en-2 expression.

The regional neural markers for *Xenopus* embryos discussed above only cover one region of the neural plate. This region, called the chordal plate, lies over the region of embryo containing the notochord and gives rise to the portion of the nervous system posterior to the midbrain. In contrast, markers have yet to be reported for the portion of the neural plate, called the prechordal neural plate, that gives rise to the nervous system anterior to the midbrain boundary. The absence of markers for the prechordal plate may reflect an emphasis of current research on genes whose structure and expression is conserved during evolution. Such genes are more likely to be expressed in the chordal neural plate, since this region gives rise to the portion of the nervous system that is more conserved anatomically among different vertebrates species than the portion formed by the pre-chordal plate. Therefore, different approaches may be required to isolate markers for the prechordal nervous system. A distinction between the chordal and prechordal neural plates may also be justified for embryo-logical reasons. For example, in chick embryos, the neural plate is first laid down in both the posterior and anterior directions beginning at a point corresponding to the midbrain, thus suggesting that the prechordal and chordal regions of the plate form separately. Thus, the first regional division of the neural plate may be into chordal (posterior) and prechordal (anterior); these regions may then be further subdivided along the A/P axis. The known regional markers are more likely to be useful in marking the A/P axis of the chordal plate.

THE EXPRESSION OF EPIDERMAL AND NEURAL MARKERS DURING NEURAL INDUCTION

Early studies of neural induction led to models that posit how the inter-action between the organizer and ectoderm leads to the formation and patterning of the neural plate. The gist of these models is that the organizer involutes underneath the ectoderm during gastrulation and forms dorsal mesodermal derivatives. This dorsal mesodermal mantle then imprints on the overlying ectoderm the region that will form the neural plate and, within the neural plate, regions that will form different areas of the nervous system (see Figure 1). From these models, one can predict the pattern by which various molecular markers should be expressed during induction. The markers have not always behaved as expected, however, in testing these predictions. When the markers fail to confirm our expectations, they reveal features of neural induction not previously appreciated. The full implication of these new results in terms of induction mechanisms is still not clear, so it is inappropriate to consider alternative models of induction until this line of investigation has been pursued further. Nonetheless, discussion of a few cases in which markers have given unexpected results illustrates how this approach has generated new information about the induction process.

Expression of Epidermal Markers During Neural Induction

Epidermal markers are expressed in the epidermis but are absent from neural tissue in late stages of development. During gastrulation, when the ectoderm separates into epidermal and neural tissue, one might have predicted that epidermal markers would be absent from ectoderm destined to be neural but present autonomously in all ectoderm outside the neural plate. The results of recent studies analyzing the expression of epidermal markers during neural induction are not consistent with this simple prediction, however. Instead, the results suggest that epidermal markers can be divided into two categories on the basis of two unexpected expression patterns during neural induction: (*a*) epidermal markers, including DG-81 and E-cadherin, that are expressed during gastrulation, and (*b*) markers, including Epi-1, that are expressed relatively late. The expression of markers in each category says something new about how inducing signals from the organizer diverts the ectoderm toward a neural pathway and away from epidermal differentiation.

DG-81, the epidermal keratin, and E-cadherin are molecules associated with the differentiation of epidermal tissue, yet the transcripts encoding these two molecules appear extremely early in development. Transcripts for both molecules are first detected in embryos before gastrulation begins (Jamrich et al 1987). Moreover, the DG81 RNA present in embryos before gastrulation has been localized by in situ hybridization to ectoderm on both the prospective dorsal and ventral sides of the blastula. The expression of DG81 RNA is later lost from ectoderm that comes into contact with involuting organizer tissue during gastrulation (Jamrich et al 1987). Similar observations suggest that E-cadherin is also expressed in ectoderm on both the prospective dorsal (neural) and ventral (epidermal) side of the blastula (Angres et al 1991, Choi & Gumbiner 1989, Levi et al 1991). Thus, these observations lead to the surprising conclusion that the expression of at least two epidermal-specific genes occurs in all ectoderm before gastrulation begins and is later turned off in the region of ectoderm that forms the neural plate. The expression of these genes suggests that the ectoderm initiates at least part of the transcriptional program required for epidermal differentiation before gastrulation, then somehow turns this program off following induction.

The Role of Epidermal Differentiation in Ectodermal Competence

The early expression of epidermal markers in the ectoderm may bear on a phenomenon observed by embryologists when they determined the stage in development when the ectoderm responds to neural induction. Ectoderm from pregastrulating embryos responds optimally to neural induction, whereas ectoderm from late-gastrulating embryos responds poorly if at all (Holtfreter & Hamburger 1955). Because ectodermal "competence" ends approximately when the ectoderm is thought to interact with inducing tissue, this phenomenon may limit the area of ectoderm that responds to induction (Albers 1987). The unexpected expression of epidermal genes in blastula ectoderm suggests that the basis for the loss of competence could be at the level of transcription. One possibility is that the transcription factors responsible for epidermal differentiation are first expressed before gastrulation at low levels, then steadily increase in levels of expression during gastrulation, perhaps by a positive feedback mechanism. The expression levels of transcription factors may determine whether the ectoderm is still able to respond to induction and go down the neural pathway. For example, in *Drosophila*, the levels of different transcription factors of the helix-loop-helix family can be a determining factor in controlling cell fate, presumably by competing for the control of gene expression (Jones 1990). Similarly, the choice between the neural and epidermal pathways may be determined by the relative levels of regulatory transcription factors required for these two fates. By the end of gastrulation, the expression of the epidermal regulatory factors may define a point of no return, when ectoderm no longer responds to neural induction. This model will be testable when more is known about the transcription factors that control epidermal gene expression.

Control of Epidermal Gene Expression

The expression of epidermal genes in early embryos suggests that neural induction can be studied as a problem in gene regulation. In other words, one way to elucidate the induction pathway is to analyze the mechanisms that control the transcription of genes whose expression is turned off or on during induction. The most complete analysis of this type has been carried out with one of the epidermal keratin genes in the XK81 family described above, called XK81A1 (Jonas et al 1989). The first step in this analysis was to identify the *cis*-acting elements within the upstream sequences of the XK81A1 gene that are required for correct expression of this gene in early embryos. Accordingly, a 500 base pair (bp) region of the XK81A1 gene around the start of transcription was placed upstream of a reporter gene and introduced back into embryos. Expression of the hybrid gene was found to occur in epidermal tissue but not in other regions of the embryo; the same developmental time course was observed with the endogenous gene (Jonas et al 1989). The next step was to identify the transcription factors that bind this 500-base-pair region and confer proper regulated expression of the XK81A1 gene. Toward this end, a protein called KTF-1 appears to bind, by gel shift and DNA footprinting analysis, a site within the 500-bp-promoter region of the XK81A1 gene, about 150 bp upstream from the start of transcription (Snape et al 1990). Removing the KTF-1 binding site from the 500-bp-promoter fragment reduces but does not eliminate transcriptional activity. This result indicates that transcription factors other than KFT-1 are apparently required for full transcriptional activity of the XK81A1 gene. In addition, the results of other experiments suggest that the KFT-1 might be considered a more general transcription factor required for the activity of this tissue-specific promoter. First, the KFT-1 binding site, attached to a heterologous core promoter, enhances transcription without any apparent tissue specificity. Second, the KFT-1 binding activity can be found by gel shift analysis in extracts prepared from tissues other than epidermis. Thus, the KFT-1 transcription factor may not explain either the tissue specificity or the developmental time course of XK81A1 gene expression. Other, yet to be discovered, factors presumably explain how the XK81A1 promoter is turned on autonomously and specifically in epidermis and off in the ectoderm in response to neural induction. Although the analysis of the XK81A1 promoter is still in early stages, this line of analysis should be important in the future for understanding how changes in cell fate in response to induction are controlled at the level of gene expression.

Epi-1 Expression in Isolated Ectoderm

The expression of another epidermal marker, Epi-1, in the ectoderm during neural induction is quite different from the pattern of DG81 and E-cadherin expression discussed above. Epi-1 expression first appears in the ectoderm after the formation of the neural plate. Moreover, the most striking features of Epi-1 expression are observed when the ectoderm is isolated from blastula embryos before gastrulation begins (London et al 1988). Ectoderm isolated and cultured from the prospective dorsal side of blastula embryo stains less intensely with the Epi-1 antibody than ectoderm taken from the prospective ventral side (see Figure 2). This observation indicates that not all ectoderm in a blastula has the same ability to express Epi-1, and that the reduction of Epi-1 expression can occur in ectoderm isolated from the embryo before neural tissue is induced. Moreover, the expression of Epi-1 in ventral ectoderm can be reduced to the levels observed in dorsal ectoderm by placing it edge-on with organizer tissue for a brief period (Savage & Phillips 1989). This result indicates that the dorsal/ventral difference in the ability of the ectoderm to express Epi-1 appears to be imposed by signals from the organizer.

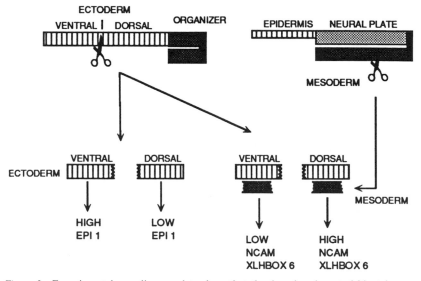

Figure 2   Experimental paradigm used to show that the dorsal and ventral blastula ecto-derms are not equivalent.

Further Evidence for a Predisposition Signal

A difference in blastula ectoderm between the dorsal and ventral sides was also reported in a series of experiments by Sharpe et al (1987). In these experiments, the ectoderm was isolated from the prospective dorsal or ventral side of a blastula and then combined with posterior axial mesoderm, a form of organizer tissue (see Figure 2). The extent to which the ectoderm responded to this inducer was then assessed by measuring the levels of two neural transcripts, X1HBox6 and NCAM, by RNAase protection analysis. The neural transcripts were not expressed in ectoderm isolated and cultured alone but were induced in ectoderm cultured in contact with axial mesoderm; however, the levels to which the two transcripts were induced by mesodermal tissue differed depending on what region of the blastula ectoderm was used for the explant (see Figure 2). Induction of ectoderm isolated from the prospective dorsal side of the blastula produced more neural tissue, as monitored by the expression of neural transcripts, than did ectoderm isolated from the prospective ventral side. These observations suggest that the dorsal ectoderm is more capable than the ventral ectoderm of responding to neural induction.

The bias in the dorsal ectoderm observed in the experiments of Sharpe et al has been termed *predisposition*. How predisposition arises is unknown, but it is tempting to think that a signal from the organizer is involved. Indeed, the Epi-1 results suggest that a signal from the organizer affects the ability of dorsal ectoderm to undergo epidermal differentiation, as measured by the expression of Epi-1, while not actually inducing neural tissue. Perhaps the same organizer signal is also responsible for generating predisposition. For example, since competence is a limiting factor in the ability of the ectoderm to respond to induction (see preceding section), an organizer signal that extended competence, or slowed the loss of competence, in dorsal ectoderm would have the desired effect of both decreasing epidermal differentiation and increasing the response to neural induction. Obviously, the mechanism underlying predisposition is currently a matter of speculation. This mechanism is likely to be an increasingly important issue in the field, since evidence for predisposition has recently been found in other examples of embryonic induction, such as the one required for the formation of the lens (Henry & Grainger 1990) or of the mesoderm (Sokol & Melton 1991).

A predisposition signal such as the one discussed above has raised several important issues that are likely to be considered more in the future. First, when does neural induction begin? Previously, it was assumed that induction occurs during gastrulation; in contrast, recent results such as those with Epi-1 suggest that signals from the organizer pass into the ectoderm even before gastrulation begins. Induction, therefore, could begin as early as the blastula stages when the organizer itself first forms. Second, by what route do signals pass from the organizer to the ectoderm during neural induction? Previously, these signals were thought to travel between the organizer and ectoderm when the two tissues come into contact as apposed cell layers during gastrulation. Recent results suggest instead that signals can also pass edgewise between the organizer and ectoderm across the boundary they share in the blastula. The basis of edgewise signaling is discussed further below. Third, what is the nature of the signals underlying neural induction? The formation of neural tissue may be viewed more accurately as a two-step process. In this view, the first step might consist of a predisposition signal that biases ectoderm toward neural and away from epidermal differentiation. The second step might consist of a neuralizing signal that evokes an actual response in terms of new tissue differentiation. A two-signal process raises the question of which signal should be rightly considered the determining factor during the formation of neural tissue. This question can be illustrated by making the distinction between permissive and instructive inducing signals suggested by Gurdon (1987). For example, the early predisposition signal underlying the first step in induction could be an instructive factor in generating neural tissue. This would be the case if only the ectoderm that received this signal (i.e. dorsal ectoderm) before and during gastrulation were then capable of undergoing neural development in response to a neuralizing signal after the completion of gastrulation. In contrast, the neuralizing signal underlying the second step in induction could be a permissive factor in generating neural tissue, particularly if this signal was not restricted in its distribution.

The advantages afforded by using molecular markers rather than histology to measure the response of the ectoderm to induction are evident from the recent experiments described above (Gurdon 1987). Molecular markers can score changes occurring in the ectoderm during induction that in some cases cannot be measured with histology. For example, markers revealed the biased expression of Epi-1 in the ectoderm, thus suggesting that ectoderm is predisposed toward neural development before the formation of neural tissue is actually induced. Similarly, the probes to DG81 and E-cadherin revealed the transient expression of epidermal properties in dorsal ectoderm before gastrulation, a finding that suggests that the ectoderm initiates epidermal differentiation before induction occurs. Thus, the use of molecular markers to analyze neural induction has made a convincing case for assaying the earliest possible consequence of inducer activity, rather than waiting for histological differentiation to occur.

Edgewise Neural Induction

The Epi-1 experiments suggested that an inducing signal could pass from the organizer into the ectoderm before gastrulation and set up a bias in the ability of the ectoderm to undergo epidermal differentiation. This observation was contrary to the traditional view that inducing signals only pass to the ectoderm from the underlying mesoderm during gastrulation, a view primarily based on experiments carried out by Holtfreter (Hamburger 1988, Holtfreter 1933). Holtfreter showed that if amphibian embryos were cultured in high salt and stripped of their vitelline membrane, they gastrulated abnormally so that the mesoderm and endoderm came to lie outside rather than inside the ectodermal cell layer. These inside-out embryos, or exogastrulae, contained relatively normal axial mesoderm connected through a stalk of tissue to an isolated ectodermal sack. Importantly, Holtfreter concluded on histological grounds that exogastrulae were neural deficient, thereby indicating that apposition of dorsal mesoderm with the ectoderm during normal gastrulation was essential for inducing signals to pass between the two tissues.

Vertical apposition between the ectoderm and organizer tissue does not occur in an exogastrula because of abnormal gastrulation movements.

Nonetheless, contact does occur between the ectoderm and organizer tissue in an exogastrula at the boundary that the two tissues share in the blastula (see Figure 3). Therefore, exogastrulae should contain at least some neural tissue, specifically at the junction where ectoderm and organizer tissue meet. This prediction was borne out when exogastrulae were examined for the expression of neural transcripts (Dixon & Kintner 1989, Kintner & Melton 1987, Ruiz i Altaba 1990). Exogastrulae express neural transcripts, such as NCAM and the neurofilament NF-3, and at least some of these transcripts can be found by in situ hybridization at the junction between the ectodermal sack and dorsal mesoderm (Kintner & Melton 1987).

The surprising finding about exogastrulae, however, is the amount of neural tissue they contain despite the limited contact between the ectoderm and the organizer (Dixon & Kintner 1989). Exogastrulae express neural transcripts at almost the same levels expressed in stage-matched control embryos. At least some of these transcripts are present in exogastrulae in regions of the extruded ectoderm located away from the point of contact with organizer tissue (Ruiz i Altaba 1990). These results are not consistent with Holtfreter's conclusions about exogastrulae and raise the question of how neural tissue forms efficiently under these conditions. The same question is raised by results obtained with an explant of blastula tissue called a Keller sandwich (see Figure 3) (Keller & Danilchik 1988). Keller sandwiches are constructed by dissecting the dorsal marginal zone and ectoderm from two blastula embryos. By combining two such isolates together into a sandwich configuration, the tissues remain flat during gastrulation

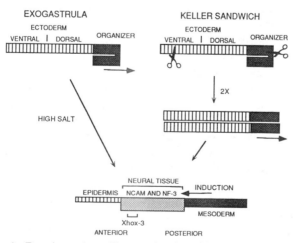

Figure 3   Experimental paradigm used to demonstrate edgewise induction.

(Figure 3). As with an exogastrula, the gastrulation movements in a Keller sandwich do not result in the involution of organizer tissue underneath the ectoderm (Keller & Danilchik 1988). Nonetheless, Keller sandwiches form copious amounts of neural tissue, as measured by the expression of NCAM and NF-3 RNA (Dixon & Kintner 1989, Keller & Danilchik 1988).

The idea of edgewise signals passing between the ectoderm and the organizer is discussed above in the context of predisposition. The observations with exogastrulae and Keller sandwiches suggest further that an edgewise interaction may be a route by which neuralizing signals pass between the organizer and ectoderm (Figure 3). An unresolved question is how an edgewise interaction is efficient enough to generate almost as much neural tissue normally found in embryos. One way an edgewise interaction could be efficient would be if the neuralizing signal were to spread within the plane of the ectoderm through the induction of neural tissue by neural tissue. Induction by this mechanism was originally described by Spemann, who called it homeogenetic induction (Hamburger 1988). Recent studies, however, show that homeogenetic induction is too inefficient in *Xenopus* embryos to explain the induction of neural tissue over long distances (Jones & Woodland 1989). Another possibility is that the edgewise transfer of a neuralizing signal occurs via gastrulating cells within neural tissue (Jacobson & Sater 1988, Keller & Danilchik 1988). Gastrulation movements in neural tissue involve a portion of the blastula called the notoplate, which lies within the ectoderm that directly abuts the organizer tissue in the dorsal marginal zone. During gastrulation and neuralation, cells in the notoplate converge and extend along the A/P axis, and thereby occupy the midline of the neural plate. The properties of the notoplate, in terms of its location in the blastula and movements during gastrulation, are strikingly similar to those associated with the prospective notochordal cells (Gordon & Jacobson 1978). The similarity between these blastula tissues ends after gastrulation, however, in that the notoplate forms part of the neural plate whereas the notochord forms part of the dorsal mesoderm. Moreover, in exogastrulae and Keller sandwiches, the movements of gastrulation bring the notoplate, but not the prospective notochord cells, into extensive contact with the ectodermal cell layer (Keller & Danilchik 1988). Thus, the edgewise induction of neural tissue could be explained if the notoplate, like the notochord, were a source of neural inducing signals.

The use of neural markers to analyze exogastrulae illustrates the advantage of using assays for gene expression to quantify the induction of neural tissue. Quantification is very useful for comparing the amount of neural tissue formed under experimental conditions to the amount of neural tissue formed normally in the embryo. Moreover, gene expression is the earliest change that can be scored quantitatively when the ectoderm responds to neural induction. These considerations might explain the discrepancy between Holtfreter's observations and more recent studies on exogastrulae. Although other explanations are conceivable, it is still a strong possibility that histology produced misleading results in the case of the exogastrulae by greatly underestimating the amount of neural tissue formed in this abnormal embryo.

Expression of Regional Markers in Response to Neural Induction

One problem faced in the early studies of neural induction was scoring the regional characteristics of neural tissue formed in experimental explants by their histological features. The regional features of neural tissue in explants were generally scored by using ectodermal derivatives that form in conjunction with the nervous system at different points along the A/P axis, such as the otic vesicle. These ectodermal placodes, however, are a rather crude measure of regional neural characteristics. Thus, the isolation of probes that mark directly an early consequence of neural regionalization, even before these consequences can be detected histologically, has been a great boon to the study of neural regionalization.

Early studies on neural induction suggested that the A/P axis of the neural plate is likely to depend on interactions between the organizer tissue and the ectoderm. The analysis of this interaction has led to several models for how regionalization occurs. For reasons of brevity, the reader is referred to several reviews that discuss the experimental bases behind these models, since they have been very influential in guiding recent studies on regionalization of the neural plate (Hamburger 1988, Saxen 1989). Although these early studies revealed some of the important aspects of regionalization, many features of this process remain obscure. For example, the results from some of the early experiments suggested that the neural plate is regionalized by a point-to-point imprinting of regional characteristics from the underlying mesoderm formed by organizer tissue during gastrulation. In other experiments, the results suggested that regionalization could arise through the action of as little as two signals emanating from the underlying mesoderm in opposing gradients along the A/P axis. Finally, some experiments suggested that interactions between cells within the neural plate may also contribute to regionalization. Regionalization is largely an unsolved problem, but recent studies demonstrate how molecular markers will help in studying this complex process.

Harland and colleagues have studied the mechanisms underlying the expression of the en-2, the engrailed-like molecule, within the anterior regions of the chordal neural plate. First, they examined en-2 expression in exogastrulae, where neural tissue forms in the absence of underlying dorsal mesoderm (Dixon & Kintner 1989). En-2 expression was greatly reduced in exogastrulae, thus suggesting that the underlying mesoderm is necessary for full en-2 expression in neural tissue (Hemmati Brivanlou & Harland 1989). Second, this idea was tested directly by combining blastula ectoderm with a portion of the underlying mesoderm, the notochord, isolated from gastrulated embryos (Hemmati Brivanlou et al 1990). Notochord isolated from both the anterior and posterior regions of an early tadpole induced blastula ectoderm to express NCAM, the general neural marker. In contrast, the anterior notochord, but not the posterior notochord, induced blastula ectoderm to express en-2 (see Figure 4). Importantly, this result was obtained even when the responding ectoderm had been isolated from ventralized embryos. Ventralized embryos are generated when the early steps in axis determination are blocked by irradiating fertilized eggs on the vegetal pole with ultraviolet light (Gerhart & Keller 1986). Because ventralized embryos lack an organizer, the ectoderm from these embryos rules out the possibility that the anterior notochord induces en-2 expression by simply revealing an inherent regional organization established in the ectoderm by edgewise signals from the organizer. The conclusion from these studies is that the expression of en-2 within a restricted region of the nervous system is dependent on a regionalizing signal emanating locally from a restricted region of the underlying anterior mesoderm. This conclusion is consistent with the results of Sharpe and

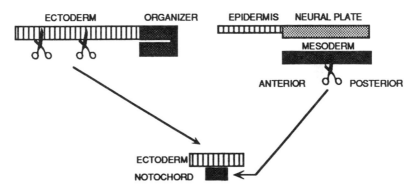

Figure 4 Experiment paradigm used to demonstrate the notochordal signals that induce the expression of en-2.

Gurdon, who showed that the expression of XIF-3, an anterior chordal marker, is also dependent on the apposition of ectoderm with the appropriate region of underlying mesoderm along the A/P axis (Sharpe & Gurdon 1990, Sharpe et al 1989).

Although the results with the en-2 marker suggest that some regional characteristics of neural plate are imprinted by the underlying mesoderm, quite different results were obtained with the Xhox-3 marker. At late stages of embryonic development, Xhox-3 expression becomes restricted to anterior regions of the chordal nervous system (Ruiz i Altaba & Melton 1989). To determine whether this expression pattern in neural tissue is dependent on the underlying mesoderm, Xhox-3 expression was examined in exogastrulae (Ruiz i Altaba 1990). In contrast to en-2, exogastrulae express Xhox-3 transcripts, and these transcripts are localized to the neural tissue that forms in the isolated ectodermal sack. Surprisingly, the expression of Xhox-3 transcripts within the neural tissue of an exogastrula is localized along an axis (Figure 3). The A/P axis in an exogastrula consists of a posterior point defined as the junction between the ectoderm and marginal zone tissue, and an anterior point defined by the position of a cement gland that forms at the tip of the ectodermal sack. Xhox-3 expression is localized along this axis toward the anterior end, that is, to a position that more or less corresponds to the normal position of Xhox-3 expression in the nervous system of normal embryos. Thus, in contrast to en-2, localized expression of Xhox-3 appears to occur in neural tissue without the benefit of regionalizing signals from the underlying mesoderm. The signal that turns on the appropriate expression of Xhox-3 in exogastrulae is unknown, but one source of regionalizing signals could be the notoplate, the gastrulating tissue along midline of the neural plate discussed above in the context of edgewise neural induction. In any case, the results with Xhox-3 complement those with en-2, suggesting that regionalizing signals may come both from the underlying mesoderm and from within the plane of the ectoderm. The expression of more regional markers will have to be examined before it is clear whether these two mechanisms work in parallel or redundantly to establish the A/P expression of these markers.

The markers isolated thus far cannot be used to determine the mechanisms underlying the patterning of the anterior-most end of the neural plate. Nonneural markers have been isolated that recognize anterior structures in the *Xenopus* body plan, of which the most common are cDNAs encoding transcripts expressed in the cement gland (Jamrich & Sato 1989, Sive et al 1989). Because the cement gland forms at the anterior-most point in the embryo, these cDNAs have been useful as markers to investigate how A/P patterning of the *Xenopus* body axis occurs. Although these markers have been used to investigate the mechanisms underlying the specification of the cement gland, how cement gland specification relates to the mechanisms underlying neural A/P patterning is not yet clear. For example, both exogastrulae and Keller sandwiches form a cement gland (Ruiz i Altaba 1990), even though these embryos lack anterior neural structures such as eyes. One interpretation of this result is that the cement gland is a by-product formed whenever neural tissue is generated regardless of how much of the A/P neural axis is present. Apparently, the specification of the cement gland can occur independently of signals required for generating the anterior portion of the nervous system.

ANALYSIS OF GENE FUNCTION DURING EARLY NEURAL DEVELOPMENT

Several complex changes occur in the ectoderm as it diverts along a neural pathway and forms neural tissue. These early changes include the appearance of new tissue and cell morphologies, the formation of progenitor cells for neural cell types, and the division of neural plate into different regions along the body axis. In *Xenopus* embryos, these changes occur simultaneously over a window of about six hours, thus making the processes underlying these changes difficult to sort out and study independently. One simplifying hypothesis, however, is to assume that these changes are initiated by the expression of a few regulatory molecules whose activity in the ectoderm is in turn controlled by induction. An appropriate analogy might be made with neurogenesis in *Drosophila*, in which a genetic analysis suggests that the specification of a neural rather than a hypodermal fate can be dissected into a pathway containing a relatively small number of genes (Hartley 1990). Similarly, a relatively few number of molecules may establish the neural fate of the ectoderm in amphibian embryos when they are expressed in the ectoderm as a consequence of induction.

A number of the molecules expressed in the neural plate have been proposed to have a pivotal role in the formation of neural tissue. Cell adhesion molecules, such as NCAM and the cadherins, are thought to mediate the morphological changes underlying neural histogenesis. Transcription regulatory factors such as the homeobox gene products are thought to mediate the new patterns of gene expression underlying the formation of different regions of the nervous system. A direct demonstration of gene function, however, is required to distinguish molecules involved in establishing a neural anlage from those required for later steps in neural differentiation. *Xenopus* is a particularly attractive vertebrate system for functional studies, in part because gene expression can be manipulated in early embryos by microinjecting fertilized eggs with cloned genes, synthetic RNA transcripts, or antibodies (Gurdon & Melton 1981). Two examples of gene manipulation in the context of early neural development are discussed below to illustrate the type of information that can be gleaned from this experimental approach.

Cell Adhesion Molecules in Early Neural Development

The idea that differential cell recognition or adhesion plays a role in morphogenesis dates back to the experiments of Townes & Holtfreter (1955). In their classic study, epidermal and neural tissues were isolated from early embryos, dissociated, mixed together, and allowed to reassociate. Upon reassociation, the epidermal cells segregated and reformed an epidermis whereas the neural cells segregated and reformed a neural tube. From this result, it was suggested that the first step in morphogenesis of epidermal or neural tissue requires a special class of cell surface molecules that permit cells of the same type to recognize each other. Cell recognition is then followed by cell segregation and tissue morphogenesis. The implication of this notion is that during development, the differential expression of these recognition molecules in presumptive neural or epidermal cells would be the first step in tissue histogenesis.

Cell adhesion molecules have been considered the most likely candidates to underlie cell recognition and morphogenesis in vertebrate embryos (Edelman 1986, Takeichi 1991). The two cell adhesion molecules, NCAM and N-cadherin, are expressed in the ectoderm when it is induced to form neural tissue. Therefore, they are likely to mediate cell recognition and morphogenesis in early neural development. One test of this idea is to determine whether the expression of these adhesion molecules has any dominant effect on the development of the ectoderm. This test could be carried out in *Xenopus* by injecting synthetic RNA transcripts encoding NCAM or N-cadherin into fertilized eggs, with the expectation of obtaining embryos in which these molecules are expressed in the ectoderm in the absence of neural induction.

Effects of NCAM and N-cadherin Expression on Ectodermal Cell Boundaries

One characteristic of a cell recognition molecule as suggested by the experiments of Townes and Holtfreter is that its expression in the ectoderm should cause changes consistent with a role in cell segregation and the formation of tissue boundaries. To test the effects of expressing NCAM and N-cadherin in the ectoderm, mosaic embryos were generated by injecting RNA encoding NCAM or N-cadherin into just one animal blastomere of an embryo at the 16-cell stage (Detrick et al 1990). The resulting embryos then contained a subpopulation of ectodermal cells ectopically expressing

N-cadherin or NCAM that were then followed through development. One question of particular interest was whether these ectodermal cells mixed with the ectodermal cells derived from neighboring blastomeres (Wetts & Fraser 1989). When a blastomere was injected with NCAM RNA, its descendants expressed NCAM on their cell surface (Kintner 1988). Surprisingly, ectodermal cell expressing NCAM mixed with cells derived from neighboring blastomeres as well as ectodermal cells injected with control RNAs. This result indicates that the expression of NCAM in a subpopulation of ectodermal cells does not generate cell boundaries. In contrast, ectodermal cells expressing N-cadherin by RNA injection mixed very little with ectodermal cells from neighboring blastomeres (Detrick et al 1990). In some cases, N-cadherin expression suppressed cell mixing to the point that a sharp cell boundary arose between the descendants of the injected and neighboring blastomeres. This result indicates that ectodermal cell boundaries can be generated by the differential expression of N-cadherin (Detrick et al 1990).

Abnormal Neural Tube Formation

The dominant effects of N-cadherin expression on ectodermal cell mixing implied that the tissue boundary between the neural plate and the surrounding nonneural ectoderm may arise by a boundary of N-cadherin expression. This model predicted that if the normal boundary of N-cadherin expression were to be erased in embryos by RNA injection, the segregation of neural from nonneural ectoderm should fail to occur. This prediction was borne out when embryos injected with N-cadherin were examined for morphological defects after neural tube formation (Detrick et al 1990, Fujimora et al 1990). Neural tube defects were evident in all embryos in which RNA injection had resulted in the misexpression of N-cadherin across the boundary between neural and nonneural ectoderm. In a large fraction of the embryos, the morphological defect consisted of a continuous cell layer between the epithelium in the neural tube and the surrounding ectoderm, thus suggesting that the ectoderm had failed to segregate into neural and nonneural tissue. In contrast, the morphology of the neural tube was normal in embryos injected with NCAM RNA and in embryos injected with a variety of control RNAs (Detrick et al 1990). The results from this analysis indicates that N-cadherin, but not NCAM, has at least some of the properties expected of a cell recognition molecule that underlies morphogenesis during early neural development.

The misexpression experiments with NCAM and N-cadherin illustrate how the dominant effects of a gene product can be analyzed in *Xenopus* embryos by RNA injection. A complementary analysis is to ask whether a molecule is necessary for a particular aspect of early neural development to occur normally. For example, while the expression of NCAM on its own does not appear to generate cell boundaries, NCAM may still be required for these boundaries to form. To test this possibility, the expression of NCAM would have to be blocked in early embryos to determine whether morphogenesis would still be normal in the absence of NCAM.

Blocking Gene Function in Early Neural Development

The most promising method to block gene function in *Xenopus* is to introduce antisense RNA or oligonucleotides into embryos in order to remove RNA transcripts encoding the gene product of interest (Gieblehaus et al 1988, Harland & Weintraub 1985, Izant & Weintraub 1984, Melton 1985). Unfortunately, antisense methodology has not always been effective when applied to *Xenopus*, although why antisense experiments fail is still not clear (Dagle et al 1990, Rebagliati & Melton 1987, Woolf et al 1990). For these reasons, alternative approaches have been taken to knock out gene function in *Xenopus*, including injecting blocking antibodies into embryos. One of the more dramatic examples of this approach was reported by Wright et al (1989) in their studies of the *Xenopus* Hox gene, X1HBox1. In these studies, antibodies were raised against the long form of X1HBox1, a homeobox-containing protein expressed in cervical regions of the central nervous system. Wright et al hypothesized that the expression of this putative transcription factor plays a role in patterning this region of the nervous system in the same way that homeobox-containing genes play a role in patterning the *Drosophila* body plan. To test this hypothesis, the function of the X1HBox1 protein was blocked in embryos by injecting X1HBox1 antibodies into fertilized eggs. The injected embryos were then examined for developmental defects.

The most significant result in these experiments is that embryos injected with the X1HBox1 antibody contained developmental defects primarily confined to the region of the embryo where the long form of X1HBox1 is normally expressed, that is, the anterior portions of the spinal cord. The defects in the anterior spinal cord of these embryos consisted of a thinning of the roof plate and altered morphology in the spinal roots. Although these defects could be explained in several ways, the most intriguing explanation is that the portion of the neural tube in these embryos that would have normally formed anterior spinal cord has formed posterior hindbrain instead. A transformation of the spinal cord (posterior) to hindbrain (anterior) was predicted by the hypothesis that the loss of X1HBox1 function should mirror the posterior-to-anterior transformations observed in *Drosophila* when genes in the ANTP-UBX complex are mutant. Thus, these results with X1HBox1 are the best evidence at present that the homeobox-containing genes play a role in establishing different regions of the nervous system along the A/P axis.

SUMMARY

Genes that are differentially expressed in the ectoderm as it diverts along the neural and epidermal pathways of differentiation can be used to study the inducing signals underlying induction as well as how ectoderm responds to these signals by forming neural tissue. Although these genes have provided, and will continue to provide, new information about the induction process, they are not likely to provide the whole story. For example, *Notch* is a gene required for neurogenesis in *Drosophila* embryos. When the *Notch* gene product is absent, the embryo forms far too many neuroblasts at the expense of the hypodermal cell layer (Artavanis-Tsakonis 1988, Campos-Ortega 1988). Even though *Notch* appears to play a role in deciding the fate of neural/hypodermal cells, the *Notch* gene product is expressed ubiquitously in early embryos in the neurogenic region (Hartley et al 1987, Kidd et al 1989). Thus, one possibility is that *Notch* function does not necessarily depend on differential expression of the *Notch* gene product within cells in the neurogenic regions [although an alternative view has been suggested by Greenspan (1990)]. Thus, some of the molecules controlling early neural development may not be expressed differentially when the ectoderm forms the neural plate. Obviously, other approaches will have to be taken to isolate and characterize these molecules. In this light, it is noteworthy that a molecule has been identified in *Xenopus* that is remarkably similar to *Drosophila Notch* in both structure and developmental expression (Coffman et al 1990). One hope is that the analysis of this molecule in combination with the molecules that are differentially expressed during neural induction will eventually lead to a molecular understanding of early neural development in vertebrate embryos.

ACKNOWLEDGMENTS

I would like to thank Drs. Greg Lemke, Nancy Papalopulu, and Barbara Ranscht for critical reading of the manuscript and the National Institutes of Health for support. The author is a McKnight Scholar and a fellow of the Sloan Foundation.

Literature Cited

Akam, M. 1987. The molecular basis for Metameric pattern in the *Drosophila* embryo. *Development* 101: 1–22

Akam, M. 1989. Hox and HOM: Homologous gene clusters in insects and vertebrates. *Cell* 57: 347–49

Akers, R. M., Phillips, C. R., Wessels, N. K. 1986. Expression of an epidermal antigen used to study tissue induction in the early *Xenopus laevis* embryo. *Science* 231: 613–16

Albers, B. 1987. Competence as the main factor determining the size of the neural plate. *Develop. Growth Differ.* 29: 535–45

Angres, B., Muller, A. H. J., Kellermann, J., Hausen, P. 1991. Differential expression of two cadherins in *Xenopus laevis*. *Development* 111: 829–44

Artavanis-Tsakonis, S. 1988. The molecular biology of the *Notch* locus and the fine tuning of differentiation in *Drosophila*. *Trends Neurosci.* 4(4): 95–100

Campos-Ortega, J. A. 1988. Cellular interactions during early neurogenesis of *Drosophila melanogaster*. *Trends Neurosci.* 11: 400–3

Carrasco, A. E., Malacinski, G. 1987. Localization of *Xenopus* homeo-box gene transcripts during embryogenesis and in the adult nervous system. *Dev. Biol.* 121: 69–81

Carrasco, A. E., McGinnis, W., Gehring, W. J., DeRobertis, E. M. 1984. Cloning of a *X. laevis* gene expressed during early embryogenesis that codes for a peptide region homologous to *Drosophila* homeotic genes. *Cell* 37: 409–14

Charnas, L., Richter, K., Sargent, T., Dawid, I. 1987. Complementary DNA cloning of a nervous system specific intermediate filament from *Xenopus laevis* with homology to mammalian neurofilament. *Soc. Neurosci. Abstr.* 450: 15

Cho, K. W. Y., Goetz, J., Wright, C. V. E., Fritz, A., Hardwicke, J., DeRobertis, E. M. 1989. Differential utilization of the same reading frame in a *Xenopus* homeobox gene encodes two related proteins sharing the same DNA-binding specificity. *EMBO J.* 7: 2139–49

Choi, Y.-S., Sehgal, R., McCrea, P., Gumbiner, B. 1990. A cadherin-like protein in eggs and cleaving embryos of *Xenopus laevis* is expressed in oocytes in response to progesterone. *J. Cell Biol.* 110: 1575–82

Choi, Y.-S., Gumbiner, B. 1989. Expression of cell adhesion molecule E-cadherin in *Xenopus* embryos begins at gastrulation and predominates in the ectoderm. *J. Cell Biol.* 108: 2449–85

Coffman, C., Harris, W., Kintner, C. 1990. *Xotch*, the *Xenopus* homolog of *Drosophila Notch*. *Science* 249: 1438–41

Cunningham, B. A., Hemperly, J. J., Murray, B. A., Prediger, E. A., Brackenbury, R., Edelman, G. M. 1987. Neural cell adhesion molecule: Structure, immunoglobulin-like domains, cell surface modulation and alternative RNA splicing. *Science* 236: 799–806

Dagle, J. M., Walder, J. A., Weeks, D. L. 1990. Targeted degradation of mRNA in *Xenopus* oocytes and embryos mediated by modified antisense oligonucleotide. *Nucleic Acids Res.* 18: 4751–58

Detrick, R. J., Dickey, D., Kintner, C. R. 1990. The effect of N-cadherin misexpression on morphogenesis in *Xenopus* embryos. *Neuron* 4: 493–506

Dixon, J. E., Kintner, C. R. 1989. Cellular contacts required for neural induction in *Xenopus* embryos: Evidence for two signals. *Development* 106: 749–57

Doe, C. Q., Scott, M. P. 1988. Segmentation and homeotic gene function in the developing nervous system. *Trends Neurosci.* 11: 101–6

Duboule, D., Dolle, P. 1989. The murine *Hox* gene network: Its structural and functional organization resembles that of *Drosophila* homeotic genes. *EMBO J.* 8: 1507–8

Dworkin-Rastl, E., Kelley, D. B., Dworkin, M. B. 1986. Localization of specific mRNA sequences in *Xenopus laevis* embryos by *in situ* hybridization. *J. Embryol. Exp. Morph.* 91: 153–68

Edelman, G. M. 1986. Cell adhesion molecules in the regulation of animal form and tissue pattern. *Annu. Rev. Cell Biol.* 2: 81–116

Edelman, G. M. 1988. Morphoregulatory molecules. *Biochemistry* 27: 3533–43

Edelman, G. M., Gallin, W. J., Delovee, A., Cunningham, B. A. 1983. Early epochal maps of two different cell adhesion molecules. *Proc. Natl. Acad. Sci. USA* 80: 4384–88

Fujimora, T., Miyatani, S., Takeichi, M. 1990. Ectopic expression of N-cadherin perturbs histogenesis in *Xenopus* embryos. *Development* 110: 97–104

Gehring, W. J. 1987. Homeo boxes in the study of development. *Science* 236: 1245–52

Gerhart, J., Keller, R. 1986. Region-specific cell activities in amphibian gastrulation. *Annu. Rev. Cell Biol.* 2: 201–29

Gieblehaus, D. H., Eib, D. W., Moon, R. T. 1988. Antisense RNA inhibits expression of membrane skeleton protein 4.1 during

282   KINTNER

embryonic development of *Xenopus*. *Cell* 53: 601–15

Ginsberg, D., DeSimone, D., Geiger, B. 1991. Expression of a novel cadherin (EP-cadherin) in unfertilized eggs and early *Xenopus* embryos. *Development* 111: 315–26

Godsave, S. F., Anderton, B. H., Wylie, C. C. 1986. The appearance and distribution of intermediate filament proteins during differentiation of the central nervous system, skin and notochord of *Xenopus laevis*. *J. Embryol. Exp. Morph*. 97: 201–23

Gordon, R., Jacobson, A. G. 1978. The shaping of tissues in embryos. *Sci. Am*. 238: 106–13

Graham, A., Papalopulu, N., Krumlauf, R. 1989. The murine and *Drosophila* gene complexes have common features of organization and expression. *Cell* 57: 367–78

Greene, L. A. 1989. A new neuronal intermediate filament protein. *Trends Neurosci*. 12: 228–30

Greenspan, R. J. 1990. The *Notch* gene, adhesion, and developmental fate in the *Drosophila* embryo. *New Biol*. 2(7): 595–600

Gurdon, J. B. 1987. Embryonic induction-molecular prospects. *Development* 99: 285–306

Gurdon, J. B., Melton, D. A. 1981. Gene transfer in amphibian eggs and oocytes. *Annu. Rev. Genet*. 15: 189–208

Hamburger, V. 1988. *The Heritage of Experimental Embryology: Hans Spemann and the Organizer*. Oxford: Oxford Univ. Press

Harland, R., Weintraub, H. 1985. Translation of mRNA injected into *Xenopus* oocytes is specifically inhibited by antisense RNA. *J. Cell Biol*. 101: 1094–99

Hartenstein, V. 1989. Early neurogenesis in *Xenopus*: The spatio-temporal pattern of proliferation and cell lineages in the embryonic spinal cord. *Neuron* 3: 399–411

Hartley, D. A., Xu, T., Artavanis-Tsakonas, S. 1987. The embryonic expression of the *Notch* locus of *Drosophila melanogaster* and the implications of point mutations in the extracellular EGF-like domain of the predicted protein. *EMBO J*. 6(11): 3407–17

Harvey, R. P., Tabin, C. J., Melton, D. A. 1986. Embryonic expression and nuclear localization of *Xenopus Homeobox* (Xhox) gene products. *EMBO J*. 5: 1237–44

Hatta, K., Takeichi, M. 1986. Expression of N-cadherin adhesion molecules associated with early morphogenetic events in chick development. *Nature* 320: 447–49

Hemmati Brivanlou, A., de la Torre, J. R., Holt, C., Harland, R. M. 1991. Cephalic expression and molecular characterization of *Xenopus* EN-2. *Development* 111: 715–24

Hemmati Brivanlou, A., Harland, R. M. 1989. Expression of an *engrailed*-related protein is induced in the anterior neural ectoderm of early *Xenopus* embryos. *Development* 106: 611–17

Hemmati Brivanlou, A., Stewart, R., Harland, R. 1990. Region-specific neural induction of an engrailed protein by anterior notochord in *Xenopus*. *Science* 250: 800–2

Henry, J. J., Grainger, R. M. 1990. Early tissue interactions leading to embryonic lens formation in *Xenopus laevis*. *Dev. Biol*. 141: 149–63

Holland, P. W. H., Hogan, B. L. M. 1986. Phylogenetic distribution of *Antennapedia*-like homeo boxes. *Nature* 321: 251–53

Holtfreter, J. 1933. Die totale Exogastrulation, eine Selbstablosung des Ekotdrems von Entomesoderm. *Roux's Arch. Entw. Mech. Org*. 129: 669–793

Holtfreter, J., Hamburger, V. 1955. *Analysis of Development*. Philadelphia, PA: Saunders

Hunkapiller, T., Hood, L. 1986. The growing immunoglobin gene superfamily. *Nature* 323: 15

Itoh, K., Yamashita, A., Kubota, H. Y. 1988. The expression of epidermal antigens in *Xenopus laevis*. *Development* 104: 1–14

Izant, J. G., Weintraub, H. 1984. Inhibition of thymidine kinase gene expression by antisense RNA: A molecular approach to genetic analysis. *Cell* 36: 1007–15

Jacobson, A. G., Sater, A. K. 1988. Features of embryonic induction. *Development* 104: 341–57

Jacobson, M., Rutishauser, U. 1986. Induction of neural cell adhesion molecule (N-CAM) in *Xenopus* embryos. *Dev. Biol*. 116: 524–31

Jamrich, M., Sargent, T. D., Dawid, I. B. 1987. Cell-type-specific expression of epidermal cytokeratin genes during gastrulation of *Xenopus laevis*. *Genes Develop*. 1: 124–32

Jamrich, M., Sato, S. 1989. Differential gene expression in the anterior neural plate during gastrulation of *Xenopus laevis*. *Development* 105: 779–86

Jessell, T. M. 1988. Adhesion molecules and the hierarchy of neural development. *Neuron* 1: 3–13

Jonas, E., Sargent, T. D., Dawid, I. 1985. Epidermal keratin gene expressed in embryos of *Xenopus laevis*. *Proc. Natl. Acad. Sci. USA* 82: 5413–17

Jonas, E. A., Snape, T. D., Sargent, T. D.

1989. Transcriptional regulation of a *Xenopus* embryonic keratin gene. *Development* 106: 399–405

Jones, E. A., Woodland, H. R. 1986. Development of the ectoderm in *Xenopus*: Tissue specification and the role of cell association and division. *Cell* 44: 345–55

Jones, E. A., Woodland, H. R. 1989. Spatial aspects of neural induction in *Xenopus laevis*. *Development* 107: 785–91

Jones, N. 1990. Transcriptional regulation by dimerization: Two sides to an incestuous relationship. *Cell* 61: 9–11

Joyner, A. L., Martin, G. R. 1987. En-1 and En-2, two mouse genes with sequence homology to the *Drosophila engrailed* gene: Expression during embryogenesis. *Genes Dev.* 1: 29–38

Kay, B. K., Schwartz, L. M., Rutishauser, U., Qui, T. H., Peng, H. B. 1988. Patterns of NCAM expression during myogenesis in *Xenopus laevis*. *Development* 103: 463–72

Keller, R., Danilchik, M. 1988. Regional expression, pattern and timing of convergence and extension during gastrulation of *Xenopus laevis*. *Development* 103: 193–209

Kidd, S., Baylies, M. K., Gasic, G. P., Young, M. W. 1989. Structure and distribution of the *Notch* protein in developing *Drosophila*. *Genes Devel.* 3: 1113–29

Kintner, C. R. 1988. Effects of altered expression of the neural cell adhesion molecule, NCAM, on early neural development in *Xenopus* embryos. *Neuron* 1: 545–55

Kintner, C. R., Melton, D. M. 1987. Expression of *Xenopus* N-CAM RNA is an early response of ectoderm to induction. *Development* 99: 311–25

Levi, G., Crossin, K. L., Edelman, G. 1987. Expression sequences and distribution of two primary cell adhesion molecules during embryonic development of *Xenopus laevis*. *J. Cell Biol.* 105: 2359–72

Levi, G., Gumbiner, B., Thiery, J. P. 1991. The distribution of E-cadherin during *Xenopus laevis* development. *Development* 111: 145–58

London, C., Akers, R., Phillips, C. 1988. Expression of Epi 1, an epidermis-specific marker in *Xenopus laevis* embryos, is specified prior to gastrulation. *Dev. Biol.* 129: 380–89

McGinnis, W., Levine, M., Hafen, E., Kuroiwa, A., Gehring, W. J. 1984a. A homologous protein coding sequence in *Drosophila* homeotic genes and its conservation in other metazoans. *Nature* 308: 428–33

McGinnis, W., Levine, M. S., Hafen, E., Kuroiwa, A., Gehring, W. 1984b. A conserved DNA sequence in homeotic genes of the *Drosophila* antennapedia and bithorax complexes. *Nature* 308: 428–33

Melton, D. A. 1985. Injected antisense RNAs specifically block messenger RNA translation *in vivo*. *Proc. Natl. Acad. Sci. USA* 82: 144–48

Miyatani, S., Winkles, J. A., Sargent, T. D., Dawid, I. B. 1986. Stage-specific keratins in *Xenopus laevis* embryos and tadpoles: The XK81 gene family. *J. Cell Biol.* 103: 1957–65

Nagafuchi, A., Takeichi, M. 1988. Cell binding function of E-cadherin is regulated by the cytoplasmic domain. *EMBO J.* 7(12): 3679–84

Nagafuchi, A., Takeichi, M. J. 1989. Transmembrane control of cadherin-mediated cell adhesion: A 94 kDa protein functionally associated with a specific region of the cytoplasmic domain of E-cadherin. *Cell Reg.* 1: 37–44

Oliver, G., Wright, C. V. E., Hardwicke, J., DeRobertis, E. M. 1988. Differential anteroposterior expression of two proteins encoded by a homeobox gene in *Xenopus* and mouse embryos. *EMBO J.* 7: 3199–3209

Osborn, M., Weber, K. 1982. Intermediate filaments: Cell-type specific markers in differentiation and pathology. *Cell* 31: 303–6

Ozawa, M., Baribault, H., Kemler, R. 1989. The cytoplasmic domain of the cell adhesion molecule uvomorulin associates with three independent proteins structurally related in different species. *EMBO J.* 8: 1711–17

Ozawa, M., Kemler, R. 1990. Correct proteolytic cleavage is required for the cell adhesive function of Uvomorulin. *J. Cell Biol.* 111: 1645–50

Patel, N. H., Martin-Blanco, E., Coleman, K. G., Poole, S. J., Ellis, M. C., et al. 1989. Expression of *engrailed* proteins in arthropods, annelids and chordates. *Cell* 58: 955–68

Rebagliati, M. R., Melton, D. A. 1987. Antisense RNA injections in fertilized frog eggs reveals an RNA duplex unwinding activity. *Cell* 48: 599–605

Richter, K., Grunz, H., Dawid, I. B. 1988. Gene expression in the embryonic nervous system of *Xenopus laevis*. *Proc. Natl. Acad. Sci. USA* 85: 8086–90

Ruiz i Altaba, A. 1990. Neural expression of *Xenopus* homeobox gene Xhox3 during embryonic development. *Development* 108: 595–604

Ruiz i Altaba, A., Melton, D. A. 1989. Bimodal and graded expression of the *Xenopus* homeobox gene *Xhox3* during embryonic development. *Development* 106: 173–83

Savage, R., Phillips, C. R. 1989. Signals from the dorsal blastopore lip region during gastrulation bias the ectoderm toward a nonepidermal pathway of differentiation in *Xenopus laevis*. *Dev. Biol.* 133: 157–68

Saxen, L. 1989. Neural induction. *Int. J. Dev. Biol.* 33: 21–48

Sharpe, C. R. 1988. Developmental expression of a neurofilament-M and two vimentin-like genes in *Xenopus laevis*. *Development* 103: 269–77

Sharpe, C. R., Fritz, A., DeRobertis, E. M., Gurdon, J. B. 1987. A homeobox-containing marker of posterior neural differentiation shows the importance of predetermination in neural induction. *Cell* 49: 749–58

Sharpe, C. R., Gurdon, J. B. 1990. The induction of anterior and posterior neural genes in *Xenopus laevis*. *Development* 109: 765–74

Sharpe, C. R., Pluck, A., Gurdon, J. B. 1989. XIF3, a *Xenopus* peripherin gene, requires an inductive signal for enhanced expression in anterior neural tissue. *Development* 107: 701–14

Sive, H. L., Hattori, K., Weintraub, H. 1989. Progressive determination during formation of the anteroposterior Axis in *Xenopus laevis*. *Cell* 58: 171–80

Snape, A. M., Jonas, E. A., Sargent, T. D. 1990. KTF-1, a transcriptional activator of *Xenopus* embryonic keratin expression. *Development* 109: 157–66

Sokol, S., Melton, D. A. 1991. Preexistent pattern in *Xenopus* animal pole revealed by induction with Activin. *Nature* 351: 409–11

Spemann, H. 1938. *Embryonic Development and Induction*. New York: Hafner

Steinert, P. M., Roop, D. R. 1988. Molecular and cellular biology of intermediate filaments. *Annu. Rev. Biochem.* 57: 593–625

Takeichi, M. 1988. The cadherins: Cell-cell adhesion molecules controlling animal morphogenesis. *Development* 102: 639–55

Takeichi, M. 1991. Cadherin cell adhesion receptors as a morphogenetic regulator. *Science* 251: 1451–55

Tapscott, S. J., Bennett, G. S., Holtzer, H. 1981a. Neuronal precursor cells in the chick neural tube express neurofilament proteins. *Nature* 292: 836–38

Tapscott, S. J., Bennett, G. S., Toyama, Y., Kleinbart, F., Holtzer, H. 1981b. Intermediate filament proteins in developing chick spinal cord. *Dev. Biol.* 86: 40–54

Townes, P. L., Holtfreter, J. 1955. Directed movements and selective adhesion of embryonic amphibian cells. *J. Exp. Zool.* 128: 53–120

Wetts, R., Fraser, S. E. 1989. Slow intermixing of cells during *Xenopus* embryogenesis contributes to the consistency of the blastomere fate map. *Development* 105: 9–15

Woolf, T. M., Jennings, C. G., Rebagiati, M., Melton D. A. 1990. The stability, toxicity, effectiveness of unmodified and phosphorothioate antisense oligodeoxynucleotides in *Xenopus* oocytes and embryos. *Nucleic Acids Res.* 18: 1763–69

Wright, C. V. E., Cho, K. W. Y., Hardwicke, J., Collins, R. H., DeRobertis, E. M. 1989. Interference with function of a homeobox gene in *Xenopus* embryos produces malformations of the anterior spinal cord. *Cell* 59: 81–93

Wright, C. V. E., Morita, E. A., Wilkin, D. J., DeRobertis, E. M. 1990. The *Xenopus* X1HBox6 homeo protein, a marker of posterior neural induction is expressed in proliferating neurons. *Development* 109: 225–34

Annu. Rev. Neurosci. 1992. 15:285–320

THE NEOSTRIATAL MOSAIC:
Multiple Levels of Compartmental Organization in the Basal Ganglia[1]

Charles R. Gerfen

Laboratory of Cell Biology, National Institute of Mental Health, Bethesda, Maryland 20892

KEY WORDS: Basal ganglia, striatum, dopamine receptors, Parkinson's disease

INTRODUCTION

The basal ganglia provide a major neural system through which the cortex effects behavior. Most notable among these effects are those related to the voluntary control of movement, which is compromised by neurodegenerative diseases that involve the basal ganglia. Two such diseases, Parkinson's disease and Huntington's chorea, display a spectrum of movement impairment (Albin et al 1989). Parkinson's disease, which results in the degeneration of dopaminergic systems in the basal ganglia, produces a disability to initiate desired movements. On the other hand, Huntington's chorea, which results in the degeneration of the major projection neurons of the basal ganglia, is characterized by uncontrolled movements. The complexity of these and other disorders that accompany basal ganglia dysfunction suggest its broad role in the subtlest components of voluntary movement. That memory, motivational, and emotional aspects of movement behavior are affected by this neural system is related to the fact that the striatum, which is the principal component of the basal ganglia, receives inputs from virtually all cortical areas (Carman et al 1965; Kemp & Powell 1970; Webster 1961), including limbic-related areas (Heimer & Wilson 1975). How the striatum processes cortical inputs is central to the function of the basal ganglia.

---

[1] The US Government has the right to retain a nonexclusive, royalty-free license in and to any copyright covering this paper.

285

The organization of corticostriatal systems has been described in terms of parallel processing mechanisms (Alexander et al 1986). According to this scheme, segregated parallel circuits connect limbic, prefrontal, oculomotor, and motor cortical areas through subregions of the basal ganglia with ventral tier thalamic nuclei that feed back to those same cortical areas. These functional circuits provide a conceptual framework for relating subregions of the basal ganglia to specific aspects of behavior, much as cytoarchitecturally defined cortical areas are defined. Disinhibition has been proposed as the basic mechanism by which these basal ganglia circuits affect behavior (Chevalier et al 1985, Deniau & Chevalier 1985, Hikosaka & Wurtz 1983b). Basal ganglia output nuclei, the entopeduncular nucleus (internal globus pallidus in primates) and the substantia nigra pars reticulata, provide tonic inhibition of ventral tier thalamic nuclei and the superior colliculus. Disinhibition of these inhibitory pathways results from cortical excitation of inhibitory striatal projections to the entopeduncular nucleus and substantia nigra (Chevalier et al 1985, Deniau & Chevalier 1985). Thus, movements initiated in response to sensory, memory, or motivationally contingent cues have been correlated with pauses in the tonic activity of substantia nigra output neurons (Hikosaka & Wurtz 1983a,b). The opposed tonic activity of these neurons is regulated, in part, by the striatal outputs to the globus pallidus (external segment of the globus pallidus in primates). Increased activity in this pathway results, by way of polysynaptic connections through the globus pallidus and subthalamic nucleus (Kita et al 1983, Kita & Kitai 1987a,b), in increased activity of entopeduncular (internal globus pallidus in primates) and substantia nigra neurons. The important contribution of increased tonic activity of these neurons through this indirect pathway to the execution of movements has been suggested (Alexander & Crutcher 1990). Balanced opposition of cortically driven striatal output systems thus appears to be responsible for the generation of normal movements.

The neuroanatomical substrates that underlie the processing of cortical inputs to produce this balanced opposition reflect organizational schemes that are common to most regions of the striatum. Unlike the cortex, the striatum lacks the clear cytoarchitectural definition that has aided analysis of information processing through layer-specific cortical connections (Jones 1984). Although the striatum lacks cytoarchitectural definition, subpopulations of striatal neurons are organized in functional compartments. This compartmental organization can be shown in some cases to be comparable to cytoarchitectural features of the cortex and in other cases to represent an organization unique to the striatum. Two such levels of compartmental organization are described that are not exclusive but represent overlapping sets of subpopulations of striatal output neurons.

The first level of compartmental organization is defined by the segregation of striatal output neurons into patch and matrix compartments (Bolam et al 1988, Gerfen 1985, Kawaguchi et al 1989), which are related to laminar and regional aspects of cortical organization (Donoghue & Herkenham 1986, Gerfen 1984, 1989). The second level of compartmental organization is the separation of projections to the external segment of the globus pallidus and to the substantia nigra (Albin et al 1989, Gerfen et al 1990, Kawaguchi et al 1990). The major output neurons of the basal ganglia are the GABAergic neurons of the substantia nigra pars reticulata and the entopeduncular nucleus (internal segment of the globus pallidus in primates). For this review these two cell groups are considered part of the same extended nucleus and striatonigral projections refer to inputs to this extended nucleus. Neurons giving rise to striatopallidal and striatonigral projections are separate and intermingled in both striatal patch and matrix compartments. The organization of striatopallidal and striatonigral systems provides for the conversion of the excitatory inputs to the striatum into balanced opposed modulation of output neurons of the basal ganglia (Albin et al 1989, Gerfen et al 1990). The cellular and molecular mechanisms responsible for regulating the balance between these systems establish the capacity for the basal ganglia to affect the selection of specific behaviors.

GENERAL ORGANIZATION OF THE BASAL GANGLIA

The principal component of the basal ganglia is the striatum, which comprises the caudate, putamen, and accumbens nuclei. One neuron cell type, the medium spiny neuron, accounts for 90–95% of the striatal neuron population. These neurons have a medium-sized cell body, approximately 20–25 μm in diameter, from which radiate branched dendrites that are densely laden with spines (Kemp & Powell 1971, Wilson & Groves 1980). The dendritic arbors extend in a domain approximately 150–250 μm in diameter such that neighboring neurons share common inputs. As they are the predominant neuron cell type in the striatum, medium spiny neurons are the major target of extrinsic afferents. Cortical and thalamic inputs provide excitatory inputs that make asymmetric synaptic contact mainly with the heads of the spines (Bouyer et al 1984, Hattori et al 1978, Somogyi et al 1981). Dopamine fibers from the midbrain cell groups are the other major source of extrinsic input to these neurons, which make symmetric synaptic contact primarily with the necks of dendritic spines and on the interspine dendritic shafts (Bouyer et al 1984, Freund et al 1984). Other inputs to medium spiny neurons come primarily from two major sources within the striatum, from the local axon collaterals of other medium spiny neurons or from striatal interneurons (Wilson & Groves 1980).

Intrinsic neurons, whose axons do not exit the striatum, comprise about 10% of the striatal neuronal population and have a profound role in striatal organization. Among the neurons that have been clearly identified in this class are the large aspiny neurons that utilize acetylcholine as a transmitter (Bolam et al 1984), and several types of medium aspiny neurons, which include those that contain somatostatin (DiFiglia & Aronin 1982) and neuropeptide Y (Vincent & Johansson 1983) and those that contain the calcium-binding protein parvalbumin (Cowan et al 1990, Gerfen et al 1985).

The basic organizational scheme of the basal ganglia is diagrammed in Figure 1. Pyramidal cortical neurons located primarily in layer 5, but also some in layers 2 and 3 and 6, provide inputs to striatum (Ferino et al 1987, Jones et al 1977, Royce 1982, Wilson 1987). These inputs utilize the amino acid glutamate as a neurotransmitter (Spencer 1976), and thus provide excitatory inputs to the striatal medium spiny neurons (Kitai et al 1976). Neurons in the intralaminar thalamic nuclei and the parafascicular nucleus

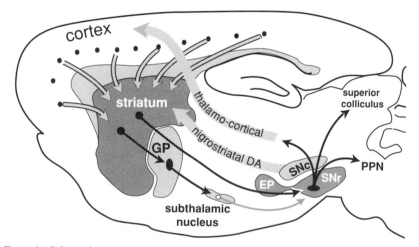

Figure 1   Schematic representation of the major connections of the basal ganglia. The main component of the basal ganglia, the striatum, receives inputs from the cortex. Two major striatal output pathways target the globus pallidus (GP) and entopeduncular (EP)–substantia nigra pars reticulata (SNr) complex. Dopamine (DA) neurons in the substantia nigra pars compacta (SNc) receive inputs from the striatum (not diagrammed) and provide feedback via the nigrostriatal DA pathway. EP and SNr neurons provide inhibitory inputs to the pedunculopontine nucleus (PPN), superior colliculus, and thalamus. Nigrothalamic inputs target intralaminar nuclei that provide feedback to the striatum (not shown) and ventral tier thalamic nuclei that provide inputs to the frontal cortex.

(Gerfen et al 1982, Herkenham & Pert 1981, Krettek & Price 1977) also provide excitatory inputs to striatal medium spiny neurons (Kitai et al 1976). Striatal medium spiny neurons are the primary source of striatal projections (Grofova 1975). These neurons are normally relatively quiescent, and activity in the output pathways is generated as a result of cortical and thalamic excitatory inputs (Kitai et al 1976, Wilson 1981). Another major source of inputs to the striatum is from the midbrain dopaminergic cell groups in the ventral tegmental area, substantia nigra, and retrorubral area (Freund et al 1984, Gerfen et al 1987, Jimenez-Castellanos & Graybiel 1987). As is discussed in some detail below, this dopaminergic input appears to modulate the responsiveness of striatal output neurons to cortical and thalamic inputs. Medium spiny striatal neurons utilize γ-amino butyric acid (GABA) as their principal transmitter (Kita & Kitai 1988) and thus provide inhibition to the targets of striatal output in the globus pallidus, entopeduncular nucleus (the internal segment of the globus pallidus in primates), and the substantia nigra (Chevalier et al 1985, Deniau & Chevalier 1985, Kita & Kitai 1988). The principal neuron type in the globus pallidus is GABAergic and provides inhibitory inputs to the subthalamic nucleus and to the substantia nigra (Smith et al 1990). Neurons in the subthalamic nucleus provide an excitatory input to the substantia nigra (Kita & Kitai 1987a,b, Nakanishi et al 1988). This latter input to the substantia nigra represents an indirect pathway, which is responsible, in part, for the tonic activity of GABAergic neurons in the substantia nigra pars reticulata. The direct striatonigral pathway provides principally inhibitory inputs to both the dopaminergic and GABAergic neurons in the substantia nigra (Chevalier et al 1985, Deniau & Chevalier 1985). As mentioned, nigral dopaminergic neurons are the source of the feedback pathway to the striatum. Nigral GABAergic neurons, located in the substantia nigra pars reticulata and in the entopeduncular nucleus, provide inhibitory inputs to the intermediate layers of the superior colliculus, the pedunculopontine nucleus, and the thalamus (Gerfen et al 1982).

Many connections have been omitted from the general scheme described so as to focus on the basic elements detailed in this review. Two levels of compartmental organization that may be overlain on the general scheme of basal ganglia connections are described. The first is the patch-matrix compartmental organization of the striatum and the second is the organization of separate striatopallidal and striatonigral systems.

PATCH-MATRIX STRIATAL COMPARTMENTS

The striatal patch and matrix compartments are defined on the basis of specific neurochemical markers and the connections of the underlying striatal neurons. The striatal patch compartment is defined by areas of dense μ-opiate receptor binding (Herkenham & Pert 1981) and areas of low acetylcholinesterase labeling, also referred to as striosomes (Graybiel & Ragsdale 1978). The striatal matrix compartment, which is complementary to the patches, is composed of neurons that contain a 28 kD calcium-binding protein (calbindin) and a rich plexus of somatostatin immunoreactive fibers (Gerfen 1985, Gerfen et al 1985). These neurochemical markers display a consistent complementary pattern in the rat throughout the majority of both the dorsal and ventral striatum, although within the medial aspect of the nucleus accumbens these patterns are not as distinct (Voorn et al 1989). Thus, these markers may be used to define the mosaic organization of the striatum into distinct patch and matrix compartments. These compartments have been well characterized in the rat, and a similar organization obtains in the primate and cat. Other neurochemical markers, most notably the immunohistochemical distribution of the peptides enkephalin and substance P (Beckstead & Kersey 1985, Gerfen 1984, Graybiel et al 1981), and some of the input (Malach & Graybiel 1986) and output (Gimenez-Amaya & Graybiel 1990) connections of the striatum, display patterns of heterogeneity in the striatum that sometimes are and sometimes are not consistent with the patch and matrix compartments as defined here. As is discussed below, these patterns reveal multiple levels of compartmental organization, only some of which are related to the patch-matrix compartments.

Striatal Patch-matrix Output Systems

Although the neurochemical markers described above have been useful in defining the patch-matrix compartments, this organization appears to be related to the segregation of separate populations of striatal medium spiny neurons that have distinct input-output connections (Gerfen 1984, 1985, Gerfen et al 1985, Kawaguchi et al 1989). In rats, retrograde axonal tracing studies show that both patch and matrix neurons project to the substantia nigra, but that patch neurons provide inputs to the location of dopaminergic cells, most specifically to the ventral tier of dopaminergic neurons in the pars compacta and dopaminergic cell islands in the pars reticulata, whereas matrix neurons provide inputs to the location of the GABAergic neurons in the substantia nigra pars reticulata (Gerfen 1984, 1985, Gerfen et al 1985). Calbindin immunoreactivity, which labels striatonigral projection neurons in the matrix, confirms, by its specific distribution in terminals in the non-dopaminergic parts of the substantia nigra pars reticulata, the patch-matrix organization of the striatonigral pathway (Gerfen et al 1985). A similar pattern of calbindin immunoreactivity in the striatonigral pathway of primates (Gerfen et al 1985) and a later study in the cat (Jimenez-Castellanos & Graybiel 1989) suggests that a similar dichotomy of patch and matrix striatonigral pathways occurs in these species as well. Multiple studies have demonstrated that the dendrites of patch and matrix medium spiny neurons remain restricted, for the most part, to the compartment of the parent neuron (Bolam et al 1988, Gerfen 1985, Herkenham et al 1984, Kawaguchi et al 1989). This is an important characteristic, as it suggests that inputs that are confined to the patch compartment will affect only patch output neurons, whereas inputs directed to the matrix will affect matrix output neurons.

Dopaminergic Patch-matrix Input Systems

Dopaminergic neurons in the midbrain ventral tegmental area, substantia nigra, and retrorubral area provide a massive input to the striatum that is compartmentally organized (Gerfen et al 1987a, Herkenham et al 1984a, Jimenez-Castellanos & Graybiel 1987). Matrix-directed dopaminergic projections arise from a continuous group of neurons that are located in the ventral tegmental area, the dorsal tier of the substantia nigra pars compacta, and the retrorubral area. Patch-directed dopaminergic neurons arise from ventral substantia nigra pars compacta neurons and from dopaminergic cells clustered in islands in the pars reticulata. Moreover, the matrix-directed dopaminergic neurons also express calbindin immunoreactivity, whereas dopaminergic neurons projecting to the patches do not (Gerfen et al 1987b). These studies establish the existence of separate dorsal and ventral tier dopaminergic mesostriatal systems that are directed to the striatal patch and matrix compartments and that are also biochemically distinct. An additional feature of this organization is that the dorsal tier dopaminergic neurons have dendrites that remain confined to the area of the dopaminergic neurons, whereas the ventral tier, patch-directed neurons have dendrites that extend into the parts of the substantia nigra pars reticulata where GABAergic neurons are located (Gerfen et al 1987). This is of possible interest, given the reports that the latter dopaminergic dendrites have been reported to release dopamine (Cheramy et al 1981). As described above, the output of the striatal patch compartment specifically targets this ventral set of dopaminergic neurons. Whether the significance of this input is related to the dendritic release of dopamine by these neurons remains to be determined. It is of interest that the dorsal tier dopaminergic neurons that provide the major input to the striatal matrix receive little in the way of inputs from the striatum, but rather receive inputs from other areas, such as the amygdala (Gonzales & Chesselet 1990).

Corticostriatal Patch-matrix Inputs

Several studies have demonstrated that corticostriatal inputs are heterogeneously distributed within the striatum (Donoghue & Herkenham 1986, Gerfen & Sawchenko 1984, Goldman-Rakic 1982, Ragsdale & Graybiel

1981). A recent study of the compartmental organization of the cortico-striatal projection in the rat suggests that the patch-matrix organization of the striatum is related to the laminar organization of the cortex (Gerfen 1989). The majority of corticostriatal neurons are located in layer 5 and in the deeper parts of layer 3. By using the anterograde axonal tracer PHA-L (Gerfen & Sawchenko 1984), it was possible to examine the projections of sublaminae of layer 5 to the striatum. Results demonstrated that cortico-striatal neurons in the deep parts of layer 5 project preferentially to the striatal patch compartment, whereas projections from upper layer 5 and from supragranular layers project primarily to the striatal matrix compartment. Although all cortical areas examined project to both striatal compartments, the relative input to each compartment varies such that periallocortical areas provide a dense input to the patch compartment, whereas the input from neocortical areas to the patch compartment is relatively sparse. Retrograde tracing studies suggest that the trends in the relative contribution of corticostriatal inputs to the two compartments is related to the numbers of deep versus superficial layer 5 corticostriatal neurons in periallo- and neocortical areas; neocortical areas have relatively few deeper layer 5 neurons that contribute inputs to the striatum (Ferino et al 1987, Wilson 1987). The relative paucity of inputs from neocortical areas to the patch compartment may explain reports from tracing studies in primates in which injections into neocortical areas appear to label inputs primarily to the matrix compartment (Goldman-Rakic 1982). In this regard, it may be significant that the area of the primate putamen that receives inputs from the highly evolved neocortical areas, the primary motor and sensory cortices, appears to have a very diminished patch compartment. Thus, in primates, in which neocortical areas are greatly expanded as compared with rodents, the relative contribution of cortical inputs to the striatal patch compartment may be diminished. Nonetheless, the laminar organization of corticostriatal inputs would appear to be a fundamental determinant of the functional significance of the striatal patch and matrix compartments (Figure 2).

Relating the patch-matrix organization of the striatum to the laminar organization of the cortex merely begs the question as to the functional significance of cortical output organization. The laminar organization of the cerebral cortex is related to the aggregation of pyramidal neurons that have common axonal projection targets (Gilbert & Kelly 1975, Jones 1984). Thus, layer 6 cortical output neurons provide inputs to the thalamus, layer 5 neurons provide inputs to other subcortical structures, including the striatum, brainstem, and spinal cord, and supragranular layer neurons are the major source of corticocortical projections. In this regard a number of different subtypes of corticostriatal neurons have been described that

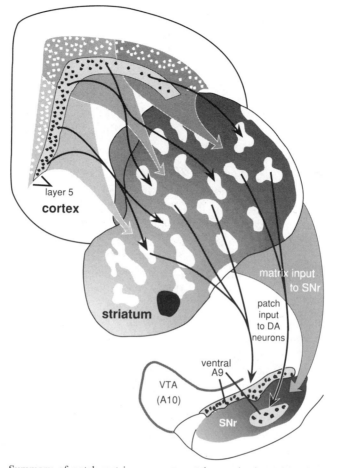

Figure 2 Summary of patch-matrix compartmental organization of corticostriatal and striatonigral pathways. Corticostriatal neurons in the deep parts of layer 5 provide inputs to the striatal patch compartment, whereas superficial layer 5 neurons provide inputs to the striatal matrix. Patch neurons provide inputs to the location of dopaminergic neurons in the ventral tier of the substantia nigra pars compacta and islands of dopamine neurons in the pars reticulata. Striatal matrix neurons provide inputs to the location of GABAergic neurons in the substantia nigra pars reticulata (SNr).

differ on the basis of axon collaterals to other brain sites, such as to the cerebellum, pyramidal tract, thalamus, contralateral cortex, and contralateral striatum (Donoghue & Kitai 1981, Jinnai & Matsuda 1979, Royce & Bromley 1984, Wilson 1987). Corticostriatal neurons also show differences in local axon collaterals with different patterns of laminar and regional spread (Wilson 1987). Although the laminar and sublaminar distributions of subtypes of corticostriatal neurons have not been established, these types of connectional distinctions presumably underlie the functional significance of the laminar origins of corticostriatal projections to the striatal patch and matrix compartments. There are other aspects of cortical organization, such as the radial columnar organization, that may contribute to heterogeneously distributed striatal afferents. Such organization is not to be confused, however, with the laminar relationship to the striatal patch-matrix organization.

Functional Significance of Patch-matrix Compartmentation

Previous studies have suggested a relationship between limbic-related areas, such as the amygdala (Ragsdale & Graybiel 1988) and prelimbic cortex (Donoghue & Herkenham 1986, Gerfen 1984), and the striatal patches and between sensorimotor cortical areas and the striatal matrix (Donoghue & Herkenham 1986, Gerfen 1984). In the rat, however, it appears that both prelimbic and sensorimotor cortical areas provide inputs to both compartments (Gerfen 1989). Nonetheless, the relative contribution of inputs to the striatal compartments from limbic, or periallocortical, and neocortical areas varies markedly, such that the limbic association with the patch compartment and the neocortical association with the matrix compartment may have some functional utility. On the one hand, such functions may be considered in terms of the regional connections of the striatum. Allo- and periallocortical areas project principally to the ventral striatum, and neocortical areas project principally to the dorsal striatum (Heimer & Wilson 1975). It is on the basis of this regional organization that the dichotomy in limbic- and nonlimbic-related striatal regions have been considered (Haber et al 1985, Heimer & Wilson 1975). Yet such generalizations raise the difficulties of defining limbic and nonlimbic functions. Although the concept of the limbic system has evolved, the original neuroanatomical, functional, and phylogenetic dichotomy of medial, allo-, and periallocortical areas subserving emotional and visceral functions and lateral, neocortical areas subserving cognitive and sensorimotor functions is generally still accepted (Swanson 1987). As Swanson points out, however, recent appreciation of neuroanatomical interrelationships have blurred the simple dichotomy of such functions. As both allo- and periallocortical and neocortical areas provide inputs to the striatum, in this structure ascribing a functional role to particular inputs is most difficult. Another approach is suggested. Rather than attempt to define the type of information conveyed to the striatum in behavioral terms, it is suggested that regional transitions in both cortical and basal ganglia organization be examined.

Although discussion of the issues of cortical evolution is beyond the scope of this review, it is probably noncontroversial to suggest that a major transition from allo- to neocortical areas is characterized by increased specialization of function typified by increased laminar organization. This perhaps has been best discussed in the context of the evolution of thalamo-cortical connections, i.e. nonspecific projection systems have been supplanted through evolution with more specific connections (Herkenham 1986). In the context of corticostriatal systems, as discussed above, there is a transition from periallocortical to neocortical areas of deep layer 5 projections to the patch compartment to superficial layer 5 projections to the matrix compartment (Gerfen 1989). In the striatum, this transition of corticostriatal patch-matrix systems is exemplified by the relative paucity of the patch compartment in striatal areas receiving neocortical inputs, particularly in the putamen of the primate. There are other indices of regional transition in the striatum. For example, two neurochemical markers of the matrix compartment, somatostatin and calbindin immuno-reactivity, are highest in the ventral striatum and diminish dorsally (Gerfen et al 1985). Other peptides, such as substance P and dynorphin, also show dorsal-ventral regional variations, which are discussed below. A regional transition occurs in the composition of the neurons of the globus pallidus: The dorsal pallidum is composed of mainly GABAergic neurons with infrequent cholinergic neurons, whereas in the ventral pallidum this mixture of neurons is reversed (Grove et al 1986). Peptides expressed in striatopallidal afferents show regional variations; enkephalin is densest in terminals in the dorsal pallidum, whereas substance P is densest in terminals in the ventral pallidum (Alheid & Heimer 1988, Haber & Nauta 1983). Finally, in the substantia nigra, dopaminergic neurons are concentrated in the dorsal pars compacta and ventral tegmental area, whereas GABAergic neurons are distributed in the ventrally located pars reticulata. These regional variations in basal ganglia organization are shown in a diagram, including the patch-matrix compartmental connections, in Figure 3. One of the connections in this diagram is hypothesized. Although it is known that the striatum provides inputs to cholinergic neurons in the pallidum (Grove et al 1986), it is not known whether these arise from patch compartment neurons.

The regional variations in basal ganglia organization are suggested to be an extension of the transition of periallocortical to neocortical subcortical circuits. Alheid & Heimer (1988) have proposed that the subcortical connections of allo- and neocortical areas share common general organizational schemes whose specific elements reflect a transition in the final targets of these systems and in the feedback mechanisms they employ. The regional transitions in the basal ganglia may be considered in this context.

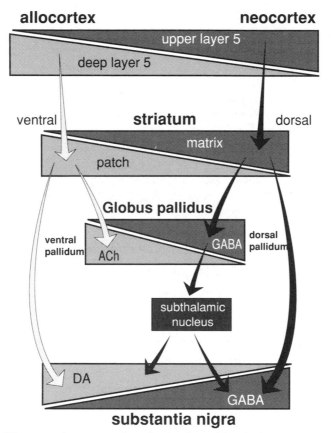

Figure 3   Diagrammatic representation of regional transitions in the components of patch-matrix compartmental connections. See text for details. (ACh, acetylcholine; DA, dopamine; ENK, enkephalin; SP, Substance P.)

Some outputs of the ventral striatum target cholinergic (Grove et al 1986) and dopaminergic neurons (Nauta et al 1978), two neurochemical systems that provide direct feedbacks to the cortex and striatum. Although there appears to be some cholinergic input to the reticular nucleus of the thalamus from the basal forebrain cell groups (Hallanger et al 1987), the majority of cholinergic neurons in the ventral and dorsal pallidum appear to provide principally ascending cortical projections (Grove 1988, McKinney et al 1983, Saper 1984). Conversely, the main outputs of the dorsal striatum target GABAergic neurons in the globus pallidus and substantia nigra and, through the thalamic connections of the latter, provide a more indirect feedback to the cortex and striatum (Wilson 1990). It is suggested that these two types of circuits are contained in the connections of the patch and matrix compartments, respectively (Gerfen 1984, 1985). In many regions of the striatum, both of these circuits exist. Rather than specify that the patch compartment conveys limbic information in dorsolateral patches, it is suggested that the type of neuroanatomical circuitry that is typical of the "limbic-ventral" striatum is retained in the dorsal striatum in the connections of the patch compartment. This is analogous to the emergence of neocortical organization in which the type of connections of older cortical areas is retained to some extent but supplanted by new features of organization (Herkenham 1986). Thus striatal patch-matrix compartments may be viewed as two phylogenetically distinct neuroanatomical circuits through which cortical information is processed. Regionally, the mix of these two circuit systems varies such that in the ventromedial striatum, allo- and periallocortical circuitry dominates, whereas in the dorsolateral striatum, neocortical circuitry dominates. In much of the striatum the two circuits coexist, and interactions between them may provide mechanisms for regulating the balance in the striatopallidal and striatonigral systems.

STRIATOPALLIDAL AND STRIATONIGRAL SYSTEMS

As described above, striatal medium spiny neurons are segregated into separate populations that form the basis of the patch and matrix compartments, whose connections are related to the laminar (Gerfen 1989) and regional (Donoghue & Herkenham 1986, Gerfen 1984) organization of the cortex. Medium spiny neurons may also be categorized on the basis of their projections to the globus pallidus, entopeduncular nucleus, and substantia nigra. In a recent study (Kawaguchi et al 1990), medium spiny neurons were intracellularly injected with biocytin, and both the local axon collateral and projection axons of these neurons were traced. Many neurons extend axon collaterals to each of the target nuclei of striatal outputs; however, the relative extent of the arborization in a particular target area provides the basis for defining two basic striatal output neurons. One type, the striatopallidal neurons, extends an axon that forms a dense arborization within the globus pallidus. These neurons do not appear to extend an axon collateral past the globus pallidus. A second type, the striatonigral neurons (including those with projections to the entopeduncular nucleus), extends an axon collateral in the globus pallidus that does not ramify in this nucleus, and axon collaterals that extend to and arborize extensively in the entopeduncular nucleus and/or substantia nigra. The local axon collateral of most striatopallidal and striatonigral neurons extends in a domain approximately equal to their dendritic domain, which is on average 250–350 μm in diameter. A subtype of striatopallidal neuron, however, extends a local axon collateral over a distance of more than 1 mm. Although these studies examined a relatively limited number of striatal neurons, striatopallidal neurons with extended local axon collaterals appear to be infrequent.

Studies employing injections of retrogradely transported markers into the globus pallidus and substantia nigra have provided estimates of the relative numbers of striatopallidal and striatonigral neurons (Loopuijt & Kooy 1985). These numbers must be considered estimates, given the findings that striatonigral neurons extend an axon collateral into the globus pallidus, albeit a small one (Kawaguchi et al 1990). Since determining the extent of uptake of any retrogradely transported marker by either fibers of passage or small collaterals is impossible, these techniques cannot provide exact numbers of the striatal output types. Nonetheless, there is some consensus that the numbers of striatopallidal neurons and striatonigral neurons are approximately equal. Moreover, both types of neurons appear to be somewhat intermingled (Gimenez-Amaya & Graybiel 1990) and to be present in roughly equal numbers in both striatal patch and matrix compartments (Gerfen & Young 1988). As discussed above, striatonigral neurons in the patch and matrix compartments project differentially to the location of dopaminergic and GABAergic neurons in the substantia nigra (Gerfen 1985). The question is raised as to whether there is a similar difference in the projection targets of striatopallidal patch and matrix neurons.

Biochemical Characterization of Striatal Output Neurons

NEUROTRANSMITTER AND PEPTIDE CONTENT   Most, and perhaps all, striatal output neurons express glutamic acid decarboxylase (GAD) immunoreactivity (Kita & Kitai 1988) and thus presumably utilize GABA as a neurotransmitter. In addition, striatopallidal and striatonigral neurons contain different sets of neuropeptides. The first suggestion of the biochemical distinction between these output neurons was provided by immunohistochemical studies. Enkephalin immunoreactivity is densely distributed in terminals in the globus pallidus but only sparsely in the substantia nigra, whereas substance P immunoreactivity is sparse in terminals in the dorsal globus pallidus, though dense in terminals in the ventral pallidum and in the substantia nigra (Alheid & Heimer 1988, Haber & Nauta 1983). Similarly, dynorphin immunoreactivity is also dense in the substantia nigra but sparse in the globus pallidus (Vincent et al 1982). Thus, immunohistochemical studies suggest that striatopallidal neurons express enkephalin, whereas striatonigral neurons express substance P and dynorphin. That the differential expression of peptides by these output neurons is not absolute is suggested by the finding that a certain percentage of striatal neurons co-express substance P and enkephalin immuno-reactivity (Penny et al 1986).

The patterns of peptide immunoreactivity in the striatum have led to some confusion regarding the distribution of neurons expressing a given peptide relative to the patch-matrix compartmental organization of the striatum. A number of early studies described the patterns of both enke-phalin and substance P immunoreactivity as displaying heterogeneous patterns that correlated, to some extent, with the patch-matrix com-partments of the striatum (Beckstead & Kersey 1985, Gerfen 1984, Gray-biel et al 1981). These studies revealed little in the way of distinct cellular labeling, and the compartmental patterns were primarily of immuno-reactivity of indistinct neuropil labeling. More recent studies have shown peptidergic immunoreactive neurons to be more evenly distributed in both compartments (Penny et al 1986). Most likely these different patterns of immunoreactive labeling are related to the parameters of fixation (Graybiel & Chesselet 1984). Nonetheless, the patterns of immunoreactive labeling suggest that the relative expression of different peptides differs regionally in the patch and matrix compartments.

As the determinants of immunoreactive labeling of striatal cells remain unclear, in situ hybridization histochemical procedures have been used to characterize the biochemical phenotype of striatal neurons (Gerfen & Young 1988). Results show that the majority of striatopallidal neurons express enkephalin mRNA, whereas the majority of striatonigral neurons express both dynorphin and substance P mRNA. Additionally, approxi-mately 50–60% of both striatal patch and matrix neurons express each peptide mRNA; however, the relative expression of dynorphin mRNA is higher per cell in patch than in matrix neurons in the dorsal striatum, whereas in the ventral striatum there is a more equal relative expression in both compartments. Conversely, substance P mRNA shows an opposite relative expression pattern, as it is higher in ventral striatal patch neurons than in matrix neurons and is roughly equally expressed in both com-partments in the dorsal striatum. The patterns of peptide mRNA in situ hybridization histochemistry labeling are consistent with immunoreactive patterns of labeling.

Striatonigral and striatopallidal neurons may thus be generally charac-terized by their respective expression of substance P/dynorphin and enke-phalin. Although these expression patterns characterize the majority of striatal output neurons, some neurons express combinations of peptides that do not strictly adhere to this simplified scheme. Estimates of between 10 and 25% of the striatal neuron population have been suggested for such neurons. One subtype of striatal neuron that projects principally to the globus pallidus and expresses the mRNA encoding the tachykinin neurokinin B has been shown to also contain both enkephalin and substance P (Burgunder & Young 1989). This neuron type represents approximately 15% of the striatal population and accounts for some of the neurons that co-express enkephalin and substance P.

D_1 AND D_2 DOPAMINE RECEPTOR EXPRESSION   The rapid advances that have been made in the past few years in cloning G-protein-coupled receptors, which include dopamine receptors, has provided the ability to further characterize striatal neurons with in situ hybridization histochemistry (ISHH) (Figure 4). The D_1 and D_2 dopamine receptors, which are characterized by their respective stimulation and inhibition of adenylyl cyclase activity (Kebabian & Calne 1979, Stoof & Kebabian 1981), have been recently cloned (Bunzow et al 1988, Dearry et al 1990, Monsma et al 1990, Sunahara

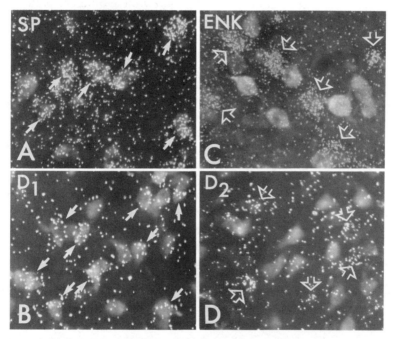

Figure 4   Striatal neurons retrogradely labeled with the fluorescent dye fluorogold after injection into the substantia nigra combined with darkfield illumination of silver grains produced by ISHH labeling with ^{35}S-labeled oligonucleotide probes for (A) substance P (SP), (B) the D_1 dopamine receptor (D_1), (C) enkephalin (ENK), and (D) the D_2 dopamine receptor (D_2). Striatonigral neurons show ISHH labeling for both substance P (A, *solid arrows*) and the D_1 dopamine receptor (B, *solid arrows*). Striatal neurons that are unlabeled by fluorogold, and presumably project to the globus pallidus, show ISHH labeling for both enkephalin (C, *open arrows*) and the D_2 dopamine receptor (D, *open arrows*).

et al 1990, Zhou et al 1990). In situ hybridization histochemical localization of mRNA encoding these receptors has shown that, for the most part, the D_1 dopamine receptor is expressed by striatonigral neurons that also contain substance P and dynorphin mRNAs (Gerfen et al 1990), whereas the D_2 dopamine receptor is expressed by striatopallidal neurons that also contain enkephalin mRNA (Gerfen et al 1990, Le Moine et al 1990). The somewhat specific expression of the D_1 and D_2 dopamine receptors by the two striatal output pathways is consistent with receptor binding studies that have shown D_1 receptor binding in the substantia nigra and D_2 receptor binding in the globus pallidus that originates from striatal output pathways (Beckstead 1988, Harrison et al 1990, Richfield et al 1989). Thus, both in situ hybridization histochemical studies and receptor-binding studies suggest the respective expression of D_1 and D_2 dopamine receptors in striatonigral and striatopallidal neurons. These studies do not suggest an exclusive distribution in these neurons, however. A subset of striatal output neurons appears to express both the D_1 and D_2 dopamine receptors. The 15–20% of striatal neurons that express the peptide neurokinin B appear to express both receptors. Thus, neurokinin B provides a marker of neurons that co-express the D_1 and D_2 dopamine receptors.

Although mRNAs encoding both D_1 and D_2 dopamine receptors have been cloned and used to characterize striatal neuronal populations, there exists the possibility that further subtypes of these receptors may be identified in the future. Already, two subtypes of the D_2 receptor have been cloned, one of which differs by the insertion of an extra piece in the third cytoplasmic domain (Monsma et al 1989), and another that differs sufficiently from the D_2 receptor to be designated as the D_3 receptor (Sokoloff et al 1990). This latter receptor is localized principally in the ventral striatum and nucleus accumbens and shares many of the pharmacologic properties of the D_2 receptor. Additionally, evidence suggests that a D_1 receptor is coupled to the phosphoinositide second messenger pathway (Felder et al 1989, Mahan et al 1990, Undie & Friedman 1990) and so is probably distinct from that which has already been cloned. Fitting these other subtypes of dopamine receptors into the currently described characterization of striatonigral and striatopallidal neurons will obviously have to await their identification.

Dopaminergic Regulation of Striatal Output Systems

The recognition that Parkinson's disease results from the degeneration of the nigrostriatal dopaminergic system is one of the cornerstones of the neurochemical basis of behavior (Hornykiewicz 1966). In the 1970s, studies demonstrated that dopamine regulates the expression of peptide levels in the striatum with the finding that chronic neuroleptic treatments result in an elevation of striatal enkephalin levels (Hong et al 1978). These studies initiated a line of research that has provided a strategy for examining the role of dopamine in modulating striatal output neurons. Following the work of Hong et al (1978), numerous studies have shown that a decrease in dopamine action in the striatum, with either neuroleptic treatments or with 6-hydroxydopamine (6-OHDA)- induced dopamine striatal deafferentation, results in an increase in enkephalin peptide and mRNA levels in the striatum but a decrease in substance P immunoreactivity and mRNA levels (Bannon et al 1986, Gerfen et al 1991, Hanson et al 1981a, Hong et al 1983, Mocchetti et al 1985, Normand et al 1988, Sivam et al 1986, Tang et al 1983, Young et al 1986). Pharmacologic treatments that increase dopaminergic action with either amphetamine or dopamine agonists result in an elevation of striatal levels of both substance P- and dynorphin-immunoreactivity and mRNA (Gerfen et al 1991, Hanson et al 1981b, Li et al 1987, 1988). Taken together with the localization of enkephalin in striatopallidal neurons and substance P and dynorphin in striatonigral neurons, these lesion and pharmacologic data suggest that dopamine oppositely modulates neurons contributing to the striatopallidal and striatonigral pathways.

The differential localization of D_1 and D_2 dopamine receptors on striatonigral and striatopallidal neurons provides a reasonable explanation of the effect of dopamine on these neurons. The direct effect of D_1 and D_2 receptor activation has been studied with the unilateral 6-OHDA-lesioned animal as a model to examine regulation through D_1 and D_2 dopamine receptors (Gerfen et al 1990). With this paradigm, the regulation of D_1 and D_2 dopamine receptor, enkephalin, substance P, and dynorphin mRNA levels were examined in response to either intermittent (21 daily injections) or continuous (injections delivered through an osmotic mini-pump) systemic administration of D_1 and D_2 selective agonists. These studies show that reduced levels of substance P and D_1 receptor mRNA levels resulting from 6-OHDA-induced nigrostriatal dopamine deafferentation are selectively reversed and dynorphin mRNA levels are significantly elevated with intermittent treatments with the selective D_1 agonist SKF-38393. Neither continuous SKF-38393 nor D_2 selective agonist treatment affects levels of these mRNAs. Conversely, the elevation of both enkephalin and D_2 receptor mRNA levels following 6-OHDA lesions is selectively reversed with continuous treatment with the D_2 selective agonist quinpirole (1 mg/kg/day for 21 days). The other drug treatment regimens had no significant effect on these mRNA levels. These results provide direct evidence that dopamine differentially regulates gene expression in striatonigral and striatopallidal neurons through the respective selective expression of D_1 and D_2 dopamine receptors by these neurons. Moreover, the critical relationship between the time course of drug delivery and the selective action through the two receptors suggests that molecular processes by which receptor occupancy alters gene expression is fundamentally distinct for the two dopamine receptors.

Studies employing D_1 and D_2 dopamine receptor agonist treatments in intact animals have yielded more complex results. For example, D_1 and not D_2 antagonist treatments result in elevation of enkephalin levels (Jiang et al 1990), and D_2 and not D_1 agonist treatments have been reported to elevate substance P levels (Haverstick et al 1989). The differences between these results and those in the unilateral lesioned model may reflect interactions between subpopulations of neurons, which are discussed in detail below.

FUNCTIONAL ASPECTS OF PARKINSON'S DISEASE   Changes in peptide mRNA levels in striatonigral and striatopallidal neurons provide an excellent paradigm for examining receptor-specific mechanisms of dopaminergic regulation of these neurons. The functions of the specific peptides examined remain unknown, however. Moreover, as all of these neurons utilize GABA as a transmitter, the physiologic consequence of dopaminergic modulation of striatal neurons is probably mediated at least in part by GABA (Chevalier et al 1985). Rather, it is suggested that the specific changes in peptide levels provide a means of assaying the relative activity level of these neurons. Several lines of evidence would appear to substantiate making this assumption. First, it has been reported that D_1 agonist treatments of 6-OHDA lesioned animals, which specifically elevate peptide levels in striatonigral neurons, also induce expression of the immediate early oncogene product *c-fos* (Robertson et al 1990). Second, studies employing comparable lesion and pharmacologic treatments as those used in examining peptide mRNA expression have revealed parallel changes in 2-deoxyglucose utilization in the globus pallidus and substantia nigra (Engber et al 1990, Trugman & Wooten 1987). Taken together, these studies suggest that dopamine deafferentation in the striatum results in both elevated enkephalin mRNA levels and increased activity in striatopallidal neurons (Figure 5). Furthermore, in lesioned animals D_1 agonist treatments selectively increase substance P and dynorphin mRNA levels and activity in striatonigral neurons.

Elevation of activity in the striatopallidal pathway resulting from striatal dopamine deafferentation has been proposed to underlie the bradykinesia of Parkinson's disease (Albin et al 1989, Mitchell et al 1989). Consistent with this are the concurrent decrease in 2-deoxyglucose utilization in the external pallido-subthalamic pathway and increased utilization in the internal pallido-thalamic pathway (Mitchell et al 1989). Bradykinesia is

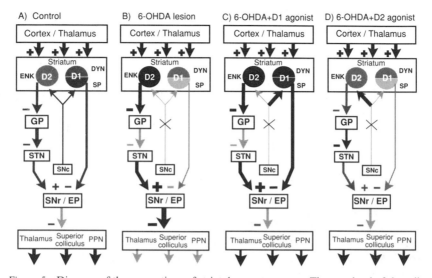

Figure 5   Diagram of the connections of striatal output neurons. The grey level of the cells denotes relative peptide mRNA levels and the thickness of the lines indicates relative activity measured in studies with 2-deoxyglucose (Trugman & Wooten 1987). (A) *Control*: The cortex and thalamus provide excitatory input to the striatum. Striatal neurons that contain enkephalin (ENK) and the D_2 dopamine receptor (D_2) provide an inhibitory input to the globus pallidus (GP). Pallidal neurons inhibit the subthalamic nucleus (STN), which provides an excitatory input to the substantia nigra pars reticulata (SNr). Striatal neurons that express the D_1 dopamine receptor (D_1), dynorphin (DYN), and substance P (SP) provide an inhibitory input to the substantia nigra/entopeduncular nucleus (SNr/EP). SNr/EP GABA-ergic neurons inhibit neurons in the thalamus, superior colliculus, and pedunculopontine nucleus (PPN). Normal behavioral activity (*arrows* at the bottom of the diagram) is dependent on coordinated striatonigral and striatopallidal outputs that regulate SN output. (B) *6-OHDA*: Dopamine lesions result in increased enkephalin expression and activity in striatopallidal neurons. This results in increased firing of SN GABAergic neurons and in diminished behavioral activity (*arrows* at bottom of diagram). (C) *6-OHDA+D_1 agonist*: D_1 agonist treatment after 6-OHDA lesions does not alter the lesion-induced increase in enkephalin in the striatopallidal pathway but reverses the decrease in substance P and significantly increases dynorphin in striatonigral neurons. (D) *6-OHDA+D_2 agonist*: Continuous D_2 agonist treatment after 6-OHDA lesions has no effect on the striatonigral pathway but reverses the lesion-induced increase in enkephalin in the striatopallidal pathway. This reverses the increased excitatory input from the subthalamic nucleus to the substantia nigra pars reticulata. From Gerfen et al (1990).

thus thought to be a result of increased activity in the indirect striatonigral output system, which results in increased tonic activity of the inhibitory inputs to the thalamus and brainstem targets of the internal segment of the globus pallidus and substantia nigra pars reticulata. Verification of this idea comes from the report that lesions of the subthalamic nucleus, designed to disrupt the indirect striatonigral pathway, result in profound reversal of bradykinesia in monkeys made Parkinsonian with MPTP lesions (Bergman et al 1990).

Therapeutic approaches to the treatment of Parkinson's disease might be suggested based on the understanding of D_1 and D_2 receptor regulation of striatal output pathways (Gerfen et al 1990). As discussed, the principal deficit produced by striatal dopamine depletion in Parkinson's disease is an increase in striatopallidal activity, which is paralleled by increased enkephalin expression in striatopallidal neurons. To overcome this lesion-induced increase in striatopallidal function, D_2 receptor activation is necessary, and to be most effective appears to require a continuous treatment protocol with a D_2 selective agonist. Supplemental treatment with a pharmacologic agent that acts on D_1 receptors would be necessary to overcome the diminished function in this pathway caused by dopamine deafferentation. It is suggested, however, that the dosage required for a D_1 agent would be less if given in conjunction with a D_2 selective agonist. Interestingly, L-DOPA administered in twice daily injections to rats with 6-OHDA nigrostriatal lesions appears to affect striatonigral and not striatopallidal pathways in a manner similar to a D_1 selective agonist (C. R. Gerfen and T. M. Engber, unpublished findings). Continuous treatment with L-DOPA appears to have little effect on either pathway. To have a behavioral effect, both D_1 selective agonists and L-DOPA given on an intermittent schedule would appear to increase striatonigral function to an abnormally high level in order to overcome the increased function of striatopallidal neurons caused by dopamine deafferentation. Maintaining such abnormally high levels of function with long-term treatments might underlie the deterioration of L-DOPA as an effective therapeutic agent in the treatment of Parkinson's disease.

BALANCED OPPOSITION OF STRIATOPALLIDAL AND STRIATONIGRAL OUTPUTS

The disinhibitory processes of striatal outputs have provided a coherent model to explain the movement disorders that result from the loss of striatal dopamine in Parkinson's disease and animal models of this disease. Reversal of bradykinesia in MPTP-treated monkeys with lesions of the subthalamic nucleus has provided perhaps the best verification of this model (Bergman et al 1990). Extrapolating this model to the generation of normal behavior undoubtedly oversimplifies the processes involved. The generation of eye movements appears to be clearly related to disinhibitory processes whereby pauses in nigrotectal activity are correlated with eye movements (Hikosaka & Wurtz 1983a,b), yet single unit recording studies of internal globus pallidus activity in relationship to limb movements in primates reveals a more complex pattern of activity. In such studies, specific limb movements are correlated with both decreases and increases in the activity of internal pallidal firing. Such results are suggested to demonstrate that although disinhibitory mechanisms are involved in the generation of specific movements, the opposed mechanism, which results in increased inhibitory processes, may suppress antagonistic movements (Alexander & Crutcher 1990). These data are not incompatible with the model of Parkinson's disease discussed above but are introduced to illustrate the positive aspect that increased activity in the striatopallidal pathway may have in the generation of normal behavior. Thus, the balanced opposition in the relative activity of the striatopallidal and striatonigral pathways provides mechanisms for both allowing and disallowing specific muscle activity. In Parkinson's disease, the mechanisms for disallowing behavior appear to prevail, but in the normal condition, such processes are an integral part of the generation of normal behavior.

D_1 and D_2 Receptor Interactions

As discussed above, the failure of long-term pharmacologic treatments in the management of Parkinson's disease may be due to the difficulty of reconstituting the normal complex dynamics of combined D_1 and D_2 receptor activation. D_1 receptor stimulation–induced increases in striatonigral function and D_2 receptor stimulation–induced decreases in striatopallidal function reflect a unique situation in which one receptor is activated in the absence of activation of the other. These rather uniform actions on the striatonigral and striatopallidal neurons represent the direct action of activation of their receptors. Combined activation of these receptors, however, brings into play intra- and intercellular mechanisms that function to generate the complex patterns of striatal output activity required for the generation of normal behavior. For example, coactivation of D_1 and D_2 receptors enacts mechanisms whereby D_1 receptor activation may result in increased striatopallidal activity.

Evidence for the synergistic effects of D_1 and D_2 receptor activation is provided from the work of Walters and coworkers (Carlson et al 1990, Walters et al 1987). Their studies have shown that D_1 agonist pretreatment potentiates D_2 agonist induced increases in globus pallidus activity (Walters et al 1987). Other studies have shown that in the 6-OHDA lesioned animal, combinations of D_1 agonist and D_2 agonist treatments result in a broader range of responses in pallidal neurons than occurs with treatments of only one of the agonists; some units show increases and some decreases in activity (Carlson et al 1990). Other studies have shown that both D_1 and D_2 receptor activation is required to inhibit the activity of $(Na^+ + K^+)$ATPase in striatal neurons (Bertorello et al 1990). These latter studies suggest that synergistic effects of D_1 and D_2 receptor activation may occur within an individual neuron. Based on current knowledge of the distribution of D_1 and D_2 receptors primarily by separate populations of striatal neurons, multiple mechanisms for the synergistic effects of D_1 and D_2 receptor activation are proposed. One mechanism may occur in neurons that coexpress D_1 and D_2 receptors, a second mechanism may be mediated through the local axon collaterals that interconnect neurons that express different receptors, a third mechanism may involve striatal interneurons (Figure 6), and a fourth may involve the patch-matrix compartmental organization of the striatum.

D_1 AND D_2 RECEPTOR EXPRESSION IN NEUROKININ B NEURONS   Although the majority of striatal output neurons appear to express only one of the D_1 and D_2 dopamine receptors. Some striatal neurons may express both

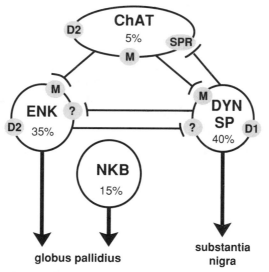

Figure 6   Schematic diagram of the major types of striatal neurons (approximate percentage of the total striatal population is listed) and the possible interactions between them that provide mechanisms for balancing striatopallidal and striatonigral outputs. Cholinergic interneurons (ChAT) express the D_2 dopamine-, muscarinic-, and substance P receptor (SPR) mRNAs. Striatonigral neurons contain the peptides dynorphin (DYN) and substance P (SP) and express the D_1 dopamine- and muscarinic (M) receptor mRNAs. The majority of striatopallidal neurons contain enkephalin (ENK) and express the D_2 dopamine- and muscarinic receptor mRNAs. A minority of striatopallidal neurons contain neurokinin B (NKB) and do not express either the D_1 or D_2 dopamine receptor mRNAs. Interactions between D_1 and D_2 receptors may occur in NKB neurons by direct interactions between ENK and DYN neurons through an unknown receptor mechanism (?), or via substance P mediated regulation of cholinergic neurons. Intrastriatal connections of the NKB neurons are still relatively unknown.

receptors, but another subset of output neurons, which contain the tachy-kinin neurokinin B, do not, for the most part, express either the D_1 or D_2 subtype (C. R. Gerfen, unpublished observations). Interestingly, neuro-kinin B mRNA levels are regulated by both D_1 and D_2 dopamine receptor-mediated mechanisms. Nigrostriatal dopaminergic deafferentation results in an increase in neurokinin B mRNA, which is reversed with continuous treatment with the D_2 selective agonist quinpirole (1 mg/kg/day for 5 days). The lesion-induced elevation of neurokinin B mRNA expression is further increased following intermittent D_1 selective agonist treatments (SKF-38393, 10 mg/kg/day for 5 days). Thus, these neurons represent an interesting subpopulation of striatal neurons in which effects of D_1 and D_2 dopamine receptor activation may occur within an individual neuron, but by indirect mechanisms.

Neurokinin B is contained in 15–20% of striatal output neurons (Bur-gunder & Young 1989). These neurons project principally to the globus pallidus and thus represent a subset of striatopallidal neurons. Although some of these neurons are grouped in clusters of three to five, for the most part they are distributed in a dispersed pattern that does not show any clear grouping such as evidenced by neurons that form striatal patches. The cellular localization of the neurokinin B receptor has yet to be determined; however, neurokinin B has been shown to be the most effective of the striatal tachykinin peptides in inducing the release of acetylcholine (Arenas et al 1991). As is discussed below, striatal cholinergic neurons are critically involved in the dopaminergic modulation of striatal output neurons.

LOCAL AXON COLLATERALS OF MEDIUM SPINY NEURONS   Striatal medium spiny output neurons have a local axon collateral that for most neurons spreads in a domain approximately equivalent to that of its dendritic arbor (Wilson & Groves 1980). These local axon collaterals appear to make synaptic contact with other medium spiny neurons through asymmetric synapses (Wilson & Groves 1980). As mentioned above, these neurons use GABA as a neurotransmitter and also express GABA and benzodiazepine receptor mRNAs (Lolait et al 1989). Thus, neighboring striatal output neurons are connected in a manner, and express the appropriate receptors, that may provide for their affecting one another through a GABA-mediated inhibitory synaptic process (Park et al 1980). This is one potential mechanism whereby a striatonigral neuron that is specifically modulated through its expression of a D_1 receptor might influence the activity of a neighboring striatopallidal neuron that expresses a D_2 receptor. Such mechanisms are still only postulated, as there is as yet no direct evidence for such interactions. The peptides that are co-expressed by these neurons also are potential neuroactive substances through which such interactions may occur. For the most part, however, the localization of receptors to which the opiate peptides bind has not been determined, so the role of these peptides is unclear.

The substance P receptor, a G-protein-coupled receptor that has recently been cloned and characterized (Hershey & Krause 1990, Yokota et al 1989), is localized specifically on striatal cholinergic neurons (Gerfen 1991b). Because substance-P-containing local axon collaterals of medium spiny neurons make synaptic contact with cholinergic neurons (Bolam et al 1986), a specific mechanism whereby cholinergic neurons are involved in interactions between striatal output neurons is suggested. This is discussed below.

STRIATAL CHOLINERGIC INTERNEURONS   A small number of striatal neurons have large cell bodies that extend aspiny dendrites over relatively large domains (up to 1000 μm in extent) and extend an even more widespread axon collateral that remains confined to the striatum (Wilson et al 1990). These aspiny interneurons express choline acetyltransferase and are referred to as striatal cholinergic interneurons. These neurons are distinct from other striatal neurons in exhibiting irregular but tonic activity (Wilson et al 1990). Although the neurons are relatively sparse, representing less than 5% of the striatal population in rats, they exert a profound influence on striatal function. The mechanisms of this influence are complex and are now being made clear with studies elucidating the mechanisms that regulate acetylcholine release and the post-synaptic action of acetylcholine through muscarinic receptors expressed on striatal neurons.

Dopamine appears to inhibit acetylcholine release through a D_2-receptor-mediated process (Lehmann & Langer 1983, Stoof & Kebabian 1982) that is modulated in part by muscarinic autoreceptors expressed on cholinergic neurons (Drukarch et al 1990). Increased acetylcholine release that follows dopamine deafferentation of the striatum may be partly responsible for the increased activity of striatopallidal neurons. This is supported by the finding that the muscarinic antagonist scopolamine partially blocks neuroleptic-induced elevation of striatal enkephalin levels (Hong et al 1985). Subtypes of muscarinic cholinergic G-protein-coupled receptors have been cloned (Bonner et al 1987). In situ hybridization localization has shown that the m_1 subtype is expressed by most striatal projection neurons, the m_2 subtype is selectively expressed by striatal cholinergic neurons, and the m_4 subtype is expressed by approximately half of the striatal output neurons (Weiner et al 1990). The functional significance of the differential pattern of muscarinic receptor subtype distribution in the striatum is not yet clear; however, their localization does place the subtypes in a position to mediate cholinergic regulation of striatal output neurons.

The possible complexity of this regulation is suggested by a recent study that showed the physiologic effect of muscarinic receptor activation, mediated through a K^+ channel, to depend on the state of the membrane polarization of the post-synaptic neuron (Akins et al 1990). Akins et al propose that the effect of muscarinic receptor activation is to stabilize the resting membrane potential of medium spiny neurons, which normally fluctuates between two levels of hyperpolarization (Wilson 1981).

The tachykinin peptides neurokinin B, substance K, and substance P have been shown to induce release of acetylcholine in the striatum (Arenas et al 1991). This raises the possibility that a functional role of these peptides, which are expressed by different sets of striatal output neurons, may be to regulate the release of acetylcholine. Such a function is dependent on the localization of the appropriate tachykinin receptors. In this context it is of interest that the substance P receptor (neurokinin 1 receptor) is selectively localized to the cholinergic neurons in the striatum and does not appear to be expressed by other striatal neurons (Gerfen 1991). This suggests another possible mechanism for the interaction between D_1 and D_2 dopamine receptors. Stimulation of D_1 receptors appears to increase substance P expression and may increase its release, which, acting on substance P receptors on cholinergic neurons, could increase the release of acetylcholine. An increase in acetylcholine release may then affect striatopallidal neurons, whose response to muscarinic receptor activation is modulated by D_2 dopamine mechanisms (C. R. Gerfen, unpublished observations). At this time a mechanism can only be postulated, but it is suggested to illustrate a possible intercellular mechanism that may underlie interaction between D_1 and D_2 dopamine receptors.

Although the functional role of the selective expression of the substance P receptor by cholinergic neurons remains speculative, it is appropriate to comment on the possible significance of the role of peptides in the basal ganglia. Substance P is expressed by striatonigral neurons and is localized in terminals in the substantia nigra by immunohistochemical techniques (Haber & Nauta 1983), yet substance P receptor binding studies have failed to demonstrate appropriate binding sites within the substantia nigra (Danks et al 1986). This has been cited as one of numerous cases of a mismatch between receptor binding sites and the localization of endogenous ligands (Herkenham 1987). Striatonigral neurons also utilize GABA as a transmitter, and for this there is evidence of a physiologic role in that stimulation of striatal outputs results in GABA-mediated inhibition of neurons in the substantia nigra (Chevalier et al 1985). Thus, striatonigral neurons may affect striatal medium spiny neurons and substantia nigra neurons by a GABA receptor mediated process and simultaneously induce release of acetylcholine from striatal cholinergic neurons by a substance

P receptor mediated process. This suggests that receptor binding and endogenous ligand mismatches may reflect the synthesis and transport of a specific neuropeptide in all axon collaterals of a neuron but that the actual physiologic action of that peptide may occur only at those synaptic sites at which the postsynaptic neuron expresses the appropriate receptor. This organization provides a neuron with the capability of differentially affecting different target neurons dependent on the type of receptor expressed by those target neurons.

PATCH-MATRIX COMPARTMENTS AND STRIATOPALLIDAL AND STRIATONIGRAL SYSTEMS   The subpopulations of striatal output neurons described above—namely striatopallidal neurons that express enkephalin and the D_2 dopamine receptor and striatonigral neurons that express substance P, dynorphin, and the D_1 dopamine receptor—appear to be evenly distributed in both patch and matrix compartments (Gerfen & Young 1988). Nonetheless, the relative expression of some of these markers in individual neurons appears to be different in the patch-matrix compartments. For example, substance P mRNA levels appear to be higher in patch than in matrix neurons in the ventral striatum, whereas dynorphin mRNA levels appear to be higher in patch than in matrix neurons in the dorsal striatum (Gerfen & Young 1988). That these patterns of relative expression are under dopaminergic regulation is suggested by the findings that in the intact striatum, dopamine agonist treatments elevate dynorphin in dorsal and substance P in ventral patch neurons, somewhat specifically (Gerfen et al 1991, Li et al 1987, 1988).

The dopaminergic regulation of the expression of dynorphin in the dorsal striatum may serve as a model of the functional significance of the patch-matrix compartments as they relate to the regulation of the balanced opposition of striatopallidal and striatonigral output systems (Figure 7, Gerfen et al 1991). In the normal striatum, complex interactions may occur between subpopulations of striatal output neurons and interneurons that establish a balance in the regulation of the two output systems. The pattern of expression of dynorphin mRNA in the dorsal striatum delineates these interactions. Dynorphin is expressed by equal numbers of patch and matrix neurons, but its expression in patch neurons is relatively higher than in matrix neurons. In rats treated with the nonselective dopamine agonist apomorphine, this relative difference in dynorphin expression in the patch compartment is augmented. In the dopamine deafferented striatum, however, the same apomorphine treatment or D_1 selective agonist treatment results in a dramatic elevation of dynorphin mRNA expression in both patch and matrix neurons. In this situation, the normal regulatory mechanisms that establish a balance in the striatonigral and striatopallidal

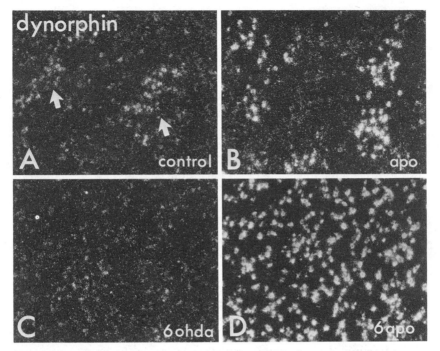

Figure 7 Darkfield photomicrographs of autoradiographically generated grains (seen as *white dots*) produced by ISHH labeling of striatal sections via an oligonucleotide probe complementary to dynorphin mRNA (A–D). Sample areas in the dorsolateral striatum are shown from a control animal (A), from an animal that received 10 days of twice daily injections of 5 mg/kg apomorphine (B), from the lesioned side of an animal that had a 6-OHDA lesion of the nigrostriatal pathway (C), and from the lesioned side of an animal that received a 6-OHDA lesion followed 7 days later by 10 days of twice daily apomorphine treatments (D). Note that in the controls, neurons in the patches (indicated with *arrows* and identified in adjacent sections with ^{3}H-naloxone binding, not shown) show more labeling than do cells in the matrix. From Gerfen et al (1991).

output systems may become inoperative. In addition to the possible cellular and molecular mechanisms regulating the balanced opposition of these output systems discussed above, several relating to patch-matrix compartmental organization may also be involved. One is the dual mesostriatal dopaminergic system originating from dorsal and ventral tier dopamine neurons that differentially targets the matrix and patch compartments, respectively (Gerfen et al 1987). In the normal animal, these systems may provide some mechanism of differentially altering peptide expression in the striatum. Another is the possible role of a striatal interneuronal system. Somatostatin striatal interneurons distribute fibers that are somewhat restricted to the matrix compartment (Chesselet & Graybiel 1985, Gerfen 1984, 1985) and could mediate the differential effect on peptides in the compartments via a dopaminergic mediated process (Chesselet & Reisine 1983). Unfortunately, we can still only speculate as to the regulatory mechanisms involved. Nonetheless, patch-matrix compartmental organization appears to provide mechanisms for differentially regulating striatopallidal and striatonigral output neurons, and these influences appear to vary regionally.

CONCLUSIONS

Balanced opposition in the cortically driven activity of striatopallidal and striatonigral pathways is proposed to underlie the complex manner in which the basal ganglia affect behavior. This review has described aspects of the compartmental organization of the striatum that provide both cellular and molecular mechanisms responsible for regulating this balance. Although all striatal neurons utilize GABA as a transmitter, the majority of striatopallidal neurons express the D_2 dopamine receptor and enkephalin, whereas the majority of striatonigral neurons express the D_1 dopamine receptor and the peptides dynorphin and substance P. In the dopamine-depleted striatum, D_2 agonists decrease and D_1 agonists increase peptide expression in striatopallidal and striatonigral neurons, respectively. This suggests that the direct action of dopamine is to oppositely modulate these output pathways; however, the rather uniform responses of striatonigral and striatopallidal neurons would appear to occur on the unique condition that one dopamine receptor subtype is activated in the absence of the other. In the normal striatum, intra- and intercellular interactions occur that provide for more complex patterns of regulation of striatonigral and striatopallidal outputs. Four such mechanisms of D_1 and D_2 receptor interaction are described. One type occurs in a subpopulation of striatopallidal neurons that express both D_1 and D_2. A second mechanism of interaction may occur by way of the local axon collaterals of medium spiny neurons that may interconnect neurons that express different dopamine receptors. A third mechanism may involve striatal interneurons. Expression of the tachykinin substance P in striatonigral neurons is elevated by D_1 agonist stimulation. Striatal cholinergic interneurons express the substance P receptor and release acetylcholine in response to its activation. Since cholinergic muscarinic receptor activation regulates striatopallidal neurons, a cholinergic mediated linkage of striatonigral and striatopallidal neurons is proposed. A fourth mechanism of regulation of striatonigral and striatopallidal output neurons may involve the patch-matrix compartmental organization of the striatum. Peptide expression in these output neurons varies regionally in patch-matrix compartments in a manner that suggests dopaminergic and possibly striatal interneuronal involvement. These mechanisms of D_1 and D_2 receptor interactions are not exhaustive but are presented to illustrate possible mechanisms for the generation of complex patterns of regulation of the balanced opposition of striatonigral and striatopallidal outputs.

In conclusion, perhaps the fundamental function of the basal ganglia is to select from myriad possibilities a specific behavioral action. Balanced opposition of the activity of striatonigral and striatopallidal neurons provides the mechanism whereby a specific behavior is facilitated or disfacilitated. Studies examining gene regulation of subpopulations of striatal output neurons reveal the cellular and molecular mechanisms involved in this process. The critical processes whereby select subpopulations of striatal neurons are coordinated are dependent on the integrative properties of corticostriatal systems (Wilson 1990) that interact with the intrastriatal organization of mechanisms regulating the balanced opposition of striatonigral and striatopallidal output systems.

ACKNOWLEDGMENTS

I would like to thank Charlie Wilson, Miles Herkenham, Lennart Heimer, Gary Alexander, Roger Albin, Mahlon DeLong, and Steve Wise for discussions during the course of writing this review. This work was supported by the Intramural Research Program of the National Institute of Mental Health.

Literature Cited

Akins, P. T., Surmeier, D. J., Kitai, S. T. 1990. Muscarinic modulation of a transient K$^+$ conductance in rat neostriatal neurons. *Nature* 344: 240–42

Albin, R. L., Young, A. B., Penney, J. B. 1989. The functional anatomy of basal ganglia disorders. *Trends Neurosci.* 12: 366–75

Alexander, G. E., Crutcher, M. D. 1990. Functional architecture of basal ganglia circuits: Neural substrates of parallel processing. *Trends Neurosci.* 13: 266–71

Alexander, G. E., DeLong, M. R., Strick, P. L. 1986. Parallel organization of functionally segregated circuits linking basal ganglia and cortex. *Annu. Rev. Neurosci.* 9: 357–81

Alheid, G. F., Heimer, L. 1988. New perspectives in basal forebrain organization of special relevance for neuropsychiatric disorders: The striatopallidal, amygdaloid, and corticopetal components of the substantia innominata. *Neuroscience* 27: 1–39

Arenas, E., Alberch, J., Perez-Navarro, E., Solsona, C., Marsal, J. 1991. Neurokinin receptors differentially mediate endogenous acetylcholine release evoked by tachykinins in the neostriatum. *J. Neurosci.* 11: 2332–38

Bannon, M. J., Lee, J.-M., Girand, P., Young, A., Affolter, J.-U., et al. 1986. Dopamine antagonist haloperidol decreases substance P, substance K and preprotachykinin mRNAs in rat striatonigral neurons. *J. Biol. Chem.* 261: 6640–42

Beckstead, R. M. 1988. Association of dopamine D1 and D2 receptors with specific cellular elements in the basal ganglia of the cat: The uneven topography of dopamine receptors in the striatum is determined by intrinsic striatal cells, not nigrostriatal axons. *Neuroscience* 27: 851–63

Beckstead, R. M., Kersey, K. S. 1985. Immunohistochemical demonstration of differential substance P-, Met-enkephalin-, and glutamic acid decarboxylase-containing cell and axon distributions in the corpus striatum of the cat. *J. Comp. Neurol.* 232: 481–98

Bergman, H., Whitman, T., DeLong, M. R. 1990. Reversal of experimental Parkinsonism by lesions of the subthalamic nucleus. *Science* 249: 1436–38

Bertorello, A. M., Hopfield, J. F., Aperia, A., Greengard, P. 1990. Inhibition by dopamine of $(Na^+ + K^+)ATPase$ activity in neostriatal neurons through D_1 and D_2 dopamine receptor synergism. *Nature* 347: 386–88

Bolam, J. P., Ingham, C. A., Izzo, P. N., Levey, A. I., Rye, D. B., et al. 1986. Substance P-containing terminals in synaptic contact with cholinergic neurons in the neostriatum and basal forebrain: A double immunocytochemical study in the rat. *Brain Res.* 397: 279–89

Bolam, J. P., Izzo, I. N., Graybiel, A. M. 1988. Cellular substrate of the histochemically defined striosome/matrix system of the caudate nucleus: A combined Golgi and immunocytochemical study in cat and ferret. *Neuroscience* 24: 853–75

Bolam, J. P., Wainer, B. H., Smith, A. D. 1984. Characterization of cholinergic neurons in the rat neostriatum. A combination of choline acetyltransferase immunocytochemistry, Golgi-impregnation and electron microscopy. *Neuroscience* 12: 711–12

Bonner, T. I., Buckley, N. J., Young, A. C., Brann, M. R. 1987. Identification of a family of muscarinic acetylcholine receptor genes. *Science* 237: 527–32

Bouyer, J. J., Park, D. H., Joh, T. H., Pickel, V. M. 1984. Chemical and structural analysis of the relation between cortical inputs and tyrosine hydroxylase-containing terminals in rat neostriatum. *Brain Res.* 302: 267–75

Bunzow, J. R., Tol, H. H. M. V., Grandy, D. K., Albert, P., Salon, J., et al. 1988. Cloning and expression of a rat D2 dopamine receptor. *Nature* 336: 783–87

Burgunder, J. M., Young, W. S. 1989. Distribution, projection and dopaminergic regulation of the neurokinin B mRNA-containing neurons of the rat caudate-putamen. *Neuroscience* 32: 323–35

Carman, J. B., Cowan, W. M., Powell, T. P. S. 1965. The organization of corticostriate connexions in the rabbit. *Brain* 86: 525–62

Carlson, J. H., Bergstrom, D. A., Demo, S. D., Walters, J. R. 1990. Nigrostriatal lesion alters neurophysiological responses to selective and nonselective D-1 and D-2 dopamine agonists in rat globus pallidus. *Synapse* 5: 83–93

Cheramy, A., Leviel, V., Glowinski, J. 1981. Dendritic release of dopamine in the substantia nigra. *Nature* 289: 537–42

Chesselet, M.-F., Graybiel, A. M. 1985. Striatal neurons expressing somatostatin-like immunoreactivity: Evidence for a peptidergic interneuronal system in the cat. *Neuroscience* 17: 547–71

Chesselet, M.-F., Reisine, T. D. 1983. Somatostatin regulates dopamine release in striatal slices and cat caudate nuclei. *J. Neurosci.* 3: 232–36

Chevalier, G., Vacher, S., Deniau, J. M., Desban, M. 1985. Disinhibition as a basic process in the expression of striatal function. I. The striato-nigral influence on tecto-spinal/tecto-diencephalic neurons. *Brain Res.* 334: 215–26

Cowan, R. C., Wilson, C. J., Emson, P. C., Heizmann, C. W. 1990. Parvalbumin containing GABAergic interneurons in the rat neostriatum. *J. Comp. Neurol.* 302: 197–205

Danks, J. A., Rothman, R. B., Cascieri, M. A., Chicchi, G. G., Liang, T., et al. 1986. A comparative autoradiographic study of the distribution of substance P and eledoisin binding sites in rat brain. *Brain Res.* 385: 273–81

Dearry, A., Gingrich, J. A., Faladrdeau, P., Fremeau, J. R. T., Bates, M. D., et al. 1990. Molecular cloning and expression of the gene for a human D1 dopamine receptor. *Nature* 347: 72–76

Deniau, J. M., Chevalier, G. 1985. Disinhibition as a basic process in the expression of striatal functions. II. The striato-nigral influence on thalamocortical cells of the ventromedial thalamic nucleus. *Brain Res.* 334: 227–33

DiFiglia, M., Aronin, N. 1982. Ultrastructural features of immunoreactive somatostatin neurons in the rat caudate-nucleus. *J. Neurosci.* 2: 1267–74

Donoghue, J. P., Herkenham, M. 1986. Neostriatal projections from individual cortical fields conform to histochemically distinct striatal compartments in the rat. *Brain Res.* 365: 397–403

Donoghue, J. P., Kitai, S. T. 1981. A collateral pathway to the neostriatum from corticofugal neurons of the rat sensory-motor cortex: An intracellular HRP study. *J. Comp. Neurol.* 210: 1–13

Drukarch, B., Schepens, E., Stoof, J. C. 1990. Muscarinic receptor activation attenuates D2 dopamine receptor mediated inhibition of acetylcholine release in rat striatum: Indications for a common signal transduction pathway. *Neuroscience* 37: 1–9

Engber, T. M., Susel, Z., Kuo, S., Chase, T. N. 1990. Chronic levodopa treatment alters basal and dopamine agonist-stimulated cerebral glucose utilization. *J. Neurosci.* 10: 3889–95

Felder, C. C., Jose, P. A., Axelrod, J. 1989. The dopamine-1 agonist, SKF 82526, stimulates phospholipase-C activity independent of adenylate cyclase. *J. Pharmacol. Exp. Ther.* 248: 171–75

Ferino, F., Thierry, A. M., Saffroy, M., Glowinski, J. 1987. Interhemispheric and subcortical collaterals of medial prefrontal cortical neurons in the rat. *Brain Res.* 417: 257–66

Freund, T. F., Powell, J. F., Smith, A. D. 1984. Tyrosine hydroxylase-immunoreactive boutons in synaptic contact with identified striatonigral neurons with particular reference to dendritic spines. *Neuroscience* 13: 1189–1215

Gerfen, C. R. 1984. The neostriatal mosaic: Compartmentalization of corticostriatal input and striatonigral output systems. *Nature* 311: 461–64

Gerfen, C. R. 1985. The neostriatal mosaic: I. Compartmental organization of projections from the striatum to the substantia nigra in the rat. *J. Comp. Neurol.* 236: 454–76

Gerfen, C. R. 1989. The neostriatal mosaic: Striatal patch-matrix organization is related to cortical lamination. *Science* 246: 385–88

Gerfen, C. R. 1991. Substance P receptor mRNA selectively expressed by cholinergic neurons in the striatum and basal forebrain. *Brain Res.* 556: 165–70

Gerfen, C. R., Baimbridge, K. G., Miller, J. J. 1985. The neostriatal mosaic: Compartmental distribution of calcium binding protein and parvalbumin in the basal ganglia of the rat and monkey. *Proc. Natl. Acad. Sci. USA* 82: 8780–84

Gerfen, C. R., Baimbridge, K. G., Thibault, J. 1987a. The neostriatal mosaic. III. Biochemical and developmental dissociation of dual nigrostriatal dopaminergic systems. *J. Neurosci.* 7: 3935–44

Gerfen, C. R., Engber, T. M., Mahan, L. C., Susel, Z., Chase, T. N., et al. 1990. D1 and D2 dopamine receptor-regulated gene expression of striatonigral and striatopallidal neurons. *Science* 250: 1429–32

Gerfen, C. R., Herkenham, M., Thibault, J. 1987b. The neostriatal mosaic. II. Patch-

and matrix-directed mesostriatal dopaminergic and non-dopaminergic systems. *J. Neurosci.* 7: 3915–34

Gerfen, C. R., McGinty, J. F., Young, I. W. S. 1991. Dopamine differentially regulates dynorphin, substance P and enkephalin expression in striatal neurons: In situ hybridization histochemical analysis. *J. Neurosci.* 11: 1016–31

Gerfen, C. R., Sawchenko, P. E. 1984. An anterograde neuroanatomical tracing method that shows the detailed morphology of neurons, their axons and terminals: Immunohistochemical localization of an axonally transported plant lectin, *Phaseolus vulgaris-leucoagglutinin* (PHA-L). *Brain Res.* 290: 219–38

Gerfen, C. R., Staines, W. A., Arbuthnott, G. W., Fibiger, H. C. 1982. Crossed connections of the substantia nigra in the rat. *J. Comp. Neurol.* 207: 283–303

Gerfen, C. R., Young, W. S. 1988. Distribution of striatonigral and striatopallidal peptidergic neurons in both patch and matrix compartments: An in situ hybridization histochemistry and fluorescent retrograde tracing study. *Brain Res.* 460: 161–67

Gilbert, C. D., Kelly, J. P. 1975. The projection of cells in different layers of the cat's visual cortex. *J. Comp. Neurol.* 163: 81–105

Gimenez-Amaya, J. M., Graybiel, A. M. 1990. Compartmental origins of the striatopallidal projections in the primate. *Neuroscience* 34: 111–26

Goldman-Rakic, P. S. 1982. Cytoarchitectonic heterogeneity of the primates neostriatum: Subdivision into island and matrix cellular compartments. *J. Comp. Neurol.* 205: 398–413

Gonzales, C., Chesselet, M.-F. 1990. Amygdalonigral pathway: An anterograde study in the rat with phaseolus vulgaris leucoagglutinin (PHA-L). *J. Comp. Neurol.* 297: 182–200

Graybiel, A. M., Ragsdale, J. C. W. 1978. Histochemically distinct compartments in the striatum of human, monkey and cat demonstrated by acetylcholinesterase staining. *Proc. Natl. Acad. Sci. USA* 75: 5723–26

Graybiel, A. M., Chesselet, M. F. 1984. Compartmental distribution of striatal cell bodies expressing [Met]enkephalin-like immunoreactivity. *Proc. Natl. Acad. Sci. USA* 81: 7980–84

Graybiel, A. M., Ragsdale, J. C. W., Yoneika, E. S., Elde, R. P. 1981. An immunohistochemical study of enkephalins and other neuropeptides in the striatum of the cat with evidence that the opiate peptides are arranged to form mosaic patterns in register with the striosomal compartments visible with acetylcholinesterase staining. *Neuroscience* 6: 377–97

Grofova, I. 1975. The identification of striatal and pallidal neurons projecting to substantia nigra. An experimental study by means of retrograde axonal transport of horseradish peroxidase. *Brain Res.* 91: 286–91

Grove, E. A. 1988. Efferent connections of the substantia innominata in the rat. *J. Comp. Neurol.* 277: 347–64

Grove, E. A., Domesick, V. B., Nauta, W. J. H. 1986. Light microscopic evidence of striatal input to intrapallidal neurons of cholinergic cell group Ch4 in the rat: A study employing the anterograde tracer *Phaseolus vulgaris leucoagglutinin* (PHA-L). *Brain Res.* 367: 379–84

Haber, S. N., Groenewegen, H. J., Grove, E. A., Nauta, W. J. H. 1985. Efferent connections of the ventral pallidum: Evidence of a dual striato pallidofugal pathway. *J. Comp. Neurol.* 235: 322–35

Haber, S. N., Nauta, W. J. H. 1983. Ramifications of the globus pallidus in the rat as demonstrated by patterns of immunohistochemistry. *Neuroscience* 9: 245–60

Hallanger, A. E., Levey, A. I., Lee, H. J., Rye, D. B., Wainer, B. H. 1987. The origins of cholinergic and other subcortical afferents to the thalamus in the rat. *J. Comp. Neurol.* 262: 105–24

Hanson, G. R., Alphs, L., Pradham, S., Lovenberg, W. 1981a. Haloperidol-induced reduction of nigral substance P-like immunoreactivity: A probe for the interactions between dopamine and substance P neuronal systems. *J. Pharmacol. Exp. Ther.* 218: 568–74

Hanson, G. R., Alphs, L., Pradham, S., Lovenberg, W. 1981b. Response of striatonigral substance P systems to a dopamine receptor agonist and antagonist. *Neuropharmacology* 20: 541–48

Harrison, M. B., Wiley, R. G., Wooten, G. F. 1990. Selective localization of striatal D1 receptors to striatonigral neurons. *Brain Res.* 528: 317–22

Hattori, T., McGeer, E. G., McGeer, P. L. 1978. Fine structural analysis of corticostriatal pathway. *J. Comp. Neurol.* 185: 347–54

Haverstick, D. M., Rubenstein, A., Bannon, M. J. 1989. Striatal tachykinin gene expression regulated by interaction of D-1 and D-2 dopamine receptors. *J. Pharmacol. Exp. Ther.* 248: 858–62

Heimer, L., Wilson, R. D. 1975. The subcortical projections of the allocortex: Similarities in the neural associations of the hippocampus, the piriform cortex, and the neocortex. In *Golgi Centennial Symposium*, ed. M. Santini, pp. 177–93. New York: Raven

Herkenham, M. 1986. New perspectives on the organization and evolution of nonspecific thalamocortical projections. In *Cerebral Cortex*, ed. E. G. Jones, A. Peters, pp. 403–45. New York: Plenum

Herkenham, M. 1987. Mismatches between neurotransmitter and receptor localizations in brain: Observations and implications. *Neuroscience* 23: 1–38

Herkenham, M., Edley, S. M., Stuart, J. 1984. Cell clusters in the nucleus accumbens of the rat, and the mosaic relationship of opiate receptors, acetylcholinesterase and subcortical afferent terminations. *Neuroscience* 11: 561–93

Herkenham, M., Pert, C. B. 1981. Mosaic distribution of opiate receptors, parafascicular projections and acetylcholinesterase in rat striatum. *Nature* 291: 415–18

Hershey, A. D., Krause, J. E. 1990. Molecular characterization of a functional cDNA encoding the rat substance P receptor. *Science* 247: 958–62

Hikosaka, O., Wurtz, R. H. 1983a. Visual and occulomotor functions of monkey substantia nigra pars reticulata. III. Memory contingent visual and saccade responses. *J. Neurophysiol.* 49: 1268–84

Hikosaka, O., Wurtz, R. II. 1983b. Visual and occulomotor functions of monkey substantia nigra pars reticulata. IV. Relation of substantia nigra to superior colliculus. *J. Neurophysiol.* 49: 1285–1301

Hong, J. S., Tilson, H. A., Yoshikawa, K. 1983. Effects of lithium and haloperidol administration on the rat brain levels of substance P. *J. Pharmacol. Exp. Ther.* 224: 590

Hong, J. S., Yang, H.-Y. T., Fratta, W., Costa, E. 1978. Rat striatal methionine-enkephalin content after chronic treatment with cataleptogenic and noncataleptogenic drugs. *J. Pharmacol. Exp. Ther.* 205: 141–47

Hong, J. S., Yoshikawa, K., Kanamatsu, T., Sabol, S. L. 1985. Modulation of striatal enkephalinergic neurons by antipsychotic drugs. *Fed. Proc.* 44: 2535–39

Hornykiewicz, O. 1966. Dopamine (3-hydroxytryptamine) and brain function. *Pharmacol. Rev.* 18: 925–64

Jiang, H.-K., McGinty, J. F., Hong, J. S. 1990. Differential modulation of striatonigral dynorphin and enkephalin by dopamine receptor subtypes. *Brain Res.* 507: 57–64

Jimenez-Castellanos, J., Graybiel, A. M. 1987. Subdivisions of the dopamine-containing A8-A9-A10 complex identified by their differential mesostriatal innervation of striosomes and extrastriosomal matrix. *Neuroscience* 23: 223–42

Jimenez-Castellanos, J., Graybiel, A. M. 1989. Compartmental origins of striatal efferent projections in the cat. *Neuroscience* 32: 297–321

Jinnai, K., Matsuda, Y. 1979. Neurons of the motor cortex projecting commonly on the caudate nucleus and the lower brainstem in the cat. *Neurosci. Lett.* 13: 121–26

Jones, E. G. 1984. Laminar distribution of cortical efferent cells. In *Cerebral Cortex,* Vol. 1: *Cellular Components of the Cerebral Cortex,* ed. A. P. Jones, E. G. Jones, pp. 521–53. New York: Plenum

Jones, E. G., Coulter, J. D., Burton, H., Porter, R. 1977. Cells of origin and terminal distribution of corticostriatal fibers arising in the sensory-motor cortex of monkeys. *J. Comp. Neurol.* 173: 53–80

Kawaguchi, Y., Wilson, C. J., Emson, P. 1989. Intracellular recording of identified neostriatal patch and matrix spiny cells in a slice preparation preserving cortical inputs. *J. Neurophysiol.* 62: 1052–68

Kawaguchi, Y., Wilson, C. J., Emson, P. 1990. Projection subtypes of rat neostriatal matrix cells revealed by intracellular injection of biocytin. *J. Neurosci.* 10: 3421–38

Kebabian, J. W., Calne, D. B. 1979. Multiple receptors for dopamine. *Nature* 277: 93–96

Kemp, J. M., Powell, T. P. S. 1971. The structure of the caudate nucleus of the cat: Light and electron microscopic study. *Philos. Trans. R. Soc. London* (Biol.) 262: 383–401

Kemp, J. M., Powell, T. P. S. 1970. The cortico-striate projection in the monkey. *Brain* 93: 525–46

Kita, H., Chang, H. T., Kitai, S. T. 1983. Pallidal inputs to subthalamus: Intracellular analysis. *Brain Res.* 264: 255–65

Kita, H., Kitai, S. T. 1987a. Efferent projections of the subthalamic nucleus in the rat: Light and electron microscopic analysis with the PHA-L method. *J. Comp. Neurol.* 260: 435–52

Kita, H., Kitai, S. T. 1987b. Subthalamic inputs to the globus pallidus and the substantia nigra in the rat: Light and electron microscopic studies using PHA-L methods. *J. Comp. Neurol.* 260: 435–52

Kita, H., Kitai, S. T. 1988. Glutamate decarboxylase immunoreactive neurons in rat neostriatum: Their morphological types and populations. *Brain Res.* 447: 346–52

Kitai, S. T., Koscis, J. D., Preston, R. J., Sugimori, M. 1976. Monosynaptic inputs to caudate neurons identified by intracellular injection of horseradish peroxidase. *Brain Res.* 109: 601–6

Krettek, J. E., Price, J. L. 1977. The cortical projections of the mediodorsal nucleus and adjacent thalamic nuclei in the rat. *J. Comp. Neurol.* 171: 157–92

Le Moine, C., Normand, E., Guitteny, A. F., Fouque, B., Teoule, R., et al. 1990. Dopamine receptor gene expression by enkephalin neurons in rat forebrain. *Proc. Natl. Acad. Sci. USA* 87: 230–34

Lehmann, J., Langer, S. Z. 1983. The striatal cholinergic interneuron: Synaptic target of dopaminergic terminals? *Neuroscience* 10: 1105–20

Li, S. J., Sivam, S. P., McGinty, J. F., Douglass, J., Calavetta, L., et al. 1988. Regulation of the metabolism of striatal dynorphin by the dopaminergic system. *J. Pharmacol. Exp. Ther.* 246: 403–8

Li, S. J., Sivam, S. P., McGinty, J. F., Huang, Y. S., Hong, J. S. 1987. Dopaminergic regulation of tachykinin metabolism in the striatonigral pathway. *J. Pharmacol. Exp. Ther.* 243: 792–98

Lolait, S. J., O'Carroll, A.-M., Kusano, K., Mahan, L. C. 1989. Pharmacological characterization and region-specific expression in brain of the B2- and B3-subunits of the rat GABA$_A$ receptor. *FEBS Lett.* 258: 17–21

Loopuijt, L. D., van der Kooy, D. 1985. Organization of the striatum: Collateralization of its efferent axons. *Brain Res.* 348: 86–99

Mahan, L. C., Burch, R. M., Monsma, J. F. J., Sibley, D. R. 1990. Expression of striatal D1 dopamine receptors coupled to inositol phosphate production and Ca^{2+} mobilization in *xenopus* oocytes. *Proc. Natl. Acad. Sci. USA* 87: 2196–2200

Malach, R., Graybiel, A. M. 1986. Mosaic architecture of the somatic sensory-recipient sector of the cat's striatum. *J. Neurosci.* 6: 3436–10

McKinney, M., Coyle, J. T., Hedreen, J. C. 1983. Topographic analysis of the innervation of the rat neocortex and hippocampus by the basal forebrain cholinergic system. *J. Comp. Neurol.* 217: 103–21

Mitchell, I. J., Clarke, C. E., Boyce, S., Robertson, R. G., Peggs, D., et al. 1989. Neural mechanisms underlying Parkinsonian symptoms based upon regional uptake of 2-deoxyglucose in monkeys exposed to 1-methyl-4-phenyl-1,2,3,6-tetrahyropyridine. *Neuroscience* 32: 213–26

Mocchetti, I., Schwartz, J. P., Costa, E. 1985. Use of mRNA hybridization and radioimmunoassay to study mechanisms of drug-induced accumulation of enkephalins in rat brain structures. *Mol. Pharmacol.* 28: 86–91

Monsma, F. J., McVittie, L. D., Gerfen, C. R., Mahan, L. C., Sibley, D. R. 1989.

Alternative RNA splicing produces multiple D2 dopamine receptors. *Nature* 342: 926–29

Monsma, F. J., McVittie, L. D., Gerfen, C. R., Mahan, L. C., Sibley, D. R. 1990. Molecular cloning and expression of a D1 dopamine receptor linked to adenylyl cyclase activation. *Proc. Natl. Acad. Sci. USA* 87: 6723–27

Nakanishi, H., Kita, H., Kitai, S. T. 1988. An N-methyl-D-aspartate receptor mediated excitatory postsynaptic potential evoked in subthalamic neurons in an in vitro slice preparation in the rat. *Neurosci. Lett.* 95: 130–36

Nauta, W. J. H., Smith, G. P., Faull, R. L. M., Domesick, V. B. 1978. Efferent connections and nigral afferents of the nucleus accumbens septi in the rat. *Neuroscience* 3: 385–401

Normand, E., Popovich, T., Onteniente, B., Fellmann, D., Piatier-Tonneau, D., et al. 1988. Dopaminergic neurons of the substantia nigra modulate preproenkephalin gene expression in rat striatal neurons. *Brain Res.* 439: 39–46

Park, M. R., Lighthall, J. W., Kitai, S. T. 1980. Recurrent inhibition in the rat neostriatum. *Brain Res.* 194: 359–69

Penny, G. R., Afsharpour, S., Kitai, S. T. 1986. The glutamate decarboxylase-, leucine enkephalin-, methionine enkephalin-and substance P-immunoreactive neurons in the neostriatum of the rat and cat: Evidence for partial population overlap. *Neuroscience* 17: 1011–45

Ragsdale, C. W. J., Graybiel, A. M. 1981. The fronto-striatal projection in the cat and monkey and its relationship to inhomogeneities established by acetylcholinesterase histochemistry. *Brain Res.* 208: 259–66

Ragsdale, C. W. J., Graybiel, A. M. 1988. Fibers from the basolateral nucleus of the amygdala selectively innervate striosomes in the caudate nucleus of the cat. *J. Comp. Neurol.* 269: 506–22

Richfield, E. K., Penney, J. B., Young, A. B. 1989. Anatomical and affinity state comparisons between dopamine D1 and D2 receptors in the rat central nervous system. *Neuroscience* 30: 767–77

Robertson, G. S., Vincent, S. R., Fibiger, H. C. 1990. Striatonigral projection neurons contain D1 dopamine receptor-activated c-*fos*. *Brain Res.* 523: 288–90

Royce, G. J. 1982. Laminar origin of cortical neurons which project upon the caudate nucleus: A horseradish peroxidase investigation in the cat. *J. Comp. Neurol.* 205: 8–29

Royce, G. J., Bromley, S. 1984. Fluorescent double labeling studies of thalamostriatal

and corticostriatal neurons. In *The Basal Ganglia*, ed. J. S. McKenzie, R. E. Kemm, L. N. Wilcock, pp. 131–46. New York: Plenum

Saper, C. B. 1984. Organization of cerebral cortical afferent systems in the rat. II. Magnocellular basal nucleus. *J. Comp. Neurol.* 222: 313–42

Sivam, S. P., Strunk, C., Smith, D. R., Hong, J.-S. 1986. Preproenkephalin-A gene regulation in the rat striatum: Influence of Lithium and Haloperidol. *Mol. Pharmacol.* 30: 186–91

Smith, Y., Bolam, J. P., Krosigk, M. V. 1990. Topographical and synaptic organization of the GABA-containing pallidosubthalamic projection in the rat. *Eur. J. Neurosci.* 2: 500–11

Sokoloff, P., Giros, B., Martres, M.-P., Bouthenet, M.-L., Schwarz, J.-C. 1990. Molecular cloning and characterization of a novel dopamine receptor (D_3) as a target for neuroleptics. *Nature* 347: 146

Somogyi, P., Bolam, J. P., Smith, A. D. 1981. Monosynaptic cortical input and local axon collaterals of identified striatonigral neurons. A light and electron microscopic study using the Golgi-peroxidase-degeneration procedure. *J. Comp. Neurol.* 195: 567–84

Spencer, H. J. 1976. Antagonism of cortical excitation of striatal neurons by glutamic acid diethylester: Evidence for glutamic acid as an excitatory transmitter in the rat striatum. *Brain Res.* 102: 91–101

Stoof, J. C., Kebabian, J. W. 1981. Opposing roles for the D-1 and D-2 dopamine receptors in efflux of cAMP from rat neostriatum. *Nature* 294: 366–68

Stoof, J. C., Kebabian, J. W. 1982. Independent in vitro regulation by the D2 dopamine receptor of dopamine-stimulated efflux of cyclic AMP and K+-stimulated release of acetylcholine from rat neostriatum. *Brain Res.* 250: 263–70

Sunahara, R. K., Niznik, H. B., Weiner, D. M., Stormann, T. M., Brann, M. R., et al. 1990. Human dopamine D1 receptor encoded by an intronless gene on chromosome 5. *Nature* 347: 80–83

Swanson, L. W. 1987. Limbic system. In *Encyclopedia of Neuroscience*, ed. G. Adelman, pp. 589–91. Boston: Birkhauser

Tang, F., Costa, E., Schwartz, J. P. 1983. Increase of proenkephalin mRNA and enkephalin content of rat striatum after daily injection of haloperidol for 2 to 3 weeks. *Proc. Natl. Acad. Sci. USA* 80: 3841–44

Trugman, J. M., Wooten, G. F. 1987. Selective D1 and D2 dopamine agonists differentially alter basal ganglia glucose utilization in rats with unilateral 6-hydroxydopamine substantia nigra lesions. *J. Neurosci.* 7: 2927–35

Undie, A. S., Friedman, E. 1990. Stimulation of a dopamine D1 receptor enhances inositol phosphates formation in rat brain. *J. Pharmacol. Exp. Ther.* 253: 987–92

Vincent, S. R., Hökfelt, T., Christensson, I., Terenius, L. 1982. Immunohistochemical evidence for a dynorphin immunoreactive striatonigral pathway. *Eur. J. Pharmacol.* 85: 251–52

Vincent, S. R., Johansson, O. 1983. Striatal neurons containing both somatostatin and avian pancreatic polypeptide (APP)-like immunoreactivities and NADPH diaphorase activity: A light and electron microscopic study. *J. Comp. Neurol.* 217: 252–63

Voorn, P., Groenewegen, H., Gerfen, C. R. 1989. *J. Comp. Neurol.* 289: 189–201

Walters, J. R., Bergstrom, D. A., Carlson, J. H., Chase, T. N., Braun, A. R. 1987. D1 dopamine receptor activation required for postsynaptic expression of D2 agonist effects. *Science* 236: 719–22

Webster, K. E. 1961. Cortico-striate inter-relations in the albino rat. *J. Anat.* 95: 532–44

Weiner, D. M., Levey, A. I., Brann, M. R. 1990. Expression of muscarinic acetyl-choline and dopamine receptor mRNAs in rat basal ganglia. *Proc. Natl. Acad. Sci. USA* 87: 7050–54

Wilson, C. J. 1981. Spontaneous firing patterns of identified spiny neurons in the rat neostriatum. *Brain Res.* 220: 67–80

Wilson, C. J. 1987. Morphology and synaptic connections of crossed cortico-striatal neurons in the rat. *J. Comp. Neurol.* 263: 567–80

Wilson, C. J., Chang, H. T., Kitai, S. T. 1990. Firing patterns and synaptic potentials of identified giant aspiny interneurons in the rat neostriatum. *J. Neurosci.* 10: 508–19

Wilson, C. J., Groves, P. M. 1980. Fine structure and synaptic connections of the common spiny neuron of the rat neostriatum: A study employing intracellular injection of horseradish peroxidase. *J. Comp. Neurol.* 194: 599–615

Yokota, Y., Sasai, Y., Tanaka, K., Fujiwara, T., Tsuchida, K., et al. 1989. Molecular characterization of a functional cDNA for rat substance P receptor. *J. Biol. Chem.* 264: 17649–52

Young, W. S. III, Bonner, T. I., Brann, M. R. 1986. Mesencephalic dopaminergic neurons regulate the expression of neuro-peptide mRNAs in the rat forebrain. *Proc. Natl. Acad. Sci. USA* 83: 9827–31

Zhou, Q.-Y., Grandy, D. K., Thambi, L., Kushner, J. A., Tol, H. H. M. V., et al. 1990. Cloning and expression of human and rat D1 dopamine receptors. *Nature* 347: 76–80

Annu. Rev. Neurosci. 1992. 15:321–51
Copyright © 1992 by Annual Reviews Inc. All rights reserved

VOLTAGE-SENSITIVE DYES AND FUNCTIONAL ACTIVITY IN THE OLFACTORY PATHWAY

Angel R. Cinelli* and John S. Kauer

Section of Neuroscience, Department of Neurosurgery,* and Department of Cell Biology and Anatomy, Tufts Medical School and New England Medical Center, Boston, Massachusetts 02111

KEY WORDS: information processing, odors, olfactory bulb, olfactory epithelium, ensemble recording

BACKGROUND AND INTRODUCTION

Many of the difficulties encountered in studying how information is handled in the CNS relate to problems with observing and characterizing activities of large numbers of neurons, yet it is undoubtedly the interactions between multitudes of cellular elements that underlie the complexity of processing in neuronal circuits. One of the greatest challenges facing the study of circuitry function, then, is to be able to observe the anatomical and physiological characteristics of many cells simultaneously in minute detail. The large numbers and great diversity of neurons, the segregation of processing functions within them, and the enormous complexity of their interconnections require methods that permit simultaneous observations to be made from many neuronal sites with high spatial and temporal resolution.

This review summarizes results obtained with voltage-sensitive dye recording, a method having good spatial and temporal resolution for observing distributed neuronal activity, as it has been applied to the olfactory pathway. Since spatially and temporally distributed activity patterns appear to be basic components of odorant encoding, the olfactory system has been useful for analyzing general mechanisms, possibly relevent to other brain regions, by which such patterns might carry information. In the present review we describe results obtained with voltage-sensitive dye

321

0147–006X/92/0301–0321$02.00

recording in the context of physiological properties about which information is already available, as a first step in assessing the fidelity of optical recording in this system. We focus on the olfactory pathway because several reviews have recently been published covering the use of optical methods in other systems (Blasdel 1989, Cohen et al 1989, Lieke et al 1989, Ross 1989, Salzberg 1989). We do not discuss methodological details at length, since they also have been described elsewhere (Cohen et al 1978, Grinvald 1985, Grinvald et al 1988, Salzberg 1983).

The difficult problem of exploring how individual neurons functionally interact in neural circuits has historically been approached via acute and chronic multiple-site EEG and single unit recording techniques, or methods that measure the distribution of metabolic activity (such as 2-deoxyglucose), protein synthesis, or expression of RNA message associated with activity (such as *c-fos*). Recently, optical methods employing voltage-sensitive (and other) dyes have also been used for these purposes, usually focusing on two aspects of the problem: (*a*) to analyze, in detail, events occurring simultaneously in identifiable, often in vitro, single neurons or neuronal processes, and (*b*) to study the structure of spatiotemporal activity patterns in large neuronal populations in which signals from single cells are not necessarily individually resolved.

Optical recordings from single neurons were first carried out on the squid giant axon; this preparation is still useful for screening new optical probes, although certain dyes are found to work better in some preparations than in others (Cohen & Lesher 1986). To assess the usefulness of a dye for investigating a particular question, then, the dye should be tested in the preparation of interest. In addition to the squid axon, single-cell resolution has also been obtained in somata of intact invertebrate ganglia (Krauthamer & Ross 1984, Ross & Krauthamer 1984, Salzberg et al 1973) and from single neurons and their processes in tissue culture (Grinvald et al 1981, 1983). In a similar way, exogenously applied optical probes have been used for recording membrane potential changes in fine nerve terminals of the neurohypophysis in *Xenopus* (Salzberg et al 1983) and mouse (Gainer et al 1986, Obaid et al 1983, Salzberg et al 1984). In intact preparations, the method has been applied to the analysis of individual neuronal interactions involved in simple behaviors (London et al 1987, Zecevic et al 1989). Voltage-sensitive dyes have also been injected intracellularly for analysis of events in single identified neurons, although signals obtained under these conditions thus far are small (Grinvald et al 1987), and better results have usually been obtained with optical probes for calcium, which often give larger signals (Connor 1986, Regehr et al 1989, Ross 1989, Tank et al 1989).

Many of these studies have had the important advantage of being able directly to control for toxicity or for unwanted interactions of the dye with the tissue through conventional physiological recording methods. Comparisons of optical signals with electrophysiological events are more difficult when recording from neuron ensembles in which cellular resolution is not achieved or from which recording the same transmembrane phenomena with electrodes may not be possible. In this context, studies on layered brain structures either in situ or in slices have been particularly useful for analyzing the origins of the optical signals by relating them to their spatial distributions. In addition to the olfactory structures described here, these kinds of studies have been carried out in slices of hippocampus (Grinvald et al 1982, Saggau et al 1986) and cerebellum (Konnerth et al 1987).

In studies of in vivo optical responses, defining the relationship between the signals and their origins in particular neurons, processes, or laminae has been more difficult, although the major goal of these studies has often not been to assign physiological signals to specific cellular components but rather to define their overall spatio-temporal properties. With this approach, in vivo optical responses have been studied in mammalian visual (Blasdel & Salama 1986, Grinvald et al 1986, Lieke et al 1989, Orbach et al 1985), somatosensory (London et al 1989, Orbach et al 1985), gustatory (London 1990), and olfactory (Cattarelli & Cohen 1989) cortices, in the optic tectum of the frog (Grinvald et al 1984), and in the olfactory bulb of the salamander (Kauer 1988, Kauer et al 1987, Orbach & Cohen 1983, Cinelli et al, submitted). Recently, Ts'o et al (1990) have reported that responses to visual stimuli can also be obtained in mammalian occipital cortex without using dyes. Such "intrinsic" optical signals, which probably relate to light scatter during transmembrane voltage changes, may prove helpful for clinical situations in which the use of dyes is difficult.

Methodology

The technical aspects of voltage-sensitive dye recording have been extensively reported in previous reviews (Cohen & Lesher 1986, Cohen & Salzberg 1978, Grinvald 1985, Salzberg 1983). Generally, these dyes are thought to function as voltage transducers by binding to the plasma membrane in ways that allow their fluorescence or absorption to be modulated by changes in transmembrane voltage. Changes in the optical properties of dye itself can occur in microseconds (Cohen & Salzberg 1978, Salzberg 1989). Therefore, temporal resolution of neuronal signals is normally limited not by the dyes themselves but by the need to reduce the bandwidth of the photo-detectors to achieve adequate signal to noise ratios.

Extracellular delivery of voltage-sensitive dyes usually consists of little more than a topical application of them, but observations of the signals require specialized sets of photodetectors or an especially sensitive video camera, analog to digital converters, and a computer. Both epi- and trans-illumination of fluorescent and absorption dyes have been successfully used to make potential measurements (Orbach & Cohen 1983). The apparent time courses of signals in optical records are critically dependent on the time constants of the capacitive coupling in diode detector recording systems and on detector lag in video camera based systems (Connor 1986, Lasser-Ross et al 1990, Kauer & Cinelli, in preparation). Photodiode array detectors are generally required for measuring rapid signals such as action potentials, although spatial resolution has been limited to 124 or, more recently to 464 (Senseman et al 1990a) recording sites. Improved spatial resolution (up to 256,000 sites) can be obtained by using video camera detectors (Blasdel & Salama 1986, Kauer 1988), but here, temporal resolution is usually limited by the standard video frame rate of 33 ms/frame (although see Figure 3B). A promising combination of reasonable sensitivity and rapid response with the high spatial resolution of solid state video chips may arise from the development of specialized partially parallel read-out diode arrays or specialized video systems (Matsumoto & Ichikawa 1990, Lasser-Ross et al 1991).

The following discussion summarizes results that have been obtained using optical recording methods in the olfactory pathway. Although these studies are in their early stages of development and many of the data are still in preliminary form, the potential utility of the approach has become apparent.

OPTICAL METHODS APPLIED TO THE OLFACTORY SENSORY EPITHELIUM

Individual olfactory receptor cells have broad odor response spectra (Revial et al 1978a,b, 1982, 1983), and odorants having different qualities generate different gradients of activity over the olfactory mucosa (Kauer & Moulton 1974, Mackay Sim et al 1982, Mustaparta 1971, Thommesen & Doving 1977). These findings suggest that at the level of the olfactory epithelium, odors are encoded by patterns of activity distributed across the receptor cell population (see Kauer 1991 for review).

To examine these distributed properties, Kent (1990) and Kent & Mozell (1988) have recently used a photodiode array and the potentiometric probe WW781 to study odor responses from frog and salamander olfactory epithelia. The distributions and time courses of these signals suggest they arise from the same sources as the electrically recorded electro-olfactogram

AMYL ACETATE   d-LIMONENE   BUTANOL

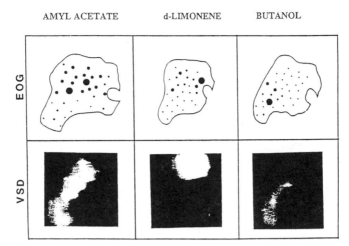

Figure 1 Comparisons of the spatial distributions of responses to three different odorants applied to the salamander olfactory mucosa via electro-olfactogram (EOG) and voltage-sensitive dye (VSD) recording. In the EOG maps, the size of each dot is proportional to the intensity of the response (from Mackay Sim et al 1982 with permission). Although only maximum activities are shown in the VSD maps, lower-level responses were found throughout the mucosa (from Kent 1990 with permission).

(EOG) (Ottoson 1956). As shown in Figure 1 for the salamander, the spatial distribution of the optical responses agrees well with activity patterns obtained in previous EOG studies for a number of odorants (Mackay Sim et al 1982). Consistent with earlier findings, each pure odorant evoked larger responses in some regions and smaller responses in most, if not all, other regions of the mucosa. These results confirm the presence of a heterogeneous, widespread distribution of receptor responsivity and provide additional evidence for relatively low odor response specificity in single receptor cells. A significant, additional step in using optical recording will be to image the olfactory epithelium at magnifications sufficient to observe activity in individually identifiable cells, especially in the context of the recent evidence for the possibility of a large multigene family that may code for odorant receptor proteins (Buck & Axel 1991).

OPTICAL METHODS APPLIED TO THE OLFACTORY BULB

The first synaptic relay in the olfactory pathway, the olfactory bulb, is a highly laminated structure having well-defined circuitry in which activity can be triggered not only by synchronous electrical pulses but also by physiologically appropriate odorant stimuli. Although the morphology and functional properties of the bulb have been studied in detail in a number of species (for reviews see Halasz 1990, Mori 1987, Shepherd 1972), specific data on how the chemical properties of odors are encoded by this structure are still lacking.

As at the epithelial level, bulbar responses to single odorant stimuli are widely distributed (Adrian 1950, Kauer 1991, Moulton 1976, Sharp et al 1975, Stewart et al 1979) and thus optical methods would appear to be particularly well-suited for examining them. Orbach & Cohen (1983) first demonstrated that optical signals could be recorded from the salamander olfactory bulb in response to electrical stimulation. Figure 2 shows a

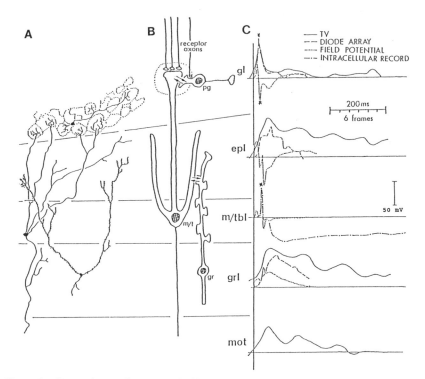

Figure 2 Comparisons of events evoked by orthodromic electrical stimulation in different layers of the salamander olfactory bulb with field potential (· · ·) and intracellular (– · – ·) electrophysiological recording, and photodiode (- - - -) and video camera VSD (——) recordings. The distribution of major cell types (periglomerular, mitral/tufted, and granule cells) in the bulbar layers are shown in A, a highly simplified circuit diagram in B, and a comparison of the time courses from the different recording methods in C. Abbreviations: pg, periglomerular cell; gl, glomerular layer; epl, external plexiform layer; m/tbl, mitral/tufted body layer; grl, granule layer; m/t, mitral/tufted cell; mot, medial olfactory tract (from Kauer 1988 with permission).

comparison among various electrophysiological measures, including field potentials and intracellular recordings from mitral/tufted cells, and optical records by using both photodiode and video camera detectors from single sites in different layers of the salamander bulb. Figures 3D, 8, and 9 illustrate distribution patterns of these responses. These data also demonstrate that the time courses of optical signals are, in general, consistent with those observed using electrophysiological measures.

Several merocyanine and styryl voltage-sensitive dyes have been tested in the salamander olfactory bulb in vivo and in vitro (Orbach & Cohen 1983). In the fluorescence mode, RH160, RH414, and RH795 (Grinvald et al 1982a) have been used most extensively. Similar patterns of depolarization and hyperpolarization after orthodromic electrical stimulation have been seen with each one, as well as with the pyrazo-oxonal dyes RH482 and RH155 in the absorbance mode (Cinelli & Salzberg 1987). A summary of some of the properties of these dyes is shown in Table 1.

The largest fluorescence and absorption signals have been obtained with the styryl dyes RH414 and RH795 (Grinvald et al 1982a, Kauer 1988, London 1990, Orbach & Cohen 1983) and the pyrazo-oxonol dye RH155 (Cinelli & Salzberg 1988, Konnerth et al 1987), making possible single trial experiments that do not require several runs to be averaged. At moderate electrical stimulus intensities (2–4 × threshold), contributions of intrinsic signals (Cohen et al 1968, 1972, Lieke et al 1989) observed in the absorption mode were negligible, although with stronger electric stimuli they became more apparent (Cinelli & Salzberg 1988, 1990, in preparation). These intrinsic changes had characteristics similar to those studied in greater detail in other systems (Salzberg et al 1985) in that they were slow, long-lasting, and represented an increase in tissue transparency that was manifest as a decrease in absorbance upon stimulation. Both intrinsic and extrinsic responses were lost when 100–150 mM KCl was applied to the tissue.

Optical recordings from the bulb after electric stimulation have been obtained in skate (*Raja erinacea*) (Cinelli & Salzberg 1988, 1990), salamander (*Ambystoma tigrinum*) (Kauer et al 1987, Kauer 1988, Orbach & Cohen 1983, Senseman et al 1990), and mouse (Cinelli & Salzberg 1987). Although there are some differences in architectonic structure among bulbs of these species, certain anatomical features appear to be relatively constant; that is, the ubiquitous presence of glomeruli, the structure of the major laminae, and the relationships among interneurons and output cell types (Andres 1970, Halasz 1990, Northcutt & Kicliter 1980). There also are similarities in a number of physiological responses, including the presence of long-lasting inhibition (Halasz 1990) after both electrical and odorant stimulation.

Table 1 Selected properties of some voltage-sensitive dyes

Volt. Dye	Other Name	Description	MW	Dye Ref.	Opt. Resp.	Peak Abs.	Exc/Em.	S/N ratio	Struc.	Species
XVII		merocyanine		(4)	abs	750		++	OB	sal.
XXV		mer-oxonol		(4)	abs	660			OB	sal.
XXII	NK2367 (NK)	mer-oxonol		(4)	abs	675			OB	sal.
RGA 452		oxonol		(3)	abs/flu	510	510/660		OB	sal.
RGA 461		oxonol		(3)	abs/flu	510	510/660		OB	sal.
RH 155	NK3041 (NK) R1114 (MP)	pyra-oxonol	941	(1)	abs	705		++++	OB,PC	ska/sal. /mse.
RH 160	S1107 (MP)	styryl	470	(1)	abs/flu	526	514/705	++	OB	sal.
RH 237		styryl		(1)	abs/flu	544	544/714		OB	sal.
RH 292		styryl		(1)	abs/flu	550	550/715	+	OB	sal.
RH 364		styryl		(1)	abs	500		++	OB	sal.
RH 414	T1111 (MP)	styryl	569	(2)	abs/flu	531	531/714	+++	OE, OB	frog/sal.
RH 482	NK2761 (NK) JPW1132(LB)	pyra-oxonol	768	(1)	abs	705			OB	sal./mse
RH 795	R649 (MP)	styryl	569	(3)	abs/flu	530	530/712	++++	OB	sal.
WW 781	W 435 (MP)	merocyanine	759	(4)	abs/flu	605	605/639	++	OE	frog/sal.
DiSC3 (5)	D 306 (MP)	carbocyanine	546	(5)	abs/flu	622	622/675		OE	frog

(1) Grinvald et al 1982a.
(2) Grinvald et al 1984.
(3) Orbach et al 1985.
(4) Waggoner 1976.
(6) Waggoner 1979.
(7) Shapiro 1981.
(MP) Molecular Probes, Eugene, OR.
(NK) Nippon Kankoh-Shikiso Kenkyusho, Okayama, Japan.

Compared to recordings in the living animal, in vitro experiments have the obvious advantage of having fewer mechanical artifacts due to circulation and respiration, although cardiac pulsations in vivo can be removed by subtraction after acquiring optical signals that have been synchronized to them (Kauer 1988, Orbach et al 1985, Orbach & Cohen 1983). In addition, in vitro preparations permit control over and manipulation of the ionic and pharmacological environments (Cinelli & Salzberg 1990, in preparation) and allow simultaneous intracellular and patch clamp records to be obtained (Wellis & Kauer 1991). One difference noted between in vivo and in vitro preparations is that the degree of bulbar inhibition seen in the external plexiform layer is often reduced in vitro, perhaps because centrifugal fibers from the brain have been eliminated (Cinelli & Salzberg 1990).

Optical Responses to Electrical Stimulation

FAST DEPOLARIZING RESPONSES   Electrical stimulation of olfactory afferents evokes a rapid, early peak in the olfactory nerve and glomerular regions and a later, rapid peak in the deeper bulbar layers near the region of mitral/tufted cell bodies and axons (see individual traces in Figure 3A–C and spatial distribution in Figure 3D). The early, fast component (*large arrows* in Figure 3) seems to be presynaptic, related to synchronous action potential discharges in olfactory nerve afferents in the peripheral bulbar layers. The later, fast component appears to be postsynaptic, related to action potentials in mitral/tufted cells (or interneurons) of deeper bulbar laminae (Orbach & Cohen 1983) (*small arrows* in Figure 3). This interpretation is supported by the fact that the presynaptic, olfactory nerve component remains constant after synaptic blockade with Cd^{2+} (Cinelli & Salzberg 1987, 1990), or low Ca^{2+}/high Mg^{2+} (Cinelli & Salzberg 1988), and in condition/test shock experiments (Figure 3B), whereas the second fast and slow components are lost. Tetrodotoxin abolishes both rapid and slow components completely. The superficial component also has been localized to the peripheral bulb layers by 16 ms/frame video data acquisition (Figure 3C), which affords high spatial resolution. Senseman et al (1990a) have measured the early fast component with a diode array. These studies have presented preliminary evidence on the distributed nature of olfactory nerve afferents as they spread across the bulb by recording optical signals related to the propogated potentials in the nerve with millisecond time resolution.

The later, fast component evoked by orthodromic stimuli is seen predominately in the external plexiform and the mitral/tufted body layers (Cinelli & Salzberg 1987, 1990) (see Figure 3D). This component exhibits a latency slightly longer than the fast component evoked in the olfactory

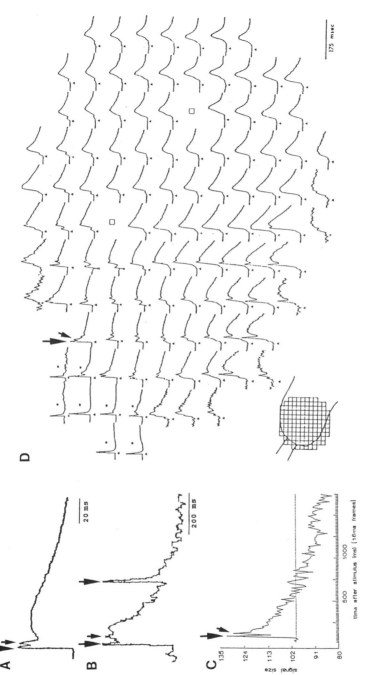

Figure 3 Optical responses from the olfactory bulb after electric orthodromic stimulation. A. Photodiode traces from one periglomerular site in a slice of mouse bulb. Early (*large arrows*) and later (*small arrows*) fast responses precede a slow depolarization (from Cinelli & Salzberg 1987 with permission). B. Photodiode recording from a single external plexiform layer (epl) site in salamander bulb slice (Cinelli & Salzberg 1988 with permission). Postsynaptic components are suppressed by preceding conditioning pulse (interstimulus interval; 400 ms). C. Plot of response at one pixel location (out of 16 K) from a video recording (16 ms/frame) at one epl location from an in vivo salamander olfactory bulb also shows two fast depolarizing responses (*arrows*) (from Kauer et al 1991 with permission). D. Optical responses from photodiode array showing responses at 111 different bulbar locations. *Inset* shows position of the array over the bulb; note the two fast, early depolarizing components seen near the glomerular region (from Orbach & Cohen 1983 with permission). Calibrations: A. 25 ms, C. 400 ms, D. 200 ms, E. 175 ms.

nerve layer, and its refractory period differs from the olfactory nerve compound action potential, appearing more like that of the first negative-going (N1) wave of the orthodromically evoked field potential (Cinelli & Salzberg 1990) (Figure 4 IIA). This component may thus be a manifestation of depolarization in mitral/tufted cells.

SLOW DEPOLARIZING RESPONSES Electrical olfactory nerve stimulation also evokes a slower, relatively long-lasting depolarizing response (Orbach & Cohen 1983) both in vitro and in vivo (see Figures 2–8). These signals had a time course lasting 100–250 msec in the skate (Cinelli & Salzberg 1990), 200–400 msec in the salamander (A. R. Cinelli and B. M. Salzberg, in preparation; A. R. Cinelli et al, in preparation), and 30–80 msec in the mouse (Cinelli & Salzberg 1987), with durations inversely proportional to stimulus intensity (Cinelli & Salzberg 1988, Cinelli et al, in preparation). As described below, various experimental conditions can generate both suppression and/or facilitation of this signal (Figure 4I, II).

In each of the species tested, this response had its onset as early as the N1 wave of the evoked field potential. At bulbar locations, where fast components were seen deep to the glomerular level, the fast signals were partially obscured by the beginning of the slow component (Cinelli & Salzberg 1987, 1990), thus suggesting that the slow signals begin early in the response. Due to its slow rise time, its peak appeared to be approximately coincident with the second negative-going (N2) wave of the field potential, depending on temperature and stimulus intensity (see Figures 2 and 4 IIA). Unlike the two fast responses, this response was small in the olfactory nerve layer and mitral/tufted axonal pathways, and was observed predominantly in the external plexiform layer, where it reached maximum amplitude (Cinelli & Salzberg 1990, in preparation; A. R. Cinelli et al, submitted). The external plexiform layer is the location where branches of mitral/tufted secondary dendrites and granule cell dendrites arborize most profusely (see Figures 2 and 3D), and the slow response may thus be related to depolarization of one or both of these neuron populations.

Suppression and facilitation of these signals have been tested via a double shock paradigm as shown in Figure 4. The size, latency, time course, and spatial spread of the signals were related to stimulus intensity in a nonlinear way. In all species studied, the slow responses to test stimuli were partially suppressed when conditioning volleys were delivered at more than three times threshold. The degree of suppression was related not only to stimulus intensity but also to the condition/test interval and to the bulbar sector in which they were monitored (see Figure 4 IIB) (Cinelli & Kauer 1991, Cinelli & Salzberg 1990, Orbach & Cohen 1983). At short condition/test intervals (Figure 4 IA), the slow responses to the test stimuli

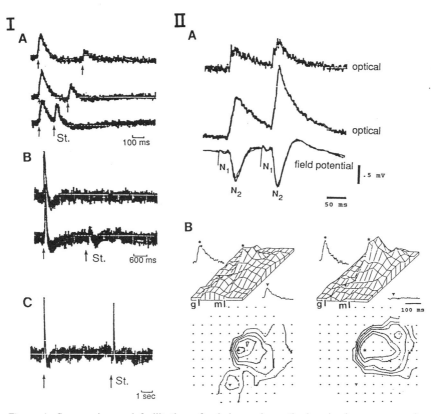

Figure 4 Suppression and facilitation of subglomerular orthodromic slow responses in slices from skate olfactory bulb.

I. Suppressive effects evoked by paired pulses. IA shows the stronger suppression on the test responses evoked by conditioning volleys at interstimulus intervals of 300, 200, and 100 ms. IB shows the degree of suppression at longer intervals of 1.8 and IC shows that at 5 secs.

II. Facilitatory effects of stimuli applied close to the threshold level. In IIA, facilitation of a subsequent response on the optical activity at the mitral/tufted layer (*top trace*), and especially in the region (cpl) just above this (*middle trace*) (120 ms interstimulus interval); observe that a similar effect was obtained on the field potential (*bottom trace*). Calibration marks represent 50 ms for IIA and B; gl and ml stand for glomerular and mitral/tufted layers, respectively; N1 and N2 for the field potential waves. From Cinelli & Salzberg 1990, with permission. In IIB, three-dimensional plots (*upper diagrams*) and iso-optic contours (*lower diagrams*) illustrate the area of activity at 30 msec, at two intensities (1.5 and 2.0 × threshold for left and right, respectively); *asterisks* and *triangles* indicate the actual positions of the isolated traces (*left* and *right* of each diagram) taken from the areas of maximal activity during the weak stimulus; note as the stimulus intensity increases, the spread of the activated area diminishes, but peak response increases.

were reduced either in size or duration, since they occurred on the abrupt falling phase of the conditioning signal. At longer condition/test intervals, a more dramatic period of suppression appeared (Figure 4 IB, C), the duration and magnitude of which depended on stimulus intensity.

In contrast to these results, orthodromic stimuli just above threshold evoked a period of facilitation in the slow signals (see Figure 4 II). Similar behavior was observed in the N2 wave of the field potential (Cinelli & Salzberg 1990, in preparation) (Figure 4 IIA). The magnitude of test response facilitation depended on the condition/test intervals and on stimulus intensity. In the skate (Cinelli & Salzberg 1990) and salamander (Cinelli & Kauer 1991, A. R. Cinelli and B. M. Salzberg, in preparation), the degree of facilitation of the test responses was inversely related to stimulus strength. This facilitation may arise from an enhancement of mitral/tufted cell dendritic depolarization, which, in turn, may increase excitatory transmitter release and thus enhance granule dendritic EPSPs, leading to an increase in the N2 wave (Cinelli & Salzberg 1990). Although differences in the time courses of these events were evident in the different species studied, all showed similar periods of suppression and facilitation that correlated well with events observed with conventional electrophysiological recording methods (Cinelli & Salzberg 1990, Orbach & Cohen 1983).

Low concentrations of sodium or tetrodotoxin in the bath solution abolished all slow component responses evoked by either olfactory nerve or tract stimulation in all species in vitro and in vivo (Figure 5 IA), but they continued to be elicited with direct, focal stimulation of the glomerular/external plexiform region (Cinelli & Kauer 1990, Cinelli & Salzberg 1988, in preparation) (Figure 5 IB), which does not rely on activation of bulbar circuitry via Na^+-dependent, propagated action potentials in the nerve or tract.

In each of the species tested, the slow component was dependent on extracellular Ca^{2+} concentration (Cinelli & Kauer 1990, Cinelli & Salzberg 1987, 1988, 1990) regardless of the type of stimulation used (Figure 5 II). All orthodromic slow responses at the subglomerular level were decreased in Ringer's solution with lowered Ca^{2+} or Cd^{2+}, since synaptic transmission is blocked at the glomerular level. Focal stimulation applied to the external plexiform layer under these conditions also showed reduced slow responses. These data do not differentiate between a slow component elicited by calcium-related synaptic transmission and processes related to a calcium current.

Additional evidence on the involvement of calcium is provided by the observation that the slow depolarizations were enhanced when Ca^{2+} was increased in the bath solution or when it was replaced by Ba^{2+} or Sr^{2+} (Cinelli & Salzberg 1990, in preparation) (Figure 5 III). These effects

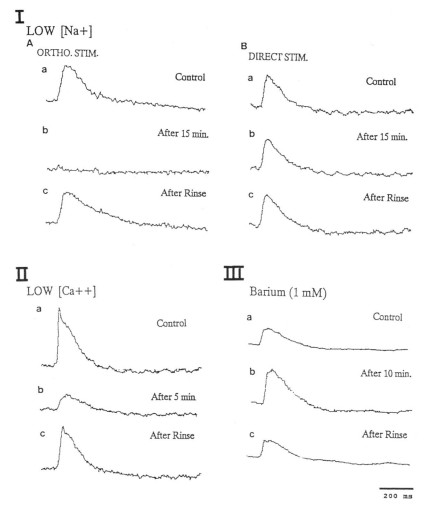

Figure 5 I. Sodium dependency of the external plexiform layer optical responses evoked by orthodromic electric volleys applied to the olfactory nerves of the salamander (A), or by direct stimulation of the glomerular level (B). In both A and B, 96% of Na^+ was replaced by choline. Under these conditions, only the response evoked by the direct stimulation remains (b). The signals returned after washing in normal Ringer's solution (c).

II and III. Calcium dependency of optical signals in salamander olfactory bulb slices. II shows the reduction on the size of the slow component evoked by direct stimuli at the glomerular layer, in a low $[Ca^{2+}]_0$ medium (from 3.6 mM to 0.72 mM Ca^{2+}, by Mg^{2+} substitution) (b). Note the recovery of the signal after wash in normal Ringer's solution (c).

III shows the effects of replacing Ca^{2+} by 1 mM Ba^{2+} on subglomerular signals elicited by orthodromic stimuli. Note the enhancement of the response and the return to baseline with subsequent Ringer wash. From Cinelli & Salzberg 1988 and 1991, in preparation, with permission.

probably originate either from a reduction of inhibitory feedback from granule cells or from deficient activation of a calcium-dependent K^+ current. It is well accepted that Ba^{2+} can effectively replace Ca^{2+} as a current carrier, but cannot serve as effectively for synaptic neurosecretion or for activation of other (more specific) currents (Hagiwara & Byerly 1981, Hagiwara et al 1974, Meech 1978). These events have been further analyzed in preliminary experiments designed to measure calcium directly with calcium-sensitive optical probes (Cinelli et al 1989, Cinelli & Salzberg 1988, Senseman et al 1990b), but the role of calcium still has not been determined conclusively.

Potassium channel blockers such as tetraethylammonium (TEA), Cs^{2+}, and Ba^{2+} produced a broadening of both the fast and slow responses evoked by either orthodromic or antidromic stimulation in all the species studied (Cinelli & Salzberg 1987, 1988, 1990) (Figure 5 III). The orthodromic slow component especially showed an increase in magnitude and duration after TEA. TEA and Ba^{2+} applied together produced a greater change in the slow response than either drug alone, even when applied in higher doses. These effects may reflect an enhancement of the depolarizing events resulting from blockade of outward currents and/or the lack of proper repolarization in the processes that generate the optical signals (Cinelli & Salzberg 1990).

Glial depolarization, resulting from extracellular potassium accumulation during neural activity, has been suggested to contribute to the slow components (Konnerth et al 1985, 1987, Lev-Ram & Grinvald 1986). However, barium at relatively high concentrations, reduces glial depolarization produced by an increase in extracellular potassium (Ballanyi et al 1987, Connors & Ransom 1986). This effect also has been observed with optical measurements of membrane potential (Astion et al 1989, Lev-Ram & Grinvald 1986). Barium does not abolish either the orthodromic (Figure 5 III) or antidromic slow optical responses in bulbar slices, but increases their size and duration (Cinelli & Kauer 1990, Cinelli & Salzberg 1987, 1990, in preparation), thus suggesting that the glial contribution to these responses is not substantial.

Much of the inhibition onto mitral/tufted cells is generally thought to be exerted through GABA released at dendrodendritic synapses with either periglomerular (Halasz & Shepherd 1983, Kosaka et al 1985) or granule cell dendrites (Margolis 1981, Mugnaini et al 1984, Ribak et al 1977). In the species studied, GABA application reduced the slow components in a dose-dependent manner without modifying the early, fast signals (see Figure 6 IA). This effect was also observed after focal stimulation of the external plexiform layer (Cinelli & Salzberg 1990, in preparation). These data suggest that GABA, acting on mitral/tufted processes, suppresses the activity in neuronal elements that generate this slow signal. That baclofen (GABA$_b$ agonist) but not muscimol (GABA$_a$ agonist) had a greater suppressive effect on the slow signal than authentic GABA suggests that a GABA$_b$ receptor may be involved (Cinelli & Salzberg 1990) (see Figure 6 IB). Phaclophen has not yet been tested.

Picrotoxin (Hamilton et al 1988, A. R. Cinelli et al, in preparation) and bicuculline (Cinelli & Salzberg 1987, 1990, in preparation), both GABA$_a$ antagonists, can enhance the slow component, although these effects are not as dramatic as with the agonists (see Figure 6 IIA, B). These compounds may act predominantly on hyperpolarizing responses close to the soma, which appear to be elicited by a GABA-mediated Cl$^-$ conductance (Jahr & Nicoll 1982, Nowycky et al 1981); thus both GABA receptor subtypes may be involved.

Glutamate appears to be a major excitatory neurotransmitter in both axonal and dendritic synapses of mitral/tufted cells (Blakely et al 1987, Halasz & Shepherd 1983). In addition, NMDA receptors may be involved in the excitatory actions from mitral to granule cell dendrites (Jacobson et al 1986, Tombley & Westbrook 1990). In optical recordings, the application of selective NMDA receptor blockers increases the size and the duration of the orthodromic slow component, as well as reducing the following period of hyperpolarization (Cinelli & Salzberg 1990, Hamilton et al 1988).

HYPERPOLARIZING RESPONSES   Under certain conditions, orthodromically applied electrical stimuli can evoke an early hyperpolarization in the glomerular region that precedes the onset of a depolarizing response, similar to that seen with odor stimulation (Kauer et al 1990, Hamilton & Kauer 1985, 1989, Wellis et al 1989, A. R. Cinelli et al, submitted). To elicit these signals in the salamander, low-intensity electric volleys at distal sites on the olfactory nerve were required (Figure 7 IA, B). This response component

---

Figure 6   Pharmacological actions on the orthodromic slow responses. I shows the effects of GABA and its agonists on optical signals. IA shows that GABA directly applied to the bath did not affect the fast component observed at the olfactory nerves (*upper traces*) but selectively and reversibly reduced the amplitude and increased the latency of the slow component (*lower traces*). A reduction in the size of the slow component was also reversibly produced by 50 μM baclofen, a GABA$_b$ agonist (IB trace), without affecting its latency. IC shows that muscimol (a GABA$_a$ agonist) did not significantly affect the size of the slow component but increased its latency after the postsynaptic fast component observed in the region of mitral/tufted somata region. II shows the effects of the GABA$_a$ antagonists bicuculline (100 μM) (A) and picrotoxin (100 μM) (B), which enhanced the size and duration of the slow responses. The traces in A were recorded with a diode array; those in B with a video camera. Note the blocking of the long-lasting hyperpolarization with picrotoxin (*dotted line* in Bb).

I

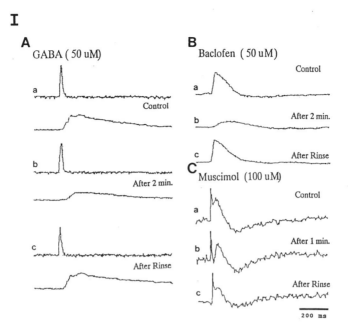

A

GABA (50 uM)

a   Control

b   After 2 min.

c   After Rinse

B

Baclofen (50 uM)

a   Control

b   After 2 min.

c   After Rinse

C

Muscimol (100 uM)

a   Control

b   After 1 min.

c   After Rinse

200 ms

II

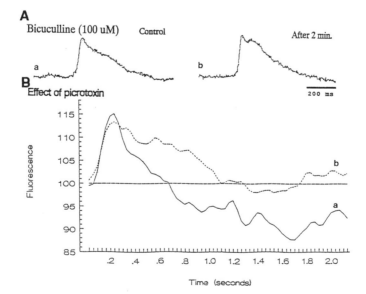

A

Bicuculline (100 uM)   Control   After 2 min.

a   b

200 ms

B

Effect of picrotoxin

b

a

Fluorescence — 115, 110, 105, 100, 95, 90, 85

Time (seconds) — .2 .4 .6 .8 1.0 1.2 1.4 1.6 1.8 2.0

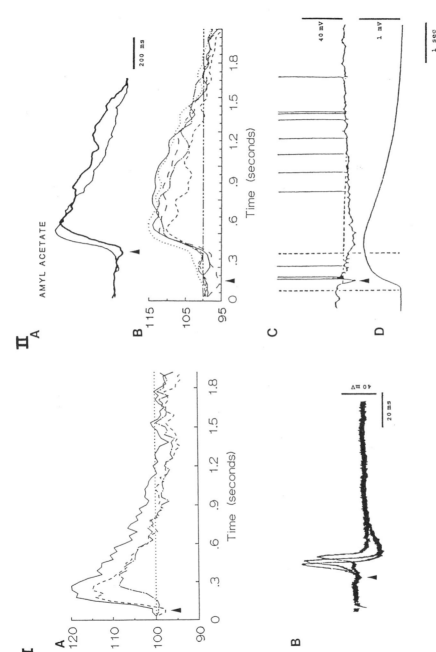
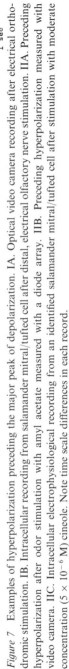

Figure 7 Examples of hyperpolarization preceding the major peak of depolarization. IA. Optical video camera recording after electrical orthodromic stimulation. IB. Intracellular recording from salamander mitral/tufted cell after distal, electrical olfactory nerve stimulation. IIA. Preceding hyperpolarization after odor stimulation with amyl acetate measured with a diode array. IIB. Preceding hyperpolarization measured with video camera. IIC. Intracellular electrophysiological recording from an identified salamander mitral/tufted cell after stimulation with moderate concentration (5×10^{-6} M) cineole. Note time scale differences in each record.

was observed both in vitro and in vivo and was especially prominent after stimulation with odors, whereupon it characteristically appeared before any spikes (see arrows in Figure 7 II). Its amplitude was maximum at subglomerular levels above the region of mitral somata. These observations suggest (Kauer et al 1990) that this response may be a manifestation of inhibition of mitral/tufted cells generated by periglomerular cells (Getchell & Shepherd 1975a,b, Halasz & Shepherd 1983). The responses were rather widespread spatially within the layer, although, interestingly, increased stimulus intensity produced a more restricted distribution. That this response is elicited only with low-intensity, distal electrical stimulation or with odors suggests that it may depend on relatively asynchronous activation of the bulbar circuitry via patterned activation of the afferents.

Following the early hyperpolarization and the fast and slow depolarizations, a long-lasting hyperpolarization, with a peak amplitude close to the mitral layer, has been observed with both electrical and odor stimulation (Figure 6 IIB, *solid line*). Optical recordings with video imaging show that the duration can be as long as 5 sec in the salamander bulb (A. R. Cinelli et al, in preparation, Kauer 1988). In general, this response is evoked with intense electrical orthodromic stimulation, often greater than three times the threshold necessary for eliciting the slow depolarizing response. The magnitude and duration of the hyperpolarization increase with stimulus intensity, and its size is directly related to the preceding slow depolarizing response (Cinelli & Salzberg 1990). The duration of this component correlates with the refractory period for the depolarizing responses (see Figure 3C, 6 IIB, 7 II). Maximal suppression of the depolarization coincides with the maximal amplitude of the hyperpolarization (Cinelli & Salzberg 1990). This hyperpolarization also resembles the long-lasting hyperpolarization seen in intracellular mitral cell records (Hamilton & Kauer 1988) (see Figure 7 IIC).

Antidromic stimulation evokes patterns of depolarization and hyperpolarization similar to those with orthodromic electrical stimulation. In salamander (Cinelli & Salzberg 1988, in preparation, A. R. Cinelli et al, submitted) and mouse (Cinelli & Salzberg 1987) olfactory bulbs, electric volleys applied to the output olfactory tract evoked a fast component that propagated to the mitral/tufted somata region, following a period of slow depolarization and a long-lasting hyperpolarization.

Optical Responses to Odor Stimulation

Early observations by Adrian (1951) first suggested that different odors evoke maximal excitation in different regions of the olfactory bulb. Subsequent studies with massed unit (Moulton 1965) and EEG (Freeman & Skarda 1985) recording showed that different odorants can evoke not only different frequency spectra at any one site in the bulb, but also that multiple loci are activated. Similar distributed odor-elicited activity patterns have been seen in 2-deoxyglucose studies (Sharp et al 1975, Stewart et al 1979).

The patterns of bulbar activity evoked by odor stimuli were first examined with optical methods in the salamander by using a photodiode array (Kauer et al 1987) (Figure 8 IA). These signals consisted of depolarizing events with latencies and time courses longer than the responses evoked by orthodromic electric volleys. Long-lasting hyperpolarizing signals were not seen at that time, probably because of the short AC coupling time constant of the diode array recording system. In general, the signals were sequentially distributed across and within the layers of the bulb, culminating in activation of fibers in the output medial olfactory tract, in a sequence consistent with electrophysiological recordings (Hamilton & Kauer 1989) (see Figure 8A).

Video-rate imaging of fluorescence with potentiometric dye (RH414 or RH795) has permitted further study of the distribution of these events and provides a way to test hypotheses about odor encoding (A. R. Cinelli et al, submitted, Kauer 1988). For example, distinctively different temporal patterns of depolarization and hyperpolarization have been observed within the bulbar layers after stimulation with various odorants (Figure 8 IB–D). The temporal patterns were characterized by brief, small amplitude hyperpolarizations, periods of depolarization, followed by longer duration hyperpolarizations that were similar to those observed in electrophysiological observations (see Hamilton & Kauer 1989 and Figure 7 IIC).

In general, higher odor concentrations shortened the response latency, increased the size and duration of the depolarization, and increased the size and duration of the subsequent after-hyperpolarization. The activity was more widely distributed within the layers with increased intensity and had poorly defined borders. Substantial overlap was seen between patterns (Figure 8 B–D) elicited by different odors similar to what has been found in 2DG experiments (Stewart et al 1979). These data provide additional evidence that odor information is encoded by patterns of activity distributed not only in space but also in time. They also indicate the additional need to develop methods for acquiring information about the distribution of these signals throughout the depths of the bulb and for making quantitative assessments of similarity among activity patterns elicited by different odors (A. R. Cinelli et al, in preparation). These kinds of data should permit careful correlations of odorant molecular structure with the detailed structure of the responses.

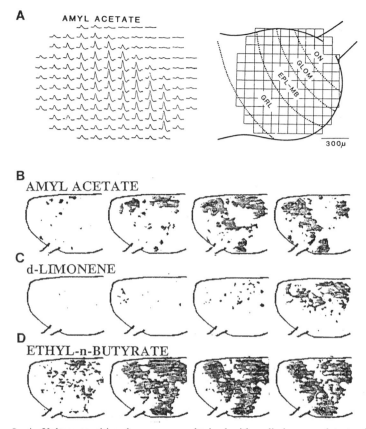

Figure 8   A. Voltage-sensitive dye response obtained with a diode array detector in the salamander olfactory to odor stimulation with amyl acetate (10^{-2} of saturated vapor, about 10^{-6} M) (Kauer et al 1987). Orientation of the array on the bulb is shown to the *right*. B.C.D. Distribution of voltage sensitive dye responses obtained with a video camera detector (Kauer 1988) after stimulation with three odorants in the same salamander olfactory bulb, each at a concentration of about 10^{-2} of saturated vapor. Only the first four of 256 frames are shown to illustrate the differences in latency and distribution of these early signals.

One of the most consistent and striking anatomical characteristics of the olfactory bulb is its glomerular organization. The occurrence of these structures in many species has suggested that the glomerulus may serve as a functional unit (Adrian 1953, Shepherd 1972). Experiments with 2-deoxyglucose (Coopersmith & Leon 1984, Jourdan et al 1980, Sharp et al 1975, Skeen 1977, Stewart et al 1979, Teicher et al 1980) have provided general support for this notion, although the precise relationships between odorant structure and glomerular activity are still not clear, since different pure odorants can elicit patterns showing great overlap (Stewart et al 1979).

Patterns of activity in response to specific odorants suggest the existence of modular elements in the bulb consisting of restricted groups of glomeruli and their associated mitral/tufted and granule cell circuitry (Kauer & Cinelli 1991). Evidence for these functional groups has been obtained via computerized analyses of 2DG autoradiographic images and by video imaging voltage-sensitive dyes (see Figure 9). Although the data are still

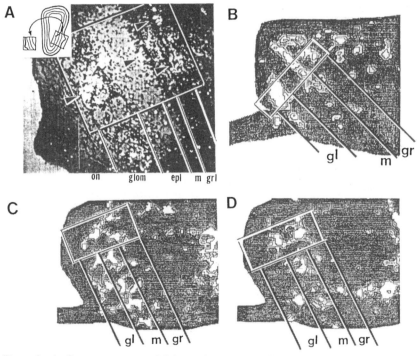

Figure 9   A. Computer-processed 2-deoxyglucose autoradiograph from the olfactory bulb of the mouse superimposed on the histological section (Kauer & Cinelli 1991) from which it was obtained. The highest levels of 2DG uptake are shown by the *white spots*. The distribution of activity as shown by 2DG in the box is characterized by the highest uptake in a periglomerular region surrounding several discrete glomeruli, in a small group of mitral/tufted cell bodies directly beneath these glomeruli, and in three groups of granule cell bodies, two on either side of the mitral focus and a smaller one directly beneath it. The group of active cells in the region encompassed by the box has been termed a module (Kauer & Cinelli 1991). B.C.D. Single video frames of voltage-sensitive dye signals from the bulb of the salamander after orthodromic electrical (B) and odor stimulation (C, amyl acetate; D, ethyl-*n*-butyrate). Note "modules" of activity in these examples showing a similar involvement of glomeruli, mitral/tufted cells, and granule cells.

preliminary, such modular elements could be candidates for functional components underlying the generation of different response patterns by different odorants. It is in the study of these kinds of functional properties that voltage-sensitive dye recording has major advantages over other ensemble assay methods.

Optical Responses in the Contralateral Olfactory Bulb

The olfactory bulb receives rich centrifugal projections from a number of higher brain regions (Davis & Macrides 1981, Luskin & Price 1983, Macrides & Davis 1983, Pinching & Powell 1972, Price 1969, Shipley et al 1985), as well as projections from homotopic regions of the contralateral bulb (Davis et al 1978, Schoenfeld et al 1985, Scott et al 1985).

These connections may correlate activity between corresponding regions of the two bulbs via projections to at least two levels, the glomerular and granule cell layers (Luskin & Price 1983, Macrides & Davis 1983). Stimuli applied to the contralateral olfactory bulb are reported to exert strong inhibitory effects on ipsilateral mitral/tufted cells via activation of granule cells (Nakashima et al 1978). Optical recordings have shown that olfactory nerve stimulation, in addition to generating the ipsilateral responses discussed above, also evokes longer latency activity in the contralateral bulb (A. R. Cinelli et al, submitted, Kauer 1988) as shown in Figure 10. In general, these events are characterized by an initial period of hyperpolarization (not shown in Figure 10) with a duration considerably longer than that evoked on the ipsilateral side, followed by a period of depolarization that resembles the slow component described above, followed by a second hyperpolarization. The origins of these signals are not yet defined.

OPTICAL METHODS APPLIED TO HIGHER OLFACTORY STRUCTURES

A major projection target for output fibers from the bulb is the piriform cortex, which receives the complex spatio-temporal patterns generated by odors in the bulb (Ferreyra-Moyano & Cinelli 1986, Haberly & Bower 1984, Moulton 1976). Morphological data show that the connections within the piriform cortex have widespread horizontal and vertical components. This organization suggests that the arrangement of the fiber systems may be optimized for extraction of information from odor-specific spatial patterns (Haberly & Bower 1989, Haberly & Price 1978).

Optical methods offer an approach for examining the distribution of activity in the piriform cortex. Voltage-sensitive dye signals have been recorded in mouse slices (Cinelli & Salzberg 1987) and in vivo in rats

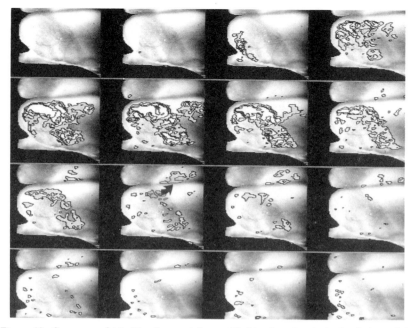

Figure 10 Sequence of 16 video frames taken at 30 fr/s after electrical stimulation of the olfactory nerve of an in vivo salamander olfactory bulb (from Kauer 1988 with permission). The contour lines encircle regions of depolarization shown by changes in the voltage-sensitive dye RH 414 one (*darker shading*) and two (*lighter shading*) SDs above background. The *curved arrow* in the tenth frame shows activity in the contralateral bulb.

(Cattarelli & Cohen 1989). So far, only electrical stimulation has been used. In these studies, electrical pulses applied to the lateral olfactory tract evoked first a fast component, which had characteristics similar to that seen in the nerve layer of the bulb (Cinelli & Salzberg 1987, 1990, in preparation), in hippocampal slices at the level of the stratum radiatum (Ginvald et al 1982b), and in parallel fibers in cerebellar slices (Konnerth et al 1987). This component probably represents afferent presynaptic compound action potentials. In deeper layers (Cinelli & Salzberg 1987), a second fast component was observed, which could be differentiated from the more superficial one by its postsynaptic characteristics and refractory period.

After the fast responses, a relatively long-lasting, postsynaptic depolarization was also reported (Cattarelli & Cohen 1989, Cinelli & Salzberg 1987); however, its characteristics have not yet been fully examined. Preliminary data suggest (Cinelli & Salzberg 1987) that glial depolarizations were not involved in these responses. As in the olfactory bulb, there was

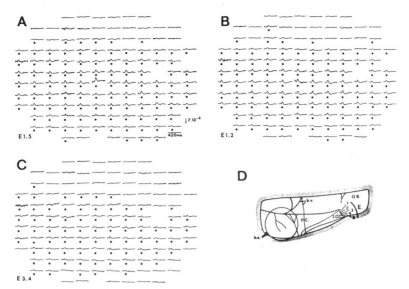

Figure 11 Distribution of piriform cortex responses after electrical stimulation of various sites in the lateral-olfactory tract (LOT) (shown in D). Note the wide overlap of response with stimulation at different sites. Stimulation = single 80 V, 300 us shock. A, B, and C are responses to LOT stimulation delivered through electrodes 1–5, 1–2, and 3–4 respectively, b.e., ball electrode; b.v., blood vessel; E, electrode; P.C., piriform cortex; *circle* delineates region of recording (from Cattarelli & Cohen 1989 with permission).

also a period of after-hyperpolarization (Cattarelli & Cohen 1989, Cinelli & Salzberg 1987).

When different groups of axons were stimulated (see Figure 11), the spatial patterns of the responses showed significant overlap, and increases in stimulation intensity enlarged the area of activation similar to results seen with 2DG measures of activity in the bulb (Greer et al 1981). The response latencies increased with increasing distance from the stimulating electrode, and distributions of the activity pattern were more restricted than others. Other differences in the waveforms of the responses were also observed (Cattarelli & Cohen 1989) (see Figure 11).

CONCLUSION

We now have good reason for optimism about beginning to understand how information is processed in complex neuronal assemblies by the use of ensemble optical recording methods that permit observation of events distributed throughout the network. With such methods, general rules about how information is manipulated by neuronal interactions may emerge, just as rules have emerged about how action potentials and synaptic events are generated in single cells. Among the first steps required for the observation of such basic properties of neuronal interactions are careful analyses of the recording systems in order to assess how well these data compare with information previously obtained with conventional electrophysiological methods.

We have reviewed such comparisons for data obtained with voltage-sensitive dye recording in the olfactory system. This modality appears to encode unitary, "monomolecular" odorant stimuli by neural events distributed across many cells at each level of the system, from the receptor epithelium to the olfactory cortex. Thus, the olfactory system may function as a kind of neural parallel distributed processor (see Kauer 1991), even at the stage of primary odor transduction. Since these distributed events appear at the most peripheral level in the olfactory pathway, study of this system may reveal algorithms related to distributed processing that would be observed only at higher levels in other sensory pathways.

We still do not have a clear picture of the mechanisms involved in coding of odor information. This is due partly to the lack of information about the primary transduction process at the receptor level, although new information has recently become available on a multigene family that might encode receptor proteins (Buck & Axel 1991). In addition, how molecular information about single odorant molecules is distributed among many neural elements is difficult to observe. Eventually, with knowledge of the detailed interconnectivity of the system and the ability to observe multitudes of elements at once, a general understanding of parallel distributed processing events in nervous systems may emerge.

ACKNOWLEDGMENTS

This work was carried out with support from the National Institutes of Health, the Pew Freedom Trust, the Office of Naval Research, and the Department of Neurosurgery, Tufts/New England Medical Center. We thank Walter Dent for expert technical assistance and Dr. Barbara Talamo for reading the manuscript.

Literature Cited

Adrian, E. D. 1950. Sensory discrimination with some recent evidence from the olfactory organ. *Br. Med. Bull.* 6: 330–33

Adrian, E. D. 1951. Olfactory discrimination. *L'Annee Psychol.* 50: 107–13

Adrian, E. D. 1953. Sensory messages and sensation. The response of the olfactory organ to different smells. *Acta Physiol. Scand.* 29: 5–14

Andres, K. H. 1970. Anatomy and ultrastructure of the olfactory bulb in fish, amphibia, reptiles, birds and mammals. In *Taste and Smell in Vertebrates*, ed. G. E. Wolstenholme, J. Knight, pp. 177–96. London: Churchill

Astion, M. L., Obaid, A. L., Orkand, R. K. 1989. Effects of barium and bicarbonate on glial cells of *Necturus* optic nerve. Studies with microelectrodes and voltage-sensitive dyes. *J. Gen. Physiol.* 93: 731–44

Ballanyi, K., Grafe, P., ten Bruggencate, G. 1987. Ion activities and potassium uptake mechanisms of glial cells in guinea-pig olfactory slices. *J. Physiol.* 382: 159–74

Blakely, R. D., Ory Lavollee, L., Grzanna, R., Koller, K. J., Coyle, J. T. 1987. Selective immunocytochemical staining of mitral cells in rat olfactory bulb with affinity purified antibodies against *N*-acetyl-aspartyl-glutamate. *Brain Res.* 402: 373–78

Blasdel, G. G. 1989. Visualization of neuronal activity in monkey striate cortex. *Annu. Rev. Physiol.* 51: 561–81

Blasdel, G. G., Salama, G. 1986. Voltage-sensitive dyes reveal a modular organizations in monkey striate cortex. *Nature* 321: 579–85

Buck, L., Axel, R. 1991. A novel multigene family may encode odorant receptors: A molecular basis for odor recognition. *Cell* 65: 175–87

Cattarelli, M., Cohen, L. B. 1989. Optical recording of the in vivo piriform cortex responses to electrical stimulation of the lateral olfactory tract in the rat. *Chem. Senses* 14: 577–86

Cinelli, A. R., Hamilton, K. A., Kauer, J. S. 1989. Evidence for Ca^{++} influx involvement in the long-lasting depolarization of mitral/tufted cell dendrites in the salamander olfactory bulb. *Neurosci. Abstr.* 15: 927 (Abstr.)

Cinelli, A. R., Kauer, J. S. 1990. Analysis of calcium involvement in optical signals from the in vitro salamander olfactory bulb. *Neurosci. Abstr.* 16: 403 (Abstr.)

Cinelli, A. R., Kauer, J. S. 1991. Spatiotemporal characterization of facilitated optical responses during paired orthodromic stimuli in the salamander olfactory bulb. *Neurosci. Abstr.* 17: (Abstr.)

Cinelli, A. R., Salzberg, B. M. 1987. Optical recording of electrical activity in slices of mammalian olfactory structures: Extrinsic signals from olfactory bulb and pyriform and sulcal cortices of the mouse. *Neurosci. Abstr.* 13: 1411 (Abstr.)

Cinelli, A. R., Salzberg, B. M. 1988. Early and late responses in slices of salamander olfactory bulb: Optical recording of electrical events that depend on Ca^{++}. *Neuro-*

sci. Abstr. 14: 1187 (Abstr.)

Cinelli, A. R., Salzberg, B. M. 1990. Multiple site optical recording of transmembrane voltage (MSORTV), Single unit recording, and evoked field potentials from the olfactory bulb of skate (Raja erinacea). *J. Neurophysiol.* 64: 1767–90

Cohen, L. B., Hille, B., Keynes, R. D. 1968. Light scattering and birefringence changes during nerve activity. *Nature* 218: 438–41

Cohen, L. B., Hopp, H.-P., Wu, J.-Y., Xiao, C., London, J., Zecevic, D. 1989. Optical measurement of action potential activity in invertebrate ganglia. *Annu. Rev. Physiol.* 51: 527–41

Cohen, L. B., Keynes, R. D., Landowne, D. 1972. Changes in light scattering that accompany the action potential in squid giant axon: Current dependent components. *J. Physiol.* 224: 727–32

Cohen, L. B., Lesher, S. 1986. Optical monitoring of membrane potential: Methods of multisite optical measurement. *Soc. Gen. Physiol. Ser. B* 40: 71–99

Cohen, L. B., Salzberg, B. M. 1978. Optical measurement of membrane potential. *Rev. Physiol. Biochem. Pharmacol.* 83: 35–88

Cohen, L. B., Salzberg, B. M., Grinvald, A. 1978. Optical methods for monitoring neuron activity. *Annu. Rev. Neurosci.* 1: 171–82

Connor, J. A. 1986. Measurement of free calcium levels in tissue culture neurons using change coupled device (CCD) camera technology. In *Imaging Function in the Nervous System: Optical Methods in Cellular Neurobiology*, pp. 14–24. Washington, DC: Soc. Neurosci.

Connors, B. W., Ransom, B. R. 1986. Electrophysiological properties of ependymal cells (radial glia) in dorsal cortex of the turtle pseudemys scripta. *J. Physiol.* 385: 287–306

Coopersmith, R., Leon, M. 1984. Enhanced neural response to familiar olfactory cues. *Science* 225: 849–51

Davis, B. J., Macrides, F. 1981. The organization of centrifugal projections from the anterior olfactory nucleus, ventral hippocampal rudiment, and piriform cortex to the main olfactory bulb in the hamster: an autoradiographic study. *J. Comp. Neurol.* 203: 475–93

Davis, B. J., Macrides, F., Young, W. M., Schneider, S. P., Rosene, D. L. 1978. Efferents and centrifugal afferents of the main and accessory olfactory bulbs in hamsters. *Brain Res. Bull.* 3: 59–72

Ferreyra-Moyano, H., Cinelli, A. R. 1986. Axonal projections and conduction properties of olfactory peduncle neurons in the armadillo (*Chaetophractus vellerosus*). *Exp. Brain Res.* 64: 527–34

Freeman, W. J., Skarda, C. A. 1985. Spatial EEG patterns, non-linear dynamics and perception: The neo-Sherringtonian view. *Brain Res.* 357: 147–75

Gainer, H., Wolfe, S. A. J., Obaid, A. L., Salzberg, B. M. 1986. Action potentials and frequency—dependent secretion in the mouse hypophysis. *Neuroendocrinology* 43: 557–63

Getchell, T. V., Shepherd, G. M. 1975a. Synaptic actions on mitral and tufted cells elicited by olfactory nerve volleys in the rabbit. *J. Physiol.* 251: 497–522

Getchell, T. V., Shepherd, G. M. 1975b. Short axon cells in the olfactory bulb: Dendrodendritic synaptic interactions. *J. Physiol.* 251: 523–48

Greer, C. A., Stewart, W. B., Kauer, J. S., Shepherd, G. M. 1981. Topographical and laminar localization of 2-deoxyglucose uptake in rat olfactory bulb induced by electrical stimulation of olfactory nerves. *Brain Res.* 217: 279–93

Grinvald, A. 1985. Real-time optical mapping of neuronal activity: From single growth cones to the intact mammalian brain. *Annu. Rev. Neurosci.* 8: 263–305

Grinvald, A., Anglister, L., Freeman, J. A., Hildesheim, R., Manker, A. 1984. Real-time optical imaging of naturally evoked electrical activity in intact frog brain. *Nature* 308: 848–50

Grinvald, A., Fine, A., Farber, I. C., Hildesheim, R. 1983. Fluorescence monitoring of electrical responses from small neurons and their processes. *Biophys. J.* 42: 195–98

Grinvald, A., Frostig, R. D., Lieke, E. E., Hildesheim, R. 1988. Optical imaging of neuronal activity. *Physiol. Rev.* 68: 1285–1366

Grinvald, A., Hildesheim, R., Farber, I. C., Anglister, L. 1982a. Improved fluorescent probes for the measurement of rapid changes in membrane potential. *Biophys. J.* 39: 301–8

Grinvald, A., Lieke, E., Frostig, R. D., Gilbert, C. D., Wiesel, T. N. 1986. Functional architecture of cortex revealed by optical imaging of intrinsic signals. *Nature* 324: 361–64

Grinvald, A., Manker, A., Segal, M. 1982b. Visualization of the spread of electrical activity in rat hippocampal slices by voltage sensitive optical probes. *J. Physiol.* 333: 269–91

Grinvald, A., Ross, W. N., Farber, I. C. 1981. Simultaneous optical measurements of electrical activity from multiple sites on processes of cultured neurons. *Proc. Natl. Acad. Sci. USA* 78: 3245–49

Grinvald, A., Salzberg, B. M., Lev-Ram, V., Hildesheim, R. 1987. Optical recording of synaptic potentials from processes of single neurons using intracellular potentiometric dyes. *Biophys. J.* 51: 643–51

Haberly, L. B., Bower, J. M. 1984. Analysis of association fiber system in piriform cortex with intracellular recording and staining techniques. *J. Neurophysiol.* 51: 90–112

Haberly, L. B., Bower, J. M. 1989. Olfactory cortex: Model circuit for study of associative memory. *Trends Neurosci.* 12: 258–64

Haberly, L. B., Price, J. L. 1978. Association and commissural fiber systems of the olfactory cortex of the rat. *J. Comp. Neurol.* 178: 711–40

Hagiwara, S., Byerly, L. 1981. Calcium channels. *Annu. Rev. Neurosci.* 4: 69–125

Hagiwara, S., Fukada, J., Eaton, D. S. 1974. Membrane current carried by Ca^{++}, Sr^{++}, and Ba^{++} in barnacle muscle fibre during voltage clamp. *J. Gen. Physiol.* 63: 564–78

Halasz, N. 1990. *The Vertebrate Olfactory System.* Budapest: Akademiai Kiado. 281 pp.

Halasz, N., Shepherd, G. M. 1983. Neurochemistry of the vertebrate olfactory bulb. *Neuroscience* 10: 579–619

Hamilton, K. A., Kauer, J. S. 1985. Intracellular potentials of salamander mitral/tufted neurons in response to odor stimulation. *Brain Res.* 338: 181–85

Hamilton, K. A., Kauer, J. S. 1988. Responses of mitral/tufted cells to orthodromic and antidromic electrical stimulation in the olfactory bulb of the tiger salamander. *J. Neurophysiol.* 59: 1736–55

Hamilton, K. A., Kauer, J. S. 1989. Patterns of intracellular potentials in salamander mitral/tufted cells in response to odor stimulation. *J. Neurophysiol.* 62: 609–25

Hamilton, K. A., Neff, S. R., Kauer, J. S. 1988. Evidence for GABA-mediated inhibition in the salamander olfactory bulb. *Soc. Neurosci. Abstr.* 14: 1187 (Abstr.)

Jacobson, I., Butcher, S., Hamberger, A. 1986. An analysis of the effects of excitatory amino acid receptor antagonists on evoked field potentials in the olfactory bulb. *Neuroscience* 19: 267–73

Jahr, C. E., Nicoll, R. A. 1982. An intracellular analysis of dendrodentric inhibition in the turtle in vitro olfactory bulb. *J. Physiol.* 326: 213–34

Jourdan, F., Duveau, A., Astic, L., Holley, A. 1980. Spatial distribution of 14C 2-deoxyglucose uptake in the olfactory bulbs of rats stimulated with two different odours. *Brain Res.* 188: 139–54

Kauer, J. S. 1988. Real-time imaging of evoked activity in local circuits of the salamander olfactory bulb. *Nature* 331: 166–68

Kauer, J. S. 1991. Contributions of topography and parallel processing to odor coding in the vertebrate olfactory pathway. *Trends Neurosci.* 14: 79–85

Kauer, J. S., Cinelli, A. R. 1991. Are there structural and functional modules in the vertebrate olfactory bulb? *J. Electron Microsc. Tech.* In press

Kauer, J. S., Hamilton, K. A., Neff, S. R., Cinelli, A. R. 1990. Temporal patterns of membrane potential in the olfactory bulb observed with intracellular recording and voltage-sensitive dye imaging: early hyperpolarization. In *Chemosensory Information Processing*, ed. D. Schild, pp. 305–14. Berlin: Springer-Verlag

Kauer, J. S., Moulton, D. G. 1974. Responses of olfactory bulb neurones to odour stimulation of small nasal areas in the salamander. *J. Physiol.* 243: 717–37

Kauer, J. S., Neff, S. R., Hamilton, K. A., Cinelli, A. R. 1991. The salamander olfactory pathway: Visualizing and modelling circuit activity. In *Olfaction as a Model System for Computational Neuroscience*, ed. J. Davis, H. Eichenbaum. Cambridge: MIT Press. In press

Kauer, J. S., Senseman, D. M., Cohen, L. B. 1987. Odor-elicited activity monitored simultaneously from 124 regions of the salamander olfactory bulb using a voltage-sensitive dye. *Brain Res.* 418: 255–61

Kent, P. F. 1990. The recording of odorant-induced mucosal activity patterns with a voltage-sensitive dye. PhD dissertation, SUNY, Syracuse, NY

Kent, P. F., Mozell, M. M. 1988. The recording of odorant-induced mucosal activity patterns with a voltage-sensitive dye. *AChemS* 10: 195 (Abstr.)

Konnerth, A., Obaid, A. L., Salzberg, B. M. 1985. Elasmobranch cerebellar slices in vitro: Selective binding of potentiometric probes allows optical recording of electrical activity from different cell types. *Biol. Bull.* 169: 553–54

Konnerth, A., Obaid, A. L., Salzberg, B. M. 1987. Optical recording of electrical activity from parallel fibres and other cell types in skate cerebellar slices in vitro. *J. Physiol.* 393: 681–702

Kosaka, T., Hataguchi, Y., Hama, K., Nagatsu, I., Wu, J. Y. 1985. Coexistence of immunoreactivities for glutamate decarboxylase and tyrosine hydroxylase in some neurons in the periglomerular region of the rat main olfactory bulb: Possible coexistence of gamma-aminobutyric acid (GABA) and dopamine. *Brain Res.* 343: 166–71

Krauthamer, V., Ross, W. N. 1984. Regional variations in excitability of barnacle neurons. *J. Neurosci.* 4: 673–82

Lasser-Ross, N., Miyakawa, H., Lev-Ram, V., Young, S., Ross, W. N. 1991. High time resolution fluorescence imaging with a ccd camera. *J. Neurosci. Meth.* 36: 253–61

Lev-Ram, V., Ginvald, A. 1986. Ca^{2+}- and K^+-dependent communication between central nervous system myelinated axons and oligodendrocytes revealed by voltage-sensitive dyes. *Proc. Natl. Acad. Sci. USA* 83: 6651–55

Lieke, E. E., Frostig, R. D., Arieli, A., Ts'o, D. Y., Hildesheim, R., Grinvald, A. 1989. Optical imaging of cortical activity: Real-time imaging using extrinsic dye-signals and high resolution imaging based on slow intrinsic signals. *Annu. Rev. Physiol.* 51: 543–59

London, J. A. 1990. Optical recording of activity in the hamster gustatory cortex elicited by electrical stimulation of the tongue. *Chem. Senses* 15: 137–43

London, J. A., Cohen, L. B., Wu, J. Y. 1989. Optical recordings of the cortical response to whisker stimulation before and after the addition of an epileptogenic agent. *J. Neurosci.* 9: 2182–90

London, J. A., Zecevic, D., Cohen, L. B. 1987. Simultaneous optical recording of activity from many neurons during feeding in Navanax. *J. Neurosci.* 7: 649–61

Luskin, M. B., Price, J. L. 1983. The topographic organization of associational fibers of the olfactory system in the rat, including centrifugal fibers to the olfactory bulb. *J. Comp. Neurol.* 216: 264–91

Mackay Sim, A., Shaman, P., Moulton, D. G. 1982. Topographic coding of olfactory quality: Odorant-specific patterns of epithelial responsivity in the salamander. *J. Neurophysiol.* 48: 584–96

Macrides, F., Davis, B. J. 1983. The olfactory bulb. In *Chemical Anatomy*, ed. P. C. Emson, pp. 391–426. New York: Raven

Margolis, F. L. 1981. Neurotransmitter biochemistry of the mammalian olfactory bulb. In *Biochemistry of Taste and Olfaction*, pp. 369–94. New York: Academic

Matsumoto, G., Ichikawa, M. 1990. Optical system for real-time imaging of electrical activity with a 128×128 photopixel array. *Neurosci. Abstr.* 16: 490 (Abstr.)

Meech, R. W. 1978. Calcium-dependent potassium activation in nervous tissues. *Annu. Rev. Biophys. Bioeng.* 7: 1–18

Mori, K. 1987. Membrane and synaptic properties of identified neurons in the olfactory bulb. *Prog. Neurobiol.* 29: 275–320

Moulton, D. G. 1965. Differential sensitivity to odors. *Cold Spring Harbor Symp. Quant. Biol. Transduct. Meeting* 30: 201–6

Moulton, D. G. 1976. Spatial patterning of response to odors in the peripheral olfactory system. *Physiol. Rev.* 56: 578–93

Mugnaini, E., Wouterlood, F. G., Dahl, A. L., Oertel, W. H. 1984. Immunocytochemical identification of GABAergic neurons in the main olfactory bulb of the rat. *Arch. Ital. Biol.* 122: 83–113

Mustaparta, H. 1971. Spatial distribution of receptor responses to stimulation with different odours. *Acta Physiol. Scand.* 82: 154–66

Nakashima, M., Mori, K., Takagi, S. F. 1978. Centrifugal influence on olfactory bulb activity in the rabbit. *Brain Res.* 154: 301–16

Northcutt, R. G., Kicliter, E. 1980. Organization of the amphibian telencephalon. In *Comparative Neurology of the Telencephalon*, ed. S. O. E. Ebbesson, pp. 203–55. New York: Plenum

Nowycky, M. C., Mori, K., Shepherd, G. M. 1981. GABAergic mechanisms of dendrodendritic synapses in isolated turtle olfactory bulb. *J. Neurophysiol.* 46: 639–48

Obaid, A. L., Gainer, H., Salzberg, B. M. 1983. Optical recording of action potentials from mammalian nerve terminals in vitro. *Biol. Bull.* 165: 530 (Abstr.)

Orbach, H. S., Cohen, L. B. 1983. Optical monitoring of activity from many areas of the in vitro and in vivo salamander olfactory bulb: A new method for studying functional organization in the vertebrate central nervous system. *J. Neurosci.* 3: 2251–62

Orbach, H. S., Cohen, L. B., Grinvald, A. 1985. Optical mapping of electrical activity in rat somatosensory and visual cortex. *J. Neurosci.* 5: 1886–95

Ottoson, D. 1956. Analysis of the electrical activity of the olfactory epithelium. *Acta Physiol. Scand.* 122: 1–83

Pinching, A. J., Powell, T. P. S. 1972. The termination of centrifugal fibres in the glomerular layer of the olfactory bulb. *J. Cell Sci.* 10: 637–55

Price, J. L. 1969. The origin of the centrifugal fibers to the olfactory bulb. *Brain Res.* 14: 542–45

Regehr, W. G., Connor, J. A., Tank, D. W. 1989. Optical imaging of calcium accumulation in hippocampal pyramidal cells during synaptic activation. *Nature* 341: 533–36

Revial, M. F., Duchamp, A., Holley, A. 1978a. Odour discrimination by frog olfactory receptors: A second study. *Chem. Senses* 3: 7–21

Revial, M. F., Duchamp, A., Holley, A., MacLeod, P. 1978b. Frog olfraction: Odour groups, acceptor distribution and

receptor categories. *Chem. Senses* 3: 23–33

Revial, M. F., Sicard, G., Duchamp, A., Holley, A. 1982. New studies on odour discrimination in the frog's olfactory receptor cells. I. Experimental results. *Chem. Senses* 7: 175–91

Revial, M. F., Sicard, G., Duchamp, A., Holley, A. 1983. New studies on odour discrimination in the frog's olfactory receptor cells. II. Mathematical analysis of electrophysiological responses. *Chem. Senses* 8: 179–94

Ribak, C. E., Vaughn, J. E., Saito, K., Barber, R., Roberts, E. 1977. Glutamate decarboxylase localization in neurons of the olfactory bulb. *Brain Res.* 126: 1–18

Ross, W. N. 1989. Changes in intracellular calcium during neuron activity. *Annu. Rev. Physiol.* 51: 491–506

Ross, W. N., Krauthamer, V. 1984. Optical measurements of potential changes in axons and processes of neurons of a barnacle ganglion. *J. Neurosci.* 4: 659–72

Saggau, P., Galvan, M., ten Bruggencate, G. 1986. Long-term potentiation in guinea pig hippocampal slices monitored by optical recording of neuronal activity. *Neurosci. Lett.* 69: 53–58

Salzberg, B. M. 1989. Optical recording of voltage changes in nerve terminal and in fine neuronal processes. *Annu. Rev. Physiol.* 51: 507–26

Salzberg, B. M. 1983. Optical recording of electrical activity in neurons using molecular probes. In *Current Methods in Cellular Neurobiology*, ed. J. Barker, J. McKelvy, pp. 139–87. New York: Wiley

Salzberg, B. M., Davila, H. V., Cohen, L. B. 1973. Optical recording of impulses in individual neurones in an invertebrate central nervous system. *Nature* 246: 508–9

Salzberg, B. M., Obaid, A. L., Gainer, H. 1984. Optical methods monitor action potentials and secretory activity at the nerve terminals of vertebrate hypophysis. *J. Gen. Physiol.* 84: 3a–4a (Abstr.)

Salzberg, B. M., Obaid, A. L., Gainer, H. 1985. Large and rapid changes in light scattering accompany secretion by nerve terminals in the mammalian neurohypophysis. *J. Gen. Physiol.* 86: 395–411

Salzberg, B. M., Obaid, A. L., Senseman, D. M., Gainer, H. 1983. Optical recording of action potentials from vertebrate nerve terminals using potentiometric probes provides evidence for sodium and calcium components. *Nature* 306: 36–40

Schoenfeld, T. A., Marchand, J. E., Macrides, F. 1985. Topographic organization of tufted cell axonal projections in the hamster main olfactory bulb: An intrabulbar associational system. *J. Comp. Neurol.* 235: 503–18

Scott, J. W., Ranier, E. C., Pemberton, J. L., Orona, E., Mouradian, L. E. 1985. Pattern of rat olfactory bulb mitral and tufted cell connections to the anterior olfactory nucleus pars externa. *J. Comp. Neurol.* 242: 415–24

Senseman, D. M., Vasquez, S., Nash, P. L. 1990a. Animated pseudocolor activity maps (PAM'S): scientific visualization of brain electrical activity. In *Chemosensory Information Processing, NATO ASI Ser. H*, Vol. 39, ed. D. Schild, pp. 329–47. Berlin/Heidelberg: Springer-Verlag

Senseman, D. M., Wu, J. Y., Cohen, L. B. 1990b. Evoked neural activity in the salamander olfactory bulb monitored with calcium, pH, and voltage-sensitive probes. *Neurosci. Abstr.* 16: 403 (Abstr.)

Shapiro, H. M. 1981. Flow cytometric probes of early events in cell activation. *Cytometry* 1: 301–14

Sharp, F. R., Kauer, J. S., Shepherd, G. M. 1975. Local sites of activity-related glucose metabolism in rat olfactory bulb during olfactory stimulation. *Brain Res.* 98: 596–600

Shepherd, G. M. 1972. Synaptic organization of the mammalian olfactory bulb. *Physiol. Rev.* 52: 864–917

Shipley, M. T., Halloran, F. J., de la Torre, J. 1985. Surprisingly rich projection from locus coeruleus to the olfactory bulb in the rat. *Brain Res.* 329: 294–99

Skeen, L. C. 1977. Odor-induced patterns of deoxyglucose consumption in the olfactory bulb of the tree shrew, *Tupia Glis. Brain Res.* 124: 147–53

Stewart, W. B., Kauer, J. S., Shepherd, G. M. 1979. Functional organization of rat olfactory bulb analysed by the 2-deoxyglucose method. *J. Comp. Neurol.* 185: 715–34

Tank, D. W., Sugimori, M., Connor, J. A., Llinas, R. R. 1989. Spatially resolved calcium dynamics of mammalian Purkinje cells in cerebellar slice. *Science* 242: 773–77

Teicher, M. H., Stewart, W. B., Kauer, J. S., Shepherd, G. M. 1980. Suckling pheromone stimulation of a modified glomerular region in the developing rat olfactory bulb revealed by the 2-deoxyglucose method. *Brain Res.* 194: 530–35

Thommesen, G., Doving, K. B. 1977. Spatial distribution of the EOG in the rat; a variation with odour quality. *Acta Physiol. Scand.* 99: 270–80

Trombley, P. Q., Westbrook, G. L. 1990. Excitatory synaptic transmission in cultures of rat olfactory bulb. *J. Neurophysiol.* 64: 598–606

Ts'o, D. Y., Frostig, R. D., Lieke, E. E., Grinvald, A. 1990. Functional organization of primate visual cortex revealed by high resolution optical imaging. *Science* 249: 417–20

Waggoner, A. S. 1976. Optical probes of membrane potential. *J. Membr. Biol.* 27: 317–34

Wellis, D. P., Kauer, J. S. 1991. Whole cell patch and optical recordings of synaptic responses in salamander olfactory bulb. *Neurosci. Abstr.* 17: (Abstr.)

Wellis, D. P., Scott, J. W., Harrison, T. A. 1989. Discrimination among odorants by single neurons of the rat olfactory bulb. *J. Neurophysiol.* 61: 1161–77

Zecevic, D., Wu, J. Y., Cohen, L. B., London, J. A., Hopp, H. P., Falk, C. X. 1989. Hundreds of neurons in the Aplysia abdominal ganglion are active during the gill-withdrawal reflex. *J. Neurosci.* 9: 3681–89

Annu. Rev. Neurosci. 1992. 15:353–75

THE ROLE OF THE AMYGDALA IN FEAR AND ANXIETY

Michael Davis

Ribicoff Research Facilities of the Connecticut Mental Health Center, Department of Psychiatry, Yale University School of Medicine, New Haven, Connecticut 06508

KEY WORDS: learning, memory, conditioning

INTRODUCTION

Converging evidence now indicates that the amygdala plays a crucial role in the development and expression of conditioned fear. Conditioned fear is a hypothetical construct used to explain the cluster of behavioral effects produced when an initially neutral stimulus is consistently paired with an aversive stimulus. For example, when a light, which initially has no behavioral effect, is paired with an aversive stimulus such as a footshock, the light alone can elicit a constellation of behaviors that are typically used to define a state of fear in animals. To explain these findings, it is generally assumed (cf. McAllister & McAllister 1971) that during light-shock pairings (training session), the shock elicits a variety of behaviors that can be used to infer a central state of fear (unconditioned responses—Figure 1). After pairing, the light can produce the same central fear state and thus the same set of behaviors formerly produced by the shock. Moreover, the behavioral effects that are produced in animals by this formerly neutral stimulus (now called a conditioned stimulus—CS) are similar in many respects to the constellation of behaviors that are used to diagnose generalized anxiety in humans (Table 1). This chapter summarizes data supporting the idea that the amygdala, and its many efferent projections, may represent a central fear system involved in both the expression and acquisition of conditioned fear.

353

0147–006X/92/0301–0353$02.00

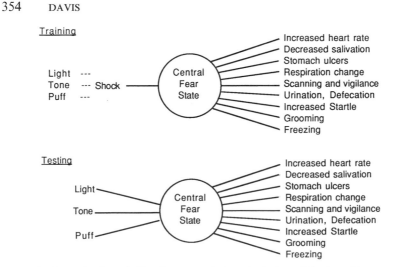

Figure 1 General scheme believed to occur during classical conditioning with an aversive conditioned stimulus. During training, the aversive stimulus (e.g. shock) activates a central fear system that produces a constellation of behaviors generally associated with aversive stimuli (unconditioned responses). After consistent pairings of some neutral stimulus such as a light or tone or puff of air with shock during the training phase, the neutral stimulus is capable of producing a similar fear state and hence the same set of behaviors (conditioned responses), formerly only produced by the shock.

Table 1 Comparison of measures in animals typically used to index fear and those in the DSM-III manual to index generalized anxiety in people

Measures of fear in animal models	DSM-III criteria—generalized anxiety
Increased heart rate	Heart pounding
Decreased salivation	Dry mouth
Stomach ulcers	Upset stomach
Respiration change	Increased respiration
Scanning and vigilance	Scanning and vigilance
Increased startle	Jumpiness, easy startle
Urination	Frequent urination
Defecation	Diarrhea
Grooming	Fidgeting
Freezing	Apprehensive expectation— something bad is going to happen

FEAR, ANXIETY, AND THE AMYGDALA

A variety of animal models have been used to infer a central state of fear or anxiety. In some models fear is inferred when an animal freezes, thus interrupting some ongoing behavior such as pressing a bar or interacting socially with other animals. In other models, fear is measured by changes in autonomic activity, such as heart rate, blood pressure, or respiration. Fear can also be measured by a change in simple reflexes or a change in facial expressions and mouth movements. Thus fear appears to produce a complex pattern of behaviors that are highly correlated with each other.

Anatomical Connections Between the Amygdala and Brain Areas Involved in Fear and Anxiety

The suggestion was made in several previous reviews (Gray 1989, Gloor 1960, Kapp et al 1984, 1990, Kapp & Pascoe 1986, Sarter & Markowitsch 1985) and supported by work in many laboratories that the central nucleus of the amygdala has direct projections to hypothalamic and brainstem areas that may be involved in many of the symptoms of fear or anxiety (summarized in Figure 2). Direct projections from the central nucleus of the amygdala to the lateral hypothalamus (Krettek & Price 1978a, Price & Amaral 1981, Shiosaka et al 1980) appear to be involved in activation of the sympathetic autonomic nervous system during fear and anxiety (cf. LeDoux et al 1988). Direct projections to the dorsal motor nucleus of the vagus nerve (Hopkins & Holstege 1978, Schwaber et al 1982, Takeuchi et al 1983, Veening et al 1984) may be involved in several autonomic measures of fear or anxiety, since the vagus nerve controls many different autonomic functions.

Projections of the central nucleus of the amygdala to the parabrachial nucleus (Hopkins & Holstege 1978, Krettek & Price 1978a, Price & Amaral 1981, Takeuchi et al 1982) may be involved in respiratory changes during fear, as electrical stimulation (Cohen 1971, 1979, Bertrand & Hugelin 1971, Mraovitch et al 1982) or lesions (Baker et al 1981, Von Euler et al 1976) of the parabrachial nucleus are known to alter various measures of respiration.

Projections from the amygdala to the ventral tegmental area (Beckstead et al 1979, Phillipson 1979, Simon et al 1979, Wallace et al 1989) may mediate stress-induced increases in dopamine metabolites in the prefrontal cortex (Thierry et al 1976). Direct amygdalar projections to the locus coeruleus (e.g. Cedarbaum & Aghajanian 1978, Wallace et al 1989), or indirect projections via the paragigantocellularis nucleus (Aston-Jones et al 1986) or perhaps via the ventral tegmental area (e.g. Deutch et al 1986), may mediate the response of cells in the locus coeruleus to conditioned

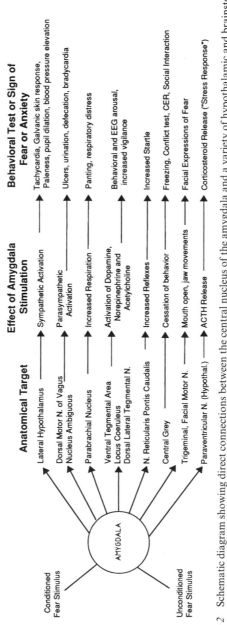
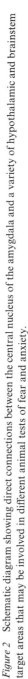

Figure 2   Schematic diagram showing direct connections between the central nucleus of the amygdala and a variety of hypothalamic and brainstem target areas that may be involved in different animal tests of fear and anxiety.

fear stimuli (Rasmussen & Jacobs 1986), as well as being involved in other actions of the locus coeruleus linked to fear and anxiety (cf. Redmond 1977). Direct projections of the amygdala to the lateral dorsal tegmental nucleus (e.g. Hopkins & Holstege 1978) and parabrachial nuclei (see above), which have cholinergic neurons that project to the thalamus (cf. Pare et al 1990), may mediate increases in synaptic transmission in thalamic sensory relay neurons (Pare et al 1990, Steriade et al 1990) during states of fear. This cholinergic activation, along with increases in thalamic transmission accompanying activation of the locus coeruleus (Rogawski & Aghajanian 1980), may thus lead to increased vigilance and superior signal detection in a state of fear or anxiety. In addition, release of norepinephrine onto motoneurons via amygdala activation of the locus coeruleus, or via amygdalar projections to serotonin-containing raphe neurons (Magnuson & Gray 1990), could lead to enhanced motor performance during a state of fear, because both norepinephine and serotonin facilitate excitation of motoneurons (e.g. McCall & Aghajanian 1979, White & Neuman 1980).

Projections of the amygdala to the nucleus reticularis pontis caudalis (Inagaki et al 1983, Rosen et al 1991) probably are involved in fear-potentiation of the startle reflex (Hitchcock & Davis 1991). The central nucleus of the amygdala projects to a region of the central grey (Beitz 1982, Gloor 1978, Hopkins & Holstege 1978, Krettek & Price 1978a, Post & Mai 1980) that has been implicated in conditioned fear in a number of behavioral tests (Borszcz et al 1989, Hammer & Kapp 1986, LeDoux et al 1988, Liebman et al 1970) and is thought to be a critical part of a general defense system (cf. Adams 1979, Bandler & Depaulis 1988, Blanchard et al 1981, Fanselow 1991, Graeff 1988, LeDoux et al 1988, Zhang et al 1990). Direct projections to the trigeminal and facial motor nuclei (Holstege et al 1977, Post & Mai 1980, Ruggiero et al 1982) may mediate some of the facial expressions of fear. Finally, direct projections of the central nucleus of the amygdala to the paraventricular nucleus of the hypothalamus (Gray 1989, Silverman et al 1981, Tribollet & Dreifuss 1981), or indirect projections by way of the bed nucleus of the stria terminalis and preoptic area, which receive input from the amygdala (De Olmos et al 1985, Krettek & Price 1978a, Weller & Smith 1982) and project to the paraventricular nucleus of the hypothalamus (Sawchenko & Swanson 1983, Swanson et al 1983), may mediate the prominent neuroendocrine responses to fearful or stressful stimuli.

The identity of the transmitters released onto these target sites by amygdaloid neurons is just beginning to emerge. Gray (1989) estimates that 25% of the neurons in the central nucleus of the amygdala and bed nucleus of the stria terminalis contain known neuropeptides. The main output neurons of the amygdala contain corticotropin-releasing factor, somatostatin, and neurotensin, with smaller contributions from substance P and galanin-containing cells.

Elicitation of Fear by Electrical Stimulation of the Amygdala

Electrical stimulation of the amygdala can produce a complex pattern of behavioral and autonomic changes that highly resembles a state of fear. Stimulation of the amygdala can alter heart rate and blood pressure, both measures used to study cardiovascular changes during fear conditioning (Anand & Dua 1956, Applegate et al 1983, Bonvallet & Gary Bobo 1972, Cox et al 1987, Faiers et al 1975, Frysinger et al 1984, Galeno & Brody 1983, Gelsema et al 1987, Harper et al 1984, Heinemann et al 1973, Hilton & Zbrozyna 1963, Iwata et al 1987, Kaada 1951, Kapp et al 1982, Koikegami et al 1957, Morgenson & Calaresu 1973, Pascoe et al 1989, Reis & Oliphant 1964, Schlor et al 1984, Stock et al 1978, 1981, Timms 1981). These effects are often critically dependent on the state of the animal and level of anesthesia (e.g. Frysinger et al 1984, Galeno & Brody 1983, Harper et al 1984, Iwata et al 1987, Stock et al 1978, Timms 1981) and in some instances may result from stimulation of fibers of passage rather than cell bodies (cf. Lewis et al 1989). Amygdala stimulation can also produce gastric ulceration (Henke 1980b, 1982, Innes & Tansy 1980, Sen & Anand 1957), which may result from chronic fear or anxiety.

Electrical stimulation of the amygdala also alters respiration (Anand & Dua 1956, Applegate et al 1983, Bonvallet & Gary Bobo 1972, Harper et al 1984), a prominent symptom of fear, especially in panic disorders. Electrical stimulation of the central nucleus of the amygdala produces a cessation of ongoing behavior (Applegate et al 1983, Gloor 1960, Kaada 1972, Ursin & Kaada 1960). Cessation of ongoing behavior is the critical measure of fear or anxiety in several animal models, such as freezing (Blanchard & Blanchard 1969, Bolles & Collier 1976, Fanselow & Bolles 1979), the operant conflict test (Geller & Seifter 1960), the conditioned emotional response (Estes & Skinner 1941) that correlates with freezing (e.g. Bouton & Bolles 1980, Mast et al 1982), and the social interaction test (File 1980). Electrical stimulation of the amygdala also elicits jaw movements (Applegate et al 1983, Gloor 1960, Kaku 1984, Ohta 1984) and activation of facial motoneurons (Fanardjian & Manvelyan 1987), both of which may be included in the facial expressions seen during the fear reaction. These motor effects may be indicative of a more general effect of amygdala stimulation, namely that of modulating brainstem reflexes such as the massenteric (Gary Bobo & Bonvallet 1975, Bonvallet & Gary Bobo 1975), baroreceptor (Lewis et al 1989, Schlor et al 1984, Pascoe et al 1989), nictitating membrane (Whalen & Kapp 1991), and startle reflex (Rosen & Davis 1988a,b). In most cases, stimulation of the amygdala facilitates these reflexes, although whether it does or not may depend on the exact amygdala sites being stimulated (see Whalen & Kapp 1991 for a discussion of this point). In humans, electrical stimulation of the amygdala elicits feelings of fear or anxiety, as well as autonomic reactions indicative of fear (Chapman et al 1954, Gloor et al 1981). Some of the emotional content of dreams may result from activation of the amygdala, stimulation of which increases ponto-geniculo-occipital activity that occurs during paradoxical (dream) sleep (cf. Calvo et al 1987).

Finally, electrical stimulation of the amygdala has been shown to increase plasma levels of corticosterone, thus indicating an excitatory effect of the amygdala on the hypothalamo-pituitary-adrenal axis (Dunn & Whitner 1986, Feldman et al 1982, Mason 1959, Matheson et al 1971, Redgate & Fahringer 1973, Smelik & Vermes 1980, Setekleiv et al 1961, Yates & Maran 1974). As mentioned above, some of these excitatory effects may be mediated through the preoptic area and bed nucleus of the stria terminalis, which receive input from the amygdala (De Olmos et al 1985, Krettek & Price 1978a, Weller & Smith 1982) and project to the paraventricular nucleus of the hypothalamus (Sawchenko & Swanson 1983, Swanson et al 1983). Electrical stimulation of these nuclei increases plasma corticosterone levels (Dunn 1987, Saphier & Feldman 1986). Elevated plasma levels of corticosterone produced by amygdala stimulation can be attenuated by bilateral lesions of the stria terminalis, medial preoptic area, and the bed nucleus of the stria terminalis (Feldman et al 1990). Direct projections from the medial nucleus of the amygdala to the hypothalamus exist as well (Gray et al 1989, Silverman et al 1981, Tribollet & Dreifuss 1981), and these projections may also mediate some of the excitatory effects of the amygdala on the hypothalamic-pituitary axis.

The highly correlated set of behaviors seen during fear may result from activation of a single area of the brain (the amygdala, especially its central nucleus), which then projects to a variety of target areas, each of which is critical for specific symptoms of fear and the perception of anxiety. Moreover, it must be assumed that all of these connections are already formed in an adult organism, because electrical stimulation produces these effects in the absence of prior explicit fear conditioning. Thus, much of the complex behavioral pattern seen during fear conditioning has already been "hard wired" during evolution. For a formerly neutral stimulus to produce the constellation of behavioral effects used to define a state of fear or anxiety, it is only necessary for that stimulus to activate the amygdala, which in turn will produce the complex pattern of behavioral changes by virtue of its innate connections to different brain target sites. Plasticity during fear conditioning probably results from a change in synaptic inputs prior to or in the amygdala, rather than from a change in its efferent target areas. The ability to produce long-term potentiation (Clugnet & LeDoux 1990, Chapman et al 1990) in the amygdala and the finding that local infusion of NMDA antagonists into the amygdala blocks the acquisition (Miserendino et al 1990) and extinction (Falls et al 1992) of fear conditioning is consistent with this hypothesis.

The Role of the Amygdala in Fear Elicited by a Conditioned Stimulus

If fear conditioning results from an activation of the amygdala, one would expect that a conditioned stimulus would activate units in the amygdala and that lesions of the amygdala would prevent a conditioned stimulus from producing fear. Thus far, recording single unit activity in the amygdala has been difficult because many of the cells are small and have very low spontaneous rates of firing. Nonetheless, several studies have shown that a neutral stimulus paired with aversive stimulation will alter neural firing in the amygdala (Applegate et al 1982, Henke 1983, Pascoe & Kapp 1985b, Umemoto & Olds 1975). In addition, many studies indicate that lesions of the amygdala block the effects of a conditioned stimulus in a variety of behavioral test situations. Lesions of the amygdala eliminate or attenuate freezing normally seen in response to a stimulus formerly paired with shock (Blanchard & Blanchard 1972, LeDoux et al 1988, 1990); in the presence of a dominant male rat (Bolhuis et al 1984, Luiten et al 1985); or in a continuous passive avoidance test (Slotnick 1973). Lesions of the amygdala counteract the normal reduction of bar pressing in the operant conflict test (Shibata et al 1986) or the conditioned emotional response paradigm (Kellicut & Schwartzbaum 1963, Spevack et al 1975). In birds, lesions of the archistriatum, believed to be homologous with the mammalian amygdala, block the development of a conditioned emotional response (Dafters 1976) or heart rate acceleration in response to a cue paired with a shock (Cohen 1975). In both adult (Gentile et al 1986, Kapp et al 1979) and infant mammals (Sananes & Campbell 1989), lesions of the central nucleus block conditioned changes in heart rate. Ibotenic acid lesions of the central nucleus of the amygdala (Iwata et al 1986) or localized cooling of this nucleus (Zhang et al 1986) also block conditioned changes in blood pressure. Lesions of the lateral amygdala (basal, lateral, and accessory basal nuclei) attenuate and lesions of the medial amygdala (corticomedial and central nucleus) eliminate negative contrast following sucrose reduction (Becker et al 1984), a measure of emotionality sensitive to anxiolytic drugs (cf. Flaherty 1990). Perhaps similarly, lesions of the amygdala block the effects of positive behavioral contrast (Henke 1972, Henke et al 1972), decreased responsiveness to shifts in reward magnitude in monkeys (Schwartzbaum 1960) and rats (Kemble & Beckman 1970), and effects of frustrative nonreward (e.g. Henke 1977, 1973). Lesions of the central nucleus or of the lateral and basal nuclei of the amygdala block fear-potentiated startle (Hitchcock & Davis 1991, 1987, Sananes & Davis 1992). This, along with a large literature implicating the amygdala in many other measures of fear such as active and passive avoidance (for reviews see Kaada 1972, Sarter & Markowitsch 1985, Ursin et al 1981) and evaluation and memory of emotionally significant sensory stimuli (Bennett et al 1985, Bresnahan & Routtenberg 1972, Ellis & Kesner 1983, Gallagher et al 1980, Gallagher & Kapp 1981, 1978, Gold et al 1975, Handwerker et al 1974, Kesner 1982, Liang et al 1985, Liang et al 1986, McGaugh et al 1990, Mishkin & Aggleton 1981) provide strong evidence for a crucial role of the amygdala in fear.

The Role of the Amygdala in Unconditioned Fear

Lesions of the amygdala are known to block several measures of innate fear in different species (cf. Blanchard & Blanchard 1972, Ursin et al 1981). Lesions of the cortical amygdaloid nucleus and perhaps the central nucleus markedly reduce emotionality in wild rats measured in terms of flight and defensive behaviors (Kemble et al 1984, 1990). Large amygdala lesions or those which damaged the cortical, medial, and, in several cases, the central nucleus dramatically increase the number of contacts a rat will make with a sedated cat (Blanchard & Blanchard 1972). Some of these lesioned animals crawl all over the cat and even nibble its ear, a behavior never shown by the nonlesioned animals. Following lesions of the archistriatum, believed to be homologous with the mammalian amygdala, birds become docile and show little tendency to escape from humans (Phillips 1964, 1968), consistent with a general taming effect of amygdala lesions reported in many species (cf. Goddard 1964). Finally, lesions of the amydaloid complex inhibit adrenocortical responses following olfactory or sciatic nerve stimulation (Feldman & Conforti 1981) and attenuate the compensatory hypersecretion of ACTH that normally occurs following adrenalectomy (Allen & Allen 1974). Lesions of the central nucleus have been found to attenuate ulceration significantly (Henke 1980a) and elevated levels of plasma corticosterone produced by restraint stress (Beaulieu et al 1986, 1987). Moreover, lesions of the medially projecting component of the ventroamygdalofugal pathway, which carries the fibers connecting the central nucleus of the amygdala to the hypothalamus, attenuate the increase in ACTH secretion following adrenalectomy, whereas lesions of the stria terminalis do not (Allen & Allen 1974). Finally, lesions of the amygdala have been reported to block the ability of high levels of noise, which may be an unconditioned fear stimulus (cf. Leaton & Cranney 1990), to produce hypertension (Galeno et al 1984) or activation of tryptophan hydroxylase (Singh et al 1990).

Other measures that have been used to index innate fear have produced less consistent data concerning amygdala lesions, however. Large electrolytic lesions of the amygdaloid complex (Bresnahan et al 1976, Corman et al 1967, Eclancher & Karli 1979, Greidanus et al 1979, Jonason & Enloe 1971, Schwartzbaum & Gay 1966), or electrolytic or ibotenic acid lesions of the central nucleus of the amygdala (e.g. Grijalva et al 1990, Jellestad et al 1986, Werka et al 1978) or of the lateral and basal nuclei (Jellestad & Cabrera 1986) produce an increase in exploratory behavior in the open field test. This does not seem to occur, however, when (*a*) open field testing is preceded by other tests on the same animals (Grossman et al 1975), (*b*) test conditions are especially familiar (e.g. McIntyre & Stein 1973), (*c*) testing occurs after considerable handling and a long time after surgery (Kemble et al 1979), or (*d*) the lesions are very small (Riolobos et al 1987), and may depend on the age of the animal when the lesions are performed (Eclancher & Karli 1979). Because increased exploratory behavior is not always associated with changes in corticosterone (Jellestad & Cabrera 1986, Jellestad et al 1986) or other measures usually associated with a loss of fear of the open field, these authors have concluded that increased locomotor activity cannot easily be explained by a general loss of fear after amygdala lesions. Moreover, other measures of neophobia, such as the time to begin eating in a novel environment, do not show consistent changes with lesions of the amygdala, as one might expect from a lesion that reduced fear (cf. Aggleton et al 1989). However, the exact way neophobia is measured may determine whether it is a valid measure of fear, at least based on the measurement of corticosterone (e.g. Misslin & Cigrang 1986).

Other data indicate that the amygdala appears to be involved in some types of aversive conditioning, but this depends on the exact unconditioned aversive stimulus that is used. For example, electrolytic lesions of the basal nucleus (Pellegrino 1968), or fiber-sparing chemical lesions of most of the amygdaloid complex (Cahill & McGaugh 1990), attenuate avoidance of thirsty rats to approach an electrified water spout through which they were previously accustomed to receiving water. Importantly, however, these same lesioned animals did not differ from controls in the rate at which they found the water spout over successive test days or their avoidance of the water spout when quinine was added to the water (Cahill & McGaugh 1990). This led Cahill and McGaugh to suggest that "the degree of arousal produced by the unconditioned stimulus, and not the aversive nature per se, determined the level of amygdala involvement" (p. 541). Although many studies have shown that electrolytic lesions of the amygdala can interfere with taste aversion learning, an elegant series of experiments have now shown that these effects result from an interruption of gustatory fibers passing through the amygdala on route to the insular cortex (Dunn &

Everitt 1988). In these studies, ibotenic acid lesions of the amygdala fail to block taste aversion learning, whereas ibotenic acid lesions of the gustatory insular cortex do. Once again, the amygdala does not seem critical for all types of aversive conditioning but only conditioning that involves an obvious fear component such as that produced by aversive shocks.

Finally, the amygdala may also be importantly involved in stimulus-response associations that do not obviously involve aversive conditioning (e.g. Aggleton & Mishkin 1986, Cador et al 1989, Everitt et al 1989, Gallagher et al 1990, Kesner et al 1989, Murray 1990, Murray & Mishkin 1985, Peinado-Manzano 1990; but see Zola-Morgan et al 1989). Hence, some of the deficits in aversive conditioning following alterations of amygdala function may be part of a more general deficit in attention (Gallagher et al 1990, Kapp et al 1990).

Conditioned Fear vs Anxiety

Clinically, fear is regarded to be more stimulus-specific than anxiety, despite very similar symptoms. Figure 2 suggests that spontaneous activation of the central nucleus of the amygdala would produce a state resembling fear in the absence of any obvious eliciting stimulus. In fact, fear and anxiety often precede temporal lobe epileptic seizures (Gloor et al 1981), which are usually associated with abnormal electrical activity of the amygdala (Crandall et al 1971). An important implication of this distinction is that treatments that block conditioned fear might not necessarily block anxiety. For example, if a drug decreased transmission along a sensory pathway required for a conditioned stimulus to activate the amygdala, that drug might be especially effective in blocking conditioned fear; however, if anxiety resulted from activation of the amygdala not involving that sensory pathway, that drug might not be especially effective in reducing anxiety. On the other hand, drugs that act specifically in the amygdala should affect both conditioned fear and anxiety. Moreover, drugs that act at various target areas might be expected to provide selective actions on some but not all of the somatic symptoms associated with anxiety.

Effects of Drugs Infused into the Amygdala on Fear and Anxiety

The central nucleus of the amygdala is known to have high densities of opiate receptors (Goodman et al 1980), whereas the basal nucleus, which projects to the central nucleus (Aggleton 1985, Krettek & Price 1978b, Millhouse & DeOlmos 1983, Nitecka et al 1981, Ottersen 1982, Smith & Millhouse 1985, Russchen 1982), has high densities of benzodiazepine receptors (Niehoff & Kuhar 1983). Local infusion of opiate agonists into the central nucleus of the amygdala blocks the acquisition of conditioned bradycardia in rabbits (Gallagher et al 1981, 1982) and has anxiolytic effects in the social interaction test (File & Rogers 1979). Furthermore, local infusion of benzodiazepines into the amygdala has anxiolytic effects in the operant conflict test (Hodges et al 1987, Nagy et al 1979, Petersen & Scheel-Kruger 1982, Petersen et al 1985, Scheel-Kruger & Petersen 1982, Shibata et al 1982, 1989, Thomas et al 1985), and in the light-dark box measure in mice (Costall et al 1989) and antagonizes the discriminative stimulus properties of pentylenetetrazol (Benjamin et al 1987). The anticonflict effect can be reversed by systemic administration of the benzodiazepine antagonist flumazenil (Hodges et al 1987, Petersen et al 1985, Shibata et al 1989) or co-administration into the amygdala of the GABA antagonist bicuculline (Scheel-Kruger & Petersen 1982) and mimicked by local infusion into the amygdala of GABA (Hodges et al 1987) or the GABA agonist muscimol (Scheel-Kruger & Petersen 1982). In general, anticonflict effects of benzodiazepines occur after local infusion into the lateral and basal nuclei (Petersen & Scheel-Kruger 1982, Petersen et al 1985, Scheel-Kruger & Petersen 1982, Thomas et al 1985) (the nuclei of the amygdala that have high densities of benzodiazepine receptors) and not after local infusion into the central nucleus (Petersen & Scheel-Kruger 1982, Scheel-Kruger & Petersen 1982). Shibata et al (1982) found just the opposite effect, however, perhaps because of local anesthetic effects, which can occur when high doses of these compounds are infused into the central nucleus (e.g. Heule et al 1983). More recently, it has been shown that the anterior parts of the basal and central nucleus are especially important for conflict performance based on both lesion and local infusion of benzodiazepines (Shibata et al 1989). Taken together these results suggest that drug actions in the amygdala may be sufficient to explain both fear-reducing and anxiety-reducing effects of various drugs given systemically. Local infusion into the amygdala of the benzodiazepine antagonist flumazenil significantly attenuated the anticonflict effect of the benzodiazepine agonist chlordiazepoxide given systemically (Hodges et al 1987). However, in a very important recent experiment, Yadin et al (1991) found that chlordiazepoxide actually had a more potent anticonflict effect in animals previously given lesions of the amygdala, even though the amygdala lesion itself released punished behavior. Clearly, more work has to be done to locate the site of the anxiolytic action of benzodiazepines given systemically.

Recently, a new class of anxiolytic compounds acting as 5-HT_3 receptor subtype antagonists have been shown to produce anxiolytic effects after local infusion into the amygdala (Costall et al 1989). Such infusions also can block some of the signs of withdrawal following subchronic administration of diazepam, ethanol, nicotine, or cocaine (Costall et al 1990) or increases in levels of dopamine or the serotonin metabolite 5-HIAA in the amygdala after activation of dopamine neurons in the ventral tegmental area (Hagan et al 1990). The latter effects, which may relate to how sensory information is gated in the amygdala (Maeda & Maki 1986), were more pronounced in the right amygdala vs the left (Hagan et al 1990), consistent with other lateralized effects reported previously (Costall et al 1987). Measures of emotionality, including fear-potentiated startle in humans, also show lateralization (cf. Lang et al 1990), consistent with a greater participation of the right vs the left hemisphere and hence perhaps the right amygdala. Future studies employing local infusion of benzodiazepine or opiate antagonists into the amygdala, coupled with systemic administration of various agonists, may be able to determine whether local binding to receptors in the amygdala is necessary to explain their anxiolytic effects. Eventually, local infusion of various drugs into specific target areas may be used to evaluate whether highly specific anxiolytic actions are produced. These results could then serve as a guide for eventually producing more selective anxiolytic compounds.

CONCLUSIONS

An impressive amount of evidence from many laboratories using a variety of experimental techniques indicates that the amygdala plays a crucial role in conditioned fear and probably anxiety. Many of the amygdaloid projection areas are critically involved in specific signs that are used to measure fear and anxiety. Electrical stimulation of the amygdala elicits a pattern of behaviors that mimic natural or conditioned states of fear. Lesions of the amygdala block innate or conditioned fear and local infusion of drugs into the amygdala have anxiolytic effects in several behavioral tests. Finally, the amygdala may be a critical site of plasticity that mediates both the acquisition and extinction of conditioned fear. A better understanding of brain systems that inhibit the amygdala and of the role of the amygdala's very high levels of peptides (cf. Gray 1989) may eventually lead to the development of more effective pharmacological strategies for treating clinical anxiety disorders.

ACKNOWLEDGMENTS

Research reported in this chapter was supported by NIMH Grant MH-25642, MH-47840, NINCDS Grant NS-18033, Research Scientist Development Award MH-00004, a grant from the Air Force Office of Scientific Research, and the State of Connecticut. My sincere thanks are extended to Bruce Kapp for helpful discussions about the brainstem projections of the amygdala and to Leslie Fields for help in typing the paper.

Literature Cited

Adams, D. B. 1979. Brain mechanisms for offense, defense and submission. *Behav. Brain Sci.* 2: 201–41

Aggleton, J. P. 1985. A description of intraamygdaloid connections in the old world monkeys. *Exp. Brain Res.* 57: 390–99

Aggleton, J. P., Blindt, H. S., Rawlins, J. N. P. 1989. Effect of amygdaloid and amygdaloid-hippocampal lesions on object recognition and spatial working memory in rats. *Behav. Neurosci.* 103: 962–74

Aggleton, J. P., Mishkin, M. 1986. The amygdala, sensory gateway to the emotions. In *Emotion: Theory, Research and Experience*, ed. R. Plutchik, H. Kellerman, pp. 281–99. New York: Academic

Allen, J. P., Allen, C. F. 1974. Role of the amygdaloid complexes in the stress-induced release of ACTH in the rat. *Neuroendocrinology* 15: 220–30

Anand, B. K., Dua, S. 1956. Circulatory and respiratory changes induced by electrical stimulation of limbic system (visceral brain). *J. Neurophysiol.* 19: 393–400

Applegate, C. D., Frysinger, R. C., Kapp, B. S., Gallagher, M. 1982. Multiple unit activity recorded from amygdala central nucleus during Pavlovian heart rate conditioning in rabbit. *Brain Res.* 238: 457–62

Applegate, C. D., Kapp, B. S., Underwood, M. D., McNall, C. L. 1983. Autonomic and somatomotor effects of amygdala central n. stimulation in awake rabbits. *Physiol. Behav.* 31: 353–60

Aston-Jones, G., Ennis, M., Pieribone, V. A., Nickell, W. T., Shipley, M. T. 1986. The brain nucleus locus coeruleus: Restricted afferent control of a broad efferent network. *Science* 234: 734–37

Baker, T., Netick, A., Dement, W. C. 1981. Sleep-related apneic and apneustic breathing following pneumotaxic lesion and vagotomy. *Resp. Physiol.* 46: 271–94

Bandler, R., Depaulis, A. 1988. Elicitation of intraspecific defence reactions in the rat from midbrain periaqueductal grey by microinjection of kainic acid, without neurotoxic effects. *Neurosci. Lett.* 88: 291–96

Beaulieu, S., DiPaolo, T., Barden, N. 1986. Control of ACTH secretion by central nucleus of the amygdala: Implication of the serotonergic system and its relevance to the glucocorticoid delayed negative feed-back mechanism. *Neuroendocrinology* 44: 247–54

Beaulieu, S., DiPaolo, T., Cote, J., Barden, N. 1987. Participation of the central amygdaloid nucleus in the response of adrenocorticotropin secretion to immobilization stress: Opposing roles of the noradrenergic and dopaminergic systems. *Neuroendocrinology* 45: 37–46

Becker, H. C., Jarvis, M. F., Wagner, G. C., Flaherty, C. F. 1984. Medial and lateral amygdalectomy differentially influences consummatory negative contrast. *Physiol. Behav.* 33: 707–12

Beckstead, R. M., Domesick, V. B., Nauta, W. J. H. 1979. Efferent connections of the substantia nigra and ventral tegmental area in the rat. *Brain Res.* 175: 191–217

Beitz, A. J. 1982. The organization of afferent projections to the midbrain periaqueductal gray of the rat. *Neuroscience* 7: 133–59

Benjamin, D., Emmett-Oglesby, M. W., Lal, H. 1987. Modulation of the dicriminative stimulus produced by pentylenetetrazol by centrally administered drugs. *Neuropharmacology* 26: 1727–31

Bennett, C., Liang, K. C., McGaugh, J. L. 1985. Depletion of adrenal catecholamines alters the amnestic effect of amygdala stimulation. *Behav. Brain Res.* 15: 83–91

Bertrand, S., Hugelin, A. 1971. Respiratory synchronizing function of the nucleus parabrachialis medialis: Pneumotaxic mechanisms. *J. Neurophysiol.* 34: 180–207

Blanchard, D. C., Blanchard, R. J. 1969. Crouching as an index of fear. *J. Comp. Physiol. Psychol.* 67: 370–75

Blanchard, D. C., Blanchard, R. J. 1972. Innate and conditioned reactions to threat in rats with amygdaloid lesions. *J. Comp. Physiol. Psychol.* 81: 281–90

Blanchard, D. C., Williams, G., Lee, E. M. C., Blanchard, R. J. 1981. Taming of wild *Rattus norvegicus* by lesions of the mesencephalic central gray. *Physiol. Psychol.* 9: 157–63

Bolhuis, J. J., Fitzgerald, R. E., Dijk, D. J., Koolhaas, J. M. 1984. The corticomedial amygdala and learning in an agonistic situation in the rat. *Physiol. Behav.* 32: 575–79

Bolles, R. C., Collier, A. C. 1976. Effects of

predictive cues on freezing in rats. *Anim. Learn. Behav.* 4: 6–8

Bonvallet, M., Gary Bobo, E. 1972. Changes in phrenic activity and heart rate elicited by localized stimulation of the amygdala and adjacent structures. *Electroenceph. Clin. Neurophysiol.* 32: 1–16

Bonvallet, M., Gary Bobo, E. 1975. Amygdala and masseteric reflex. II. Mechanisms of the diphasic modifications of the reflex elicited from the "Defence Reaction Area." Role of the spinal trigeminal nucleus (pars oralis). *Electro-encephal. Clin. Neurophysiol.* 39: 341–52

Borszcz, G. S., Cranney, J., Leaton, R. N. 1989. Influence of long-term sensitization on long-term habituation of the acoustic startle response in rats: Central gray lesions, preexposure, and ectinction. *J. Exp. Psychol.: Anim. Behav. Process* 15: 54–64

Bouton, M. E., Bolles, R. C. 1980. Conditioned fear assessed by freezing and by the suppression of three different baselines. *Anim. Learn. Behav.* 8: 429–34

Bresnahan, E., Routtenberg, A. 1972. Memory disruption by unilateral low level, subseizure stimulation of the medial amygdaloid nucleus. *Physiol. Behav.* 9: 513–25

Bresnahan, J. C., Meyer, P. M., Baldwin, R. B., Meyer, D. R. 1976. Avoidance behavior in rats with amygdala lesions in the septum, fornix longus, and amygdala. *Physiol. Psychol.* 4: 333–40

Cador, M., Robbins, T. W., Everitt, B. J. 1989. Involvement of the amygdala in stimulus-reward associations: Interaction with the ventral striatum. *Neuroscience* 30: 77–86

Cahill, L., McGaugh, J. L. 1990. Amygdaloid complex lesions differentially affect retention of tasks using appetitive and aversive reinforcement. *Behav. Neurosci.* 104: 532–43

Calvo, J. M., Badillo, S., Morales-Ramirez, M., Palacios-Salas, P. 1987. The role of the temporal lobe amygdala in pontogeniculo-occipital activity and sleep organization in cats. *Brain Res.* 403: 22–30

Cedarbaum, J. M., Aghajanian, G. K. 1978. Afferent projections to the rat locus coeruleus as determined by a retrograde tracing technique. *J. Comp. Neurol.* 178: 1–16

Chapman, P. F., Kairiss, E. W., Keenan, C. L., Brown, T. H. 1990. Long-term synaptic potentiation in the amygdala. *Synapse* 6: 271–78

Chapman, W. P., Schroeder, H. R., Guyer, G., Brazier, M. A. B., Fager, C., Poppen, J. L., Solomon, H. C., Yakolev, P. I. 1954. Physiological evidence concerning the importance of the amygdaloid nuclear region in the integration of circulating

function and emotion in man. *Science* 129: 949–50

Clugnet, M. C., LeDoux, J. E. 1990. Synaptic plasticity in fear conditioning circuits: Induction of LTP in the lateral nucleus of the amygdala by stimulation of the medial geniculate body. *J. Neurosci.* 10: 2818–24

Cohen, D. H. 1975. Involvement of the avian amygdala homologue (archistriatum posterior and mediale) in defensively conditioned heart rate change. *J. Comp. Neurol.* 160: 13–36

Cohen, M. I. 1971. Switching of the respiratory phases and evoked phrenic responses produced by rostral pontine electrical stimulation. *J. Physiol. London* 217: 133–58

Cohen, M. I. 1979. Neurogenesis of respiratory rhythm in the mammal. *Physiol. Rev.* 59: 1105

Corman, C. D., Meyer, P. M., Meyer, D. R. 1967. Open-field activity and exploration in rats with septal and amygdaloid lesions. *Brain Res.* 5: 469–76

Costall, B., Domeney, A. M., Naylor, R. J., Tyers, M. B. 1987. Effects of the 5-HT-3 receptor antagonist, GR38032F, on raised dopaminergic activity in the mesolimbic system of the rat and marmoset brain. *Br. J. Pharmacol.* 92: 881–94

Costall, B., Jones, B. J., Kelly, M. E., Naylor, R. J., Onaivi, E. S., Tyers, M. B. 1990. Sites of action of ondasetron to inhibit withdrawal from drugs of abuse. *Pharmacol. Biochem. Behav.* 36: 97–104

Costall, B., Kelly, M. E., Naylor, R. J., Onaivi, E. S., Tyers, M. B. 1989. Neuroanatomical sites of action of 5-HT-3 receptor agonist and antagonists for alteration of aversive behaviour in the mouse. *Br. J. Pharmacol.* 96: 325–32

Cox, G. E., Jordan, D., Paton, J. F. R., Spyer, K. M., Wood, L. M. 1987. Cardiovascular and phrenic nerve responses to stimulation of the amygdala central nucleus in the anaesthetized rabbit. *J. Physiol. London* 389: 541–56

Crandall, P. H., Walter, R. D., Dymond, A. 1971. The ictal electroencephalographic signal identifying limbic system seizure foci. *Proc. Amer. Assoc. Neurol. Surg.* 1: 1

Dafters, R. I. 1976. Effect of medial archistriatal lesions on the conditioned emotional response and on auditory discrimination performance of the pigeon. *Physiol. Behav.* 17: 659–65

De Olmos, J., Alheid, G. F., Beltramino, C. A. 1985. Amygdala. In *The Rat Nervous System*, ed. G. Paxinos, 1: 223–334. Orlando, FL: Academic

Deutch, A. Y., Goldstein, M., Roth, R. H.

368 DAVIS

1986. Activation of the locus coeruleus induced by selective stimulation of the ventral tegmental area. *Brain Res.* 363: 307–14

Dunn, J. D. 1987. Plasma corticosterone responses to electrical stimulation of the bed nucleus of the stria terminalis. *Brain Res.* 407: 327–31

Dunn, J. D., Whitener, J. 1986. Plasma corticosterone responses to electrical stimulation of the amygdaloid complex: cytoarchitectural specificity. *Neuroendocrinology* 42: 211–17

Dunn, L. T., Everitt, B. J. 1988. Double dissociations of the effects of amygdala and insular cortex lesions on conditioned taste aversion, passive avoidance, and neophobia in the rat using the excitotoxin ibotenic acid. *Behav. Neurosci.* 102: 3–23

Eclancher, F., Karli, P. 1979. Effects of early amygdaloid lesions on the development of reactivity in the rat. *Physiol. Behav.* 22: 1123–34

Ellis, M. E., Kesner, R. P. 1983. The noradrenergic system of the amygdala and aversive information processing. *Behav. Neurosci.* 97: 399–415

Estes, W. K., Skinner, B. F. 1941. Some quantitative properties of anxiety. *J. Exp. Psychol.* 29: 390–400

Everitt, B. J., Cador, M., Robbins, T. W. 1989. Interactions between the amygdala and ventral striatum in stimulus-reward associations: Studies using a second-order schedule of sexual reinforcement. *Neuroscience* 30: 63–75

Faiers, A. A., Calaresu, F. R., Mogenson, G. J. 1975. Pathway mediating hypotension elicited by stimulation of the amygdala in the rat. *Amer. J. Physiol.* 288: 1358–66

Falls, W. A., Miserendino, M. J. D., Davis, M. 1992. Extinction of fear-potentiated stimuli: Blockade by infusion of an NMDA antagonist into the amygdala. *J. Neurosci.* In press

Fanardjian, V. V., Manvelyan, L. R. 1987. Mechanisms regulating the activity of facial nucleus motoneurons. III. Synaptic influences from the cerebral cortex and subcortical structures. *Neuroscience* 20: 835–43

Fanselow, M. S. 1991. The midbrain periaqueductal gray as a coordinator of action in response to fear and anxiety. In *The Midbrain Periaqueductal Grey Matter: Functional Anatomical and Immunohistochemical Organization*, ed. A. Depaulis, R. Bandler. New York: Plenum

Fanselow, M. S., Bolles, R. C. 1979. Naloxone and shock-elicited freezing in the rat. *J. Comp. Physiol. Psychol.* 93: 736–44

Feldman, S., Conforti, N. 1981. Amygdalectomy inhibits adrenocortical responses to somatosensory and olfactory stimulation. *Neuroendocrinology* 32: 330–34

Feldman, S., Conforti, N., Saphier, D. 1990. The preoptic area and bed nucleus of the stria terminalis are involved in the effects of the amygdala on adrenocortical secretion. *Neuroscience* 37: 775–79

Feldman, S., Conforti, N., Siegal, R. A. 1982. Adrenocortical responses following limbic stimulation in rats with hypothalamic deafferentations. *Neuroendocrinology* 35: 205–11

File, S. E. 1980. The use of social interaction as a method for detecting anxiolytic activity of chlordiazepoxide-like drugs. *J. Neurosci. Methods* 2: 219–38

File, S. E., Rodgers, R. J. 1979. Partial anxiolytic actions of morphine sulphate following microinjection into the central nucleus of the amygdala in rats. *Pharmacol. Biochem. Behav.* 11: 313–18

Flaherty, C. F. 1990. Effect of anxiolytics and antidepressants on extinction and negative contrast. *Pharmacol. Ther.* 46: 309–20

Frysinger, R. C., Marks, J. D., Trelease, R. B., Schechtman, V. L., Harper, R. M. 1984. Sleep states attenuate the pressor response to central amygdala stimulation. *Exp. Neurol.* 83: 604–17

Galeno, T. M., Brody, M. J. 1983. Hemodynamic responses to amygdaloid stimulation in spontaneously hypertensive rats. *Amer. J. Physiol.* 245: 281–86

Galeno, T. M., VanHoesen, G. W., Brody, M. J. 1984. Central amygdaloid nucleus lesion attenuates exaggerated hemodynamic responses to noise stress in the spontaneously hypertensive rat. *Brain Res.* 291: 249–59

Gallagher, M., Graham, P. W., Holland, P. C. 1990. The amygdala central nucleus and appetitive pavlovian conditioning: Lesions impair one class of conditioned behavior. *J. Neurosci.* 10: 1906–11

Gallagher, M., Kapp, B. S. 1978. Manipulation of opiate activity in the amygdala alters memory processes. *Life Sci.* 23: 1973–78

Gallagher, M., Kapp, B. S. 1981. Effect of phentolamine administration into the amygdala complex of rats on time-dependent memory processes. *Behav. Neural. Biol.* 31: 90–95

Gallagher, M., Kapp, B. S., Frysinger, R. C., Rapp, P. R. 1980. Beta-adrenergic manipulation in amygdala central n. alters rabbit heart rate conditioning. *Pharmacol. Biochem. Behav.* 12: 419–26

Gallagher, M., Kapp, B. S., McNall, C. L.,

Pascoe, J. P. 1981. Opiate effects in the amygdala central nucleus on heart rate conditioning in rabbits. *Pharmacol. Biochem. Behav.* 14: 497–505

Gallagher, M., Kapp, B. S., Pascoe, J. P. 1982. Enkephalin analogue effects in the amygdala central nucleus on conditioned heart rate. *Pharmacol. Biochem. Behav.* 17: 217–22

Gary Bobo, E., Bonvallet, M. 1975. Amygdala and masseteric reflex. I. Facilitation, inhibition and diphasic modifications of the reflex, induced by localized amygdaloid stimulation. *Electroenceph. Clin. Neurophysiol.* 39: 329–39

Geller, I., Seifter, J. 1960. The effects of meprobamate, barbiturates, *d*-amphetamine and promazine on experimentally induced conflict in the rat. *Psychopharmacologia* 1: 482–92

Gelsema, A. J., McKitrick, D. J., Calaresu, F. R. 1987. Cardiovascular responses to chemical and electrical stimulation of amygdala in rats. *Am. J. Physiol.* 253: R712–18

Gentile, C. G., Jarrel, T. W., Teich, A., McCabe, P. M., Schneiderman, N. 1986. The role of amygdaloid central nucleus in the retention of differential pavlovian conditioning of bradycardia in rabbits. *Behav. Brain Res.* 20: 263–73

Gloor, P. 1960. Amygdala. In *Handbook of Physiology:* Sect. 1. *Neurophysiology*, ed. J. Field, pp. 1395–1420. Washington, DC: Am. Physiol. Soc.

Gloor, P. 1978. Inputs and outputs of the amygdala: What the amygdala is trying to tell the rest of the brain. In *Limbic Mechanisms: The Continuing Evolution of the Limbic System Concept*, ed. K. Livingston, K. Hornykiewicz, pp. 189–209. New York: Plenum

Gloor, P., Olivier, A., Quesney, L. F. 1981. The role of the amygdala in the expression of psychic phenomena in temporal lobe seizures. In *The Amygdaloid Complex*, ed. Y. Ben-Ari, pp. 489–507. New York: Elsevier/North-Holland

Goddard, G. V. 1964. Functions of the amygdala. *Psychol. Bull.* 62: 89–109

Gold, P. E., Hankins, L., Edwards, R. M., Chester, J., McGaugh, J. L. 1975. Memory inference and facilitation with posttrial amygdala stimulation: Effect varies with footshock level. *Brain Res.* 86: 509–13

Goodman, R. R., Snyder, S. H., Kuhar, M. J., Young, W. S. III. 1980. Differential of delta and mu opiate receptor localizations by light microscopic autoradiography. *Proc. Natl. Acad. Sci. USA* 77: 2167–74

Graeff, F. G. 1988. Animal models of aversion. In *Selected Models of Anxiety, Depression and Psychosis*, ed. P. Simon, P.

Soubrie, D. Wildlocher, pp. 115–42. Basel: Karger

Gray, T. S. 1989. Autonomic neuropeptide connections of the amygdala. In *Neuropeptides and Stress*, ed. Y. Tache, J. E. Morley, M. R. Brown, pp. 92–106. New York: Springer-Verlag

Gray, T. S., Carney, M. E., Magnuson, D. J. 1989. Direct projections from the central amygdaloid nucleus to the hypothalamic paraventricular nucleus: Possible role in stress-induced adrenocorticotropin release. *Neuroendocrinology* 50: 433–46

Greidanus, T. B. V. W., Croiset, G., Bakker, E., Bouman, H. 1979. Amygdaloid lesions block the effect of neuropeptides, vasopressin, ACTH on avoidance behavior. *Physiol. Behav.* 22: 291–95

Grijalva, C. V., Levin, E. D., Morgan, M., Roland, B., Martin, F. C. 1990. Contrasting effects of centromedial and basolateral amygdaloid lesions on stress-related responses in the rat. *Physiol. Behav.* 48: 495–500

Grossman, S. P., Grossman, L., Walsh, L. 1975. Functional organization of the rat amygdala with respect to avoidance behavior. *J. Comp. Physiol. Psychol.* 88: 829–50

Hagan, R. M., Jones, B. J., Jordan, C. C., Tyers, M. B. 1990. Effect of 5-HT-3 receptor antagonists on responses to selective activation of mesolimbic dopaminergic pathways in the rat. *Br. J. Pharmacol.* 99: 227–32

Hammer, G. D., Kapp, B. S. 1986. The effects of naloxone administered into the periaqueductal gray on shock-elicited freezing behavior in the rat. *Behav. Neural Biol.* 46: 189–95

Handwerker, M. J., Gold, P. E., McGaugh, J. L. 1974. Impairment of active avoidance learning with posttraining amygdala stimulation. *Brain Res.* 75: 324–27

Harper, R. M., Frysinger, R. C., Trelease, R. B., Marks, J. D. 1984. State-dependent alteration of respiratory cycle timing by stimulation of the central nucleus of the amygdala. *Brain Res.* 306: 1–8

Heinemann, W., Stock, G., Schaeffer, H. 1973. Temporal correlation of responses in blood pressure and motor reaction under electrical stimulation of limbic structures in the unanesthetized unrestrained cat. *Pflugers Arch. Ges. Physiol.* 343: 27–40

Henke, P. G. 1972. Amygdalectomy and mixed reinforcement schedule contrast effects. *Psychon. Sci.* 28: 301–2

Henke, P. G. 1973. Effects of reinforcement omission on rats with lesions in the amygdala. *J. Comp. Physiol. Psychol.* 84: 187–93

Henke, P. G. 1977. Dissociation of the frustration effect and the partial reinforcement extinction effect after limbic lesions in rats. *J. Comp. Physiol. Psychol.* 91: 1032–38

Henke, P. G. 1980a. The amygdala and restraint ulcers in rats. *J. Comp. Physiol. Psychol.* 94: 313–23

Henke, P. G. 1980b. The centromedial amygdala and gastric pathology in rats. *Physiol. Behav.* 25: 107–12

Henke, P. G. 1982. The telencephalic limbic system and experimental gastric pathology: A review. *Neurosci. Biobehav. Rev.* 6: 381–90

Henke, P. G. 1983. Unit-activity in the central amygdalar nucleus of rats in response to immobilization-stress. *Brain Res. Rev.* 10: 833–37

Henke, P. G., Allen, J. D., Davison, C. 1972. Effect of lesions in the amygdala on behavioral contrast. *Physiol. Behav.* 8: 173–76

Heule, F., Lorez, H., Cumin, R., Haefely, W. 1983. Studies on the anticonflict effect of midazolam injected into the amygdala. *Neurosci. Lett.* 14: S164

Hilton, S. M., Zbrozyna, A. W. 1963. Amygdaloid region for defense reaction and its efferent pathway to the brainstem. *J. Physiol. London* 165: 160–73

Hitchcock, J. M., Davis, M. 1987. Fear-potentiated startle using an auditory conditioned stimulus: Effect of lesions of the amygdala. *Physiol. Behav.* 39: 403–8

Hitchcock, J. M., Davis, M. 1991. The efferent pathway of the amygdala involved in conditioned fear as measured with the fear-potentiated startle paradigm. *Behav. Neurosci.* 105: 826–42

Hodges, H., Green, S., Glenn, B. 1987. Evidence that the amygdala is involved in benzodiazepine and serotonergic effects on punished responding but not on discrimination. *Psychopharmacology* 92: 491–504

Holstege, G., Kuypers, H. G. J. M., Dekker, J. J. 1977. The organization of the bulbar fibre connections to the trigeminal, facial and hypoglossal motor nuclei. II. An autoradiographic tracing study in cat. *Brain* 100: 265–86

Hopkins, D. A., Holstege, G. 1978. Amygdaloid projections to the mesencephalon, pons and medulla oblongata in the cat. *Exp. Brain Res.* 32: 529–47

Inagaki, S., Kawai, Y., Matsuzak, T., Shiosaka, S., Tohyama, M. 1983. Precise terminal fields of the descending somatostatinergic neuron system from the amygdala complex of the rat. *J. Hirnforsch.* 24: 345–65

Innes, D. L., Tansy, M. F. 1980. Gastric mucosal ulceration associated with electrochemical stimulation of the limbic system. *Brain Res. Bull.* 5: 33–36

Iwata, J., Chida, K., LeDoux, J. E. 1987. Cardiovascular responses elicited by stimulation of neurons in the central amygdaloid nucleus in awake but not anesthetized rats resemble conditioned emotional responses. *Brain Res.* 418: 183–88

Iwata, J., LeDoux, J. E., Meeley, M. P., Arneric, S., Reis, D. J. 1986. Intrinsic neurons in the amygdala field projected to by the medial geniculate body mediate emotional responses conditioned to acoustic stimuli. *Brain Res.* 383: 195–214

Jellestad, F. K., Cabrera, I. G. 1986. Exploration and avoidance learning after ibotenic acid and radio frequency lesions in the rat amygdala. *Behav. Neural Biol.* 46: 196–215

Jellestad, F. K., Markowska, A., Bakke, H. K., Walther, B. 1986. Behavioral effects after ibotenic acid, 6-OHDA and electrolytic lesions in the central amygdala nucleus of the rat. *Physiol. Behav.* 37: 855–62

Jonason, K. R., Enloe, L. J. 1971. Alterations in social behavior following septal and amygdaloid lesions in the rat. *J. Comp. Physiol. Psychol.* 75: 286–301

Kaada, B. R. 1951. Somatomotor, autonomic and electrophysiological responses to electrical stimulation of "rhinencephalic" and other structures in primates, cat, and dog. *Acta Physiol. Scand.* (Suppl. 24) 83: 1–285

Kaada, B. R. 1972. Stimulation and regional ablation of the amygdaloid complex with reference to functional representations. In *The Neurobiology of the Amygdala*, ed. B. E. Eleftheriou, pp. 205–81. New York: Plenum

Kaku, T. 1984. Functional differentiation of hypoglossal motoneurons during the amygdaloid or cortically induced rhythmical jaw and tongue movements in the rat. *Brain Res. Bull.* 13: 147–54

Kapp, B. S., Frysinger, R. C., Gallagher, M., Haselton, J. R. 1979. Amygdala central nucleus lesions: Effects on heart rate conditioning in the rabbit. *Physiol. Behav.* 23: 1109–17

Kapp, B. S., Gallagher, M., Underwood, M. D., McNall, C. L., Whitehorn, D. 1982. Cardiovascular responses elicited by electrical stimulation of the amygdala central nucleus in the rabbit. *Brain Res.* 234: 251–62

Kapp, B. S., Pascoe, J. P. 1986. Correlation aspects of learning and memory: Vertebrate model systems. In *Learning and Memory: A Biological View*, ed. J. L. Martinez, R. P. Kesner, pp. 399–440. New York: Academic

Kapp, B. S., Pascoe, J. P., Bixler, M. A. 1984. The amygdala: A neuroanatomical systems approach to its contribution to aversive conditioning. In *The Neuropsychology of Memory*, ed. N. Butters, L. S. Squire, pp. 473–88. New York: Guilford

Kapp, B. S., Wilson, A., Pascoe, J. P., Supple, W. F., Whalen, P. J. 1990. A neuroanatomical systems analysis of conditioned bradycardia in the rabbit. In *Neurocomputation and Learning: Foundations of Adaptive Networks*, ed. M. Gabriel, J. Moore. New York: Bradford Books

Kellicut, M. H., Schwartzbaum, J. S. 1963. Formation of a conditioned emotional response. CER following lesions of the amygdaloid complex in rats. *Psychol. Rev.* 12: 351–58

Kemble, E. D., Beckman, G. J. 1970. Runway performance of rats following amygdaloid lesions. *Physiol. Behav.* 5: 45–47

Kemble, E. D., Blanchard, D. C., Blanchard, R. J. 1990. Effects of regional amygdaloid lesions on flight and defensive behaviors of wild black rats (*Rattus rattus*). *Physiol. Behav.* 48: 1–5

Kemble, E. D., Blanchard, D. C., Blanchard, R. J., Takushi, R. 1984. Taming in wild rats following medial amygdaloid lesions. *Physiol. Behav.* 32: 131–34

Kemble, E. D., Studelska, D. R., Schmidt, M. K. 1979. Effects of central amygdaloid nucleus lesions on ingestation, taste reactivity, exploration and taste aversion. *Physiol. Behav.* 22: 789–93

Kesner, R. P. 1982. Brain stimulation: Effects on memory. *Behav. Neural Biol.* 36: 315–67

Kesner, R. P., Walser, R. D., Winzenried, G. 1989. Central but not basolateral amygdala mediates memory for positive affective experiences. *Behav. Brain Res.* 33: 189–95

Koikegami, H., Dudo, T., Mochida, Y., Takahashi, H. 1957. Stimulation experiments on the amygdaloid nuclear complex and related structures: Effects upon the renal volume, urinary secretion, movements of the urinary bladder, blood pressure and respiratory movements. *Folia Psychiat. Neurol. Jpn.* 11: 157–207

Krettek, J. E., Price, J. L. 1978a. A description of the amygdaloid complex in the rat and cat with observations on intra-amygdaloid axonal connections. *J. Comp. Neurol.* 178: 255–80

Krettek, J. E., Price, J. L. 1978b. Amygdaloid projections to subcortical structures within the basal forebrain and brain-stem in the rat and cat. *J. Comp. Neurol.* 178: 225–54

Lang, P. J., Bradley, M. M., Cuthbert, B. N. 1990. Emotion, attention, and the startle reflex. *Psychol. Rev.* 97: 377–95

Leaton, R. N., Cranney, J. 1990. Potentiation of the acoustic startle response by a conditioned stimulus paired with acoustic startle stimulus in rats. *J. Exp. Psychol. Anim. Behav. Process.* 16: 279–87

LeDoux, J. E., Cicchetti, P., Xagoraris, A., Romanski, L. M. 1990. The lateral amygdaloid nucleus: Sensory interface of the amygdala in fear conditioning. *J. Neurosci.* 10: 1062–69

LeDoux, J. E., Iwata, J., Cicchetti, P., Reis, D. J. 1988. Different projections of the central amygdaloid nucleus mediate autonomic and behavioral correlates of conditioned fear. *J. Neurosci.* 8: 2517–29

Lewis, S. J., Verberne, A. J. M., Robinson, T. G., Jarrott, B., Louis, W. J., Beart, P. M. 1989. Excitotoxin-induced lesions of the central but not basolateral nucleus of the amygdala modulate the baroreceptor heart rate reflex in conscious rats. *Brain Res.* 494: 232–40

Liang, K. C., Bennett, C., McGaugh, J. L. 1985. Peripheral epinephrine modulates the effects of post-training amygdala stimulation on memory. *Behav. Brain Res.* 15: 93–100

Liang, K. C., Juler, R. G., McGaugh, J. L. 1986. Modulating effects of posttraining epinephrine on memory: Involvement of the amygdala noradrenergic systems. *Brain Res.* 368: 125–33

Liebman, J. M., Mayer, D. J., Liebeskind, J. C. 1970. Mesencephalic central gray lesions and fear-motivated behavior in rats. *Brain Res.* 23: 353–70

Luiten, P. G. M., Koolhaas, J. M., deBoer, S., Koopmans, S. J. 1985. The corticomedial amygdala in the central nervous system organization of agonistic behavior. *Brain Res.* 332: 283–97

Maeda, H., Maki, S. 1986. Dopaminergic facilitation of recovery from amygdaloid lesions which affect hypothalamic defensive attack in cats. *Brain Res.* 363: 135–40

Magnuson, D. J., Gray, T. S. 1990. Central nucleus of amygdala and bed nucleus of stria terminalis projections to serotonin or tyrosine hydroxylase immunoreactive cells in the dorsal and median raphe nuclei in the rat. *Soc. Neurosci. Abstr.* 16: 121

Mason, J. W. 1959. Plasma 17-hydroxycorticosteroid levels during electrical stimulation of the amygdaloid complex in conscious monkeys. *Am. J. Physiol.* 196: 44–48

Mast, M., Blanchard, R. J., Blanchard, D. C. 1982. The relationship of freezing and response suppression in a CER situation. *Psychol. Record* 32: 151–67

Matheson, B. K., Branch, B. J., Taylor, A. N. 1971. Effects of amygdaloid stimulation on pituitary-adrenal activity in conscious cats. *Brain Res.* 32: 151–67

McAllister, W. R., McAllister, D. E. 1971. Behavioral measurement of conditioned fear. In *Aversive Conditioning and Learning*, ed. F. R. Brush, pp. 105–79. New York: Academic

McCall, R. B., Aghajanian, G. K. 1979. Serotonergic facilitation of facial motoneuron excitation. *Brain Res.* 169: 11–27

McGaugh, J. L., Introinicollison, I. B., Nagahara, A. H., Cahill, L., Brioni, J. D., Castellano, C. 1990. Involvement of the amygdaloid complex in neuromodulatory influences on memory storage. *Neurosci. Biobehav. Rev.* 14: 425–32

McIntyre, M., Stein, D. G. 1973. Differential effects of one- vs two-stage amygdaloid lesions on activity, exploration, and avoidance behavior in the albino rat. *Behav. Biol.* 9: 451–65

Millhouse, O. E., DeOlmos, J. 1983. Neuronal configurations in lateral and basolateral amygdala. *Neuroscience* 10: 1269–1300

Miserendino, M. J. D., Sananes, C. B., Melia, K. R., Davis, M. 1990. Blocking of acquisition but not expression of conditioned fear-potentiated startle by NMDA antagonists in the amygdala. *Nature* 345: 716–18

Mishkin, M., Aggleton, J. 1981. Multiple functional contributions of the amygdala in the monkey. In *The Amygdaloid Complex*, ed. Y. Ben-Ari, pp. 409–20. New York: Elsevier/North-Holland

Misslin, R., Cigrang, M. 1986. Does neophobia necessarily imply fear or anxiety? *Behav. Process.* 12: 45–50

Morgenson, G. J., Calaresu, F. R. 1973. Cardiovascular responses to electrical stimulation of the amygdala in the rat. *Exp. Neurol.* 39: 166–80

Mraovitch, S., Kumada, M., Reis, D. J. 1982. Role of the nucleus parabrachialis in cardiovascular regulation in cat. *Brain Res.* 232: 57–75

Murray, E. A. 1990. Representational memory in nonhuman primates. In *Neurobiology of Comparative Cognition*, ed. R. T. Kesner, D. S. Olton, pp. 127–55. Hillsdale, NJ: Erlbaum

Murray, E. A., Mishkin, M. 1985. Amygdalectomy impairs crossmodal association in monkeys. *Science* 228: 604–6

Nagy, J., Zambo, K., Decsi, L. 1979. Antianxiety action of diazepam after intraamygdaloid application in the rat. *Neuropharmacology* 18: 573–76

Niehoff, D. L., Kuhar, M. J. 1983. Benzodiazepine receptors: Localization in rat amygdala. *J. Neurosci.* 3: 2091–97

Nitecka, L., Amerski, L., Narkiewicz, O. 1981. The organization of intraamygdaloid connections: an HRP study. *J. Hirnforsch.* 22: 3–7

Ohta, M. 1984. Amygdaloid and cortical facilitation or inhibition of trigeminal motoneurons in the rat. *Brain Res.* 291: 39–48

Ottersen, O. P. 1982. Connections of the amygdala of the rat. IV. Corticoamygdaloid and intraamygdaloid connections as studied with axonal transport of horseradish peroxidase. *J. Comp. Neurol.* 205: 30–48

Pare, D., Steriade, M., Deschenes, M., Bouhassiri, D. 1990. Prolonged enhancement of anterior thalamic synaptic responsiveness by stimulation of a brain-stem cholinergic group. *J. Neurosci.* 10: 20–33

Pascoe, J. P., Bradley, D. J., Spyer, K. M. 1989. Interactive responses to stimulation of the amygdaloid central nucleus and baroreceptor afferents in the rabbit. *J. Auton. Nerv. Sys.* 26: 157–67

Pascoe, J. P., Kapp, B. S. 1985a. Electrophysiological characteristics of amygdaloid central nucleus neurons in the awake rabbit. *Brain Res. Bull.* 14: 331–38

Pascoe, J. P., Kapp, B. S. 1985b. Electrophysiological characteristics of amygdaloid central nucleus neurons during Pavlovian fear conditioning in the rabbit. *Behav. Brain Res.* 16: 117–33

Peinado-Manzano, M. A. 1990. The role of the amygdala and the hippocampus in working memory for spatial and non-spatial information. *Behav. Brain Res.* 38: 117–34

Pellegrino, L. 1968. Amygdaloid lesions and behavioral inhibition in the rat. *J. Comp. Physiol. Psychol.* 65: 483–91

Petersen, E. N., Braestrup, C., Scheel-Kruger, J. 1985. Evidence that the anticonflict effect of midazolam in amygdala is mediated by the specific benzodiazepine receptors. *Neurosci. Lett.* 53: 285–88

Petersen, E. N., Scheel-Kruger, J. 1982. The GABAergic anticonflict effect of intraamygdaloid benzodiazepines demonstrated by a new water lick conflict paradigm. In *Behavioral Models and the Analysis of Drug Action*, ed. M. Y. Spiegelstein, A. Levy. Amsterdam: Elsevier

Phillips, R. E. 1964. "Wildness" in the Mallard duck: Effects of brain lesions and stimulation on "escape behavior" and reproduction. *J. Comp. Neurol.* 122: 139–56

Phillips, R. E. 1968. Approach-withdrawal behavior of peach-faced lovebirds, *Aga-*

pornis roseicolis, and its modification by brain lesions. *Behavior* 31: 163–84

Phillipson, O. T. 1979. Afferent projections to the ventral tegmented area of Tsai and intrafascicular nucleus. A horseradish peroxidase study in the rat. *J. Comp. Neurol.* 187: 117–43

Post, S., Mai, J. K. 1980. Contribution to the amygdaloid projection field in the rat: A quantitative autoradiographic study. *J. Hirnforsch.* 21: 199–225

Price, J. L., Amaral, D. G. 1981. An autoradiographic study of the projections of the central nucleus of the monkey amygdala. *J. Neurosci.* 1: 1242–59

Rasmussen, K., Jacobs, B. L. 1986. Single unit activity of locus coeruleus in the freely moving cat: II. Conditioning and pharmacologic studies. *Brain Res.* 371: 335–44

Redgate, E. S., Fahringer, E. E. 1973. A comparison of the pituitary-adrenal activity elicited by electrical stimulation of preoptic, amygdaloid and hypothalamic sites in the rat brain. *Neuroendocrinology* 12: 334–43

Redmond, D. E. Jr. 1977. Alteration in the function of the nucleus locus coeruleus: A possible model for studies on anxiety. In *Animal Models in Psychiatry and Neurology*, ed. I. E. Hanin, E. Usdin, pp. 293–304. Oxford, UK: Pergamon

Reis, D. J., Oliphant, M. C. 1964. Bradycardia and tachycardia following electrical stimulation of the amygdaloid region in the monkey. *J. Neurophysiol.* 27: 893–912

Riolobos, A. S., Garcia, A. I. M. 1987. Open field activity and passive avoidance responses in rats after lesion of the central amygdaloid nucleus by electrocoagulation and ibotenic acid. *Physiol. Behav.* 39: 715–20

Rogawski, M. A., Aghajanian, G. K. 1980. Modulation of lateral geniculate neuron excitability by noradrenaline microiontophoresis or locus coeruleus stimulation. *Nature* 287: 731–34

Rosen, J. B., Davis, M. 1988a. Enhancement of acoustic startle by electrical stimulation of the amygdala. *Behav. Neurosci.* 102: 195–202

Rosen, J. B., Davis, M. 1988b. Temporal characteristics of enhancement of startle by stimulation of the amygdala. *Physiol. Behav.* 44: 117–23

Rosen, J. B., Hitchcock, J. M., Sananes, C. B., Miserendino, M. J. D., Davis, M. 1991. A direct projection from the central nucleus of the amygdala to the acoustic startle pathway: Anterograde and retrograde tracing studies. *Behav. Neurosci.* 105: 817–25

Ruggiero, D. A., Ross, C. A., Kumada, M.,

Reis, D. J. 1982. Reevaluation of projections from the mesencephalic trigeminal nucleus to the medulla and spinal cord: New projections. A combined retrograde and anterograde horseradish peroxidase study. *J. Comp. Neurol.* 206: 278–92

Russchen, F. T. 1982. Amygdalopetal projections in the cat. II. Subcortical afferent connections. A study with retrograde tracing techniques. *J. Comp. Neurol.* 207: 157–76

Sananes, C. B., Campbell, B. A. 1989. Role of the central nucleus of the amygdala in olfactory heart rate conditioning. *Behav. Neurosci.* 103: 519–25

Sananes, C. B., Davis, M. 1992. NMDA lesions of the lateral and basolateral nuclei of the amygdala block fear-potentiated startle and shock sensitization of startle. *Behav. Neurosci.* In press

Saphier, D., Feldman, S. 1986. Effects of stimulation of the preoptic area on hypothalamic paraventricular nucleus unit activity and corticosterone secretion in freely moving rats. *Neuroendocrinology* 42: 167–73

Sarter, M., Markowitsch, H. J. 1985. Involvement of the amygdala in learning and memory: A critical review, with emphasis on anatomical relations. *Behav. Neurosci.* 99: 342–80

Sawchenko, P. E., Swanson, L. W. 1983. The organization of forebrain afferents to the paraventricular and supraoptic nucleus of the rat. *J. Comp. Neurol.* 218: 121–44

Scheel-Kruger, J., Petersen, E. N. 1982. Anticonflict effect of the benzodiazepines mediated by a GABAergic mechanism in the amygdala. *Eur. J. Pharmacol.* 82: 115–16

Schlor, K. H., Stumpf, H., Stock, G. 1984. Baroreceptor reflex during arousal induced by electrical stimulation of the amygdala or by natural stimuli. *J. Auton. Nerv. Sys.* 10: 157–65

Schwaber, J. S., Kapp, B. S., Higgins, G. A., Rapp, P. R. 1982. Amygdaloid and basal forebrain direct connections with the nucleus of the solitary tract and the dorsal motor nucleus. *J. Neurosci.* 2: 1424–38

Schwartzbaum, J. S. 1960. Changes in reinforcing properties of stimuli following ablation of the amygdaloid complex in monkeys. *J. Comp. Physiol. Psychol.* 53: 388–95

Schwartzbaum, J. S., Gay, P. E. 1966. Interacting behavioral effects of septal and amygdaloid lesions in the rat. *J. Comp. Physiol. Psychol.* 61: 59–65

Sen, R. N., Anand, B. K. 1957. Effect of electrical stimulation of the limbic system of brain ("visceral brain") on gastric secretory activity and ulceration. *Ind. J. Med. Res.* 45: 515–21

Setekleiv, J., Skaug, O. E., Kaada, B. R. 1961. Increase of plasma 17-hydroxy-corticosteroids by cerebral cortical and amygdaloid stimulation in the cat. *J. Endocrinol.* 22: 119–26

Shibata, K., Kataoka, Y., Gomita, Y., Ueki, S. 1982. Localization of the site of the anticonflict action of benzodiazepines in the amygdaloid nucleus of rats. *Brain Res.* 234: 442–46

Shibata, K., Kataoka, Y., Yamashita, K., Ueki, S. 1986. An important role of the central amygdaloid nucleus and mam-millary body in the mediation of conflict behavior in rats. *Brain Res.* 372: 159–62

Shibata, K., Yamashita, K., Yamamoto, E., Ozaki, T., Ueki, S. 1989. Effect of benzo-diazepine and GABA antagonists on anti-conflict effects of antianxiety drugs injected into the rat amygdala in a water-lick suppression test. *Psychopharmacology* 98: 38–44

Shiosaka, S., Tokyama, M., Takagi, H., Takahashi, Y., Saitoh, T., et al. 1980. Ascending and descending components of the medial forebrain bundle in the rat as demonstrated by the horseradish per-oxidase-blue reaction. I. Forebrain and upper brainstem. *Exp. Brain Res.* 39: 377–88

Silverman, A. J., Hoffman, D. L., Zimm-erman, E. A. 1981. The descending affer-ent connections of the paraventricular nucleus of the hypothalamus (PVN). *Brain Res. Bull.* 6: 47–61

Simon, H., LeMoal, M., Calas, A. 1979. Efferents and afferents of the ventral teg-mental-A10 region studies after local injection of [3H]leucine and horseradish peroxidase. *Brain Res.* 178: 17–40

Singh, V. B., Onaivi, E. S., Phan, T. H., Boadle-Biber, M. C. 1990. The increases in rat cortical and midbrain tryptophan hydroxylase activity in response to acute or repeated sound stress are blocked by bilateral lesions to the central nucleus of the amygdala. *Brain Res.* 530: 49–53

Slotnick, B. M. 1973. Fear behavior and pas-sive avoidance deficts in mice with amyg-dala lesions. *Physiol. Behav.* 11: 717–20

Smelik, P. G., Vermes, I. 1980. The regu-lation of the pituitary-adrenal system in mammals. In *General Comparative and Clinical Endocrinology of the Adrenal Cor-tex*, ed. I. C. Jones, I. W. Henderson, pp. 1–55. London: Academic

Smith, B. S., Millhouse, O. E. 1985. The connections between basolateral and cen-tral amygdaloid nuclei. *Neurosci. Lett.* 56: 307–9

Spevack, A. A., Campbell, C. T., Drake, L.

1975. Effect of amygdalectomy on habitu-ation and CER in rats. *Physiol. Behav.* 15: 199–207

Steriade, M., Datta, S., Pare, D., Oakson, G., Dossi, R. C. 1990. Neuronal activities in brain-stem cholinergic nuclei related to tonic activation processes in thalamo-cortical systems. *J. Neurosci.* 10: 2541–59

Stock, G., Schlor, K. H., Heidt, H., Buss, J. 1978. Psychomotor behaviour and cardio-vascular patterns during stimulation of the amygdala. *Pflugers Arch. Ges. Physiol.* 376: 177–84

Stock, G., Rupprecht, U., Stumpf, H., Schlor, K. H. 1981. Cardiovascular changes during arousal elicited by stimu-lation of amygdala, hypothalamus and locus coeruleus. *J. Auton. Nerv. Syst.* 3: 503–10

Swanson, L. W., Sawchenko, P. E., Rivier, J., Vale, W. 1983. Organization of ovine corticotropin-releasing factor immuno-reactive cells and fibers in the rat brain: An immunohistochemical study. *Neuro-endocrinology* 36: 165–86

Takeuchi, Y., Matsushima, S., Matsushima, R., Hopkins, D. A. 1983. Direct amyg-daloid projections to the dorsal motor nucleus of the vagus nerve: A light and electron microscopic study in the rat. *Brain Res.* 280: 143–47

Takeuchi, Y., McLean, J. H., Hopkins, D. A. 1982. Reciprocal connections between the amygdala and parabrachial nuclei: Ultrastructural demonstration by de-generation and axonal transport of horse-radish peroxidase in the cat. *Brain Res.* 239: 538–88

Thierry, A. M., Tassin, J. P., Blanc, G., Glowinski, J. 1976. Selective activation of the mesocortical DA system by stress. *Nature* 263: 242–43

Thomas, S. R., Lewis, M. E., Iversen, S. D. 1985. Correlation of [3H]diazepam bind-ing density with anxiolytic locus in the amygdaloid complex of the rat. *Brain Res.* 342: 85–90

Timms, R. J. 1981. A study of the amyg-daloid defence reaction showing the value of althesin anesthesia in studies of the functions of the forebrain in cats. *Pflugers Arch.* 391: 49–56

Tribollet, E., Dreifuss, J. J. 1981. Local-ization of neurones projecting to the hypo-thalamic paraventricular nucleus of the rat: A horseradish peroxidase study. *Neuroscience* 7: 1215–1328

Umemoto, M., Olds, M. E. 1975. Effects of chlordiazepoxide, diazepam and chlor-promazine on conditioned emotional behaviour and conditioned neuronal activity in limbic, hypothalamic and geniculate regions. *Neuropharmacology* 14: 413–25

Ursin, H., Jellestad, F., Cabrera, I. G. 1981. The amygdala, exploration and fear. In *The Amygdaloid Complex*, ed. Y. Ben-Ari, pp. 317–29. Amsterdam: Elsevier

Ursin, H., Kaada, B. R. 1960. Functional localization within the amygdaloid complex in the cat. *Electroenceph. Clin. Neurophysiol.* 12: 1–20

Veening, J. G., Swanson, L. W., Sawchenko, P. E. 1984. The organization of projections from the central nucleus of the amygdala to brain stem sites involved in central autonomic regulation: A combined retrograde transport-immunohistochemical study. *Brain Res.* 303: 337–57

Von Euler, C., Martila, I., Remmers, J. E., Trippenbach, J. 1976. Effects of lesions in the parabrachial nucleus on the mechanisms for central and reflex termination of inspiration in the cat. *Acta Physiol. Scand.* 96: 324–37

Wallace, D. M., Magnuson, D. J., Gray, T. S. 1989. The amygdalo-brainstem pathway: Dopaminergic, noradrenergic and adrenergic cells in the rat. *Neurosci. Lett.* 97: 252–58

Weller, K. L., Smith, D. A. 1982. Afferent connections to the bed nucleus of the stria terminalis. *Brain Res.* 232: 255–70

Werka, T., Skar, J., Ursin, H. 1978. Exploration and avoidance in rats with lesions in amygdala and piriform cortex. *J. Comp. Physiol. Psychol.* 92: 672–81

Whalen, P. J., Kapp, B. S. 1991. Contributions of the amygdaloid central nucleus to the modulation of the nictitating membrane reflex in the rabbit. *Behav. Neuroscience* 105: 141–53

White, S. R., Neuman, R. S. 1980. Facilitation of spinal motoneuron excitability by 5-hydroxytryptamine and noradrenaline. *Brain Res.* 185: 1–9

Yadin, E., Thomas, E., Strickland, C. E., Grishkat, H. L. 1991. Anxiolytic effects of benzodiazepines in amygdala-lesioned rats. *Psychopharmacology* 103: 473–79

Yates, E. F., Maran, J. W. 1974. Stimulation and inhibition of adrenocorticotropin release. In *Handbook of Physiology*, Sect. 7, *Endocrinology*, *The Pituitary Gland and its Neuroendocrine Control*, ed. E. Knobil, W. H. Sawyer, 4: 367–404. Washington, DC: Am. Physiol. Soc.

Zhang, J. X., Harper, R. M., Ni, H. 1986. Cryogenic blockade of the central nucleus of the amygdala attenuates aversively conditioned blood pressure and respiratory responses. *Brain Res.* 386: 136–45

Zhang, S. P., Bandler, R., Carrive, P. 1990. Flight and immobility evoked by excitatory amino acid microinjection within distinct parts of the subtentorial midbrain periaqueductal gray of the cat. *Brain Res.* 520: 73–82

Zola-Morgan, S., Squire, L. R., Amaral, D. G. 1989. Lesions of the amygdala that spare adjacent cortical regions do not impair memory or exacerbate the impairment following lesions of the hippocampal formation. *J. Neurosci.* 9: 1922–30

Annu. Rev. Neurosci. 1992. 15:377–402

THE ORGANIZATION AND REORGANIZATION OF HUMAN SPEECH PERCEPTION

Janet F. Werker and Richard C. Tees

Department of Psychology, University of British Columbia, Vancouver, BC V6T 1Z4, Canada

KEY WORDS: infancy, speech perception, development

One of the most fundamental attributes of being human is the ability to perceive and produce language. Although humans communicate by sign and writing, by far the most common and enduring form of human communication is spoken language, or speech. The unique set of abilities that characterizes speech perception, and the early appearance of these abilities during human ontogeny, suggests that these capacities may be deeply rooted in our biology. In this chapter we review selective behavioral work on the development of speech perception in humans and try to relate these empirical findings both to the ontogeny of auditory/communicative abilities in infrahumans and to their neurobiological substrates.

We start with an overview of some of the fundamental characteristics of speech perception that make it an intriguing area of inquiry and suggest the involvement of specialized biological predispositions. We then survey research on developmental changes in speech perception. This section begins with a characterization of the "initial state" of the human infant's ability to process speech sounds (including a consideration of potential prenatal environmental influences on this state), and then reviews empirical work examining postnatal changes in speech processing, with a focus on our own work in cross-language speech perception. In the course of this analysis we relate this work, albeit selectively, to what we believe is parallel research involving the processing of auditory/communicative signals by nonhuman animals. In the final section of the chapter, we briefly speculate about possible neurobiological processes that might account for the nature of the recent developmental evidence.

377

SPEECH PERCEPTION

We typically produce approximately 12–14 individual speech sounds per second, three to four times the number of sequential arbitrary sounds that can be perceived (Warren et al 1969), and can perceive 50 or 60 segments per second with little loss in intelligibility. The speed with which speech is produced and perceived is difficult to explain, particularly given the complex computational requirements for perceiving human speech. This complexity takes two forms (Liberman et al 1967). The first is a segmentation problem. Although we think of speech as linear, there are, in reality, no clear breaks between words, syllables, and the basic elements called phonemes. Rather, the information that specifies a given phoneme is spread over several surrounding segments and often crosses syllable and even word boundaries. This is a result of coarticulation (the influence of one phoneme on another in the articulatory process).

The second problem is a lack of invariance. Simply stated, to date, no single (or even set of) invariant property(ies) in the acoustic signal has been successfully identified as necessary and sufficient for a given phoneme. Indeed, the acoustic characteristics of a given phoneme vary tremendously in different contexts (see Figure 1). Attempts to understand speech perception in terms of acoustic information have now moved to a consideration of relational information, but substantial ambiguity still exists.

The lack of equivalence between the acoustic signal and the perceived segments of an utterance has generated considerable theoretical controversy (see Mattingly & Studdert-Kennedy 1991 for discussion), which has been the catalyst for much of the empirical work. The controversy pits "specialized speech-specific" explanations against "generalized auditory" theories of speech perception. Proponents of the *speech-is-special* point of

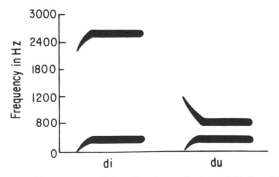

Figure 1   Spectrographic patterns sufficient for the synthesis of /d/ before /i/ and /u/. In /di/ the second formant transition rises, in /du/ it falls. (Adapted from Liberman et al 1967.)

view identify instances of different data patterns in the perception of speech vs. nonspeech sounds, whereas adherents of the *general auditory* approaches attempt to show that such differential data patterns do not exist.

One of the first pieces of evidence for specialized speech processing was the finding that human adults show "categorical" perception for speech sounds, and thus respond to only those acoustic variations which signal a difference in meaning (Liberman et al 1967). This helps to explain the perceptual invariance in light of the enormous acoustic variability. More recent work has shown that our perceptual system actually adjusts to contextual influences on speech production. For example, when one speaks rapidly, the formant transitions in a stop consonant will be shortened. Our perceptual systems calibrate for that (Miller & Liberman 1979) and other contextual influences.

Studies revealing that speech is perceived bimodally through our eyes as well as our ears provide further evidence that speech is not perceived like other acoustic signals (Summerfield 1991). For example, when shown a "talking head" producing the articulatory movements for the syllable /ba/, and presented with an acoustic /ga/, human adults typically report perceiving an instantaneous, unambiguous /da/ syllable (McGurk & MacDonald 1976). Other combinations of articulatory and acoustic information lead to different percepts, but the immediacy of the integrated speech percept is what is compelling.

The occurrence of duplex perception has also been interpreted as highlighting the independence of speech perception from the more general processing of other acoustic signals. Basically, when a speech signal is taken apart such that the base component is presented to one ear, and two of the formant transitions (representing rapid changes in frequency) to the other ear, adult listeners report two simultaneous percepts. One corresponds to the original fully integrated syllable, and the other to a nonspeech glissando—the sound that results from the transitions alone (Liberman & Mattingly 1989). Experimental manipulations indicate that when the transitional and base components are of equal intensity, only an integrated syllable is heard. The transitions have to be presented at a considerably higher intensity than the base in order for the duplexity effect to be evident (Whalen & Liberman 1987), thus suggesting that the speech percept "preempts" the non-speech percept.

Virtually all of these "special" aspects of speech perception have been shown to hold true for at least some non-speech sounds as well. The perception of non-speech, tone-onset-tone continua is partially categorical. The context effect with respect to "speaking rate" has been shown to extend to sine-wave stimuli, and duplex perception has been shown to extend to several domains, including musical stimuli and nonsignificant auditory events such as the slamming of a door (for reviews, see Fowler & Rosenblum 1991, Pisoni & Luce 1986). Thus, it is clear that these effects are not unique to the perception of speech. Speech, in particular consonant-vowel (CV) syllables, however, constitutes the only kind of signal for which all these effects can be demonstrated. It can be argued, therefore, that it is not the particuliarity of any of these effects, but rather their universality that provides the weight of evidence pointing to biological preparedness and perhaps biological specialization.

Neuropsychological studies provide more direct support for differential processing of speech sounds. Considerable evidence has shown that the rule-based aspects of language processing (phonology, syntax) are predominantly a left-hemisphere function, even, interestingly, in the signing of deaf adults with damage to one of the hemispheres (Poizner et al 1990). Well-known demonstrations of left-hemisphere processing include the early studies of aphasic patients showing a loss of speech production abilities following left, but not right, hemisphere damage, and of split-brain patients. Studies of regional brain blood-flow, auditory evoked responses, carotid sodium amytal injection, cortical stimulation, and dichotic listening indicate a right-ear, left-hemisphere advantage in the perception of consonant-vowel syllables (e.g. Kimura 1967; for a review see Kolb & Whishaw 1990).

SPEECH PERCEPTION IN HUMAN INFANTS

Perhaps the strongest evidence in support of some kind of biological preparedness comes from studies indicating that many of these behavioral and neuropsychological indices are evident even in very young infants. Prelinguistic infants as young as one month of age show categorical-like perception of speech sounds (Eimas et al 1971). When tested in a high amplitude sucking procedure, after repeated presentations of consonant-vowel syllables from one phonetic category, the infants will show an increase in sucking rate to a new stimulus, but only if it is from a contrasting (adult) phonetic category. More recent research has shown this categorical-like speech perception to be quite general (e.g. Kuhl 1987) and even to extend to phones the infant has not experienced in his/her language-learning environment (for a review, see Werker 1991), thus providing quite convincing evidence that the phonetic-relevance of early infant speech perception does not rely on specific language experience. In other words, perceptual biases exist from birth that predispose an infant to be able to discriminate acoustic variability that signals phonetic distinctiveness, and to be less able to discriminate equal-sized acoustic variability that exists between stimuli from within a single phonetic category. Such perceptual constancy is evident even across variability in intonation contour and speaker identity.

Infants also display several other quite remarkable capacities with respect to speech perception. They show context effects with respect to speaking rate (Eimas & Miller 1980). Thus the boundary between /ba/ and /wa/ shifts in accordance with the overall duration of the syllable. Perhaps even more surprisingly, infants recognize bimodal equivalences in speech. For example, in Kuhl's work, infants of 4 months of age were shown two filmed faces side by side, with a single vowel sound presented from a loudspeaker located midway between the two facial images (see Figure 2). One of the faces articulated the /a/ vowel while the other articulated the /i/ vowel (side of presentation was counterbalanced). On half the trials the vocal stimulus was /a/, and on half the trials it was /i/. These neonates clearly demonstrated recognition of auditory-visual equivalence in speech by selecting to look at the face that "matched" the vowel being spoken 73% of the time (Kuhl & Meltzoff 1982).

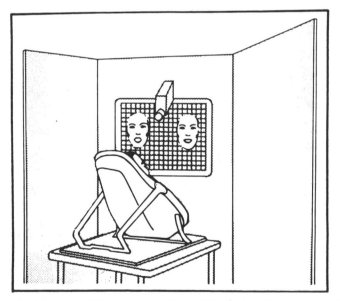

Figure 2 Infants' cross-modal speech-perception abilities are tested by presenting them with two facial images, one articulating the vowel /a/ and the other the vowel /i/. One or the other of the sounds is presented from a loudspeaker midway between the two facial images. The results show that infants look longer at the face that "matches" the speech sound they hear. (Adapted from Kuhl & Meltzoff 1982.)

These same infants also showed a rudimentary detection of auditory-motor equivalences by tending to match their vocal productions with the female speaker's vowel production. The match was apparent both in terms of prosodic characteristics (pitch control and duration) and the formant characteristics that distinguish the vowels /a/ and /i/.

Finally, neuropsychological studies reveal some evidence of left-hemispheric specialization for speech perception in early infancy. Infants display a right-ear advantage in the processing of stop consonants and a left-ear advantage for the perception of musical sounds (Best et al 1982). Both full-term and premature infants display asymmetric left-hemisphere auditory evoked responses to at least some consonant signals (Molfese & Molfese 1980). The comparability in such "test results" between infants and adults argues strongly against an interpretation that the specialized nature of speech perception is acquired through extensive experience of hearing speech in early infancy, and suggests instead a propensity that exists from birth. Whether this propensity is specific to speech or reflects more general auditory or sensorimotor abilities that are also applied to speech remains uncertain. There is no doubt, however, that the young infant is born with abilities that greatly enhance the speed, efficiency, and accuracy with which he/she can process linguistic input. At the very least, this suggests the existence of neural circuitry that is ideally suited for processing speech.

The existence of neural circuitry that is utilized rapidly to disambiguate complex signals has certainly been documented in other perceptual/motor domains. The "sophisticated" neural characteristics of the barn owl's inferior colliculus that support the animal's ability to localize sound and react appropriately (Knudsen & Konishi 1978) is perhaps the most widely cited example. The features of the neural circuitry necessary to disambiguate ambiguous visual objects, including faces (e.g. Perrett et al 1982), represent another example of a biologically significant signal that is likely to be processed by specialized central neural mechanisms in humans. The general proposition that organisms might be adapted to process, attend, or associate particular kinds of behavior with such specific stimuli is well established (e.g. Garcia & Koelling 1966, Marler 1990).

Several different neurobiological models have been proposed for explaining the specialized characteristics of human speech perception. As one example, Sussman (1989) extrapolates from the sound-localizing system of the barn owl and provides a speculative conceptualization of how a collective processing network could analyze speech sounds. Sussman proposes that hierarchically arranged arrays of biologically specialized combination-sensitive neurons might be able to account for perceptual constances both in the perception of place cues for stop consonants and for vowel normalization.

We think two points are important to keep in mind with respect to this and other auditory-based neural models of speech perception. First of all, topographical maps of auditory features such as sound localization in both cortical and subcortical structures are, in fact, sensorimotor maps. For example, both the inferior and the superior colliculus contain ordered maps of motor space that are in sensory register with visual, auditory and somatosensory maps. The little neurobiological work that has been conducted with infrahumans with respect to species-specific calls/speech suggests an intimate link between perceptual and motor processes. For example, Williams and Nottebohm (1985) report evidence that the motor system responsible in part for song production in male infant zebra finches is selectively tuned for perceiving the critical elements of song. Similar observations can be made for the colliculi of rodents, barn owls and birds in connection with the analysis of visual, auditory, and spatial location. The perception of biologically significant information across a variety of species thus seems to be organized in overlapping sensorimotor systems. As far as speech itself is concerned, electrical stimulation of the same cortical sites disrupts both the production of sequential oral facial movements and the ability to discriminate phonemes (Ojemann 1983). At some level it seems undeniable, then, that articulatory and perceptual neural systems should be highly integrated in the case of human speech (Mattingly & Liberman 1989).

Second, the evidence that is available (e.g. Tees 1990a,b) makes it clear that these perceptual-motor systems are altered by changes in auditory, visual, and tactile stimulation history. The impact of the environment is most dramatic neonatally, but there is unequivocal evidence (e.g. King & Moore 1991, Merzenich et al 1984) that perceptual-motor neural representations, and the related behavioral competences, remain somewhat plastic throughout life. Thus at the outset, we propose that both neonatal abilities and postnatal changes in speech perception competences might be best explained by reference to a specialized but flexible perceptual-motor system.

PERCEPTION OF SPECIES-SPECIFIC CALLS BY ANIMALS

An evolutionary perspective, in which one predicts continuity among phenomena and mechanisms across species, focuses attention on the evidence that many nonhuman animals also have specialized mechanisms for the perception of their own species calls.[1] This evidence can be examined

---

[1] The questions addressed by this research are distinct from those examined in investigations of the perception of human speech by nonhuman animals. That work is not reviewed here (see Kuhl 1988 for an overview).

in the context of the speech-processing characteristics mentioned above and helps put specialized speech processing into a biological context (see Petersen 1982 for a review).

One interesting set of experiments (reviewed by Ehret 1987) focuses on evidence of *categorical perception* of species-typical communicative signals. Both labeling and discrimination tests involving synthetic series of calls relevant to mice, vervet and macaque monkeys, and pygmy marmosets have been undertaken and, in each case, good evidence for categorical perception has been found.

Japanese macaque (*m. fuscata*) monkeys not only show categorical perception, but also show the kind of perceptual constancy for their species-specific calls that humans demonstrate for speech sounds. The Japanese macaque produces 80 to 90 different vocalizations. Two of these calls can be distinguished on the basis of whether a frequency-modulated sweep occurs early or late in the signal. Using operant techniques, Green (1975) found that the Japanese monkeys are able to discriminate the "linguistically distinctive" frequency-modulated segment of the calls and ignore other, noncritical acoustic information. Other, non-Japanese, "comparison" (pigtailed and bonet macaque) monkeys who do not utter these calls require extensive training to master this task. In a subsequent experiment, all monkeys were required to differentiate these same calls on the basis of a nonlinguistic acoustic cue—high vs. low fundamental frequency—instead of temporal location of frequency-modulated sweep within segments. Although the comparison monkeys quickly learned the discrimination, the Japanese monkeys had great difficulty. These findings are consistent with the possibility that Japanese monkeys use neuronal mechanisms different from those of the "comparison" monkeys in perceiving Japanese macaque cells.

Direct support for the possibility of specialized neural processing for macaque calls comes from measurements of lateralization. Japanese monkeys show a right-ear advantage, indicative of left hemisphere dominance in their ability to discriminate pairs of macaque stimuli presented alternatively on a random basis to one ear at a time. The comparison monkeys exhibit no such right-ear advantage. Although comparison and Japanese monkeys can and do use the same features of the calls when performing the discriminations (Petersen et al 1984), the cortical lesions that disrupt macaques' ability to discriminate macaque calls are specific to the left temporal cortex, and are distinct from other regions that, when lesioned, yield general hearing losses (Heffner & Heffner 1989). These results from Japanese monkeys have a parallel in comparative studies of Thai and English speakers. In Thai, changes in fundamental frequency serve a critical linguistic function by specifying semantic distinctions among words. In English such changes serve only paralinguistic functions such as providing information about the speaker's emotional state. Thai speakers demonstrate a right-ear advantage for fundamental frequency whereas English speakers may not (Van Lancker & Fromkin 1973).

Another intriguing parallel with features of human speech involves vocalizations directed primarily at infant animals by squirrel monkey mothers (Biben et al 1989). Like humans, squirrel monkeys rarely use vocalizations typical of adult "conversations" when addressing their youngsters, and instead use "caregiver" calls with several unique features (see Fernald 1984 for a review of the human work). However, whereas infant-directed speech in humans is typically higher pitched than adult-directed speech, caregiver calls in squirrel monkeys have most of their power concentrated at several kilohertz lower than most squirrel monkey calls. A special "baby talk" register is used, as in humans (Ferguson 1964).

EXPERIENCE, GENETIC PREDISPOSITION, AND THE NEONATE

If we accept the notion that there is specialization for the perception of biologically significant communicative signals, it becomes of interest to explore the ontogenetic roots of such specialization (see also Miller & Jusczyk 1989). The most obvious explanation for the remarkable capabilities of human infants is that genetic and maturational factors leave the neonate with a specialized processing system for human speech. We would argue, however, that the research to date is more consistent with the hypothesis that the abilities that exist in early infancy reflect the probabilistic outcome of both endogenous and environmental factors. The idea that we would like to entertain (as have others) is that normally inevitable pre- and postnatal experiences (i.e. species-typical) are not only influential but necessary for normal development (e.g. Gottlieb 1985a, Hebb 1980). Psychologists would characterize the resulting process as *innately guided learning* (Jusczyk & Bertoncini 1988), while neurobiologists might want to describe it in terms of *activity-dependent neurogenesis* (e.g. Greenough et al 1987). Both reflect a probabilistic epigenetic viewpoint that the phenotype or endpoint is a complex outcome of both endogenous and exogenous factors, including species-invariant sensory input. The same kind of argument for probabilistic epigenesis has been made in the case of the ontogeny of early-appearing visual competences such as stereopsis, binocularity, and avoidance of heights. For each of these basic visual abilities, the manipulation of normally "inevitable" input results in significant performance deficits (Tees 1990a,b, Mitchell 1989).

There is now strong evidence that the auditory system of the human fetus is partially functional during the last prenatal trimester. The mother's voice is transmitted to the uterus by bone conduction, as well as, perhaps, other human voices via air and fluid conduction. Although there is still considerable controversy concerning the precise acoustic information that is available to the human fetus, it is generally agreed that at the very least, low-frequency information (< 400 Hz) is available (Fifer & Moon 1988). The attenuated linguistic input that thus reaches the fetus may well play a part in sculpturing the developing speech or nonspeech-related auditory systems and facilitating the neonate's processing of speech and speech-like sounds. We know from a variety of studies (e.g. Smotherman & Robinson 1989) that nonhuman mammalian fetuses are responsive to sensory stimuli encountered during gestation, and that prenatal stimulation may alter their postnatal behavior (Turkewitz 1988). The first persuasive demonstration of prenatal influences on speech processing in humans was provided in a study by DeCasper & Fifer (1980), who demonstrated that on the first postnatal day, human babies of either gender will suck (an artificial nipple) preferentially to hear a tape of their mother's voice over that of another female. (Neonates showed no such preference for their father's voices.) There is also data indicating that speech stimuli with familiar vs. unfamiliar melodic and/or temporal characteristics elicit different cardiac patterns in human fetuses and that neonates will suck preferentially to a song or a story heard prenatally (Fifer & Moon 1988).

More recently, Mehler et al (1988) have shown that infants can discriminate global characteristics of native from nonnative speech samples within days after birth. Two-month-old American-English-learning infants in Oregon were compared to 4-day-old French-learning infants in Paris on their ability to discriminate French vs. Russian and Italian vs. English samples of speech. The four-day-old French-learning infants discriminated the French vs. Russian samples, but not the English vs. Italian speech. In contrast, the English-learning infants discriminated English from Italian, but not French from Russian. Thus, each group of infants was able to discriminate its native language from an unfamiliar language, but was unable to discriminate two unfamiliar languages. These results are interpreted by the authors as indicating that prenatal experience with a language facilitates processing of the global properties of that language, thus allowing infants to distinguish or prefer the native language over an unfamiliar one.

Together, these findings help us understand the human postnatal preference for familiar stimuli such as mother's voice and native language. We would like to suggest that they may also help explain the young infant's abilities to perceive the phonetic aspects of speech in a categorical-like fashion, the evidence of trading relations in infant perception, and perhaps even the infant's ability to discriminate nonnative sounds at birth. Our argument is as follows. Although it is very likely that detailed phonetic information is not available prenatally, some aspects of the speech signal do reach the fetus. The input that is available presumably provides enough information to enable infants to have greater facility in processing variability that corresponds to that experienced prenatally. Thus, the linguistic relevance of speech perception in the newborn may reflect an epigenetically determined processing specialization for stimuli that share global characteristics with those experienced prenatally. There is undoubtedly an innate propensity for this bias, but we are arguing that there may also be an experiential component. In terms of Greenough et al's (1987) distinction between *experience-expectant processes* and *experience-dependent processes*[2] regarding the roles played by early sensory experience/learning, we are identifying the experiential impact in terms of experience-expectant processes.

DEVELOPMENTAL STUDIES OF NONHUMANS

Although primates are *the* choice for investigating possible parallels in the ontogeny of systems for species-specific calls, data on the development of such abilities in primates is limited. In contrast, there have been many developmental studies on the effects of experience on vocal communication in birds. Most of the evidence for experiential effects on vocal communication rely on data on vocal output. Marler and his colleagues have conducted a number of studies revealing that for many songbirds, experience "hearing" a species-specific song during a sensitive period in early development is required in order to sing the species-specific song correctly as an adult (see Marler 1990). The requisite amount and timing of exposure varies tremendously across avian species, but in most cases the birds are not capable of correctly singing the song of another species even if given considerable early experience with such songs. Thus, in general, the data pattern is consistent with the notion that experience-expectant processes play an important role in the development of vocal communication.

In addition to simple exposure to the song or calls of conspecifics, other types of "nonobvious" experience are also possible (e.g. Johnston 1988). These include self-produced sounds (including calls), sounds of siblings,

---

[2] Greenough categorized two ways experience might influence the developing (and mature) organism in terms of the type of information stored and the brain mechanism involved. Experience-expectant involves incorporation of ubiquitous early environmental information by means of selective sculpturing of intrinsically overproduced synaptic connections between sensory/motor neurons. Experience-dependent involves the storage of other experiences by means of activity-associated generation of new synapses.

parents, and other conspecifics (both prenatally and postnatally), and nonvocal social stimulation. These nonobvious sources of experience can also exert a profound impact on vocal communication. For example, West and King (1988) have shown that the vocal behavior exhibited by the adult cowbird depends not only on early acoustic experiences, but also on the bird's social environment throughout its life. In illustration, an adult male cowbird can learn a new dialect as an adult when in the company of a (nonsinging, but potentially responsive) female cowbird familiar with only the new dialect.

Gottlieb (1985a,b) has also examined nonobvious experiential influences on communication development in several different species of ducklings, using responsiveness to species-specific calls rather than vocal production as the dependent variable. Gottlieb has shown that this species-specific responsiveness required relevant exposure in embryo. In the Peking duck, this experience can be either the calls of conspecifics in the environment or self-produced vocalizations—but the experience must occur during a relatively constrained period of development for the duckling to show the typical species-specific preference. Interestingly, in devocalized ducklings, the embryonic experience need not match the calls precisely. In fact, for the Peking duck, variability in the input produces greater preference than does the "ideal" repetition rate in the case of the contact call. Similarly, for the wood duck, prehatching exposure to a frequency range different from that of the assembly call is adequate for ensuring a post-hatching preference for their assembly call. These findings confirm the role that experiential influences play even in species-specific behaviors. In addition, they provide a concrete instance of our suggestion that non-identical but related exposure in embryo can canalize the perception of species-specific communicative signals in human infants.

Drawing precise parallels between most of these studies and work on human speech perception is difficult because the studies of these birds rely on production[3] and preference to reflect perceptual competence. Recent work on birdsong by Sinnott (e.g. 1987) comparing blackbirds, cowbirds, and humans after training to discriminate both birds' full songs and song elements is more promising in this regard. In this instance, species-specific coding was more apparent when birds were processing information in

---

[3] The problem with using vocal output as a sole index of perceptual processing is that it may reflect only the properties of the receptive "apparatus" that interact or guide the motor systems involved in producing vocalization. The perceptual system may well be a complex multistage one, able to process much more about the signals than is reflected in the animal's vocal behavior and only some of the stages involved in directing the motor program, e.g. for song acquisition. In any event, if we relied only on vocal output we certainly would grossly underestimate the sensory and perceptual capacities of prelinguistic human infants.

the complex, full-song context than it was when birds were specifically "tutored" with song elements out of context. These results are strikingly analogous to the pattern of data from speech perception studies with adult humans showing auditory and phonetic processes in development.

CROSS-LANGUAGE SPEECH PERCEPTION IN INFANTS

One way of collecting evidence on the interaction between biological endowment and experiential factors in human speech development is to examine speech perception in a cross linguistic framework. This allows an evaluation of the effects of naturally occurring variations in input on speech perception abilities. Developmental processes can be assessed by comparing the effects of variation in input at different ages. We began to address this question by examining language-specific influences on the perception of the phonetic aspects of speech in infants and adults.

Previous work had shown that although infants discriminate both native and nonnative speech contrasts according to phonetically relevant boundaries (Trehub 1976), adults often show difficulty discriminating nonnative speech contrasts. Thus, the existing work (Strange & Jenkins 1978) had suggested a change between infancy and adulthood in the ease with which listeners discriminate unfamiliar phonetic contrasts. Our work has been designed to investigate the validity and meaning of this claim. In a series of experiments conducted over the last ten years, we have confirmed that there is a profound developmental change between infancy and adulthood in the ease with which listeners can differentiate phonetic contrasts that are not used in their native language. Furthermore, we have shown that this developmental change from broad-based to language-specific phonetic perception is evident as early as 10–12 months of age. We briefly review this work below.

In an early experiment, English-speaking adults, Hindi-speaking adults, and English-learning infants aged six–eight months were compared on their ability to discriminate two Hindi speech contrasts: the retroflex/dental place-of-articulation contrast, /Ta/-/ta/, and the voiceless aspirated vs. breathy voiced dental stops, /t^{h}a/-/d^{h}a/. Natural rather than synthetic stimuli were used in this original study to allow us to assess discrimination and categorization of phonetic categories within the context of at least some naturally occurring variation. Infants and adults were tested in a variation of the head-turn procedure (see Figure 3). The basic logic of this procedure is that the infant is conditioned to turn her head when she detects a change in the speech stimulus. Correct (but not incorrect) head turns are reinforced with the activation of single or multiple mechanical

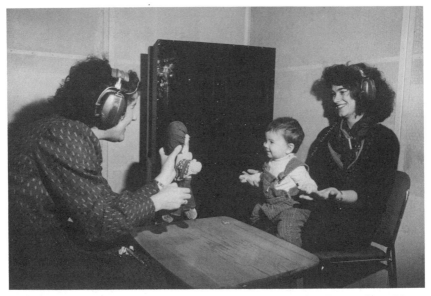

Figure 3a   An infant attending to the experimental assistant during the "no-change" intervals in the head-turn procedure.

Figure 3b   An infant turning her head toward the visual reinforcer upon detecting a change in the speech sound. (Notice that the correct head turn is "reinforced" by the activation of the toy animals as well as by clapping and praise by the experimental assistant.)

toy animals. Adults and older children indicate detection of a change by pushing a button (for fuller details of this procedure, see Kuhl 1987). The results indicated that although all the Hindi adults and most of the young English-learning infants could reach discrimination criterion on the two Hindi contrasts, the majority of the English-speaking adults could not, showing particular difficulty on the retroflex/dental distinction. Indeed, when a second group of English-speaking adults was given 25 training trials on each of the Hindi contrasts, their performance improved on the voicing distinction, but training did not affect performance of the more difficult retroflex/dental contrast (although we have subsequently developed procedures in which sensitivity to even the retroflex/dental contrast is maintained). These results confirmed that although young infants are equally sensitive to both native and nonnative phonetic contrasts, adult perception is modified by language experience, and the impact of experience is more profound for some nonnative contrasts than it is for others (Werker et al 1981).

A series of experiments was run to try to identify the age at which the developmental change in sensitivity is first apparent. After first finding that children aged 12, 8, and even as young as 4 years have difficulty with some nonnative contrasts (Werker & Tees 1983), we eventually discovered evidence of a developmental change by 10–12 months of age. Briefly, English-learning infants of 6–8, 8–10, and 10–12 months of age were compared on their ability to discriminate two non-English phonetic contrasts as well as the English bilabial/alveolar contrast, /ba/-/da/. The non-English contrasts were the Hindi retroflex/dental and an Nthlakampx (a Northwest, Interior Salish language) glottalized velar vs. glottalized uvular contrasts. The youngest English-learning infants could discriminate all three sets of contrasts, but the infants aged 10–12 months could only discriminate the native language /ba/-/da/ distinction (see Figure 4). To ensure that the performance of the older infants was not simply a general age-related performance decline, we tested a few Hindi- and Nthlakampx-learning infants aged 11–12 months and found they could quickly reach a 9 out of 10 discrimination criterion on their native contrast (Werker & Tees 1984a).

We have replicated this finding of developmental change between 6–12 months of age with a synthetically produced Hindi retroflex/dental contrast involving voiced rather than voiceless stimuli (/Da/ vs. /da/ rather than /Ta/ vs. /ta/) (Werker & Lalonde 1988). Also, Best and McRoberts have replicated the developmental change between 6 and 12 months of age for the Nthlakampx contrast by using a different procedure (C. T. Best, in preparation). They have also recently reported data showing a similar age-related reorganization for three other click contrasts.

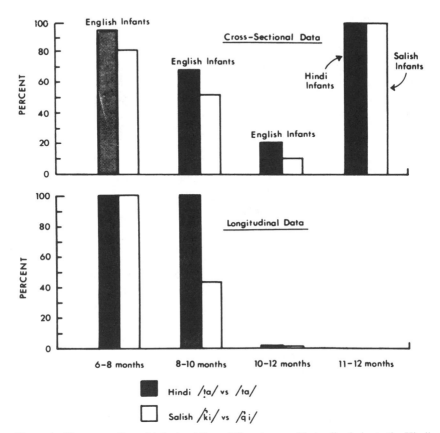

Figure 4   The proportion of infants at three different ages able to discriminate the Hindi and Nthlakampx (non-English) phonetic contrasts. (Adapted from Werker & Tees 1984a.)

One very interesting study by Best et al (1988) did not replicate our results. In this study, English-learning infants of 6–8, 8–10, 10–12, and 12–14 months of age were compared to both Zulu- and English-speaking adults on their ability to discriminate the difference between a medial vs. lateral Zulu click contrast. Not only does this contrast not occur in English, clicks do not have lexical status at all. Of interest, the English adults and the infants of all four ages were able to discriminate this nonnative contrast. Best has begun to develop a scheme for predicting which kinds of nonnative contrasts will be easy and which difficult to discriminate.

Polka is currently conducting a set of experiments with one of us that examines developmental changes in cross-language vowel perception (L. Polka and J. F. Werker, in preparation). Infants of 6–8 and 10–12 months of age and English- and German-speaking adults are being compared on their ability to discriminate two sets of German (non-English) vowel contrasts as well as one English vowel contrast. In all cases the vowels are embedded in between the consonants /d/ and /t/. The English distinction contrasts /deet/ and /dot/. The German contrasts both involve front vs. back distinctions between rounded vowels. One contrast involves two lax vowels /dʊt/ vs. /dɣt/ and the other, two tense vowels /bʊːt/ vs. /bɣːt/. The results to date show that English-speaking adults can discriminate both of these contrasts with relative ease; however, English-learning infants of 10–12 months[4] show considerably more difficulty. These data show that even when the nonnative contrast is quite acoustically distinct, if the stimuli map onto native-language phones, there will be evidence of a prelinguistic reorganization in perceptual performance.

To find out more about the nature of the discriminative abilities of young (6–8-month-old) infants and whether they "categorized" even nonnative stimuli according to phonetic identity, we utilized synthetically produced retroflex/dental stimuli. A 16-step /ba/-/da/-/Da/ continuum was synthesized varying in 16 equal steps according to the starting frequency of the first and second formant transitions (Werker & Lalonde 1988). English-learning infants of both 5–8 and 11–13 months of age were tested in three conditions (see Figure 5 for stimulus selections). In the first two, adjacent sets of three stimuli were drawn from a section of the continuum such that they were identified as /ba/ and /da/, respectively, by adult (English and Hindi) listeners. This condition would be phonemic to either a Hindi- or an English-speaking adult listener. In the second, two sets of adjacent stimuli were selected that are identified as dental /da/ and retroflex /Da/ by adult Hindi listeners but are all identified as alveolar /da/ by English listeners. This condition would be "phonemic" to a Hindi but not an English listener, and would clearly correspond to a "universal" phonetic category. In the third condition, two sets of adjacent stimuli were selected that are identified as primarily all retroflex by Hindi listeners (and as all alveolar by English listeners) and thus do not correspond to any phonetic category.

It was reasoned that if perception is phonetically relevant in early

[4]The data actually suggest that the English-learning infants of 6–8 months also have difficulty with the two German vowels, which suggests that the reorganization in vowel perception might occur earlier than that for consonant perception. P. Kuhl (personal communication) has also recently found native-language influences on vowel perception by six months.

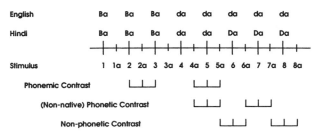

Figure 5 The top part of this figure shows the way English- vs. Hindi-speaking adults divide the synthetic /ba/-/Da/-/da/ continuum. The bottom half of the figure indicates the stimuli that were used as phonemic, phonetic, and nonphonetic contrasts in Werker & Lalonde (1988).

infancy, the infants of 6–8 months of age should be able to perform successfully in the first two, but not the third condition. That is, they should be able to categorize sets of varying stimuli according to native or "universal" phonetic boundaries, but should be unable to categorize stimuli according to an arbitrary point along the continuum that does not conform to a phonetic boundary. This is precisely what was obtained. The result confirms the phonetic relevance of speech perception at the more abstract level of categorization in infants as young as 6 months of age.

This research is similar to the research by Kuhl outlined above showing that by 6 months of age, infants can perceptually categorize vocalic stimuli differing in speaker and intonation on the basis of vowel color, and that they organize vowel categories around phonetic prototypes (e.g. Kuhl 1987). Our cross-language research complements this pattern of data by showing that young infants categorize stimuli, even stimuli that they have not heard before, in terms of phonetically relevant boundaries. The results for the 11–13-month-old infants replicated the findings of Werker & Tees (1984a). These older infants were only able to categorize stimuli according to a phonetic boundary that has functional, phonemic status in their own language but not according to a phonetic boundary that is ignored by adult speakers of their language.

Auditory experience continues to influence competence in speech perception in the months following birth. For example, although infants can detect violations of natural clausal boundaries in both native and nonnative speech by 4 months of age (Hirsh-Pasek et al 1987), by 6 months of age they are only sensitive to clausal boundaries in their native language. Further, although infants at this age can detect prosodic differences (melody, stress, tempo) between native and nonnative low-frequency words by as early as 6 months of age, they show no evidence of preferring native

Table 1 Experimental influences on speech perception

Age	Speech perception competence
Birth	Preference for mother's voice
	Preference for story and song heard prenatally
	Ability to discriminate overall prosody of native language from that of a nonfamiliar language
	Can discriminate "universal" set of phonetic contrasts
4 Months	Sensitivity to clausal boundaries in native and unfamiliar languages
6 Months	Decline in sensitivity to clausal boundaries in unfamiliar languages
	Preference for listening to words with melodic and rhythmical characteristics of native language
	Decline in sensitivity to vocalic contrasts in unfamiliar languages
9–10 Months	Preference for listening to words that conform to native-language phonotactic[a] rules
10–12 Months	Decline in sensitivity to consonant contrasts in unfamiliar languages

[a] Phonotactic refers to the phonological rules specifying the letter/sound sequences that are acceptable in a given language.

words on the basis of phonotactic information (the rules for which phonemes can occur together in a sequence) until about 9 months of age (Jusczk 1992). Overall, the data suggest that the preference of a native over a nonnative language in infants follows a developmental progression in sensitivity from the global features of language-specific sound patterning such as melody and rhythm, to sensitivity to smaller units such as phrases, clauses, individual words, and ultimately individual syllables and phonemes (see Table 1).

CROSS-LANGUAGE PERCEPTION IN ADULTS

Early research in cross-language speech perception led to the conclusion that the difficulty adults have in discriminating nonnative phonetic contrasts is quite general and irreversible (Strange & Jenkins 1978). More recent research, however, has made clear that the situation is much more complex (for a review see Werker 1991). Developmental changes in sensitivity do not apply equally to all nonnative distinctions (Best et al 1988) and do not indicate an absolute loss of the ability to discriminate nonnative distinctions (Werker & Logan 1985). Furthermore, even when adults clearly have initial difficulty with a nonnative distinction, training studies indicate that they can typically improve after practice or feedback (Logan et al 1991, Tees & Werker 1984) or after intensive study of the relevant language (MacKain et al 1981, Tees & Werker 1984). In many cases, however, the performance levels they obtain fall far short of those obtained by native speakers (Polka 1991). Finally, early (pre-linguistic) exposure to a language also seems to help maintain sensitivity to the phonetic distinctions (Tees & Werker 1984).

Of perhaps even greater interest, there is evidence that adults may show a latent capacity for discriminating even apparently quite difficult contrasts if the testing procedure is adequately sensitive. For example, we have shown that although adults most readily discriminate CV speech stimuli in terms of native language phonemic categories, sensitivity to nonnative phonetic contrasts is maintained throughout adulthood *even without training* (Werker & Logan 1985, Werker & Tees 1984b). This sensitivity to nonnative phonetic category differences exists in addition to the already known latent sensitivity to acoustic differences within a phonetic category. This led us to propose that adults can process speech in several ways, depending upon task conditions. The most readily available strategy seems to be to perceive speech in terms of native-language phonemic categories. Adults maintain a sensitivity to nonnative phonetic category differences, however, and can even be shown to be sensitive to changes within a phonetic category under some circumstances.

What do these results from cross-language studies with adults tell us about universal capabilities? They tell us that although there are clear experimentally based changes in the ease with which nonnative contrasts can be discriminated, the underlying sensitivity in both universal phonetic and to nonphonetic acoustic differences remains. Thus, it would be incorrect to conclude that lack of listening experience during the first year of life leads to some permanent "loss" in either ability. Nevertheless, these results make clear that there is a substantial developmental change in the ease with which listeners discriminate nonnative contrasts.

POSSIBLE NEUROBIOLOGICAL MECHANISMS

In previous work, we have outlined several different classes of explanations that might be useful for understanding developmental changes in speech perception (Werker 1991). These include perceptual tuning, modular recalibration, articulatory mediation, phonological development, and cognitive categorization (see also Jusczyk 1991 for an alternate cognitive model). In this chapter, we briefly identify specific processes of neurogenesis that might be related to developmental changes in human speech perception. In this endeavor, we purposely avoid a discussion of general models, such as that proposed by Edelman (1987), and focus instead on what can be said about the involvement of specific events in neurogenesis.

Before considering possible neurobiological mechanisms that might be related to developmental changes in speech perception, we briefly review what we think is a reasonable description of the key evidence.

1. Young infants, even neonates, show highly developed speech perception abilities, some of which have been influenced by prenatal experiential factors.
2. Speech perception in both adults and infants is complex, multi-dimensional, and linguistically relevant and likely to involve a highly integrated but distributed network of perceptual and motor components.
3. Postnatal linguistic experience during the first year of life influences speech perception competences—first for global characteristics of language (prosody, clausal marking) and later for vowel and consonant perception.
4. This reorganization reflects the sound patterning of the native language.
5. Developmental changes are more pronounced for some nonnative phones than for others and in all cases are reversible.
6. Even without extensive listening experience, adults are able to use different processing strategies—phonemic, phonetic, and acoustic—for categorizing speech sounds under particular task conditions.

The evidence of prenatal and life-long influence on speech processing makes clear at the outset that speech processing is unlikely to be a "closed" developmental program. The genetic program undoubtedly sets *constraints* on the kinds of stimuli that are most easily processed and that can influence the emerging perceptual and neural system, but experience undoubtedly also plays a role in influencing perceptual abilities (Jusczyk & Bertoncini 1988, Miller & Jusczyk 1989). At the neural level, *activity-dependent* and *reactive neurogenesis* must underlie the fine-tuning of the perceptual-motor network to these characteristics of the human voice that are available to the fetus. We suggest that these experience-expectant, prenatal processes yield a stable base sensitivity to the broad set of "species-specific" characteristics of human speech. The profound experiential influences evident during the first year of life reflect changes in the ease with which the child can access or use these various species-specific sensitivities. This latter tuning must be more malleable because at least some aspects of this developmental program remain "open" throughout much of the lifespan.

The behavioral evidence for high neonatal ability followed by selective postnatal decline in sensitivity suggests the involvement of regressive processes in postnatal neurogenesis (see also Cowan et al 1984, Kolb 1989). The class of regressive neuronal events that we believe to be most relevant involves competitive sparing of synapses produced in early development (Greenough et al 1987). Such *experience-expectant* plasticity has already been invoked to help explain the emergence of self-produced locomotion (Bertenthal & Campos 1987). If we invoke this process to help explain the developmental change between language-general and language-specific sensitivities seen across the first year of life, several caveats need to be considered. First, the postnatal developmental changes in language-specific sensitivity are by no means absolute or irreversible, whereas we have suggested that some aspects of prenatal tuning may be resistant to change. Thus any neural mechanism that results in irreversible loss of connections is not adequate to explain the changes seen in the first year of life. Second, it is not enough to say simply that regressive neurogenesis might be the mechanism accounting for particular developmental change. Different kinds of synaptic connections in different areas of the brain proliferate and regress at different points in development. Thus it is necessary to specify the location and timing of regressive neuronal events in relation to the appropriate developmental changes in speech-related behavior.

In considering the reversibility in sensitivity to nonnative speech contrasts, it is important to remember that competitive sparing is not irreversible in the same way that, for example, cell death is. In fact, recent research confirms that new synapses can be generated throughout much of the lifespan (e.g. Greenough et al 1987) and that dysfunctional synapses can, under certain circumstances, regain functionality. For example, when one monocularly deprives a binocular cat, the resulting apparent "loss" in "synaptic control" of binocular cortical neurons is recoverable (Mitchell 1989). Appropriate regimes of visual exposure through both eyes and each eye can reverse the neural and behavioral consequences of early competitive disadvantages. The recovery of sensitivity in adulthood to even difficult nonnative speech contrasts could also be accounted for by such *experience-dependent* neural plasticity. The importance of changes in the inhibitory connections of the neural substrates of such emerging abilities has been discussed elsewhere (e.g. Tees 1990b).

Our ability to relate behavioral changes in speech perception to specific sites of regressive neurogenesis is more problematic. This involves a consideration of the areas of the brain that might be most intimately involved in this particular skill, as well as a consideration of the developmental point in the process of neurogenesis at which behavioral consequences would most likely emerge. Data on human cortical development (Conel 1939–1967) shows different patterns of growth across brain regions. Generally, according to Kolb & Fantie (1989), postnatal changes in dendritic complexity within speech areas (though simple at birth) are among the most impressive in the brain; a dramatic increase in the density of synaptic connections of the left temporal/parietal cortex begins at 8 and declines at 20 months of age. In the frontal cortex, the peak period begins a little later; the increase in the density of synaptic connections occurs between 15 and 24 months of age, followed by a gradual decline.

To the extent that developmental changes in speech perception reflect tuning of a perceptual-motor system specialized to speech, it is likely that developmental changes in behavior will be related to neurobiological events in speech-related cortical areas. If this is the case, one would predict a tight relationship between developmental changes in speech perception and speech production. In previous writings (Werker & Pegg 1992), we have reviewed the data that show a relationship between developmental changes in babbling and in speech (e.g. Locke 1990).

At least some of the developmental changes in speech perception—particularly those showing a mapping to native-language phonological contrasts—might be related to changes in either cognitive mechanisms or to the construction of a phonological rule system (see Goldman-Rakic 1987, Werker 1991). These changes could be related to prefrontal cortical mechanisms. Goldman-Rakic (1987) reports a significant increase in synaptic proliferation in humans in the frontal and prefrontal cortex beginning around 8 months of age and continuing until about 2 years of age (see also Huttenlocher 1979). Lecours (1975) also suggests that intra- and interhemispheric cortical association bundles begin to myelinate at 7 months postnatally.

Before definitive links can be made, it is necessary to specify more precisely what kinds of changes in behavior might be related to specific synaptic proliferations and what kinds might be related to regressive events. Whether the initial reorganization and emergence of competences reflect a widespread *concurrent* period of excess synaptogenesis and subsequent pruning throughout the cortex, representing the elements of the distributed perceptual-motor network, or whether the neural unfolding is more piecemeal and hierarchical, remains uncertain. We have detailed the timing of important age-related changes in speech-related behavior. The differences in speech-related perceptual/motor competences that are environmentally influenced must, one way or another, yield changes in the dendritic fields of neurons within these neural systems. We are hopeful that the data and constructs reviewed in this chapter will help direct thinking and research toward further specification of the link between neural and behavioral events in human speech perception.

ACKNOWLEDGMENTS

This work was supported by the Natural Sciences and Engineering Research Council of Canada (OGP0001103 to J. F. Werker and OGP000179 to R. C. Tees). The stimuli for our experiments were prepared with assistance from NICHD Contract No. 1-HD-5-2910 to Haskins Laboratories.

Literature Cited

Bertenthal, B. L., Campos, J. J. 1987. New directions in the study of early experience. *Child Dev.* 58: 560–67

Best, C. T., Hoffman, H., Glanville, B. B. 1982. Development of infant ear asymmetries for speech and music. *Percept. Psychophys.* 31: 75–85

Best, C. T., McRoberts, G., Sithole, N. 1988. Examination of perceptual reorganization for nonnative speech contrasts: Zulu click discrimination by English-speaking adults and infants. *J. Exp. Psychol. Human Percept. Perform.* 14: 345–60

Biben, M., Symmes, D., Bernhards, D. 1989. Contour variables in vocal communication between squirrel monkey mothers and infants. *Dev. Psychobiol.* 22: 617–31

Conel, J. L. 1939–1967. *The Postnatal Development of the Human Cerebral Cortex*, Vol. 1–8. Cambridge, MA: Harvard Univ. Press

Cowan, W. M., Fawcett, J. W., O'Leary, D. D. M., Stanfield, B. B. 1984. Regressive events in neurogenesis. *Science* 225: 1258–65

DeCasper, A. J., Fifer, W. P. 1980. Of human bonding: Newborns prefer their mothers' voices. *Science* 208: 1174–76

Edelman, G. 1987. *Neural Darwinism: The Theory of Neuronal Group Selection*. New York: Harper & Row

Ehret, G. 1987. Categorical perception of sound signals: Facts and hypotheses from animal studies. In *Categorical Perception*, ed. S. Harnad, pp. 301–31. New York: Cambridge Univ. Press

Eimas, P. D., Miller, J. L. 1980. Contextual effects in infant speech perception. *Science* 109: 1140–41

Eimas, P. D., Siqueland, E. R., Jusczyk, P., Vigorito, J. 1971. Speech perception in infants. *Science* 171: 303–6

Ferguson, C. A. 1964. Baby talk in six languages. *Amer. Anthropol.* 66: 103–14

Fernald, A. 1984. The perceptual and affective salience of mothers' speech to infants. In *The Origins and Growth of Communication*, ed. L. Feagans, C. Garvey, R. Golinkoff, pp. 5–29. Norwood, NJ: Ablex

Fifer, W. P., Moon, C. 1988. Auditory experience in the fetus. In *Behavior of the Fetus*, ed. W. Smotherman, S. Robertson, pp. 175–88. West Caldwell, NJ: Telford Press

Fowler, C. A., Rosenblum, L. D. 1991. The perception of phonetic gestures. In *Modularity and the Motor Theory of Speech Perception*, ed. I. G. Mattingly, M. Studdert-Kennedy, pp. 33–59. Hillsdale, NJ: Erlbaum

Garcia, J., Koelling, R. 1966. Relation of cue to consequence in avoidance learning. *Psychonom. Sci.* 4: 123–24

Goldman-Rakic, P. S. 1987. Development of cortical circuitry and cognitive function. *Child. Dev.* 58: 601–22

Gottlieb, G. 1985a. On discovering significant acoustic dimensions of auditory stimulation for infants. In *Measurement of Audition and Vision in the First Year of Postnatal Life: A Methodological Overview*, ed. G. Gottlieb, N. A. Krasnegor, pp. 3–29. Norwood, NJ: Ablex

Gottlieb, G. 1985b. Development of species identification in ducklings: XI. Embryonic critical period for species-typical perception in the hatching. *Animal Behav.* 33: 225–33

Green, S. 1975. The variation of vocal pattern with social situation in the Japanese monkey (*Macaca fuscata*): A field study. In *Primate Behavior*, ed. L. Rosenblum, 4: 1–102. New York: Academic

Greenough, W. T., Black, J. E., Wallace, C. S. 1987. Experience and brain development. *Child Dev.* 58: 539–59

Hebb, D. O. 1980. *Essay on Mind*. Hillsdale, NJ: LEA Press

Heffner, H. E., Heffner, R. S. 1989. Effect of restricted cortical lesions on absolute thresholds and aphasia-like deficits in Japanese macaques. *Behav. Neurosci.* 103: 158–69

Hirsh-Pasek, K., Kemler Nelson, D. G., Jusczyk, P. W., Wright Cassidy, K., Druss, B., et al. 1987. Clauses are perceptual units for young infants. *Cognition* 26: 268–86

Huttenlocher, P. R. 1979. Synaptic density in human frontal cortex—developmental changes and effects of aging. *Brain Res.* 163: 195–205

Johnston, T. D. 1988. Developmental explanation and the ontogeny of birdsong: Nature/nurture redux. *Behav. Brain Sci.* 11: 617–63

Jusczyk, P. W. 1991. Developing phonological categories from the speech signal. In *Phonological Development: Models, Research, and Implications*, ed. C. E. Ferguson, L. Menn, C. Stoel-Gammon. Parkton, MD: York Press. In press

Jusczyk, P. W., Bertoncini, J. 1988. Viewing the development of speech perception as an innately guided learning process. *Lang. Speech* 31: 217–38

Kimura, D. 1967. Functional asymmetry of the brain in dichotic listening. *Cortex* 8: 163–78

King, A. J., Moore, D. R. 1991. Plasticity of auditory maps in the brain. *Trends Neurosci.* 14: 31–37

Knudsen, E., Konishi, M. 1978. Space and frequency are represented separately in auditory midbrain of the owl. *J. Neurophysiol.* 41: 870–84

Kolb, B. 1989. Brain development plasticity in behavior. *Amer. Psychol.* 44: 1203–12

Kolb, B., Fantie, B. 1989. Development of the child's brain and behavior. In *Handbook of Clinical Child Neuropsychology*, ed. C. R. Reynolds, E. Fletcher-Janzen, pp. 17–39. New York: Plenum

Kolb, B., Whishaw, I. A. 1990. *Fundamentals of Human Neuropsychology.* New York: Freeman. 3rd ed.

Kuhl, P. K. 1987. Perception of speech sounds in early infancy. In *Handbook of Infant Perception*, ed. P. Salapatek, L. Cohen, 2: 275–382. New York: Academic

Kuhl, P. K. 1988. Auditory perception and the evolution of speech. *Human Evol.* 3: 19–43

Kuhl, P. K., Meltzoff, A. N. 1982. The bimodal perception of speech in infancy. *Science* 218: 1138–44

Lecours, A. R. 1975. Myelogenetic correlates of the development of speech and language. In *Foundations of Language Development: A Multidisciplinary Approach*, ed. E. H. Lenneberg, E. Lenneberg, 1: 121–35. New York: Academic

Liberman, A. M., Cooper, F. S., Shankweiler, D. P., Studdert-Kennedy, M. 1967. Perception of the speech code. *Psychol. Rev.* 74: 431–61

Liberman, A. M., Mattingly, I. G. 1989. A specialization for speech perception. *Science* 243: 489–94

Locke, J. L. 1990. Structure and stimulation in the ontogeny of spoken language. *Dev. Psychobiol.* 23: 621–43

Logan, J. S., Lively, S. E., Pisoni, D. B. 1991. Training Japanese listeners to identify /r/ and /l/: A first report. *J. Acoust. Soc. Amer.* 89: 874–86

MacKain, K. S., Best, C. T., Strange, W. 1981. Categorical perceptions of English /r/ and /l/ by Japanese bilinguals. *Appl. Psycholing.* 2: 369–90

Marler, P. 1990. Innate learning preferences: Signals for communication. *Dev. Psychobiol.* 23: 557–68

Mattingly, I. G., Liberman, A. M. 1989. Speech and other auditory modules. In *Signal and Sense: Local and Global Order in Perceptual Maps*, ed. G. M. Edelman, W. E. Gall, W. W. Cowan, pp. 775–93. New York: Wiley

Mattingly, I. G., Studdert-Kennedy, M. 1991. *Modularity and the Motor Theory of Speech Perception.* Hillsdale, NJ: Erlbaum

McGurk, H., MacDonald, J. 1976. Hearing lips and seeing voices. *Nature* 264: 746–48

Mehler, J., Jusczyk, P. W., Lambert, G., Halsted, N., Bertoncini, J., et al. 1988. A precursor of language acquisition in young infants. *Cognition* 29: 143–78

Merzenich, M. M., Nelson, R. J., Stryker, M. P., Cynader, M. S., Schoppmann, A., et al. 1984. Somatosensory cortical map changes following digit amputation in adult monkeys. *J. Comp. Neurol.* 224: 592–605

Miller, J. L., Jusczyk, P. W. 1989. Seeking the neurobiological bases of speech perception. *Cognition* 33: 111–37

Miller, J. L., Liberman, A. J. 1979. Some effects of later-occurring information on the perception of stop consonant and semivowel. *Percept. Psychophys.* 25: 457–65

Mitchell, D. E. 1989. Normal and abnormal visual development in kittens: Insights into the mechanisms that underlie visual perceptual development in humans. *Canad. J. Psychol.* 43: 141–64

Molfese, D. L., Molfese, V. J. 1980. Cortical responses of preterm infants to phonetic and nonphonetic speech stimuli. *Dev. Psychol.* 16: 574–81

Ojemann, G. 1983. Brain organization for language from the perspective of electrical stimulation mapping. *Behav. Brain Sci.* 6: 189–230

Perrett, D. I., Rolls, E. T., Caan, W. 1982. Visual neurones responsive to faces in the monkey temporal cortex. *Exp. Brain Res.* 47: 329–42

Petersen, M. R. 1982. The perception of species-specific vocalizations by primates: A conceptual framework. In *Primate Communication*, ed. C. Snowden, C. Brown, M. Petersen, pp. 171–211. Cambridge, UK: Cambridge Univ. Press

Petersen, M. R., Beecher, M. D., Zoloth, S. R., Green, S., Marler, P. R., et al. 1984. Neural lateralization of vocalizations by Japanese macaques: Communicative significance is more important than acoustic structure. *Behav. Neurosci.* 98: 779–90

Pisoni, D. B., Luce, P. A. 1986. Speech perception: Research, theory, and the principal issues. In *Pattern Recognition by Humans and Machines: Speech Perception,*

Vol. 1, ed. I. C. Schwab, H. C. Nusbaum, pp. 1–50. New York: Academic

Poizner, H., Bellugi, U., Klima, E. S. 1990. Biological foundations of language: Clues from sign language. *Annu. Rev. Neurosci.* 13: 283–307

Polka, L. 1991. Cross language speech perception in adults: Phonemic, phonetic, and acoustic contributions. *J. Acous. Soc. Amer.* 89: 2961–77

Sinnott, J. M. 1987. Modes of perceiving and processing information in birdsong (*agelaius Phoeniceus, Molothrus ater*, and *Homo sapiens*). *J. Comp. Psychol.* 101: 355–66

Smotherman, W. P., Robinson, S. R. 1989. Cryptopsychobiology: The appearance, disappearance, and reappearance of a species-typical action pattern during early development. *Behav. Neurosci.* 1013: 246–53

Strange, W., Jenkins, J. J. 1978. Linguistic experience and speech perception. In *Perception and Experience*, ed. R. D. Walk, H. L. Pick, pp. 125–69. New York: Plenum

Summerfield, Q. 1991. Visual perception of phonetic gestures. In *Modularity and the Motor Theory of Speech Perception*, ed. I. G. Mattingly, M. Studdert-Kennedy, pp. 117–30. Hillsdale, NJ: Erlbaum

Sussman, H. M. 1989. Neural coding of relational invariance in speech: Human language analogs to the barn owl. *Psychol. Rev.* 96: 631–42

Tees, R. C. 1990a. Plasticity and change. In *The Cerebral Cortex of the Rat*, ed. B. Kolb, R. C. Tees, pp. 475–81. Cambridge, MA: MIT Press

Tees, R. C. 1990b. Experience, perceptual competences, and rat cortex. In *The Cerebral Cortex of the Rat*, ed. B. Kolb, R. C. Tees, pp. 507–36. Cambridge, MA: MIT Press

Tees, R. C., Werker, J. F. 1984. Perceptual flexibility: Maintenance or recovery of the ability to discriminate non-native speech sounds. *Can. J. Psychol.* 38: 579–90

Trehub, S. 1976. The discrimination of foreign speech contrasts by infants and adults. *Child Dev.* 47: 466–72

Turkewitz, G. 1988. A prenatal source for the development of hemispheric specialization. In *Brain Lateralization in Children: Developmental Implications*, ed. D. L. Molfese, S. J. Segalowitz, pp. 73–81. New York: Guildford

Van Lancker, D., Fromkin, V. A. 1973. Hemispheric specialization for pitch and "tone": Evidence from Thai. *J. Phonet.* 1: 101–9

Warren, R. M., Obusek, C. J., Farmer, R. M., Warren, R. P. 1969. Auditory sequence: Confusion of patterns other than speech and music. *Science* 196: 586–87

Werker, J. F. 1991. The ontogeny of speech perception. In *Modularity and the Motor Theory of Speech Perception*, ed. I. G. Mattingly, M. Studdert-Kennedy, pp. 91–110. Hillsdale, NJ: Erlbaum

Werker, J. F., Gilbert, J. H. V., Humphrey, K., Tees, R. C. 1981. Developmental aspects of cross-language speech perception. *Child Dev.* 52: 349–55

Werker, J. F., Lalonde, C. E. 1988. Cross-language speech perception: Initial capabilities and developmental change. *Dev. Psychol.* 24: 1–12

Werker, J. F., Logan, J. 1985. Cross-language evidence for three factors in speech perception. *Percept. Psychophys.* 37: 35–44

Werker, J. F., Pegg, J. E. 1992. Infant speech perception and phonological acquisition. In *Phonological Development: Models, Research, and Implications*, ed. C. E. Ferguson, L. Menn, C. Stoel-Gammon. Parkton, MD: York. In press

Werker, J. F., Tees, R. C. 1983. Developmental changes across childhood in the perception of non-native speech sounds. *Canad. J. Psychol.* 37: 278–86

Werker, J. F., Tees, R. C. 1984a. Cross-language speech perception: Evidence for perceptual reorganization during the first year of life. *Infant Behav. Dev.* 7: 49–63

Werker, J. F., Tees, R. C. 1984b. Phonemic and phonetic factors in adult cross-language speech perception. *J. Acous. Soc. Amer.* 75: 1866–78

West, M. J., King, A. 1988. Female visual displays affect the development of male song in the cowbird. *Nature* 334: 244–46

Whalen, D., Liberman, A. 1987. Speech perception takes precedence over nonspeech perception. *Science* 237: 169–71

Williams, H., Nottebohm, F. 1985. Auditory responses in avian vocal motor neurons: A motor theory for song perception in birds. *Science* 229: 279–82

Annu. Rev. Neurosci. 1992. 15:403–42

THE CEREBELLUM AND THE ADAPTIVE COORDINATION OF MOVEMENT

W. T. Thach,†‡§ H. P. Goodkin,* and J. G. Keating**

Departments of Anatomy and Neurobiology,* Neurology and Neurosurgery,† The Irene Walter Johnson Rehabilitation Institute,‡ and The McDonnell Center for Study of Higher Brain Function,§ Washington University Medical School, St. Louis, Missouri

KEY WORDS: motor learning, parallel fiber, climbing fiber, motor map, synergy

INTRODUCTION

Despite continuing work on the structure and function of the cerebellum, there is still no consensus as to what it does or how it does it. Its distinguishing features are well known, but they have not been fit together into any comprehensive, coherent model of function. The question remains: What is the fundamental cerebellar operation that so depends on the stylized and stereotyped circuitry, employs such highly individualistic interactions among its neurons, gives rise to characteristic pathologic signs when damaged, and is so subtly associated with the integrated motions of the vertebrate body?

Two pieces of relatively new information encouraged this attempt at a model of cerebellar function: first, the mapping of body parts and modes of motor control within the deep nuclei (Asanuma et al 1983c, Kane et al 1988, 1989, Thach et al 1982, 1990a,b); and second, the newly assessed, much longer length of the cerebellar parallel fiber (Brand et al 1976, Mugnaini 1983). Our goal has been that the model should explain both the regional differences in cerebellar function and the one generalized function that has long been implied by the stereotype of the intrinsic circuitry. To this end, we review evidence that 1. the body is mapped separately within each of the three deep cerebellar nuclei, that 2. the nuclei operate in parallel, each nucleus controlling a different mode of bodily

403

0147–006X/92/0301–0403$02.00

movement, and that 3. each mode is a function of the input and output connections of that nucleus. We show that 4. the beams of Purkinje cells, so linked by parallel fibers, project onto the nuclei and thus link the actions of the different body parts represented within each nucleus and the different modes of control across the nuclei into coordinated movement. We suggest that 5. the job of the cerebellum is thus to coordinate the elements of movement that reside in its downstream targets and 6. to adjust old and learn new movement synergies.

THE OUTPUT SIDE OF CEREBELLAR PROCESSING: MULTIPLE SOMATOMOTOR REPRESENTATIONS WITHIN THE DEEP NUCLEI

A series of anatomic pathway tracing studies on the macaque have helped clarify the connections and topographic organization of the deep cerebellar nuclei (Asanuma et al 1983a–d, Kalil 1982, Orioli & Strick 1989, Schell & Strick 1983, Stanton 1980), and are summarized as follows:

The output of the cerebellum, the cerebellar nuclei, project to a target area within the thalamus that is sufficiently free of other inputs as to be called "cerebellar thalamus" (Figure 1). The target area includes several architectonic subdivisions, VLc, VLps, VPLo, and X. The basal ganglia input to the thalamus arrives more anterior in VLo and VA, and the lemniscal arrives more posteriorl in VPL. This cerebello-thalamic target area projects to area 4 (VLc, VLps, VPLo, and parts of VLo) and to lateral area 6, the periarcuate area (X). In addition to this "specific" cerebellar-thalamic receiving area, there is a "nonspecific" thalamic target area, which projects more widely to the cerebral cortex: the centrum medianum.

Dentate and interposed nuclei each project in completely overlapping fashion to the whole width (coronal) of the contralateral thalamic receiving area; the fastigius projection is sparse, bilateral, and restricted and appears not to project to Xo. This itself suggests that at the level of cerebellar outflow, control of the thalamic target is multiple and is repeated across each nucleus. [The projections of the three nuclei arrive in register at a macro level only; at the cellular level, they appear to interdigitate in patches (interpositus) and rostrocaudal rods (dentate) (Asanuma et al 1983c).]

From the known somatotopic mapping in the cerebellar thalamus (Strick 1976), and from the topographic projection of each nucleus onto the common thalamic target, somatotopic mapping may be inferred within each of the cerebellar nuclei. In the cerebellar thalamus, the head is medial, tail lateral, trunk dorsal, and extremeties ventral; in the cerebellum, the head would then be caudal, tail rostral, trunk lateral, and extremities medial (Asanuma et al 1983c). This map has been supported by neural

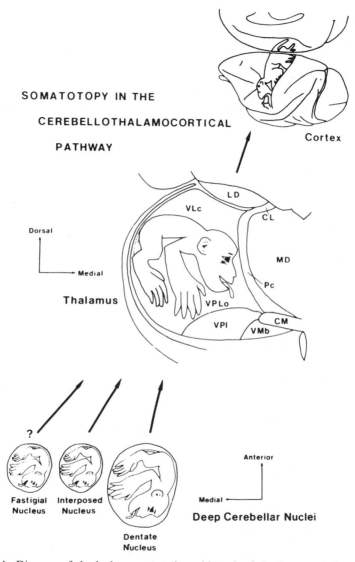

Figure 1 Diagram of the body representation with each of the deep cerebellar nuclei, thalamus, and motor cortex. VL$_c$, Ventral Lateral Nucleus, caudal division; VL$_{ps}$, Ventral Lateral Nucleus, pars postrema; VPL$_o$, Ventral Posterolateral Nucleus, oral division; X, Nucleus X; VA, Ventral Anterior Nucleus; VL$_o$, Ventral Lateral Nucleus, oral division (adapted from Asanuma et al 1983c).

recording and inactivation studies during movement (Kane et al 1988, Thach et al 1982, 1990, 1991) and is consistent with the electroanatomical *input* mapping studies in macaques of Allen and colleagues (1977, 1978).

Is the overlap complete? In the anatomical studies, the fastigius may not project to X, the medial and anterior extent of the cerebellar thalamus. This may suggest that the dentate and interpositus but not fastigius control lateral area 6, the periarcuate area. Orioli & Strick (1989) and Sasaki et al (1976) have raised the question in macaques of whether the cerebello-thalamic projection controls even further anterior cortical areas, such as area 8 and beyond.

Is there only one map per nucleus? A number of studies have raised the question of whether there is more than one body map per cerebellar nucleus (Orioli & Strick 1989, Schell & Strick 1983, Stanton 1980). If there is a separate body representation within lateral area 6 of cerebral cortex and in nucleus X of thalamus, this would seem a likely possibility.

This is not to say that the cerebellar output is exclusively or even mainly directed to the thalamus and cerebral cortex: The dentate projects to the parvocellular red nucleus, the reticular nucleus of the pontine tegmentum, and the inferior olive [principal (PO)]; the interpositus projects to the magnocellular red nucleus, the reticular nucleus of the pontine tegmentum, the inferior olive [dorsal accessory (DAO)], and the spinal cord inter-mediate grey; the fastigius projects to lateral and descending vestibular nuclei, the n. reticularis tegmenti pontis and prepositus hypoglossi, the reticular grey of the midbrain, the inferior olive [medical accessory (MAO)], and the contralateral motor neurons in the spinal cord (Asanuma et al 1983d).

CODING: WHAT DO THE MULTIPLE MAPS IN THE CEREBELLUM CONTROL?

If there is a separate body representation within each nucleus, it seems likely that each should control some different aspect of bodily function. Nevertheless, suggestions vary as to what that may be.

Parameters, Gain, or Something Else?

Holmes (1939), in his final analysis of cerebellar deficits, had factored the deficits into possible errors of start, stop, direction, acceleration, velocity, and force (cf. Brooks & Thach 1981). Evarts developed a technique for determining whether CNS neurons code for physical parameters of move-ment at single joints. The Evarts technique proved useful in testing the hypothesis of whether the causation of movement is a serial process in which each parameter is specified one after another, each in a different part of the brain (Allen & Tsukahara 1974, Eccles 1967). The predictions were that (*a*) the movement command for initiation of movement was generated in the motor cortex–dentate loop at the motor cortex or dentate; (*b*) the ongoing movement would be servo-controlled by the interpositus, which would compare the command from the motor cortex and the feedback from spinocerebellar pathways and would execute the corrected commands through interpositus projections back to the motor cortex (via the thalamus) and rubrospinal neurons (Allen & Tsukahara 1974, Eccles 1967, Evarts & Thach 1969). The results of these experiments showed in the onset of activity of the different parts of the brain—dentate, motor cortex, interpositus—slight timing differences in the order predicted. Nevertheless, just as impressive was the near-simultaneity of onsets and the great overlap of discharge patterns (e.g. Thach 1975, 1978). Moreover, the order of the timing differences varies among the different tasks, suggesting that in some instances the dentate might be the site of the movement initiating command and in others, the interpositus (Thach 1978). This question was reviewed recently (Brooks & Thach 1981), but it is important to summarize the evidence in the light of the newer evidence and interpretation that follows.

The *dentate* and the lateral cerebellar hemisphere receive input via pons from frontal and parietal association cortex (cf. Allen et al 1978, Brodal 1978); the dentate has been thought to translate mental percepts and concepts into action plans for movement (Allen & Tsukahara 1974, Brooks & Thach 1981, Eccles 1967, Evarts & Thach 1969).

In keeping with this notion, dentate neural discharge was reported to precede movement onset performed by trained monkeys, and even to lead the discharge of motor cortex neurons (Lamarre et al 1983, Thach 1975, 1978, but see Grimm & Rushmer 1974). Correspondingly, dentate inactivation was found to delay the onset of discharge of motor cortex neurons (Meyer-Lohman et al 1975, Spidalieri et al 1983) as well as the onset of movement (Meyer-Lohman et al 1977, Thach 1975, Trouche & Beaubaton 1980). This finding was interpreted to support the idea that the dentate participates in initiating volitional movement from the motor cortex (Evarts & Thach 1969, Brooks & Thach 1981). The problem with this interpretation has been that the onset of movement is only slightly delayed by 50–150 msec. Unless the purpose of the dentate is to provide that little extra speed of reaction, the paucity of the observed deficit does not appear commensurate with the phylogenetically increasing size of the dentate nucleus.

Direction of movement has been seen in several studies to correlate with the discharge of dentate, interposed, and Purkinje neurons (Fortier et al 1989, Thach 1970a,b) and in one study with dentate but not interposed neurons (Thach 1978). In other studies, in which different tasks were performed, little (MacKay 1988a,b) or no (Schieber & Thach 1985) direction signal was found in the dentate. Furthermore, even in studies in which dentate discharge did correlate with movement direction (Mink & Thach 1991a), the direction of the movement did not depend on a dentate direction signal: Sudden inactivation of the dentate during performance of a variety of tasks involving flexion or extension of the wrist produced no errors of these flexor and extensor directions (Mink & Thach 1991b).

Pathologic tremor (over 5 degrees amplitude, 3–6 Hz) has been attributed to dentate ablation in a number of animal studies (Botterell & Fulton 1938b, Brooks et al 1973, Cooke & Thomas 1976, Goldberger & Growden 1973, Vilis & Hore 1977, 1980). Indeed, in all of these studies, tremor appeared to be the salient deficit. The possibility remains, however, that the reason for the lack of greater deficit in task performance in these studies is that the neurons were controlling some process other than the movements in the tasks.

The *interpositus* receives fast feedback from movement and spinal motor programs via various spinocerebellar pathways and receives input also from the motor cortex via the pons (cf. Bloedel & Courville 1981). The theory held that the interpositus compared the command for movement and the feedback from the movement, sensed errors, and corrected them quickly during the course of the movement (Allen & Tsukahara 1974, Eccles 1967, Evarts & Thach 1969).

Timing studies in trained, visually triggered movements have shown that the order in which the interpositus begins to fire relative to the dentate and motor cortex is last (Thach 1978), but in movements triggered by a somatosensory perturbation of the part to be moved (wrist), the interpositus was first (Thach 1978). A crucial observation was made by Strick (1983): When the movement was made to oppose the perturbation, the interpositus led dentate; but when the movement was made in the direction of the perturbation, the dentate led interpositus. The interpositus therefore led when the reaction was hardwired, as in the stretch reflex ("hold a position despite displacement"); the dentate led when the reaction was counter-instinctive and learned ("go in the direction you are pushed").

In coding studies, Burton & Onoda (1977, 1978) and Soechting et al (1978) identified signals correlated with velocity and force during movements made by cats. Thach (1978) documented force and position correlations during positions held by monkeys; but in the latter study, coding was not absolute. Additions of load influenced the extent to which the neurons coded for position or force, as it did also with the muscle electromyogram (EMG) (cf. Schieber & Thach 1985). Schieber & Thach (1985) found no parameter correlations in interposed neurons during smooth-pursuit wrist tracking. They noted a pattern of increased discharge at the beginning of movement, regardless of the direction (flexor or extensor) and regardless of load or muscle pattern (increasing loaded flexors or decreasing loaded extensors) used to make the movement. This "bidirectional" discharge pattern was also seen in the discharge of Ia spindle afferent neurons (Elble et al 1984, Schieber & Thach 1985), and both the Ia's and the interposed neurons carried a signal of the animal's own tremor. Repeated penetrations into the interpositus increased the amplitude and slowed the frequency of the tremor. This result appeared compatible with the suggestions of Vilis & Hore (1977, 1980) that physiological tremor and cerebellar tremor both originate from instability of the stretch reflex. It also appeared to be consistent with the theory of MacKay & Murphy (1979) that the purpose of the cerebellum is to control the gain and stability of downstream structures (Thach et al 1986).

Since the gamma loop was active in EMG-silent, shortening, or lengthening muscle, it was inferred that gamma motoneuron discharge was present in the absence of alpha motoneuron discharge (alpha-gamma dissociation). This was discussed as a possible means of reducing tremor and small movement irregularities in the "one-muscled" movement.

What perhaps is more surprising is not that interposed and dentate neurons should resemble gamma motoneuron-stretch reflex loop activities (the two have long been known to be related), but that there should be no nuclear neural activity resembling that of alpha motoneurons, despite the many prior papers that have shown the expected load, position, and direction signals in these nuclei. Was there something about this task and what the cerebellum contributes to movement that took the alphas out and left the gammas in? As discussed below, we believe that the critical factor in this task was that the animals were trained for over two years until all EMG activity occurring in the task was confined to the loaded wrist extensor or flexor muscles. The gamma loop activity was not so constrained; indeed, it was fully present in the unloaded, EMG-silent antagonist muscle, and may have been present in muscles over other joints as well.

Heretofore, the interposed nucleus (nuclei globose and emboliform in man) has had no well-identified ablation syndrome attached to it. Uno et al (1973) reported that local cooling of the interposed nucleus did not result in the prominent errors of rate, range, and force seen following dentate cooling. Interpositus ablation has since been seen to cause an accentuation of dentate tremor (Goldberger & Growden 1973).

Others have emphasized impairment by interposed ablation of certain types of task performances, such as contact placing (Amassian et al 1972a,b, 1974, Amassian & Rudell 1978) and the learned nictitating membrane response (McCormick & Thompson 1984, Yeo et al 1984). These findings [with the observations above of Thach (1978) and Strick (1983) on discharge patterns compatible with long loop stretch reflex operation] would suggest the involvement of the interpositus in the control of some tasks and not others.

The *fastigius* has inputs from the vestibular complex, lateral reticular nucleus, and (indirectly) the spinocerebellar pathways (Brodal 1981, Jansen & Brodal 1940). It has been assumed to control proximal musculature, or stance and gait, or both, and has not been considered in models of voluntary control of limb movement (Allen & Tsukahara 1974, Eccles 1967, Evarts & Thach 1969).

Fastigial single-unit recordings during trained limb movement in monkeys have nevertheless been reported to correlate with force and its time derivatives (Bava et al 1983, Bava & Grimm 1978). The suggestion was made that the fastigius controlled the force of movement (and the time derivatives of force) and that the interpositus and dentate controlled progressively more abstract properties of movement, such as sequencing (Grimm & Rushmer 1974).

Other single-unit recording studies in the fastigial nuclei of decerebrate cats made to undergo walking and scratching movements have shown neural discharge correlated with the movements but have found little or none in the interpositus and dentate, respectively (Antziferova et al 1980, Arshavsky et al 1980). Fastigial ablation has been shown to impair stance and gait (Botterell & Fulton 1938a, Sprague & Chambers 1953; and see below). These observations would be more consistent with a role specialized for control of stance and gait.

MacKay (1988a), in recording from all three deep nuclei during visually triggered, single-jointed elbow movements, found little relation to distal movements or movement parameters, including direction, velocity, force, and muscle group. Yet small differences in timing served to distinguish the three nuclei (dentate fired earliest, interpositus next, and fastigius last, in agreement with other studies of Bava et al 1983, Bava & Grimm 1978, Thach 1970a, 1978a). MacKay concluded that "all three nuclei work to stabilize the same motor performance but at different levels." In studies of multijointed reaching movements (MacKay 1988b), neurons tended to discharge maximally at lift-off and minimally during returns to rest and decelerations in midtrajectory. Again, discharge appeared related exclusively to proximal movements, and with "no observable relation to kinematic parameters." The timing sequence was now changed, with fastigius firing first and interpositus last. Aside from timing, there was nothing else in the discharge that distinguished one nucleus from another.

In a recent study (Kane et al 1988, 1989, Thach et al 1990a,b, 1991), unit activity was recorded in the cerebellar nuclei as the monkeys performed five trained movements of the wrist. The job of the animal was to flex or extend the wrist to line up the cursor within the target window on an oscilloscope (Figure 2a) and to maintain the alignment as the window moved. The animals performed five tasks designed to dissociate hypothetical functions of the nuclei: (a) Jerk, a prompt visual triggered move; (b) Jump, identical to that above, except required to stop within the visual target; (c) Pert, return to hold position after perturbation by torque step; (d) Ramp, tracking of visual target; (e) RAM, self-paced rapidly alternating movements. All tasks were performed in two directions, under two loads. The earliest changes were seen in the dentate on the Jerk and Jump tasks. The interpositus showed the earliest changes on the Pert task, and alone showed modulation in relation to tremor on the Ramp tracking. The interpositus

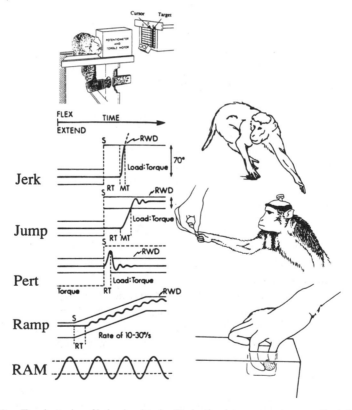

Figure 2a   Two batteries of behavioral tasks. Trained wrist movements are on the *left*. RT—reaction time. MT—movement time. RWD—reward. Nontrained multijointed movements are on the *right*.

and dentate showed equal modulation on the RAM task. Fastigial neurons occasionally showed modulations, but without the trial by trial or temporal consistency seen for the other two nuclei. The results of this unit discharge-behavioral correlation study were thus somewhat similar to those of MacKay's (1988a,b). Both the dentate and interpositus seemed to contribute to some extent to all tasks, but at different times and intensities. The fastigius lacked consistent movement correlations, thus raising serious questions about whether it was taking part in these movements.

To see whether parameters of movement or movement stability depended on the neural discharge with which it was correlated, the different nuclei were inactivated with muscimol (temporary) or kainic acid (permanent) while the proficiency of performance was monitored (Kane et al 1988, 1989, Thach et al 1990, 1992). Any impairment was documented by comparing the pre- and post-injection movement traces. Several variables were examined, including onset and termination times, movement time, direction, velocity, amplitude, tremor, and EMG patterns.

For the dentate, there was a slight (47 ms) delay in reaction time on Jerks and Jumps only. There was no overshoot of final target position. There was a tremor during Ramps only; the tremor was transient, even after permanent kainate ablation. Perts and RAMs (except for superimposition of tremor) were normal. For the interpositus, a slight (10–20 msec) delay in the reaction time on Perts occurred after one muscimol injection. A tremor appeared at the end of Jumps and Perts and during Ramps; the tremor persisted. No consistent change was seen in frequency, amplitude, or regularity of rapid alternating movements, except the superimposition of tremor. The fastigius showed no detectable abnormality in the performance of any of the five trained wrist movements. Similar small delays and irregularities of velocity and force in single-joint tracking (Beppu et al 1984, 1987) and in isometric force (Mai et al 1988) and EMG (Hallett et al 1975) have been seen in studies in cerebellar-damaged patients.

To summarize: In no instance did inactivation abolish the task performance; and the inactivation was immediate, before compensation could have occurred. Inactivation of the dentate and interpositus delayed the onset of task performances with which the neural discharge appeared best to correlate, but the impairment was often so slight as to be just barely detectable. Neither was the choice of direction or velocity of movement impaired. The most conspicuous deficit was tremor. The authors came to several conclusions:

1. the cerebellar nuclei are not significantly reponsible for the generation of any one of the modes of movement in this study;
2. the cerebellar nuclei are not significantly responsible for control of any

of the physical movement parameters underlying these movements, excepting possibly initiation and stability;

3. the actions of the cerebellar nuclei may either be responsible for the "fine control" of any and all movement; or

4. the modes of movement that the cerebellum really controls were not included in this study.

The seemingly contradictory set of findings across the many studies of (a) a good correlation of discharge with some aspects of the behavior (some studies) and (b) little or no deficit in that behavior after ablation, again raises the perplexing question of what it is that the cerebellum does. Gain control would seem to be the logical answer. In this model, the role of the cerebellum would be through tonic discharge simply to maintain the adjustment of downstream structures—thalamus and motor cortex, red, reticular, vestibular nuclei, and spinal cord—which in turn are directly responsible for the initiation and execution of movement. Nevertheless, this idea is intuitively unsatisfying. Why should there be so elaborate an apparatus for so trivial a purpose? Engineers accomplish the same task with trimpots, tiny simple potentiometers, which very accurately bias and balance high performance amplifiers. Is there some other kind of behavior we are missing?

One idea is that the behavior is purely mental, and that we miss it in animal studies and in the routine cerebellar neurological examination because we cannot or do not look for it. Time and space do not permit review of this burgeoning body of material, however. As the quest for what-it-is-the-cerebellum-does threatens to veer away from movement and toward mind, it is useful to consider again an old question:

DOES THE CEREBELLUM PREFERENTIALLY CONTROL THE COORDINATION OF MULTIJOINTED MOVEMENTS?

The Multijoint Coordination Model Versus the Single-joint Modulator Model

It is widely held that the cerebellum coordinates movement. This idea is not a new one; its foundation dates back to 1824 when Flourens concluded, following cerebellar ablations in the pigeon, "the will, the senses, the perception remained, but the coordination of movement, the ability for controlled and determined movement, was lost" (Flourens 1824).

Babinski (1899, 1906) supported this idea in his observations and inferences from a patient who, when instructed to point his toe at a target above his supine body, first flexed the hip, and only after extended the knee. Termed "asynergia," the deficit also implied of ability to coordinate the two joints simultaneously to the purpose of the task. The observation of decomposition of movement in patients with cerebellar deficits led Babinski to the same conclusion that one of the primary functions of the cerebellum is to link together the constituent, simpler movements that make up volitional, compound movements (Babinski 1899, 1906).

But the idea that the cerebellum plays a specific combining role in the coordination of movement has not gained universal acceptance, especially in the English-speaking nations. One of its earliest opponents was none other than the celebrated British neurologist, John Hughlings Jackson. Jackson wrote, "It will not, at all events, suffice to speak of coordination as a separate 'faculty.' Coordination is the function of the whole and every part of the nervous system" (Jackson 1870, as cited in Taylor 1932).

Luciani (1915) was even more explicit. After having described a variety of movement abnormalities produced by cerebellar ablation, he proposed that all could be explained by three primary deficits: atonia, astasia, and asthenia. The function of the cerebellum, he concluded, was to exert a supportive influence on the rest of the nervous system, which was necessary for its fine adjustments (cf. MacKay & Murphy 1979). Upon observing that a dog with half of its cerebellum removed was still capable of swimming "with perfect coordination," Luciani dismissed Fluorens' theory (of the cerebellum as being the seat of and necessary for coordination) as being "a fictitious entity, obscure, imperfect, and unintelligible." Further, that it "opens a false track to subsequent workers and has become a serious obstacle to advance in the physiology of the cerebellum."

Yet, it is interesting to note that Luciani's criticism of coordination comes but one page after the following statement: "when standing and walking, the cerebellum intervenes less as an organ for preserving equilibrium than as an organ which regulates tone and contraction of muscles to the right extent and in the proper combination."

Luciani's reductionist approach greatly influenced another British neurologist, Gordon Holmes. A tribute to Luciani concludes the final interpretation (Holmes 1939) of his own meticulous studies of acute cerebellar injury of man performed during WWI (Holmes 1917, 1922). Following cerebellar injury, Holmes had noted that simple movements, those that occur in one direction and at one joint, could be disturbed in rate, regularity, and force. In the last paragraph that he wrote on the subject, he ends: "In his classical contributions to the physiology of the cerebellum Luciani described as the three symptoms of cerebellar defect: atonia, astasia, and asthenia. These three symptoms occur in man as a result of acute cerebellar lesions and by them the irregularities of movement which constitute cerebellar ataxia can be fully explained" (Holmes 1939).

As for the irregularities observed in multijoint movements, Holmes concluded that they were the result of the errors in the constituent single joint movements, combining "as it were in geometric progresion so that the error of the whole movement is relatively greater than the sum of its parts" (Holmes 1939). He preferred to explain decomposition of movement not as a primary defect in combining movements, but rather as caused by the delay in initiation and the excessive range of movement of any one and all of the component movements relative to each other, and to defective postural fixation (Holmes 1939). Yet, Holmes provided no data on single joint-multijoint movement comparisons. Indeed, he presented some evidence against the hypothesis: "I have studied the relations of such simple synergies as extension of the wrist on flexion of the fingers, but though tracings show that when the fingers suddenly close wrist extension is less regular than normal, this does not appear to be an important element or specific factor of 'cerebellar ataxia'" (Holmes 1939). Is it possible that Holmes' elegant interpretations overcame his own evidence?

Whatever the case, the argument of Jackson, Luciani, and Holmes clearly asserts that the cerebellum exerts control primarily over the single joint has dominated physiology. And because other motor components, like the pyramidal system, even more conspicuously control single joints (Lawrence & Kuypers 1968, Schieber 1988), the role of the upstream cerebellum has inevitably been relegated to that of "modulation" of its downstream initiators and generators. Thus, Holmes (1939) stated: "The conclusions can be drawn that, in addition to regulating postural tone, the cerebellum reinforces or tunes up the cerebral motor apparatus, including subcortical structures with motor functions, so that they respond promptly to volitional stimuli and the impulses from them which excite muscular contractions are properly graded." And, Denny-Brown (1968) stated: "The cerebellum is not essential to any of these [pyramidal and] extrapyramidal mechanisms, but it exerts a modulating effect on all of them. . . . The cerebellum regulates the gamma discharge associated with all motor responses, but its modulating effect is still present in movements initiated by direct alpha drive in a deafferented limb." The statements foreshadow the important work of Gilman on the cerebellar control of gamma motor neuronal discharge in hypotonia (Gilman 1969) and of its independent control of alpha routes in ataxia (Gilman et al 1976, cf. also Granit et al 1955). The statements are compatible with the modern emphasis on primary cerebellar roles in stability control and tremor (Elble et al 1984, Glaser & Higgins 1966, Henatsch 1967, Matthews 1981, Schieber & Thach 1985, Vilis & Hore 1977). They point toward the reflex gain control (MacKay & Murphy 1979) and stability control (Thach et al 1986) theories of today.

We have previously interpreted the dramatic correlation of interpositus neural discharge with tremor and the production of tremor by interposed inactivation to support the argument for gain control (Elble et al 1984, Thach et al 1986) as have others before us (Glaser & Higgins 1966, Henatsch 1967). The argument runs that, because oscillation is a problem inherent in the mechanical-reflex design of the motor system, the cerebellum or other component of that system may have evolved as a specific solution to the problem: to actively damp the oscillation. The same problem of oscillation occurs in mechanical and electrical systems, and the solution of active damping is adopted in both: one uses dashpots, and the other uses resistors. If it is conceded that the cerebellum may damp oscillation, the question remains as to the mechanism: whether at the level of segmental stretch reflexes via gamma motor neuron modulation (Gilman 1969), or in a long loop at the level of motor cortex (Vilis & Hore 1977, 1980), or both. This mechanism still seems plausible because cerebellar units often have appeared to relate poorly or not at all to the various parameters of the task that the EMG and alpha motor neurons relate to; cerebellar units under certain conditions have related very specifically to activities of the gamma-reflex loop; cerebellar ablation has mainly given rise to instability and tremor, with little or no effect on other aspects of certain task performances; and cerebellar output is tonically active and, therefore, a priori influences the sensitivity of downstream targets to other inputs.

But does the cerebellum only control the gain and stability of downstream elements? Despite the physiologist's acceptance of the single joint hypothesis, many neurologists still infer from clinical observation that the function of the cerebellum is preferentially concerned with compound multijointed movement, and that the loss of this function is the major symptom of cerebellar damage (Rondot et al 1979). Even Dow (1987) in attempting to summarize Holmes' description of cerebellar deficits states, "when movements involve two or more joints acting synchronously or simultaneously, the disorders are more than the sum of each of its parts." This apparently amounts to a clinical intuitive feeling that the parts, that is, the deficits at single joints, do not add up to the whole, that is, the deficit in compound movements. Yet no one has really added up the parts to see whether the whole exceeds the sum or not.

Some other suggestions of a cerebellar mechanism for coordination can be found within the literature. Nashner has shown that in response to forward or backward platform translation, which produces primarily rotation of the body around the ankle joint, the normal subject compensates with a preprogrammed, fixed synergic pattern of rapid muscular contractions involving not only the muscles of the ankle, but hip and knee as well. In a study of patients with cerebellar diseases, primarily degenerative diseases of the cerebellar cortex, Nashner & Grimm (1978) reported abnormalities in the synergic response to perturbation. There were gross delays between normally linked pairs of muscles and trials in which there was a complete absence of response in some muscles, as if the mechanism that coordinated this group of postural muscles during stance were lost.

Electrical stimulation of the baboon's dentate nucleus produced two distinct types of movement: "simple" and "complex" movements (Rispal-Padel et al 1982, 1983). Simple movements consisted of the unidirectional displacement of a limb segment around a single joint with the cocontraction of muscles around a nearby joint. Thus, for simple movements, the dentate signal appeared to carry information not only of which limb segment to move but also the postural fixation necessary for that movement to occur. Complex movements involved the displacement of two or three joints, usually noncontiguous. These movements were stereotyped and indissociable. Both types of movements could be the result of cerebellar control of muscle synergies.

At this point, the question should be stated precisely: Does the cerebellum independently control the muscles that operate each joint, or does it have a specific role and mechanism for combining the many muscles that operate many joints in a multijoint movement? Although many studies have recorded units or ablated during single-joint tasks, and many other studies have recorded or ablated during multijoint tasks, we have found no study that has both recorded and ablated during both single-joint and multijoint tasks. Indeed, only three studies either record or ablate during both single-joint and multijoint tasks. The results of these are interesting:

Van Kan et al (1986) reported in a published abstract that interpositus cells, which responded well to free-form reaching, did not correlate as well to movements about a specific joint.

Yet, Harvey et al (1979) recorded the activity of 129 related single units in the dentate and interpositus nucleus during a whole arm reaching task. They attempted to demonstrate relation of these units to a specific joint involved in reaching through "gentle manual restraint" of specific joints to reduce the task to that of a single joint. For 50 of the 129 neurons (38,76%), there was an association with movements about a particular joint (wrist, elbow, or shoulder) or whole hand finger flexion or extension.

Kane et al (1988, 1989) and Thach et al (1990, 1992) studied the effect of nuclear inactivation across nuclei on the performance of five trained single-joint tasks, and, in the same animals in the same sessions, on the performance of five untrained tasks. The untrained activities consisted of sitting, standing, walking, reaching out for bits of food, and picking small bits of food out of deep narrow food wells with a precision pinch of the fingers. Stance, sitting, and walking were evaluated immediately after the animal was released from the primate chair in which the injection had been stereotaxically delivered. Reaching involved coordinated movement of shoulder and elbow; reaching was performed while sitting and standing on the floor, and while sitting in the primate chair. The precision pinch task required coordinated movements of the hand and fingers, including the pinch per se of thumb and forefinger and "tea-cup" posturings of the other digits to keep them out of the way of the edges of the food well. Movements were movie-filmed or video-taped and graphically reconstructed.

Each nuclear inactivation produced an incapacitating impairment of bodily movement. But for each nucleus, the type of deficit was uniquely different (Figure 2b). Fastigial inactivation prevented sitting, standing, and walking, with frequent falls to the side of the lesion. Interposed inactivation caused a severe action tremor of 3–5 Hz during reaching, but not during sitting, standing, or walking. Dentate inactivation caused excess angulation of shoulder and elbow in reaching, which resulted in overshoot of the target, and an increased use of single-digit strategies in attempting

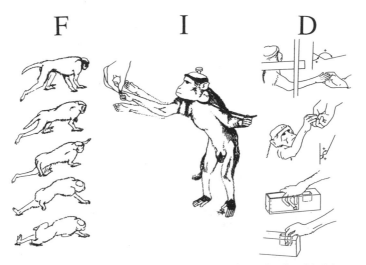

Figure 2b Major deficits produced by micro-injection of muscimol and kainic across the tasks. F represents sitting and stance after each of two fastigial muscimol injections. The figures represent sequential video frames in the course of falling that was caused by an ipsilateral muscimol injection. I represents arm position during reaching after interpositus injection of muscimol. D—deficits in reaching and pinching after dentate injection of muscimol.

to pinch and pick up food morsels, from which we inferred an incoordination of compound finger movements. The deficits were qualitatively more severe than any deficits seen in the simple, single-jointed wrist movements. Our interpretation was and is that nuclear discharge would have been primarily concerned with this last battery of tasks, and not with the first.

To summarize the studies that have specifically addressed the question of multijoint versus single-joint control, one unit recording study seemed to favor the multijoint control, whereas the second seemed to assure that single-joint control occurs. The ablation study clearly favored multijoint control over single-joint control, in addition to proposing the discrete control of different task modes for each of the nuclei.

Botterell & Fulton (1938a,b) had previously noted that midline lesions, whether the fastigial nuclei were involved or not, gave rise uniquely to difficulties in stance and gait. The animals could not stand up against gravity; if the lesions were unilateral, they fell to the side of the lesion. The essential abnormality was of much of the body musculature, but as used only in stance and gait: The midline representation was, therefore, of the whole body, but only for circumscribed functions. Lesions of the lateral hemispheres (with or without dentate involvement) and the ensuing behavioral deficits formed the corollary: Many muscles of the body were again involved, but for an entirely different set of tasks. Animals overshot the mark when reaching for an object, and showed "clumsiness" when manipulating small objects. When running headlong down the hallway, they would often bump into the wall, unable to stop in time (vision was thought to be normal).

These observations were confirmed and extended by Chambers and Sprague in cats with electrolytic lesions stereotaxically placed within the nuclei. With fastigial lesions (Sprague & Chambers 1953), the animal fell toward the side of the lesion. Anterior fastigial lesions caused the ipsilateral hindlimb to flex and the opposite to extend; posterior fastigial lesions affected the forelimb more, though the attitude (flexion or extension) was variable. Reaching and climbing were normal. In lateral hemisphere lesions (Chambers & Sprague 1955a,b), stance and gait were relatively normal, but reaching overshot and pawing movements were "clumsy."

What Is Different and What Is Common Across All Nuclei? Motor Modes and Movement Coordination

What did the rather marked changes in neural discharge relate to in all the single-joint movement studies, if it did not "generate" them? We believe that the observation that single units discharge during tasks that are not impaired by the inactivation of those neurons must mean that the neurons were indeed doing something other than controlling the monitored aspects of the task performance. We have pointed out elsewhere (Mink & Thach 1991a,b) that unit firing may correlate with parameters of movement and yet not control them: The inactivation of those units does not impair those parameters. The lesson: Correlation does not mean causation. The monkeys were trained to perform movements of the wrist, and they did so. As we have often commented (Mink & Thach 1991a, Schieber & Thach 1985, Thach 1968, 1970a, 1978), however, monkeys usually move many muscles other than those minimally necessary to operate the wrist joint. Since after nuclear lesion the wrist joint movements were themselves so little impaired, and the multijoint movements so badly impaired, we conclude that the unit activity may have been better correlated with (and helping to cause) the covert multijoint movements.

What is the meaning, if any, of the slight ablation deficits on performance of the single-joint tasks, and how do they relate to the multijoint results? Common to the single joint tasks and the multijoint tasks is the mode in which they were made. We suggest that Jerk, Jump, and Ramp have a lot in common with Reach and Pinch: They are all visually triggered or guided hand and/or limb movements. They are "volitional" movements: They are trained and/or thought about before they are made. They probably require processing in the occipital and parietal cortices, with information sent to the dentate over cerebro-ponto-cerebellar pathways. Dentate cells fire early in relation to the single-joint performances; dentate inactivation impairs onsets of both the single- and the multijoint performances. One should point out that in the visually guided single-joint movement (Ramp), the eye and the wrist are moved coordinately. Dentate inactivation is known to impair eye-hand coordination (Vercher & Gauthier 1988), whereas single movements are only slightly delayed.

What would be the "mode" of operation of the interpositus? If, as for the dentate, we take this mode to be largely determined by the type and source of input, the spinocerebellar, visual, and auditory inputs suggest behaviors for which fast input and feeback information from the part to be controlled is used to trigger movement of the part. Examples are the long-loop reflex, contact placing, and vision and audition to control the learned blink reflex (McCormick & Thompson 1984, Yeo et al 1984) and acoustic startle responses (Leaton & Supple 1986, Mortimer 1973), respectively. The mode would be "reflexes." The question remains as to how much of these are in the cerebellum and how much in the brainstem and spinal cord (cf. Bloedel & Kelly 1988).

Previous unit-recording single-joint studies (Strick 1983, Thach 1978) show that the interpositus responds at critically short latencies to perturbation from a holding position. If the "functional stretch reflex" *per se* were fully routed through the interpositus, one might expect an impairment of its performance by interposed inactivation. This result was obtained in one injection only (Thach et al 1990a,b); one wonders whether the function of the interpositus is best tested by the single-joint, muscle-loaded task. But the contact-placing reaction also consists of a coordinated action of a limb triggered by a somatosensory stimulus delivered to the limb. Ablation studies of Amassian (Amassian et al 1972a,b, 1974, Amassian & Rudell 1978) have suggested that this action is subserved by the interposed nuclear control of thalamic projections onto the cerebral motor cortex.

Are interpositus deficits also due to incoordination of numbers of muscle groups? Smith and colleagues (Frysinger et al 1984, Smith & Bourbonnais 1981, Wetts et al 1985) have specifically suggested from their own data that the cerebellar cortex overlying the interposed nucleus causally determines whether the flexor and extensor muscles acting at the wrist are reciprocally active (as in flexion and extension movements at the wrist) or coactive (as in the co-contraction that fixes the wrist when the fingers are used in a precision pinch). The proposed mechanism is simple: When the Purkinje cells turn off, the nuclear cells and the agonist and antagonist muscles at the wrist turn on. If this interpretation is correct, the interpositus would play a role in coordinating opposing muscles at a joint rather than synergist muscles across joints.

In the single-joint experiments of Schieber & Thach (1985), Kane et al (1988), and Thach et al (1990, 1992), loads applied alternately to extensors and to flexors always determined that mainly one of an antagonist pair of muscles was active. With overtraining in the Schieber experiment to the point that no other muscle activity was seen, not one interposed or dentate neuron fired in relation to the parameters that did engage the motor cortex and EMG. We thus conclude that the cerebellum was not concerned with alpha motoneurons in this "single muscle group" movement. Nuclear inactivation caused tremor even in the single joint movements: this may have been the result of inactivating the gamma loop system, as originally suggested. These studies therefore did not test the model of Smith and colleagues, which in essence deals with the coordination of agonist and antagonist through the range of reciprocal contraction to full cocontraction.

What about the fastigius? That fastigial inactivation impairs stance and gait is consistent with earlier observations (Botterell & Fulton 1938a, Sprague & Chambers 1953) and with the single-unit recording studies (Antziferova et al 1980) that show unit activity in the fastigial nucleus during fictive scratching and walking, with relatively little in the interpositus and none in dentatus, respectively (Arshavsky et al 1980), during these activities. They are also consistent with Andersson & Armstrong's (1987) observation of Purkinje cell activity related to walking (and its adaptation) in the cerebellar vermis. Thus, the fastigius is concerned with a mode of activity—stance and gait—that distinguishes it from the interpositus and dentate.

Since the acts of walking (and scratching) are by their very nature coordinated multijoint tasks, what the cerebellum adds to their control is not clear from these studies. That the essential program generators for these movements and even some measure of their adaptability lie in the brainstem and spinal cord is known (Arshavsky et al 1972a,b, Bloedel & Kelly 1988). What the cerebellum adds is therefore presumably superimposed upon these fundamental motor synergies. Knowing the magnitude and range of this control may depend on experimental strategies of the type in which Armstrong & Bloedel are engaged that allow more variety of alteration of gait.

A NEURAL NETWORK MECHANISM FOR MOVEMENT COORDINATION...

Parallel Fibers, Purkinje Cell Beams, and Coordination of Linked Nuclear Cells

One of Luciani's objections to a cerebellar role in coordination was that he saw no special feature in its structure that suggested such a function. One must remember that this view somewhat preceded the discoveries of Ramon y Cajal on the architecture of the cerebellar cortex. Is the situation still the same today?

The studies reviewed here show a somatopic representation of the body within each of the three cerebellar nuclei (Figure 3). In each representation, the mapping is of the caudo-rostral dimension of the body onto the sagittal dimension of the nucleus. The hindlimbs are represented anteriorly, the head (at least for the dentate and interposed nuclei) posteriorly; distal parts are medial, proximal parts lateral. This orientation would suggest that the myotomes, running orthogonal to the long axis of the body, run primarily in the coronal dimension and thus roughly parallel to the trajectory of the parallel fibers. Since the parallel fibers are connected to the nuclear cells by Purkinje cells, a coronal "beam" of parallel fibers would control through inhibitory modulation the nuclear cells that influence the synergistic muscles in the myotome. The parallel fiber in this way would be a single neural element spanning and coordinating the activities of multiple synergic muscles and joints.

HOW LONG IS THE PARALLEL FIBER?   In the above model, it is obvious that the longer a parallel fiber is, the more cells in the nuclei (via Purkinje cell control) that it can link together. The length of the parallel fiber then

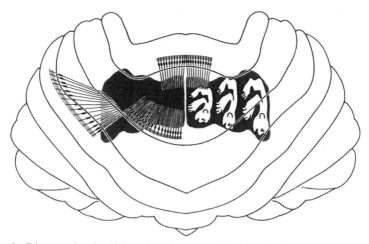

Figure 3   Diagram showing linkage into beams of Purkinje cells by parallel fibers. Beams project down onto the somatotopically organized nuclei. Purkinje cell beams thus link body parts together within each nucleus and link adjacent nuclei together. Such linkage could be the mechanism of the cerebellar role in movement coordination.

becomes a critical limit to its potential functional capacities. Ramon y Cajal believed that parallel fibers ran the full width of the cerebellum (Ramon y Cajal 1911). Since the electrophysiological studies of the Canberra group (Eccles et al 1967, Llinas 1981), the length of the parallel fibers has been assumed to be measured by the distance activity which can travel away from a local stimulus to the parallel fibers in the molecular layer. This has been about 1.5 mm. As such, this rather short parallel fiber has been modeled as a tapped delay line to generate short time intervals (Braitenberg 1967, Braitenberg & Atwood 1958).

Direct anatomical studies show parallel fibers to be much longer. In the studies of Mugnaini and colleagues (Brand et al 1976, Mugnaini 1983), cuts were made across parallel fiber beams and the fibers allowed to degenerate. Upon examination, degenerating fibers were found to extend for considerable distances: Those more superficial were longer than those deep. On the average, those for chicken were just under 10 mm, those for cat, a little over 5 mm, and those for monkey about 6 mm. The range of lengths was roughly the mean ±2 mm. Six millimeters is roughly a third the width of the macaque's cerebellar hemisphere. A 6 mm stretch of cortex projects onto about a 3 mm beam of nucleus, which is the width of one nucleus or slightly greater (Figure 3).

Thus, a beam of Purkinje cells under the influence of a set of parallel fibers of the same origin and length affects a beam of nuclear cells across an entire nucleus. Depending on the portion of the body map to which the *cortical* beam projects, that *nuclear* beam influences the synergic muscles across several joints in the limb (Figure 4), or the muscles of eye, head and neck, and arm, or whatever, depending on the pattern of projection and the folial orientation in the horizontal plane. We often caricature the folial pattern as being more or less strictly in the transverse or coronal plane; but a look at the cerebella of different animals shows how varied it actually is—some even sagittal, or nearly so. For example, in the vermis of the cat, there are near-sagittal and a variety of other orientations of cortical beams. Assuming that they project as such onto the nuclei, they group and regroup the cells of the fastigial nuclei in a variety of ways. As one reflects on the cat's unique and uncanny ability to move limbs and trunk while falling so as to land invariably standing, one may suspect a link between these uniquely coordinated movements and the uniquely configured cortical folia.

Beams also bridge and link nuclei, e.g. the two fastigial nuclei. This provides a means for coordination of the two nuclei and the two sides of

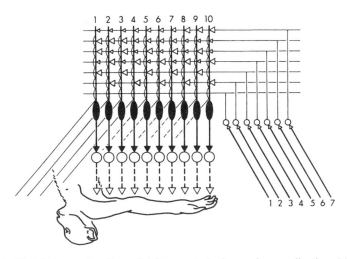

Figure 4 Model of granule cell parallel fiber control of muscular coordination: (*a*) within each nucleus, there is a use-specific (modal) representation of somatic musculature; (*b*) the orientation of the myotome is in the coronal plane; (*c*) the orientation of the parallel fibers is also in the coronal plane; (*d*) the output of the parallel fiber beam of Purkinje cells falls on the nuclear representation of the myotome; (*e*) different uses of the muscles in a limb may be coded by different subsets of parallel fibers and their differential effects on the Purkinje cells (coordination of synergist muscles); (*f*) parallel fiber beams that span the nuclei in their Purkinje cell projection may influence two or more nuclei simultaneously (coordination of modes of movement).

the body in stance and gait. Beams also link fastigius and interposed nuclei (e.g. locomotion and reflex sensitivity) and interposed and dentate nuclei (e.g. reach and reflex sensitivity), and thus coordinate their functions.

These relationships are a fact. There can be little question that parallel fibers link Purkinje cells, which inhibit nuclear cells, which in turn control the different body parts. Because of these anatomical features, linkage of body parts and modes of movement would appear to be designed into the structure. The parallel fiber is the key element of the coordination. The questions that remain are: How much does one parallel fiber control, how many parallel fibers are required to code a novel synergy, how much of coordinated behavior do parallel fibers coordinate, and what are the unique features of that type of coordination?

A Model for Controlling and Adapting Movement Synergies

1. The body is multiply represented within the deep cerebellar nuclei, with at least one body map within each nucleus.
2. Each body map controls a different mode of bodily movement, and each map and mode operate in parallel with the others.
3. Each mode has its own triggering input and its own output target (with its own inherent motor synergies), and these input-output connections determine the difference between modes.
4. The parallel fibers of the cerebellar cortex link Purkinje cells into long beams that project down onto the nuclei, which in turn link the somatotopically arranged nuclear cells into functional subgroupings.
5. These subgroupings are unique and task specific, and are the basis for the cerebellar coordination of movement.
6. The parallel fiber-Purkinje cell linkages are adjustable and are the basis of specific ad hoc learned motor synergies.
7. The learning is determined by the climbing fiber effect on the parallel fiber–Purkinje cell linkage: an error in movement activates the climbing fiber, which works to reduce the strength of connections of the parallel fiber (see below).
8. Learning occurs at synapses outside the cerebellar cortex as well, but for a different purpose: Adaptations at those closer to the motor output (or sensory input) will be generalized across all performances via those outputs (and inputs). These adaptations are useful in balancing the properties of the motor apparatus (e.g. muscles) or input (e.g. sensory organ) structures, but they are not and cannot be the mechanism for memory and control of unique task-specific synergies.
9. Memory for task-specific synergies can occur in the cerebellar cortex, where it is remote from input and output processing, and where there are adequate type and number of structures to code the many and

various synergies that make up higher vertebrate movement repertoires. The best candidate is the granule cell-parallel fiber-Purkinje cell synapse.

10. The model has predictive value.

The verification of this model will require showing the following:

1. Purkinje cells are differentially controlled in single vs. multijointed movement.
2. The parallel fiber is the agent of this control on the Purkinje cell.
3. After cerebellar cortical injury, multijointed movements are sufficiently more impaired than single-jointed movements, such that the sum of the abnormalities at the single joints cannot account for the magnitude of the abnormalities in the compound movement for all three zones and modes.
4. The climbing fibers fire along the beam when learning a synergy that involves many muscles and joints in a limb.
5. Ablation of the beam removes the learned synergy from the behavioral repertoire.

Why do climbing fibers fire in sagittal strips? Parallel fibers have been caricatured as linking together the muscles and joints in a myotomal-coronal dimension. If this is true, there remains the problem of how to coordinate the muscles and joints along the axial-sagittal dimension. One way would be to have folia slanting in a variety of ways onto the deep nuclear body map, as we have suggested may be the case in the vermis of the cat. Another way would be a time stamp across parallel fiber beams, so that learning at the elbow would reinforce contemporary and complementary patterns at the knee. Only if both are correct (or incorrect) will elbow and knee get minimum (or maximum) attention from the climbing fiber adapting both.

Relation of Output Mapping to Input Mapping

We have referred above to the common belief that the trunk is represented in the midline and the extremities laterally. This notion came from Luciani's (1811–1824) localization of abnormalities of stance and gait to midline lesions, and tremor and limb incoordination to lateral lesions (cf. also Holmes 1917, 1922). These observations have been interpreted to imply that the proximal musculature alone is used to stand and that distal musculature alone is used in reaching and manipulative movements (cf. Brown 1949). Botterell & Fulton (1938a,b), in presenting their own scheme of functional localization, commented on the lack of logic in this formulation.

Another line of work that has entrenched this belief is *input mapping* and the evoked potential studies of Adrian (1943), Snider & Stovell (1944), and Snider & Eldred (1952). These studies showed the familiar representation upside down in the anterior lobe (with vision and audition overlapping the somatosensory face) and a second representation in the paramedian lobule. The trunk was in the midline, the extremities extended out laterally into the intermediate zone cortex. Alcoholic cerebellar degeneration was found to occur first and worst in the anterior part of the anterior lobe; the chief disturbance was control of gait and the lower extremity (Victor et al 1959). This pattern fit that of the hindlimb representation in the anterior lobe of the evoked potential studies, and therefore the evoked potential *input mapping* has been reproduced in various clinical works as though generally representing localization of cerebellar functional *output mapping*. We agree with Botterell & Fulton (1938a) and Chambers & Sprague (1955a,b) that this idea may not be correct.

From the more recent studies of mossy fiber responses to tactile stimulation in the rat (Joseph et al 1978, Shambes et al 1978) and cat (Nelson & Bower 1990), it appears that that input mapping may generally be far more complex than originally thought. The input mapping consists of multiple representations of body parts in a pattern its discoverers have called "fractured somatotopy."

What are the physiological implications of such an input system? Nelson & Bower (1990) propose that it "may be involved in optimally controlling sensory receptor surfaces" (e.g. retina, fingers, whiskers) during sensory exploration. Quite another interpretation is suggested by the work on the associative conditioning of the nictitating membrane response (McCormick et al 1981, 1982, McCormick & Thompson 1984, Yeo et al 1984). The import of that work is that through learning, a movement pattern may be provided with a new and arbitrary sensory trigger. Any one of the many different sensory features represented in a mosaic patchwork could be selected, through learning, to drive the behavior.

... THAT LEARNS NEW MOVEMENTS

What is a new movement? One easily thinks of examples of skilled performances like riding a bicycle, skipping rope, serving in tennis, typing, or playing a Beethoven Sonata. What is new? Is it the novel combinations of the muscle and joint actions, or the application of old motions to novel conditions, or both?

An experimental paradigm that illustrates the learning of a synergy is the adaptation of eye-hand coordination in throwing a ball or a dart at a target while wearing wedge prism spectacles (Figure 5a; Baizer & Glickstein 1974, Kane & Thach 1989, Thach et al 1991, Weiner et al 1983). In throwing at a target, the eyes fixate the target and serve as the reference aim for the arm in throwing. The coordination between eye position and synergy of the arm throw is a skill: It has to be developed and kept up with practice. If wedge prism spectacles are placed over the eyes with the base at the right, then the optic path will be bent to the right, and the eye will have to look to the left to see the target. The arm, calibrated to the line of sight, will throw to the left of target (Figure 5b). With practice, the calibration changes, and the arm throws with each try closer to and finally on-target. Proof that gaze direction and eye position in fact comprise the reference aim for the arm throw trajectory comes when the prisms are suddenly removed and the arm throws. The eyes are now on-target, but the eye-arm calibration for the previously left-bent gaze persists; the arm throws to the right of target an amount equal to the original left error (Figure 5a). With practice, the eye position and the arm throw trajectory are recalibrated back to the original setting: The throws move closer back to and finally on-target. A good analogy is the relation between sighting

Figure 5a Throwing darts while wearing wedge prism spectacles (base to the *right*). The subject is looking directly at the target toward which she is pointing the dart, but because the prism bends the optic path 15 degrees to the right, her gaze is deviated 15 degrees to her left in order to see the target (she is looking at you). The portion of her face behind the lenses appears to the viewer to be displaced to her left, also because of the prism's bending of the optic path. The direction of throw is normally in the direction of gaze. The gaze direction has, however, been calibrated to the throw direction, and the aim of throw is true (at you).

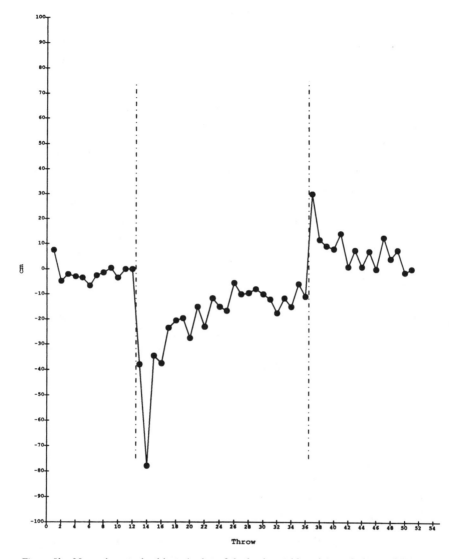

Figure 5b   Normal control subject. A plot of the horizontal locations, relative to the target center, of each successive dart hit. Before introduction of the prism, the darts hit close to the center of the target. Introduction of the prism shifts the hits to the left (*down*), and with practice they normally return toward the center of the target as the recalibration of gaze direction-throw direction is made. After removal of the prism, the hits are normally shifted to the right (*up*), a result that shows that the throw direction is still calibrated 15 degrees off the gaze direction (in actuality, the error is never quite the whole 15 degrees). With practice, the gaze-throw directions readjust back to the original value.

and shooting a gun: The linkage between the sight and the bore is calibrated by adjustment and kept true through practice.

Baizer first showed in macaques that the adjustment mechanism was abolished by cerebellar lesion (Baizer & Glickstein 1974). Weiner et al (1983) confirmed the result in patients with cerebellar disease and found that adaptation was not impaired in disease of corticospinal or basal ganglia systems. We have seen that two patients with magnetic resonance imaging-documented inferior olive hypertropy (a degenerative disease of the inferior olive) could not adapt, despite otherwise normal performance (Figure 5c; Thach et al 1991, cf. also Gauthier et al 1979). This suggests that the adaptation mechanism could be dissociated at least in degree from those of coordination and performance. We have also seen that lesions of the mossy fibers of the middle cerebellar peduncle impair motor learning. This is not to say that the cortex, the inferior olive, and the mossy fibers are equivalent or equipotential in their control of learning, but only that they are all necessary.

Vercher & Gauthier (1988) also observed impaired coordination of eye and hand after dentate lesion; movement of both members became independently saccadic. Whereas they modeled the deficit as a lack of feedback information between independent generators for eye and hand movement (Gauthier et al 1988, Gauthier & Mussa-Ivaldi 1988), we would emphasize the lack of a common feedforward control system.

A similar task has been developed for studying the correlation of Purkinje cell discharge with, and the effect of cortical inactivation on, adaptation (Keating & Thach 1990). A monkey was trained on the Jump task (Figure 2a), which during the learning phase requires coordination of the eyes in observing a moving target and the hand in tracking it. An adaptation is required when, without warning, the gain of the hand coupling to the cursor is changed (e.g. increased). Thus, when the target jumps to the same familiar position and the monkey moves the wrist to its same, familiar position sufficiently to have previously brought the cursor on target, he finds that the cursor overshoots. He has to re-scale his wrist movement and make it smaller by an amount inversely proportionate to the gain increase so that the cursor lands on target. With practice, the monkey learns to move to the new target position. If, as above, the hand points to where the eye points, a calibration of the coupling of the two is needed for this kind of performance. When the gain of the handle is changed, the hand no longer points in the line of sight: The eye-hand coordination must be recalibrated for the cursor to land where the eye is looking. As in the wedge prism task, the cerebellum may control the recalibration.

For a number of Purkinje cells so far studied during the Jump endpoint

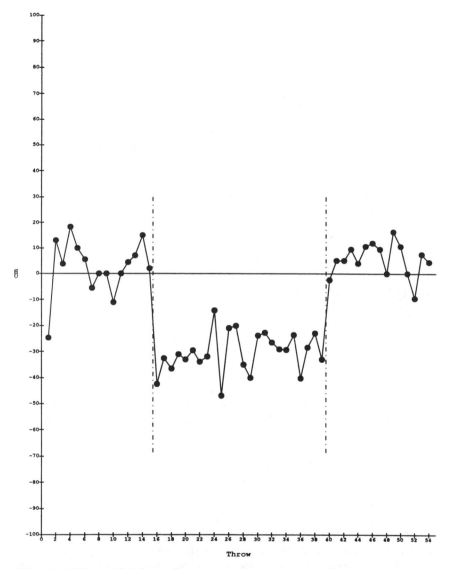

Figure 5c   Patient with inferior olive hypertrophy, a degenerative disease of the inferior olive. After introduction of the prism, there is no recalibration of the gaze-throw directions, and the throws remain to the left of target center (*down*). After removal of the prisms, the hits land where they did before the introduction of the prisms, thus indicating no adaptation to the prisms in this subject.

adaptation, the behavioral adaptation appears to be related to a transient, covarying change in complex and simple spike rates, and a persistent behavioral change to a persistent change in the simple spike rate. This is a pattern consistent with the Marr (1969)-Albus (1971) theory, and one that has been seen experimentally twice previously (Gilbert & Thach 1977, Watanabe 1984).

"Learning curves" that graphically depict the adaptation are well modeled by a simple exponential decay function. The learning can easily be distinguished from the performance in the following way. The rate of change of slope of the learning curve (a time constant) is taken to reflect the learning capacity at that particular time. The scatter of data points (each representing the actual trial-by-trial position of the wrist as it jumps to target) is measured as the variance around the mean learning curve, and is taken to reflect the performance capacity at that particular time. In previous ablation studies, a lack of distinction between an impairment of learning versus an impairment of performance has been a confounding factor.

The prediction was that different experimental manipulations would affect the one or the other (learning and performance) independently. This has been confirmed by repeated injections of muscimol into the cerebellar cortex. Cortical inactivation prolongs the adaptation time by as much as five- or six-fold with low concentrations of muscimol (1 μmol). The adaptation can be impaired in the absence of any detected change in performance. At higher concentrations (5 μmol), the adaptation is abolished for several days. The effect is localized to a point in the lateral zone of the hemisphere (which we presume to project to dentate). Injections at sites more medial, anterior, and posterior do not give the deficit.

Why Is the Cerebellar Cortex Particularly Appropriate for Task-specific Learned Motor Synergies?

The site for storage of the more unique synergies should be far away from output and input sites so that the memory for one synergy does not spill over and influence another synergy. The cerebellar cortex is such a site.

The structure must be optimized for storage of multiple separate synergies. Even with multiple connections, as in the hidden layers of a connection machine, the memory and combinatorial capacity increases with the number of the hidden units. The cerebellar cortex contains the cell type that far outnumbers all other cell types within the nervous system—the granule cell, which gives rise to the parallel fiber.

The aspect of behavior that is learned or adapted must somehow be represented within the mechanism. Since the Purkinje cell projects topically onto a small portion of a deep nucleus, and similarly onto adjacent cells without complete overlap, then a Purkinje cell "represents" its target in the nuclei. The nuclei in turn are somatotopically coded for body part and, depending on the inputs to and output targets of the nucleus, a certain "mode" of control of the body.

Purkinje cells, as representatives of the movement elements to be controlled, are linked together in large, long combinations, which then become muscle and movement pattern synergies.

Because the parallel fiber-Purkinje cell synapses are adjustable (Gilbert & Thach 1977, Ito et al 1982, Robinson 1976), the synergies can be created and eliminated. Since they are remote from input and output, creation or elimination of one synergy does not affect another synergy. There is herein the direct analogy to a look-up table with its many addresses, the addresses being fully programmable, as the memory for pre-computed commands in robotic control.

Many learnable movements are possible; the limits are set by the patterns of connection of parallel fibers, Purkinje cells, nuclear cells, downstream pattern generators (motor cortex, red nucleus, reticular and vestibular nuclei), and inter- and motoneurons. Movements are triggered in many different contexts—the limits are again set by the range of conditions represented in the mossy fibers to any given granule cell.

Some Old Puzzles Illuminated: Silent Areas, Focal Versus Diffuse Lesions

If the cerebellar cortex is viewed as the storage site that both generates old and learns new movements, some old observations are explained. One is the paradox that the cortex appears so insensitive to focal lesions and yet so vulnerable to widespread degenerative diseases. A focal lesion may wipe out one memory-movement synergy and its trigger but leave other memories and thus the means of accomplishing the same ends by similar movements. The loss or even partial damage diffusely of the whole cortex (as in cortical degenerative diseases of alcoholic thiamine deficiency, paraneoplastic degeneration, olivopontocerebellar atrophy, etc.), however, can be devastating in its effects.

Also explained is the remarkable absence of deficits after lateral hemisphere lesion in some patients, and the equally remarkable selectivity of deficits in others. Holmes (1922) commented on two musicians: the one, after removal of a left lateral hemisphere tumor, could not play left hand notes on the piano in proper sequence. The other, after a cerebellar gunshot wound, could not play the flute. We have also seen a man with a posterior inferior cerebellar artery territory infarct of the underside of the lateral cerebellum on the right who was normal (in the arm and hand) to clinical test. Yet he complained of a severe deficit. He had from youth trained himself at the sharp dealing and shuffling of cards, at card tricks, and at other sleights of hand using coins and poker chips. After the lesion, he could no longer deal the cards evenly spaced, track card order during shuffling so as to pull from the middle or bottom of the deck, put three coins on elbow or back of hand and then flip and catch them in midair, or palm a roll of poker chips and singlehandedly and serially remove the one in the middle and place it on either end of the roll. He regarded the deficits as discrete and qualitative rather than quantitative. It was as though he had lost a number of highly trained synergies of the hand and arm, whereas others were unimpaired.

THE EXECUTIVE AND THE MODULATOR MODELS, CORTICAL AND NUCLEAR LEARNING:
Either/Or, Both, All?

Often, the two sides of a long-standing dispute that has been waged on the premise that only one side can be correct are eventually seen both to be correct: The dispute has resulted from how the question was posed or studied. We believe that this may be another such situation. The question is whether the executive mechanism that initiates and controls movement is in the cerebellum (and especially in the cortex) or in the target structures to which the cerebellum projects. The answer is likely to be "both."

It is known that the motor cortex is one of these target structures, and that its repertoire is in the domain of individuated movements rather than larger synergies (Lawrence & Kuypers 1968, Schieber 1988). We have reviewed the timing studies of cerebellum-motor cortex unit recording and ablation. These have shown that the cerebellum leads and helps initiate some movements through the motor cortex. The mechanism could be as simple as an assembly of the individuated, pauci-, or single-jointed movements into compound, multijointed movements. What about other cerebellar targets?

Within the brainstem, mechanisms control the synergies responsible for the orienting of eyes, head, and neck to visual, acoustic, and somatosensory stimuli. There are also mechanisms that control tonic neck reflexes, contact placing, righting, supporting, and the labyrinthine control of stance and locomotion. The spinal cord contains basic mechanisms for muscle synergies that mediate resistance to displacement, flexor withdrawal with contralateral extension, and quadrupedal locomotion.

Since only the fastigius projects directly to motoneurons (cf. Asanuma et al 1983d), the bulk of the cerebellar outflow must use the synergies within the thalamocortical and the brainstem nuclei on which it falls. This use is likely to lie between the alternatives of "modulatory gain control only" and "executive coordinative control only." How might the "in between" condition be conceptualized and stated?

Ito showed that the tonically active Purkinje cells are inhibitory (Eccles et al 1967). A metaphor was created that is useful in thinking of how the cerebellar cortex can project onto and control downstream motor mechanisms that have their own resident programs: Cerebellar cortical outflow "sculpts" varied and individualistic patterns from monolithic stereotypes through disinhibition and inhibition (Eccles et al 1967). The essence of the metaphor is that the learned synergy may be achieved not by building it up, muscle by muscle and piece by piece, as in clay, but rather by chipping off, piece by piece, the unwanted parts of some undifferentiated, multijointed movement as in stone. Thus, a mossy fiber-triggered beam of parallel fibers activates basket-cell inhibition of off-beam flanking Purkinje cells, which releases the nuclear cells below from tonic Purkinje cell inhibition. This could release a primitive grasp, tonic neck, contact-placing, or labyrinthine reflex. The same mossy fiber-triggered beam of parallel fibers would give a learned and variable amount of activation to on-beam Purkinje cells, which would through thus controlled inhibition sculpt the released, undifferentiated synergy and give it the adaptive differentiated specifics.

Where then would initiation of movement occur? The quest has been for some single site, where sensory information triggers a motor response in the process we call *sensorimotor integration*. The cerebellum has been considered a good candidate, because so much of its input is so clearly sensory, whereas most if not all its output is motor. Timing studies support this view, in that motor cortex activity has consistently appeared to lag and depend on prior cerebellar activity. If the cerebellar cortex does indeed contain special apparatus dedicated both to building complex movements from simpler components and to carving out differentiated movements from undifferentiated movements, the initiation of a complex (or differentiated) movement would have to occur at the level of this mechanism. In either situation, the mossy fiber would provide the trigger. The parallel fiber and its beam of variably activated Purkinje cells and flanks of inhibited Purkinje cells would provide the response. The parallel fiber would have the pivotal role in combining or carving the "pieces of movement" (Marr 1969).

Clearly, not all movement is initiated by the cerebellum. The segmental stretch reflex, the vestibulo-ocular reflex, and the blink reflex can all occur without any cerebellar input. These activities are also capable of plastic adaptation without the cerebellum (Bloedel & Kelly 1988, Lisberger 1988, Wolpaw & Carp 1990). Yet in the case of multiple context-dependent learned responses of the one reflex, as in the VOR under water, in air, with and without trifocal spectacles (cf. Gauthier & Robinson 1975, Shelhamer & Robinson 1991), it seems likely that the cerebellar cortex participates and is necessary. It also seems likely that the initiation is distributed—that the primary vestibular afferent triggers responses both from the vestibular nuclei and from the cerebellar cortex at about the same time. In this situation, the vestibular nuclear response is "fundamental," and the cerebellar cortical response is "differential"—one of the several or many specially adapted responses. The fundamental response in the vestibular nuclei may precede (in time) the differential response in the cortex, as Lisberger has suggested (1988). Yet it would be the cortical component, which is based on the adaptation that is specific to a certain behavioral context, that provides both the individuality and the utility of the response. This is considerably more specific and important than "the fine control of the VOR." In the cortex there would be not just the one tuning but many, each and all necessary to make the fundamental VOR perform differentially and adequately across task requirements in the foveate animal. One wonders whether the cerebellum may prove also to extend the range of capability of the stretch reflex—under different conditions providing a variety of appropriate responses in relative stiffness, damping, length servo-assistance—when the system is adequately studied under varying conditions and requirements.

Clearly, coordination is not unique to the cerebellum. Each motor component that is downstream from the cerebellum obviously is built with its own type of coordination. As Denny-Brown stated, the cerebellum for the most part must work through them, but that fact does not exclude a role for the type of coordination we have proposed here, not does it condemn the cerebellum to the trivial role of a simple gain control of these structures. Indeed, for coordinated complex movements, we propose that the cerebellum is the executive; that it learns, initiates, continues, and stops complex movements through its actions on the downstream structures. As for the multiple learned-context-dependent performances of the one reflex, we agree that the initiation may be distributed, but propose that the specificity of the response is in the cerebellar cortex.

Just as clearly, motor learning is not unique to the cerebellar cortex, nor should it be. But, if cerebellar cortex does indeed contain special apparatus dedicated to combining simpler elements of movement into larger complex synergies, then the learning as well as the initiation of such a movement would have to occur at the level of the mechanism that provides for the combination. What is unique about cerebellar learning is the flexibility, ease, and speed of assigning the trigger-response (task specificity) and of building the tailor-made complex movement response. The triggering input may be changed from one sensory modality to another. The task performance may be programmed and adjusted without affecting other task performances. This is not so of spinal cord, vestibular, or other low-level mechanisms.

Finally, we agree with Llinas (1981) that what had been lacking in the Marr-Albus motor learning theory is how it ties in with the rest of cerebellar function. We hope this review helps to answer that objection.

SUMMARY

Based on a review of cerebellar anatomy, neural discharge in relation to behavior, and focal ablation syndromes, we propose a model of cerebellar function that we believe is both comprehensive as to the available information (at these levels) and unique in several respects. The unique features are the inclusion of new information on (a) cerebellar output—its replicative representation of body maps in each of the deep nuclei, each coding a different type and context of movement, and each appearing to control movement of multiple body parts more than of single body parts; and (b) the newly assessed long length of the parallel fiber. The parallel fiber, by virtue of its connection through Purkinje cells to the deep nuclei, appears optimally designed to combine the actions at several joints and to link the modes of adjacent nuclei into more complex coordinated acts. We review the old question of whether the cerebellum is responsible for the coordination of body parts as opposed to the tuning of downstream executive centers, and conclude that it is both, through mechanisms that have been described in the cerebellar cortex. We argue that such a mechanism would require an adaptive capacity, and support the evidence and interpretation that it has one. We point out that many parts of the motor system may be involved in different types of motor learning for different purposes, and that the presence of the many does not exclude an existence of the one in the cerebellar cortex. The adaptive role of the cerebellar cortex would appear to be specialized for combining simpler elements of movement into more complex synergies, and also in enabling simple, stereotyped reflex apparatus to respond differently, specifically, and appropriately under different task conditions. Speed of learning and magnitude of memory for both novel synergies and task-specific performance modifications are other attributes of the cerebellar cortex.

Literature Cited

Adrian, E. D. 1943. Afferent areas in the cerebellum connected with the limbs. *Brain* 66: 289–315

Albus, J. S. 1971. A theory of cerebellar function. *Math. Biosci.* 10: 25–61

Allen, G. I., Gilbert, P. F. C., Marini, R., Schultz, W., Yin, T. C. T. 1977. Integration of cerebral and peripheral inputs by interpositus neurons in monkey. *Exp. Brain Res.* 27: 81–99

Allen, G. I., Gilbert, P. F. C., Yin, T. C. T. 1978. Convergence of cerebral inputs onto dentate neurons in monkey. *Exp. Brain Res.* 32: 151–70

Allen, G. I., Tsukahara, N. 1974. Cerebrocerebellar communication systems. *Physiol. Rev.* 54: 957–1006

Amassian, V. E., Reisne, H., Wertenbaker, C. 1974. Neural pathways subserving plasticity of contact placing. *J. Physiol.* 242: 67–69

Amassian, V. E., Ross, R., Wertenbaker, C., Weiner, H. 1972b. Cerebello-thalamocortical inter-relations in contact placing and other movements in cats. In *Corticothalamic Projections and Sensorimotor Activities*, ed. T. L. Frigyesi, E. Rinvik, M. D. Yahr, pp. 395–444. New York: Raven

Amassian, V. E., Rudell, A. 1978. When does the cerebellum become important in coordinating placing movements? *J. Physiol.* 276: 35–36

Amassian, V. E., Weiner, H., Rosenblum, M. 1972a. Neural systems subserving the tactile placing reaction: A model for the study of higher level control of movement. *Brain Res.* 40: 171–78

Andersson, G., Armstrong, D. M. 1987. Complex spikes in Purkinje cells in the lateral vermis (b zone) of the cat cerebellum during locomotion. *J. Physiol.* 385: 107–34

Antziferova, L. I., Arshavsky, Yu, I., Orlovsky, G. N., Pavlova, G. A. 1980. Activity of neurons of cerebellar nuclei during fictitious scratch reflex in the cat. I. Fastigial nucleus. *Brain Res.* 200: 239–48

Arshavsky, Y. I., Berkinblit, M. B., Fuxson, O. L., Gel'fand, I. M., Orlovsky, G. N. 1972a. Recordings of neurones of the dorsal spinocerebellar tract during evoked locomotion. *Brain Res.* 43: 272–75

Arshavsky, Y. I., Berkinblit, M. B., Fuxson, O. L., Gel'fand, I. M., Orlovsky, G. N. 1972b. Origin of modulation in neurones of the ventral spinocerebellar tract during locomotion. *Brain Res.* 43: 276–79

Arshavsky, Y. I., Orlovsky, G. N., Pavlova, G. A., Perret, C. 1980. Activity of neurons of cerebellar nuclei during fictitious scratch reflex in the cat. II. The interpositus and lateral nuclei. *Brain Res.* 200: 249–58

Asanuma, C., Thach, W. T., Jones, E. G. 1983a. Cytoarchitectonic delineation of the ventral lateral thalamic region in the monkey. *Brain Res. Rev.* 5: 219–35

Asanuma, C., Thach, W. T., Jones, E. G. 1983b. Distribution of cerebellar terminations and their relation to other afferent terminations in the ventral lateral thalamic region of the monkey. *Brain Res.*

Rev. 5: 237–65

Asanuma, C., Thach, W. T., Jones, E. G. 1983c. Anatomical evidence for segregated focal groupings of efferent cells and their terminal ramifications in the cerebellothalamic pathway of the monkey. *Brain Res. Rev.* 5: 267–99

Asanuma, C., Thach, W. T., Jones, E. G. 1983d. Brainstem and spinal projections of the deep cerebellar nuclei in the monkey, with observations on the brainstem projections of the dorsal column nuclei. *Brain Res. Rev.* 5: 299–322

Babinski, J. 1899. De l'asynergie cerebelleuse. *Rev. Neurol.* 7: 806–16

Babinski, J. 1906. Asynergie et inertie cerebelleuses. *Rev. Neurol.* 14: 685–86

Baizer, J. S., Glickstein, M. 1974. Role of cerebellum in prism adaptation. *J. Physiol.* 236: 34p–35p

Bava, A., Grimm, R. J. 1978. Activity of fastigial neurons during wrist movements in primates. *Neurosci. Lett. Suppl.* 1: 141

Bava, A., Grimm, R. J., Rushmer, D. S. 1983. Fastigial unit activity during voluntary movement in primates. *Brain Res.* 288: 371–74

Beppu, H., Nagaoka, M., Tanaka, R. 1987. Analysis of cerebellar motor disorders by visually-guided elbow tracking movements: 2. Contributions of the visual cues on slow ramp pursuit. *Brain* 110: 1–18

Beppu, H., Suda, M., Tanaka, R. 1984. Analysis of cerebellar motor disorders by visually-guided elbow tracking movements. *Brain* 107: 787–809

Bloedel, J. R., Courville, J. 1981. Cerebellar afferent systems. In *Handbook of Physiology, The Nervous System*, Sect. 1, Vol. 2, ed. V. B. Brooks, pp. 877–946

Bloedel, J. R., Kelly, T. M. 1988. A proposed role for cerebellar sagittal zones in cerebellar cortical interactions. In *Neurobiology of the Cerebellar Systems: A Centenary of Ramon y Cajal's Description of the Cerebellar Circuits*, p. 30 (Abstr.)

Botterell, E. H., Fulton, J. F. 1938a. Functional localization in the cerebellum of primates. II. Lesions of midline structures (vermis) and deep nuclei. *J. Comp. Neurol.* 69: 47–62

Botterell, E. H., Fulton, J. F. 1938b. Functional localization in the cerebellum of primates. III. Lesions of hemispheres (neocerebellum). *J. Comp. Neurol.* 69: 63–87

Braitenberg, V. 1967. Is the cerebellar cortex a biological clock in the millisecond range? *Prog. Brain Res.* 25: 334–46

Braitenberg, V., Atwood, R. P. 1958. Morphological observations on the cerebellar cortex. *J. Comp. Neurol.* 109: 1–33

Brand, S., Dahl, A.-L., Mugnaini, E. 1976. The length of parallel fibers in the cat cerebellar cortex. An experimental light and electron microscope study. *Exp. Brain Res.* 26: 39–58

Brodal, P. 1978. The corticopontine projection in the rhesus monkey. Origin and principles of organisation. *Brain* 101: 251–83

Brodal, P. 1981. *Neurologic Anatomy in Relation to Clinical Medicine.* Oxford/ New York: Oxford Univ. Press

Brooks, V. B., Kozlovskaya, I. B., Atkin, A., Horvath, F. E., Uno, M. 1973. Effects of cooling dentate nucleus on tracking-task performance in monkeys. *J. Neurophysiol.* 36: 974–95

Brooks, V. B., Thach, W. T. 1981. Cerebellar control of posture and movement. In *Handbook of Physiology, The Nervous System*, Sect. 1, Vol. 2, ed. V. B. Brooks, pp. 877–46. Bethesda: Am. Physiol. Soc.

Brown, J. R. 1949. Localizing cerebellar syndromes. *J. Am. Med. Assoc.* 141: 518–21

Burton, J. E., Onoda, N. 1977. Interpositus neurons discharge in relation to a voluntary movement. *Brain Res.* 121: 167–72

Burton, J. E., Onoda, N. 1978. Dependence of the activity of interpositus and red nucleus neurons on sensory input data generated by movement. *Brain Res.* 152: 41–63

Chambers, W. W., Sprague, J. M. 1955a. Functional localization in the cerebellum. I. Organization in longitudinal cortonuclear zones and their contribution to the control of posture, both extrapyramidal and pyramidal. *J. Comp. Neurol.* 103: 105–29

Chambers, W. W., Sprague, J. M. 1955b. Functional localization in the cerebellum. II. Somatic organization in cortex and nuclei. *Arch. Neurol. Psychiat.* 74: 653–80

Cooke, J. D., Thomas, J. S. 1976. Forearm oscillations during cooling of the dentate nucleus in the monkey. *Can. J. Physiol. Pharmacol.* 54: 430–36

Denny-Brown, D. E. 1967. The fundamental organization of motor behavior. In *Neurophysiological Basis of Normal and Abnormal Motor Activities*, ed. M. D. Yahr, D. P. Purpura, pp. 414–44. New York: Raven

Dow, R. S. 1987. Cerebellum, pathology: Symptoms and signs. In *Encyclopedia of Neuroscience*, ed. G. Adelman, 1: 203–6. Boston: Birkhauser Boston, Inc.

Eccles, J. C. 1967. Circuits in the cerebellar control of movement. *Proc. Natl. Acad. Sci. USA* 58: 336–43

Eccles, J. C., Ito, M., Szentagothai, J. 1967. *The Cerebellum as a Neuronal Machine.* New York: Springer-Verlag

Elble, R. J., Schieber, M. H., Thach, W. T. 1984. Activity of spindle afferents and neurons of motor cortex and cerebellar nuclei during action tremor. *Brain Res.* 323: 330–34

Evarts, E. V., Thach, W. T. 1969. Motor mechanisms of the CNS: Cerebrocerebellar inter-relations. *Annu. Rev. Physiol.* 31: 451–98

Flourens, P. 1824. Recherches experimentales sur les propriétés et les fonctions du systéme nerveux, dans les animaux vertébres. Crevot, Paris

Fortier, P. A., Kalaska, J. F., Smith, A. M. 1989. Cerebellar neuronal activity related to whole-arm reaching movements in the monkey. *J. Neurophysiol.* 62: 198–211

Frysinger, R. C., Bourbonnais, D., Kalaska, J. F., Smith, A. M. 1984. Cerebellar cortical activity during antagonist cocontraction and reciprocal inhibition of forearm muscles. *J. Neurophysiol.* 51: 32–49

Gauthier, G. M., Hofferer, J.-M., Hoyt, W. F., Stark, L. 1979. Visual-motor adaptation: Quantitative demonstration in patients with posterior fossa involvement. *Arch. Neurol.* 36: 155–60

Gauthier, G. M., Mussa-Ivaldi, F. 1988. Oculo-manual tracking of visual targets in monkey: Role of the arm afferent information in the control of arm and eye movements. *Exp. Brain Res.* 73: 138–54

Gauthier, G. M., Robinson, D. A. 1975. Adaptation of the human vestibulo-ocular reflex to magnifying lenses. *Brain Res.* 92: 331–35

Gauthier, G. M., Vercher, J.-L., Mussa-Ivaldi, F., Marchetti, E. 1988. Oculomanual tracking of visual targets: Control learning, coordination control, and coordination model. *Exp. Brain Res.* 73: 127–37

Gilbert, P. F. C., Thach, W. T. 1977. Purkinje cell activity during motor learning. *Brain Res.* 128: 309–28

Gilman, S. 1969. The mechanism of cerebellar hypotonia. *Brain* 92: 621–38

Gilman, S., Carr, D., Hollenberg, J. 1976. Kinematic effects of deafferentation and cerebellar ablation. *Brain* 99: 311–30

Glaser, G. H., Higgins, D. C. 1966. Motor stability, stretch responses, and the cerebellum. In *Muscular Afferents and Motor Control. Proc. Nobel Symp.*, Vol. 1, ed. R. Granit, pp. 121–38. Stockholm: Almquist & Wiksell

Goldberger, M. E., Growden, J. H. 1973. Pattern of recovery following cerebellar deep nuclear lesions in monkeys. *Exp. Neurol.* 39: 307–22

Granit, R., Holmgren, B., Merton, P. A. 1955. The two routes for excitation of muscle and their subservience to the cerebellum. *J. Physiol.* 130: 213–24

Grimm, R. J., Rushmer, D. S. 1974. The activity of dentate neurons during an arm movement sequence. *Brain Res.* 71: 309–26

Hallett, M. B., Shahani, B. T., Young, R. R. 1975. EMG analysis of patients with cerebellar deficits. *J. Neurol. Neurosurg. Psychiatr.* 38: 1154–62

Harvey, R. J., Porter, R., Rawson, J. A. 1979. Discharges of intracerebellar nuclei in monkeys. *J. Physiol. London* 297: 559–80

Henatsch, H. D. 1967. Instability of the proprioceptive length servo: Its possible role in tremor phenomena. In *Neurophysiological Basis of Normal and Abnormal Motor Activities*, ed. M. D. Yahr, D. P. Purpura, pp. 75–89. New York: Raven

Holmes, G. 1917. The symptoms of acute cerebellar injuries due to gunshot injuries. *Brain* 40: 461–535

Holmes, G. 1922. Clinical symptoms of cerebellar disease and their interpretation. The Croonian lectures II. *Lancet* 1: 1177–82

Holmes, G. 1939. The cerebellum of man. The Hughlings Jackson memorial lecture. *Brain* 62: 1–30

Ito, M., Sakurai, M., Tongroach, P. 1982. Climbing induced depression of both mossy fiber responsiveness and glutamate sensitivity of cerebellar Purkinje cells. *J. Physiol. London* 324: 113–34

Jackson, J. H. 1870. A study of convulsions. In *Selected Writings of John Hughlings Jackson* (1932), ed. J. Taylor, 1: 8–36. New York: Basic Books

Jansen, J., Brodal, A. 1940. Experimental studies on the intrinsic fibers of the cerebellum. II. The cortico-nuclear projection. *J. Comp. Neurol.* 73: 267–321

Joseph, J. W., Shambes, G. M., Gibson, J. M., Welker, W. 1978. Tactile projections to granule cells in caudal vermis of the rat's cerebellum. *Brain Behav. Evol.* 15: 141–49

Kalil, K. 1982. Projections of the cerebellar and dorsal column nuclei upon the thalamus of the Rhesus monkey. *J. Comp. Neurol.* 195: 25–50

Kane, S. A., Goodkin, H. P., Keating, J. G., Thach, W. T. 1989. Incoordination in attempted reaching and pinching after inactivation of cerebellar dentate nucleus. *Abstr. Soc. Neurosci.* 15: 52

Kane, S. A., Mink, J. W., Thach, W. T. 1988. Fastigial, interposed, and dentate cerebellar nuclei: Somatotopic organization and the movements differentially controlled by each. *Soc. Neurosci. Abstr.* 14: 954

Kane, S. A., Thach, W. T. 1989. Palatal myoclonus and inferior olive function: Are they related? *Exp. Brain Res. Ser.* 17: 427–60

Keating, J. G., Thach, W. T. 1990. Cerebellar motor learning: Quantitation of movement adaptation and performance in rhesus monkeys and humans implicates cortex as the site of adaptation. *Abstr. Soc. Neurosci.* 16: 762

Lamarre, Y., Spidalieri, G., Chapman, C. E. 1983. A comparison of neuronal discharge recorded in the sensori-motor cortex, parietal cortex, and dentate nucleus of the monkey during arm movements triggered by light, sound or somesthetic stimuli. *Exp. Brain Res. Suppl.* 7: 140–56

Lawrence, D. G., Kuypers, H. G. J. M. 1968. The functional organization of the motor system in the monkey. I. The effects of bilateral pyramidal lesions. *Brain* 91: 1–14

Leaton, R. N., Supple, W. F. Jr. 1986. Cerebellar vermis: Essential for long-term habituation of the acoustic startle response. *Science* 232: 513–15

Lisberger, S. G. 1988. The neural basis for learning of simple motor skills. *Science* 242: 728–35

Llinas, R. 1981. Electrophysiology of cerebellar networks. In *Handbook of Physiology*, Sect. 1, Vol. 2, Pt. 2, ed. V. B. Brooks, pp. 831–76. Bethesda: Am. Physiol. Soc.

Luciani, L. 1915. The hindbrain. In *Human Physiology*, transl. F. A. Welby, Ch. 8, p. 467. London: Macmillan

MacKay, W. A. 1988a. Unit activity in the cerebellar nuclei related to arm reaching movements. *Brain Res.* 442: 240–54

MacKay, W. A. 1988b. Cerebellar nuclear activity in relation to simple movements. *Exp. Brain Res.* 71: 47–58

MacKay, W. A., Murphy, J. T. 1979. Cerebellar modulation of reflex gain. *Prog. Neurobiol.* 13: 361–417

Mai, N., Belsinger, P., Avarello, M., Diener, H.-C., Dichgans, J. 1988. Control of isometric finger force in patients with cerebellar disease. *Brain* 111: 973–98

Marr, D. 1969. A theory of cerebellar cortex. *J. Physiol.* 202: 437–70

Matthews, P. B. C. 1981. Muscle spindles, their messages and their fusimotor supply. In *Handbook of Physiology*, Sect. 1: *The Nervous System*, Volume 2, *Motor Control*, Pt. 1, ed. J. M. Brookhart, V. B. Mountcastle, V. B. Brooks, pp. 189–228. Bethesda, MD: Am. Physiol. Soc.

McCormick, D. A., Clark, G. A., Lavond, D. G., Thompson, R. F. 1982. Initial localization of the memory trace for a basic form of learning. *Proc. Natl. Acad. Sci. USA* 79: 2731–35

McCormick, D. A., Lavond, D. G., Clark, G. A., Kettner, R. E., Rising, C. E., Thompson, R. F. 1981. The engram found? Role of the cerebellum in classical conditioning of nictitating membrane and eyelid responses. *Bull. Psychonomic Soc.* 18: 103–5

McCormick, D. A., Thompson, R. F. 1984. Cerebellum: Essential involvement in the classically conditioned eyelid response. *Science* 223: 296–99

Meyer-Lohman, J., Conrad, B., Matsunami, K., Brooks, V. B. 1975. Effects of dentate cooling on precentral unit activity following torque pulse injections into elbow movements. *Brain Res.* 94: 237–51

Meyer-Lohman, J., Hore, J., Brooks, V. B. 1977. Cerebellar participation in generation of prompt arm movements. *J. Neurophysiol.* 40: 1038–50

Mink, J. W., Thach, W. T. 1991a. Basal ganglia motor control. 2. Late pallidal timing relative to movement onset and inconsistent pallidal coding of movement parameters. *J. Neurophysiol.* In press

Mink, J. W., Thach, W. T. 1991b. Basal ganglia motor control. 3. Pallidal ablation: Normal reaction time, muscle cocontraction, and slow movement. *J. Neurophysiol.* In press

Mortimer, J. A. 1973. Temporal sequence of cerebellar Purkinje and nuclear activity in relation to the acoustic startle response. *Brain Res.* 50: 457–62

Mugnaini, E. 1983. The length of cerebellar parallel fibers in chicken and rhesus monkey. *J. Comp. Neurol.* 220: 7–15

Nashner, L. M., Grimm, R. G. 1978. Analysis of multiloop dyscontrols in standing cerebellar patients. *Prog. Clin. Neurophysiol.* 5: 300–19

Nelson, M. E., Bower, J. M. 1990. Brain maps and parallel computers. *Trends Neurosci.* 13: 403–8

Orioli, P. J., Strick, P. L. 1989. Cerebellar connections with the motor cortex and the arcuate premotor area: An analysis employing retrograde transneuronal transport of WGA-HRP. *J. Comp. Neurol.* 288: 612–26

Ramon y Cajal, S. 1911. Histologie du Systeme Nerveux. Paris: Maloine

Rispal-Padel, L., Cicirata, F., Pons, J. C. 1982. Cerebellar nuclear topography of simple and synergistic movements in the alert baboon. *Exp. Brain Res.* 47: 365–80

Rispal-Padel, L., Cicirata, F., Pons, J.-C. 1983. Neocerebellar synergies. *Exp. Brain Res. Suppl.* 7: 213–23

Robinson, D. A. 1976. Adaptive gain control of the vestibulocular reflex by the cerebellum. *J. Neurophysiol.* 39: 954–69

Rondot, P., Bathien, N., Toma, S. 1979. Physiopathalogy of cerebellar movement. In *Cerebro-cerebellar Interactions*, ed. J. Massion, K. Sasaki, pp. 203–30. Amsterdam: Elsevier

Sasaki, K. S., Kawaguchi, S., Oka, H., Saki,

M., Mizuno, N. 1976. Electrophysiological studies on the cerebellocerebral projections in monkeys. *Exp. Brain Res.* 24: 495–507

Schell, G. R., Strick, P. L. 1983. The origin of thalamic inputs to the arcuate premotor and supplementary motor areas. *J. Neurosci.* 4: 539–60

Schieber, M. H. 1988. Motor fields of motor and premotor cortex neurons in the rhesus monkey during independent finger movements. *Abstr. Soc. Neurosci.* 14: 821

Schieber, M. H., Thach, W. T. 1985. Trained slow tracking. II. Bidirectional discharge patterns of cerebellar nuclear, motor cortex, and spindle afferent neurons. *J. Neurophysiol.* 55: 1228–70

Shambes, G. M., Gibson, J. M., Welker, W. 1978. Fractured somatotopy in granule cell tactile areas of rat cerebellar hemispheres revealed by micromapping. *Brain Behav. Evol.* 15: 94–140

Shelhamer, M. J., Robinson, D. A. 1991. Context-specific gain switching in the vestibuloocular reflex. *ARVO Abstr.* 1991

Smith, A. M., Bourbonnais, D. 1981. Neuronal activity in cerebellar cortex related to control of prehensile force. *J. Neurophysiol.* 45: 286–303

Snider, R. S., Eldred, E. 1952. Cerebro-cerebellar relationships in the monkey. *J. Neurophysiol.* 15: 27–40

Snider, R. S., Stowell, A. 1944. Receiving areas of the tactile, auditory, and visual systems in the cerebellum. *J. Neurophysiol.* 7: 331–57

Soechting, J. F., Burton, J. E., Onoda, N. 1978. Relationships between sensory input, motor output and unit activity in interpositus and red nuclei during intentional movement. *Brain Res.* 152: 65–79

Stanton, G. B. 1980. Topographical organization of ascending cerebellar projections from the dentate and interposed nuclei in macaca mulatta: An anterograde degeneration study. *J. Comp. Neurol.* 190: 699–731

Sprague, J. M., Chambers, W. W. 1953. Regulation of posture in intact and decerebrate cat. I. Cerebellum, reticular formation, and vestibular nuclei. *J. Neurophysiol.* 16: 451–63

Spidalieri, H. J., Busby, L., Lamarre, Y. 1983. Fast ballistic arm movements triggered by visual, auditory, and somesthetic stimuli in the monkey. II. Effects of unilateral dentate lesion on discharge of precentral cortical neurons and reaction. *J. Neurophysiol.* 50: 1359–79

Strick, P. L. 1976. Anatomical analysis of ventrolateral thalamic input to the primate motor cortex. *J. Neurophysiol.* 39: 1020–31

Strick, P. L. 1983. The influence of motor preparation on the response of cerebellar neurons to limb displacements. *J. Neurosci.* 3: 2007–20

Thach, W. T. 1968. Discharge of Purkinje and cerebellar nuclear neurons during rapidly alternating arm movements in the monkey. *J. Neurophysiol.* 31: 785–97

Thach, W. T. 1970a. Discharge of cerebellar neurons related to two maintained postures and two prompt movements. I. Nuclear cell output. *J. Neurophysiol.* 33: 527–36

Thach, W. T. 1970b. Discharge of cerebellar neurons related to two maintained postures and two prompt movements. II. Purkinje cell output and input. *J. Neurophysiol.* 33: 537–47

Thach, W. T. 1975. Timing of activity in the cerebellar dentate nucleus and cerebral motor cortex during prompt volitional movement. *Brain Res.* 169: 168–72

Thach, W. T. 1978. Correlation of neural discharge with pattern and force of muscular activity, joint position, and direction of intended next movement in motor cortex and cerebellum. *J. Neurophysiol.* 41: 654–76

Thach, W. T., Goodkin, H. P., Keating, J. G. 1991. Inferior olive disease in man prevents learning of novel eye-hand synergies. *Abstr. Soc. Neurosci.* 17: 1380

Thach, W. T., Kane, S. A., Mink, J. W., Goodkin, H. P. 1992. Cerebellar output: Multiple maps and motor modes in movement coordination. In *Ramon y Cajal Centenary*, ed. R. Llinas, C. Sotelo. In press

Thach, W. T., Kane, S. A., Mink, J. W., Goodkin, H. P. 1990b. Cerebellar nuclear signs: Motor modalities, somatotopy, and incoordination. Am. Acad. Neurol., Motor Control: Annual Course # 248: 109–35

Thach, W. T., Perry, J. G., Shieber, M. H. 1982. Cerebellar output: Body maps and muscles spindles. In *The Cerebellum—New Vistas*, ed. S. L. Palay, V. Chan-Palay, pp. 440–54. New York: Springer-Verlag

Thach, W. T., Schieber, M. H., Mink, J. W., Kane, S. A., Horne, M. K. 1986. Cerebellar relation to muscle spindles in hand tracking. *Prog. Brain Res.* 64: 217–24

Trouche, E., Beaubaton, D. 1980. Initiation of a goal-directed movement in the monkey. *Exp. Brain Res.* 40: 311–21

Uno, M., Kozlovskaya, I. B., Brooks, V. B. 1973. Effects of cooling interposed nuclei on tracking-task performance in monkeys. *J. Neurophysiol.* 36: 996–1003

Van Kan, P. L. E., Houk, J. C., Gibson, A. R. 1986. Body representation in the nucleus interpositus of the monkey. *Neurosci. Lett. Suppl.* 26: s231

Vercher, J.-L., Gauthier, G. M. 1988. Cerebellar involvement in the coordination control of the oculo-manual tracking systems: Effects of cerebellar dentate nucleus lesion. *Exp. Brain Res.* 73: 155–66

Victor, M., Adams, R. D., Mancall, E. L. 1959. A restricted form of cerebellar cortical degeneration occurring in alcoholic patients. *Arch. Neurol.* 1: 579–688

Vilis, T., Hore, J. 1977. Effects of changes in mechanical state of limb on cerebellar intention tremor. *J. Neurophysiol.* 43: 279–91

Vilis, T., Hore, J. 1980. Central neuronal mechanisms contributing to cerebellar tremor produced by limb perturbations. *J. Neurophysiol.* 43: 279–91

Watanabe, E. 1984. Neuronal events correlated with long-term adaptation to the horizontal vestibulo-ocular reflex in the primate flocculus. *Brain Res.* 297: 169–74

Weiner, M. J., Hallett, M., Funkenstein, H. H. 1983. Adaptation to lateral displacement of vision in patients with lesions of the central nervous system. *Neurology* 33: 766–72

Wetts, R., Kalaska, J. F., Smith, A. M. 1985. Cerebellar nuclear cell activity during antagonist cocontraction and reciprocal inhibition of forearm muscles. *J. Neurophysiol.* 54: 231–44

Wolpaw, J. R., Carp, J. S. 1990. Memory traces in spinal cord. *Trends Neurosci.* 13: 137–42

Yeo, C. H., Hardiman, M. J., Glickstein, M. 1984. Discrete lesions of the cerebellar cortex abolish classically conditioned nictitating membrane response of the rabbit. *Behav. Brain Res.* 13: 261–66

SUBJECT INDEX

CUMULATIVE INDEXES

CONTRIBUTING AUTHORS, VOLUMES 11-15

CHAPTER TITLES, VOLUMES 11–15

450